ORGANIC SYNTHESES

ORGANIC SYNTHESES

AN ANNUAL PUBLICATION OF SATISFACTORY METHODS FOR THE PREPARATION OF ORGANIC CHEMICALS

VOLUME 88
2011

JONATHAN A. ELLMAN
VOLUME EDITOR

BOARD OF EDITORS

RICK L. DANHEISER, *EDITOR-IN-CHIEF*

KAY M. BRUMMOND	DAVID L. HUGHES
HUW M. L. DAVIES	MARK LAUTENS
JONATHAN A. ELLMAN	ANDREAS PFALTZ
MARGARET M. FAUL	VIRESH H. RAWAL
TOHRU FUKUYAMA	BRIAN M. STOLTZ

CHARLES K. ZERCHER, *ASSOCIATE EDITOR*
DEPARTMENT OF CHEMISTRY
UNIVERSITY OF NEW HAMPSHIRE
DURHAM, NEW HAMPSHIRE 03824

The procedures in this text are intended for use only by persons with prior training in the field of organic chemistry. In the checking and editing of these procedures, every effort has been made to identify potentially hazardous steps and to eliminate as much as possible the handling of potentially dangerous materials; safety precautions have been inserted where appropriate. If performed with the materials and equipment specified, in careful accordance with the instructions and methods in this text, the Editors believe the procedures to be very useful tools. However, these procedures must be conducted at one's own risk. Organic Syntheses, Inc., its Editors, who act as checkers, and its Board of Directors do not warrant or guarantee the safety of individuals using these procedures and hereby disclaim any liability for any injuries or damages claimed to have resulted from or related in any way to the procedures herein.

For general information on our other products and services or for technical support, please contact our Customer Care Department within the United States at (800) 762-2974, outside the United States at (317) 572-3993 or fax (317) 572-4002.

Wiley also publishes its books in a variety of electronic formats. Some content that appears in print may not be available in electronic formats. For more information about Wiley products, visit our web site at www.wiley.com.

"John Wiley & Sons, Inc. is pleased to publish this volume of Organic Syntheses on behalf of Organic Syntheses, Inc. Although Organic Syntheses, Inc. has assured us that each preparation contained in this volume has been checked in an independent laboratory and that any hazards that were uncovered are clearly set forth in the write-up of each preparation, John Wiley & Sons, Inc. does not warrant the preparations against any safety hazards and assumes no liability with respect to the use of the preparations."

For ordering and customer service, call 1-800-CALL-WILEY.

Library of Congress Catalog Card Number: 21-17747
ISBN 978-1-118-17397-8

Printed in the United States of America
10 9 8 7 6 5 4 3 2 1

ORGANIC SYNTHESES

*Out of print.
†Deceased.

*Out of print.
†Deceased.

*Out of print.
†*Deceased.*

*Out of print.
†Deceased.

NOTICE

Beginning with Volume 84, the Editors of *Organic Syntheses* initiated a new publication protocol, which is intended to shorten the time between submission of a procedure and its appearance as a publication. Immediately upon completion of the successful checking process, procedures are assigned volume and page numbers and are then posted on the Organic Syntheses website (www.orgsyn.org). The accumulated procedures from a single volume are assembled once a year and submitted for publication. The annual volume is published by John Wiley and Sons, Inc., and includes an index. The hard cover edition is available for purchase through the publisher. Incorporation of graphical abstracts into the Table of Contents began with Volume 77. Annual volumes 70–74, 75–79 and 80–84 have been incorporated into five-year versions of the collective volumes of *Organic Syntheses*. Collective Volumes IX, X and XI are available for purchase in the traditional hard cover format from the publishers. Beginning with Volume 88, a new type of article, referred to as Discussion Addenda, appeared. In these articles submitters are provided the opportunity to include updated discussion sections in which new understanding, further development, and additional application of the original method are described. Organic Syntheses intends for Discussion Addenda to become a regular feature of future volumes. The Editors welcome comments and suggestions from users concerning the new editions.

Organic Syntheses, Inc., joined the age of electronic publication in 2001 with the release of its free web site (www.orgsyn.org). Organic Syntheses, Inc., fully funded the creation of the free website at www.orgsyn.org in a partnership with CambridgeSoft Corporation and Data-Trace Publishing Company. The site is accessible to most internet browsers using Macintosh and Windows operating systems and may be used with or without a ChemDraw plugin. Because of continually evolving system requirements, users should review software compatibility at the website prior to use. John Wiley & Sons, Inc., and Accelrys, Inc., partnered with Organic Syntheses, Inc., to develop the new database (www.mrw.interscience.wiley.com/osdb) that is available for license with internet solutions from John Wiley & Sons, Inc. and intranet solutions from Accelrys, Inc.

Both the commercial database and the free website contain all annual and collective volumes and indices of *Organic Syntheses*. Chemists can draw structural queries and combine structural or reaction transformation queries with full-text and bibliographic search terms, such as chemical name, reagents, molecular formula, apparatus, or even hazard warnings or phrases. The preparations are categorized into reaction types, allowing search by category. The contents of individual or collective volumes can be browsed by lists of titles, submitters' names, and volume and page references, with or without reaction equations.

The commercial database at www.mrw.interscience.wiley.com/osdb also enables the user to choose his/her preferred chemical drawing package, or to utilize several freely available plug-ins for entering queries. The user is also able to cut and paste existing structures and reactions directly into the structure search query or their preferred chemistry editor, streamlining workflow. Additionally, this database contains links to the full text of primary literature references via CrossRef, ChemPort, Medline, and ISI Web of Science. Links to local holdings for institutions using open url technology can also be enabled. The database user can limit his/her search to, or order the search results by, such factors as reaction type, percentage yield, temperature, and publication date, and can create a customized table of reactions for comparison. Connections to other Wiley references are currently made via text search, with cross-product structure and reaction searching to be added in the near future. Incorporations of new preparations will occur as new material becomes available.

INFORMATION FOR AUTHORS OF PROCEDURES

Organic Syntheses welcomes and encourages submissions of experimental procedures that lead to compounds of wide interest or that illustrate important new developments in methodology. Proposals for *Organic Syntheses* procedures will be considered by the Editorial Board upon receipt of an outline proposal as described below. A full procedure will then be invited for those proposals determined to be of sufficient interest. These full procedures will be evaluated by the Editorial Board, and if approved, assigned to a member of the Board for checking. In order for a procedure to be accepted for publication, each reaction must be successfully repeated in the laboratory of a member of the Editorial Board at least twice, with similar yields (generally ±5%) and selectivity to that reported by the submitters.

Organic Syntheses Proposals

A cover sheet should be included providing full contact information for the principal author and including a scheme outlining the proposed reactions (an *Organic Syntheses* Proposal Cover Sheet can be downloaded at orgsyn.org). Attach an outline proposal describing the utility of the methodology and/or the usefulness of the product. Identify and reference the best current alternatives. For each step, indicate the proposed scale, yield, method of isolation and purification, and how the purity of the product is determined. Describe any unusual apparatus or techniques required, and any special hazards associated with the procedure. Identify the source of starting materials. Enclose copies of relevant publications (attach pdf files if an electronic submission is used).

Submit proposals by mail or as e-mail attachments to:

Professor Charles K. Zercher
Associate Editor, Organic Syntheses
Department of Chemistry
University of New Hampshire
23 Academic Way, Parsons Hall
Durham, NH 03824

For electronic submissions: *org.syn@unh.edu*

Submission of Procedures

Authors invited by the Editorial Board to submit full procedures should prepare their manuscripts in accord with the Instructions to Authors which are described below or may be downloaded at orgsyn.org. Submitters are also encouraged to consult this volume of *Organic Syntheses* for models with regard to style, format, and the level of experimental detail expected in *Organic Syntheses* procedures. Manuscripts should be submitted to the Associate Editor. Electronic submissions are encouraged; procedures will be accepted as e-mail attachments in the form of Microsoft Word files with all schemes and graphics also sent separately as ChemDraw files.

Procedures that do not conform to the Instructions to Authors with regard to experimental style and detail will be returned to authors for correction. Authors will be notified when their manuscript is approved for checking by the Editorial Board, and it is the goal of the Board to complete the checking of procedures within a period of no more than six months.

Additions, corrections, and improvements to the preparations previously published are welcomed; these should be directed to the Associate Editor. However, checking of such improvements will only be undertaken when new methodology is involved.

NOMENCLATURE

Both common and systematic names of compounds are used through-out this volume, depending on which the Volume Editor felt was more appropriate. The Chemical Abstracts indexing name for each title compound, if it differs from the title name, is given as a subtitle. Systematic

Chemical Abstracts nomenclature, used in the Collective Indexes for the title compound and a selection of other compounds mentioned in the procedure, is provided in an appendix at the end of each preparation. Chemical Abstracts Registry numbers, which are useful in computer searching and identification, are also provided in these appendices.

ACKNOWLEDGMENT

Organic Syntheses wishes to acknowledge the contributions of Amgen, Inc. and Merck & Co. to the success of this enterprise through their support, in the form of time and expenses, of members of the Board of Editors.

INSTRUCTIONS TO AUTHORS

All organic chemists have experienced frustration at one time or another when attempting to repeat reactions based on experimental procedures found in journal articles. To ensure reproducibility, *Organic Syntheses* requires experimental procedures written with considerably more detail as compared to the typical procedures found in other journals and in the "Supporting Information" sections of papers. In addition, each *Organic Syntheses* procedure is carefully "checked" for reproducibility in the laboratory of a member of the Board of Editors.

Even with these more detailed procedures, the experience of *Organic Syntheses* editors is that difficulties often arise in obtaining the results and yields reported by the submitters of procedures. To expedite the checking process and ensure success, we have prepared the following "Instructions for Authors" as well as a **Checklist for Authors** and **Characterization Checklist** to assist you in confirming that your procedure conforms to these requirements. Please include completed Checklist (available at www.orgsyn.org) together with your procedure at the time of submission. Procedures submitted to *Organic Syntheses* will be carefully reviewed upon receipt and procedures lacking any of the required information will be returned to the submitters for revision.

Scale and Optimization

The appropriate scale for procedures will vary widely depending on the nature of the chemistry and the compounds synthesized in the procedure. However, some general guidelines are possible. For procedures in which the principal goal is to illustrate a synthetic method or strategy, it is expected, in general, that the procedure should result in at least 5 g and no more than 50 g of the final product. In cases where the point of the procedure is to provide an efficient method for the preparation of a useful reagent or synthetic building block, the appropriate scale may be larger, but in general should not exceed 100 g of final product. Exceptions to these guidelines may be granted in special circumstances. For example, procedures describing the preparation of reagents employed as catalysts will often be acceptable on a scale of less than 5 g.

In considering the scale for an *Organic Syntheses* procedure, authors should also take into account the cost of reagents and starting materials. In general, the Editors will not accept procedures for checking in which the cost of any one of the reactants exceeds $500 for a single full-scale run. Authors are requested to identify the most expensive reagent or starting material on the procedure submission checklist and to estimate its cost per run of the procedure.

It is expected that all aspects of the procedure will have been optimized by the authors prior to submission, and that each reaction will have been carried out at least twice on exactly the scale described in the procedure. It is appropriate to report the weight, yield, and purity of the product of each step in the procedure as a range. In any case where a reagent is employed in significant excess, a Note should be included explaining why an excess of that reagent is necessary. If possible, the Note should indicate the effect of using amounts of reagent less than that specified in the procedure.

Reaction Apparatus

Describe the size and type of flask (number of necks) and indicate how *every* neck is equipped.

"A 500-mL, three-necked, round-bottomed flask equipped with an 3-cm Teflon-coated magnetic stirbar, a 250-mL pressure-equalizing addition funnel fitted with an argon inlet, and a rubber septum is charged with...."

Indicate how the reaction apparatus is dried and whether the reaction is conducted under an inert atmosphere. This can be incorporated in the text of the procedure or included in a Note.

"The apparatus is flame-dried and maintained under an atmosphere of argon during the course of the reaction."

In the case of procedures involving unusual glassware or especially complicated reaction setups, authors are encouraged to include a photograph or drawing of the apparatus in the text or in a Note (for examples, see *Org. Syn.,* Vol. 82, 99 and Coll. Vol. X, pp 2, 3, 136, 201, 208, and 669).

Reagents and Starting Materials

All chemicals employed in the procedure must be commercially available or described in an earlier *Organic Syntheses* or *Inorganic Syntheses* procedure. For other compounds, a procedure should be included either as one or more steps in the text or, in the case of relatively straightforward preparations of reagents, as a Note. In the latter case, all requirements with regard to characterization, style, and detail also apply.

In one or more Notes, indicate the purity or grade of each reagent, solvent, etc. It is desirable to also indicate the source (company the chemical was purchased from), particularly in the case of chemicals where it is suspected that the composition (trace impurities, etc.) may vary from one supplier to another. In cases where reagents are purified, dried, "activated" (e.g., Zn dust), etc., a detailed description of the procedure used should be included in a Note. In other cases, indicate that the chemical was "used as received".

> "Diisopropylamine (99.5%) was obtained from Aldrich Chemical Co., Inc. and distilled under argon from calcium hydride before use. THF (99+%) was obtained from Mallinckrodt, Inc. and distilled from sodium benzophenone ketyl. Diethyl ether (99.9%) was purchased from Aldrich Chemical Co., Inc. and purified by pressure filtration under argon through activated alumina. Methyl iodide (99%) was obtained from Aldrich Chemical Co., Inc. and used as received."

The amount of each reactant should be provided in parentheses in the order mL, g, mmol, and equivalents with careful consideration to the correct number of significant figures.

The concentration of solutions should be expressed in terms of molarity or normality, and not percent (e.g., 1 N HCl, 6 M NaOH, not "10% HCl").

Reaction Procedure

Describe every aspect of the procedure clearly and explicitly. Indicate the order of addition and time for addition of all reagents and how each is added (via syringe, addition funnel, etc.).

Indicate the temperature of the reaction mixture (preferably internal temperature). Describe the type of cooling (e.g., "dry ice-acetone bath") and heating (e.g., oil bath, heating mantle) methods employed. Be careful to describe clearly all cooling and warming cycles, including initial and final temperatures and the time interval involved.

Describe the appearance of the reaction mixture (color, homogeneous or not, etc.) and describe all significant changes in appearance during the course of the reaction (color changes, gas evolution, appearance of solids, exotherms, etc.).

Indicate how the reaction can be monitored to determine the extent of conversion of reactants to products. In the case of reactions monitored by TLC, provide details in a Note, including eluent, R_f values, and method of visualization. For reactions followed by GC, HPLC, or NMR analysis, provide details on analysis conditions and relevant diagnostic peaks.

"The progress of the reaction was followed by TLC analysis on silica gel with 20% EtOAc-hexane as eluent and visualization with p-anisaldehyde. The ketone starting material has $R_f = 0.40$ (green) and the alcohol product has $R_f = 0.25$ (blue)."

Reaction Workup

Details should be provided for reactions in which a "quenching" process is involved. Describe the composition and volume of quenching agent, and time and temperature for addition. In cases where reaction mixtures are added to a quenching solution, be sure to also describe the setup employed.

"The resulting mixture was stirred at room temperature for 15 h, and then carefully poured over 10 min into a rapidly stirred, ice-cold aqueous solution of 1 N HCl in a 500-mL Erlenmeyer flask equipped with a magnetic stirbar."

For extractions, the number of washes and the volume of each should be indicated as well as the size of the separatory funnel.

For concentration of solutions after workup, indicate the method and pressure and temperature used.

"The reaction mixture is diluted with 200 mL of water and transferred to a 500-mL separatory funnel, and the aqueous phase is separated and extracted with three 100-mL portions of ether. The combined organic layers are washed with 75 mL of water and 75 mL of saturated NaCl solution, dried over 25 g of MgSO$_4$, filtered through a 250-mL medium porosity sintered glass funnel, and concentrated by rotary evaporation (25 °C, 20 mmHg) to afford 3.25 g of a yellow oil."

"The solution is transferred to a 250-mL, round-bottomed flask equipped with a magnetic stirbar and a 15-cm Vigreux column fitted with a short path distillation head, and then concentrated by careful distillation at 50 mmHg (bath temperature gradually increased from 25 to 75 °C)."

In cases where solid products are filtered, describe the type of filter funnel used and the amount and composition of solvents used for washes.

"... and the resulting pale yellow solid is collected by filtration on a Büchner funnel and washed with 100 mL of cold (0 °C) hexane."

When solid or liquid compounds are dried under vacuum, indicate the pressure employed (rather than stating "reduced pressure" or "dried *in vacuo*").

"... and concentrated at room temperature by rotary evaporation (20 mmHg) and then at 0.01 mmHg to provide. ..."

"The resulting colorless crystals are transferred to a 50-mL, round-bottomed flask and dried overnight in a 100 °C oil bath at 0.01 mmHg."

Purification: Distillation

Describe distillation apparatus including the size and type of distillation column. Indicate temperature (and pressure) at which all significant fractions are collected.

"... and transferred to a 100-mL, round-bottomed flask equipped with a magnetic stirbar. The product is distilled under vacuum through a 12-cm, vacuum-jacketed column of glass helices (Note 16) topped with a Perkin triangle. A forerun (ca. 2 mL) is collected and discarded, and the desired product is then obtained, distilling at 50–55 °C (0.04–0.07 mmHg). ..."

Purification: Column Chromatography

Provide information on TLC analysis in a Note, including eluent, R_f values, and method of visualization.

Provide dimensions of column and amount of silica gel used; in a Note indicate source and type of silica gel.

Provide details on eluents used, and number and size of fractions.

"The product is charged on a column (5 × 10 cm) of 200 g of silica gel (Note 15) and eluted with 250 mL of hexane. At that point, fraction collection (25-mL fractions) is begun, and elution is continued with 300 mL of 2% EtOAc-hexane (49:1 hexanes:EtOAc) and then 500 mL of 5% EtOAc-hexane (19:1 hexanes:EtOAc). The desired product is obtained in fractions 24–30, which are concentrated by rotary evaporation (25 °C, 15 mmHg). ..."

Purification: Recrystallization

Describe procedure in detail. Indicate solvents used (and ratio of mixed solvent systems), amount of recrystallization solvents, and temperature protocol. Describe how crystals are isolated and what they are washed with.

"The solid is dissolved in 100 mL of hot diethyl ether (30 °C) and filtered through a Büchner funnel. The filtrate is allowed to cool to room temperature, and 20 mL of hexane is added. The solution is cooled at −20 °C overnight and the resulting crystals are collected by suction filtration on a Büchner funnel, washed with 50 mL of ice-cold hexane, and then transferred to a 50-mL, round-bottomed flask and dried overnight at 0.01 mmHg to provide. ...."

Characterization

Physical properties of the product such as color, appearance, crystal forms, melting point, etc. should be included in the text of the procedure. Comments on the stability of the product to storage, etc. should be provided in a Note.

In a Note, provide data establishing the identity of the product. This will generally include IR, MS, ^{1}H-NMR, and ^{13}C-NMR data, and in some cases UV data. Copies of the proton and carbon NMR spectra for the products of each step in the procedure should be submitted showing integration for all resonances. Submission of copies of the NMR spectra for other nuclei are encouraged as appropriate.

In the same Note, provide quantitative analytical data establishing the purity of the product. Elemental analysis for carbon and hydrogen (and nitrogen if present) agreeing with calculated values within 0.4% is preferred. However, GC data (for distilled or vacuum-transferred samples) and/or HPLC data (for material isolated by column chromatography) may be acceptable in some cases. Provide details on equipment and conditions for GC and HPLC analyses.

In procedures involving non-racemic, enantiomerically enriched products, optical rotations should generally be provided, but enantiomeric purity must be determined by another method such as chiral HPLC or GC analysis.

In cases where the product of one step is used without purification in the next step, a Note should be included describing how a sample of the product can be purified and providing characterization data for the pure material. Copies of the proton NMR spectra of both the product both *before* and *after* purification should be submitted.

Hazard Warnings

Any significant hazards should be indicated in a statement at the beginning of the procedure in italicized type. Efforts should be made to avoid the use of toxic and hazardous solvents and reagents when less hazardous alternatives are available.

Discussion Section

The style and content of the discussion section will depend on the nature of the procedure.

For procedures that provide an improved method for the preparation of an important reagent or synthetic building block, the discussion should focus on the advantages of the new approach and should describe and reference all of the earlier methods used to prepare the title compound.

In the case of procedures that illustrate an important synthetic method or strategy, the discussion section should provide a mini-review on the new methodology. The scope and limitations of the method should be discussed, and it is generally desirable to include a table of examples. Please be sure each table is numbered and has a title. Competing methods for accomplishing the same overall transformation should be described and referenced. A brief discussion of mechanism may be included if this is useful for understanding the scope and limitations of the method.

Titles of Articles

In cases where the main thrust of the article is the illustration of a synthetic method of general utility, the title of the article should incorporate reference to that method. Inclusion of the name of the final product is acceptable but not required. In the case of articles where the objective is the preparation of a specific compound of importance (such as a chiral ligand), then the name of that compound should be part of the title.

Examples

Title without name of product: "Stereoselective Synthesis of 3-Arylacrylates by Copper-Catalyzed Syn Hydroarylation" (*Org. Synth.* **2010**, *87*, 53).

Title including name of final product (not required): "Catalytic Enantioselective Borane Reduction of Benzyl Oximes: Preparation of (S)-1-Pyridin-3-
yl-ethylamine Bis Hydrochloride" (*Org. Synth.* **2010**, *87*, 36).

Title where preparation of specific compound is the subject: "Preparation of (S)-3,3'-Bis-Morpholinomethyl-5,5',6,6',7,7',8,8'-octahydro-1,1' – bi-
2-naphthol" (*Org. Synth.* **2010**, *87*, 59).

Style and Format

Articles should follow the style guidelines used for organic chemistry articles published in the ACS journals such as *J. Am. Chem. Soc., J. Org. Chem.,
Org. Lett.*, etc. as described in the the the ACS Style Guide (3rd Ed.). The text
of the procedure should be constructed using a standard word processing
program, like MS Word, with 14-point Times New Roman font. Chemical
structures and schemes should be drawn using the standard ACS drawing
parameters (in ChemDraw, the parameters are found in the "ACS Document
1996" option) with a maximum width of 6 inches. The graphics files should
be inserted into the document at the correct location and the graphics files
should also be submitted separately. All Tables that include structures should
be entirely prepared in the graphics (ChemDraw) program and inserted into the
word processing file at the appropriate location. Tables that include multiple,
separate graphics files prepared in the word processing program will require
modification.

Acknowlegments and Author's Contact Information

Contact information (institution where the work was carried out and mailing
address for the principal author) should be included as footnote 1. This footnote
should also include the email address for the principal author. Acknowledg-
ment of financial support should be included in footnote 1.

Biographies and Photographs of Authors

Photographs and 100-word biographies of all authors should be submitted
as separate files at the time of the submission of the procedure. The format
of the biographies should be similar to those in the Volume 84 procedures
found at the orgsyn.org website. Photographs can be accepted in a number of
electronic formats, including tiff and jpeg formats.

HANDLING HAZARDOUS CHEMICALS

A Brief Introduction

General Reference: *Prudent Practices in the Laboratory*; National Academy Press; Washington, DC, 2011.

Physical Hazards

Fire. Avoid open flames by use of electric heaters. Limit the quantity of flammable liquids stored in the laboratory. Motors should be of the nonsparking induction type.

Explosion. Use shielding when working with explosive classes such as acetylides, azides, ozonides, and peroxides. Peroxidizable substances such as ethers and alkenes, when stored for a long time, should be tested for peroxides before use. Only sparkless "flammable storage" refrigerators should be used in laboratories.

Electric Shock. Use 3-prong grounded electrical equipment if possible.

Chemical Hazards

Because all chemicals are toxic under some conditions, and relatively few have been thoroughly tested, it is good strategy to minimize exposure to all chemicals. In practice this means having a good, properly installed hood; checking its performance periodically; using it properly; carrying out all operations in the hood; protecting the eyes; and, since many chemicals can penetrate the skin, avoiding skin contact by use of gloves and other protective clothing at all times.

a. Acute Effects. These effects occur soon after exposure. The effects include burn, inflammation, allergic responses, damage to the eyes, lungs, or nervous system (e.g., dizziness), and unconsciousness or death (as from overexposure to HCN). The effect and its cause are usually obvious and so are the methods to prevent it. They generally arise from inhalation or skin contact, so should not be a problem if one follows the admonition "work in a hood and keep chemicals off your hands". Ingestion is a rare route, being generally the result of eating in the laboratory or not washing hands before eating.

b. Chronic Effects. These effects occur after a long period of exposure or after a long latency period and may show up in any of numerous organs. Of the chronic effects of chemicals, cancer has received the most attention lately. Several dozen chemicals have been demonstrated to be carcinogenic in man and hundreds to be carcinogenic to animals. Although there is no simple correlation between carcinogenicity in animals and in man, there is little doubt that a significant proportion of the chemicals used in laboratories have some potential for carcinogenicity in man. For this and other reasons, chemists should employ good practices at all times.

The key to safe handling of chemicals is a good, properly installed hood, and the referenced book devotes many pages to hoods and ventilation. It recommends that in a laboratory where people spend much of their time working with chemicals there should be a hood for each two people, and each should have at least 2.5 linear feet (0.75 meter) of working space at it. Hoods are more than just devices to keep undesirable vapors from the laboratory atmosphere. When closed they provide a protective barrier between chemists and chemical operations, and they are a good containment device for spills. Portable shields can be a useful supplement to hoods, or can be an alternative for hazards of limited severity, e.g., for small-scale operations with oxidizing or explosive chemicals.

Specialized equipment can minimize exposure to the hazards of laboratory operations. Impact resistant safety glasses are basic equipment and should be worn at all times. They may be supplemented by face shields or goggles for particular operations, such as pouring corrosive liquids. Because skin contact with chemicals can lead to skin irritation or sensitization or, through absorption, to effects on internal organs, protective gloves should be worn at all times.

Laboratories should have fire extinguishers and safety showers. Respirators should be available for emergencies. Emergency equipment should be kept in a central location and must be inspected periodically.

MSDS (Materials Safety Data Sheets) sheets are available from the suppliers of commercially available reagents, solvents, and other chemical materials; anyone performing an experiment should check these data sheets before initiating an experiment to learn of any specific hazards associated with the chemicals being used in that experiment.

DISPOSAL OF CHEMICAL WASTE

General Reference: *Prudent Practices in the Laboratory* National Academy Press, Washington, D.C. 2011.

Effluents from synthetic organic chemistry fall into the following categories:

1. **Gases**

   1a. Gaseous materials either used or generated in an organic reaction.

   1b. Solvent vapors generated in reactions swept with an inert gas and during solvent stripping operations.

   1c. Vapors from volatile reagents, intermediates and products.

2. **Liquids**

   2a. Waste solvents and solvent solutions of organic solids (see item 3b).

   2b. Aqueous layers from reaction work-up containing volatile organic solvents.

   2c. Aqueous waste containing non-volatile organic materials.

   2d. Aqueous waste containing inorganic materials.

3. **Solids**

   3a. Metal salts and other inorganic materials.

   3b. Organic residues (tars) and other unwanted organic materials.

   3c. Used silica gel, charcoal, filter aids, spent catalysts and the like.

The operation of industrial scale synthetic organic chemistry in an environmentally acceptable manner* requires that all these effluent categories be dealt with properly. In small scale operations in a research or

*An environmentally acceptable manner may be defined as being both in compliance with all relevant state and federal environmental regulations *and* in accord with the common sense and good judgment of an environmentally aware professional.

academic setting, provision should be made for dealing with the more environmentally offensive categories.

1a. Gaseous materials that are toxic or noxious, e.g., halogens, hydrogen halides, hydrogen sulfide, ammonia, hydrogen cyanide, phosphine, nitrogen oxides, metal carbonyls, and the like.

1c. Vapors from noxious volatile organic compounds, e.g., mercaptans, sulfides, volatile amines, acrolein, acrylates, and the like.

2a. All waste solvents and solvent solutions of organic waste.

2c. Aqueous waste containing dissolved organic material known to be toxic.

2d. Aqueous waste containing dissolved inorganic material known to be toxic, particularly compounds of metals such as arsenic, beryllium, chromium, lead, manganese, mercury, nickel, and selenium.

3. All types of solid chemical waste.

Statutory procedures for waste and effluent management take precedence over any other methods. However, for operations in which compliance with statutory regulations is exempt or inapplicable because of scale or other circumstances, the following suggestions may be helpful.

Gases

Noxious gases and vapors from volatile compounds are best dealt with at the point of generation by "scrubbing" the effluent gas. The gas being swept from a reaction set-up is led through tubing to a large trap to prevent suck-back and into a sintered glass gas dispersion tube immersed in the scrubbing fluid. A bleach container can be conveniently used as a vessel for the scrubbing fluid. The nature of the effluent determines which of four common fluids should be used: dilute sulfuric acid, dilute alkali or sodium carbonate solution, laundry bleach when an oxidizing scrubber is needed, and sodium thiosulfate solution or diluted alkaline sodium borohydride when a reducing scrubber is needed. Ice should be added if an exotherm is anticipated.

Larger scale operations may require the use of a pH meter or starch/iodide test paper to ensure that the scrubbing capacity is not being exceeded.

When the operation is complete, the contents of the scrubber can be poured down the laboratory sink with a large excess (10–100 volumes) of water. If the solution is a large volume of dilute acid or base, it should be neutralized before being poured down the sink.

Liquids

Every laboratory should be equipped with a waste solvent container in which *all* waste organic solvents and solutions are collected. The contents of these containers should be periodically transferred to properly labeled waste solvent drums and arrangements made for contracted disposal in a regulated and licensed incineration facility.**

Aqueous waste containing dissolved toxic organic material should be decomposed *in situ*, when feasible, by adding acid, base, oxidant, or reductant. Otherwise, the material should be concentrated to a minimum volume and added to the contents of a waste solvent drum.

Aqueous waste containing dissolved toxic inorganic material should be evaporated to dryness and the residue handled as a solid chemical waste.

Solids

Soluble organic solid waste can usually be transferred into a waste solvent drum, provided near-term incineration of the contents is assured.

Inorganic solid wastes, particularly those containing toxic metals and toxic metal compounds, used Raney nickel, manganese dioxide, etc. should be placed in glass bottles or lined fiber drums, sealed, properly labeled, and arrangements made for disposal in a secure landfill.** Used mercury is particularly pernicious and small amounts should first be amalgamated with zinc or combined with excess sulfur to solidify the material.

Other types of solid laboratory waste including used silica gel and charcoal should also be packed, labeled, and sent for disposal in a secure landfill.

Special Note

Since local ordinances may vary widely from one locale to another, one should always check with appropriate authorities. Also, professional disposal services differ in their requirements for segregating and packaging waste.

**If arrangements for incineration of waste solvent and disposal of solid chemical waste by licensed contract disposal services are not in place, a list of providers of such services should be available from a state or local office of environmental protection.

PREFACE

Organic chemistry has developed into an art form where scientists produce marvelous chemical creations in their test tubes. Mankind benefits from this in the form of medicines, ever-more precise electronics and advanced technological materials.

From the press release announcing the 2010 Nobel Prize of Chemistry by the Royal Swedish Academy of Sciences.

Synthetic organic chemistry is becoming ever more powerful in its ability to efficiently create new chemical matter with enormous benefit to society. However, for the successful implementation of synthetic organic chemistry, reliable and reproducible procedures are essential whether they are to be applied to the discovery or production of new compounds. For the past 90 years *Organic Syntheses* has provided a vital mechanism for satisfying this need by publishing detailed procedures that have been carefully reproduced and validated.

Volume 88 contains thirty new procedures representing the utility, variety and creativity of synthetic organic chemistry transformations and the products that are obtained. Volume 88 for the first time also contains eight discussion addendums to procedures that were previously published in *Organic Syntheses.* These addendums provide authors with an opportunity to contribute additional insights and applications, increased scope, and most importantly, new experimental details and advances.

The procedures in this volume appear chronologically in the order that checking and editing were completed to enable immediate availability upon posting on the *Organic Syntheses* website. In contrast, my summary of the contents of this volume serves to highlight common themes of the collected protocols and therefore does not follow their sequential appearance.

Metal catalyzed processes are prominently featured in this volume with a majority of new procedures and all of the addendums including at least one metal catalyzed step. Transition-metal catalyzed carbon-carbon bond formation is in particular emphasized, which is particularly appropriate considering that the 2010 Nobel Prize in Chemistry was awarded for palladium-catalyzed cross couplings in organic synthesis. Huang, DeLuca and Hartwig describe the extremely efficient Pd-catalyzed α-arylation of esters (p. 4). Campeau and Fagnou report on a powerful method for the arylation of pyridines by the Pd-catalyzed direct arylation of pyridine N-oxides via C-H bond functionalization (p. 22). This very useful procedure clearly reflects the ingenuity and importance of the scientific contributions made by Keith Fagnou over the short time frame during which we were privileged to witness his research. Yamamoto, Sugai, Takizawa and Miyaura also provide a useful method for the synthesis of 2-aryl pyridines through the preparation of lithium 2-pyridyltriolborates for cross-coupling reactions with aryl halides (p. 79). Wender, Lesser and Sirois report an innovative convergent assembly of cyclohept-4-enones by Rh-catalyzed intermolecular [5+2] cycloaddition (p. 109). Chen, Markina, Yao and Larock provide a procedure for their extensively used synthesis of 2,3-disubstituted indoles via the Pd-catalyzed annulation of internal alkynes with 2-haloanilines (p. 377).

A majority of the discussion addendums are for procedures that describe Pd-catalyzed cross-coupling transformations. The addendum by Denmark and Liu provides important advances on the increasingly used cross-coupling of alkenylsilanols and aryl halides (p. 102). Williams has contributed a discussion addendum (p. 197) to a procedure originally authored by Stille, Echavarren, Hendrix, Albrecht and Williams to provide a perspective on the numerous advances to the Stille cross-coupling reaction over the past two decades, including expanded substrate scope, improved reaction conditions, catalyst design and applications in total synthesis. Miyaura has provided a timely addendum (p. 202) to a procedure originally authored by Miyaura and Suzuki on new conditions and boron reagents for the Pd-catalyzed cross-coupling of alkenyl boron compounds with vinyl halides. Miyaura has also provided a very useful addendum (p. 207) to a procedure originally authored by Miyaura, Ishiyama and Suzuki on the Pd-catalyzed cross-coupling of 9-alkyl-9-BBN reagents prepared by the convenient hydroboration of alkenes.

As for previous volumes, the synthesis of nitrogen heterocycles is also featured prominently and illustrates the ingenuity and variety of possible disconnections by which compounds of this class can be prepared. In addition to the aforementioned preparation of indoles by

Larock and coworkers (p. 377), an efficient synthesis of indoles by the Cu-mediated intramolecular amination of aryl halides is reported by Noji, Okano, Fukuyama and Tokuyama (p. 388). A related Cu-mediated cyclization to provide benzoxazoles is also reported by Saha, Ali and Punniyamurthy (p. 398). Counceller, Eichman, Wray, Welin and Stambuli further report on the efficient preparation of indazoles by metal free intramolecular electrophilic amination of 2-aminophenyl ketoximes (p. 33). Hein, Krasnova, Iwasaki and Fokin contribute a procedure for the singularly ubiquitous Cu-catalyzed azide-alkyne cycloaddition with their preparation of tris((1-benzyl-1H-1,2,3-triazolyl)methyl)amine (p. 238), which in turn serves as an efficient ligand for Cu-catalyzed azide-alkyne cycloadditions. Andrews and Kwon provide an innovative strategy for the preparation of pyrroline carboxylates by the phosphine catalyzed [3+2] annulation of allenes and N-tosyl imines (p. 138). Finally, two discussion addendums further illustrate the importance and relevance of indole synthesis. Ichikawa provides an addendum on more efficient strategies for the preparation of their key intermediate to 2-fluoroindoles (p. 162), and Söderberg provides an addendum for the synthesis of indoles via the Pd-catalyzed reductive N-heteroannulation of 2-nitrostyrenes (p. 291).

Asymmetric transformations are prevalent in previous volumes and this trend continues with the current volume. Noteworthy is the increase in the number of procedures describing the preparation and/or application of organocatalysts. Graham, Horning and MacMillan describe the preparation of an imidazolinone organocatalyst that has proven to be extremely versatile for mediating a variety of transformations with exceedingly high levels of enantioselectivity (p. 42). Nguyen, Oh, Henry-Riyad, Sepulveda and Romo report on an innovative catalytic enantioselective synthesis of bicyclic β-lactones via cinchona alkaloid-catalyzed aldol-lactonization (p. 121). Viózquez, Guillena, Nájera, Bradshaw, Etxebarria-Jardí and Bonjoch describe the preparation of a BINAM-prolinamide (p. 317) and its application to an efficient and highly enantioselective intramolecular aldol reaction (p. 330). Hu, Zhou, Xu, Liu and Gong report on the preparation of a chiral phosphoric acid incorporating a bis-phenanthryl-substituted binaphthol that has served as a chiral Brønsted acid for a range of enantioselective transformations (p. 406). Moreover, Zhou, Xu and Hu further report on the application of this organocatalyst for the asymmetric synthesis of β-amino-α-hydroxy esters by an innovative Rh-catalyzed three component reaction (p. 418).

Chiral auxiliaries and reagents continue to be extensively used and are also featured in this volume. Crimmins, Christie and Hughes report on the preparation of and diastereoselective aldol reaction using a thiazolidinethione chiral auxiliary that has been applied to the synthesis of a number of complex molecules (p. 364). Webster, Partridge and Aggarwal report on the asymmetric synthesis of alcohols by the clever sparteine-mediated homologation of boronic esters with lithiated primary alkyl carbamates (p. 247). Sun and Roush report two related procedures on the preparation of chiral boron reagents and their reactions with aldehydes, one procedure for (+)-B-allyldiisopinocampheylborane (p. 87) and a second procedure for (S, S)-diisopropyl tartrate (E)-crotylboronate (p. 181). These procedures provide an outstanding resource for comparing and contrasting the substrate scope and practical aspects of product isolation and purification for this important class of reagents. Finally, the discussion addendum by Bringmann, Gulder and Gulder provides a survey of the numerous applications of their 'lactone concept' for the atroposelective construction of a variety of axially chiral natural products (p. 70).

Organic Syntheses has historically showcased the preparation and use of organic reagents, and volume 88 is no exception. Snyder and Treitler contribute a procedure for the preparation of the brominating agent $Et_2SBr \cdot SbCl_5Br$ and demonstrate its particular effectiveness in biomimetic brominative polyene cyclization (p. 54). Azuma, Okano, Fukuyama and Tokuyama report the practical preparation of a Horner-Wadsworth-Emmons reagent that has seen extensive use in the preparation of (Z)-dehydroamino esters (p. 152). Eisenberger, Kieltsch, Koller, Stanek and Togni describe the preparation of a highly versatile trifluoromethyl transfer agent and demonstrate its value in the synthesis of an S-trifluoromethyl cysteine derivative (p. 168). Muchalski, Doody, Troyer and Johnston provide an efficient preparation of a diazoimide that has proven to be very useful in highly diastereoselective, Brønsted acid-catalyzed transformations (p. 212). Wein, Tong and McDonald report on the preparation of the versatile synthetic intermediate 4-trimethylsilyl-2-butyn-1-ol (p. 296). Harmata, Cai and Huang contribute an efficient route to allenic sulfones, which have proven useful for a variety of transformations (p. 309). Finally Lisboa, Hoang and Dudley report the preparation of a vinylogous acyl triflate and its fragmentation upon treatment with a dithiane-stabilized anion (p. 353).

The remaining procedures in volume 88 fall within the important category of functional group manipulation. Giguère-Bisson, Yoo and

Li contribute the efficient Cu-mediated oxidative coupling of aldehydes and amine hydrochloride salts to form amides (p. 14). Ely and Morken report the stereoselective Ni-catalyzed 1,4-hydroboration of 1,3-dienes to provide (Z)-allylic alcohols (p. 342). Leighty, Spletstoser and Georg describe an efficient procedure for the general and highly functional group compatible reduction of tertiary amides to aldehydes with Schwartz's reagent (p. 427). Finally, Cacchi, Morera and Ortar contribute a discussion addendum on the extensively used Pd-catalyzed reduction of vinyl triflates to alkenes (p. 260).

In completing this volume I first and foremost gratefully acknowledge the dedicated efforts of the submitters in the author's labs and the checkers associated with the editors' labs. These individuals deserve the primary credit for the contents of this volume. I am also very appreciative of the many contributions of my colleagues on the Board of Editors over the past eight years. Finally, I would especially like to thank Rick Danheiser (Editor in Chief) and Chuck Zercher (Associate Editor) for their guidance, collaboration and dedication that ensure the continued relevance, importance and utility of *Organic Syntheses*. As I reflect upon my term on the Board of Editors, I have come to realize that it has been one of my most fulfilling professional experiences. It has been a privilege and honor to be a part of this organization.

JONATHAN A. ELLMAN
New Haven, Connecticut

CONTENTS

SYNTHESIS OF 2-ARYL PYRIDINES BY PALLADIUM-CATALYZED DIRECT ARYLATION OF PYRIDINE *N*-OXIDES

22

Louis-Charles Campeau* and Keith Fagnou

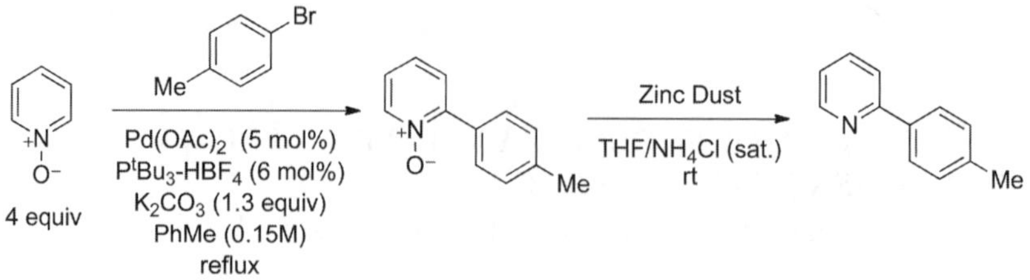

THE PREPARATION OF INDAZOLES VIA METAL FREE INTRAMOLECULAR ELECTROPHILIC AMINATION OF 2-AMINOPHENYL KETOXIMES

33

Carla M. Counceller, Chad C. Eichman, Brenda C. Wray, Eric R. Welin, and James P. Stambuli*

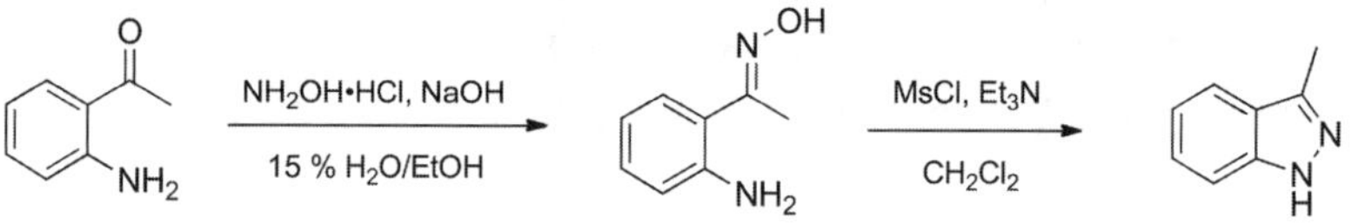

THE PREPARATION OF (2*R*,5*S*)-2-*t*-BUTYL-3,5-DIMETHYLIMIDAZOLIDIN-4-ONE

42

Thomas H. Graham, Benjamin D. Horning, and David W. C. MacMillan*

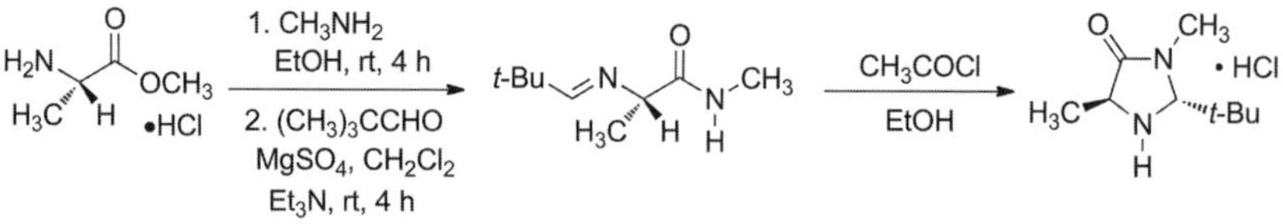

SYNTHESIS OF Et₂SBr•SbCl₅Br AND ITS USE IN BIOMIMETIC BROMINATIVE POLYENE CYCLIZATIONS

Scott A. Snyder* and Daniel S. Treitler

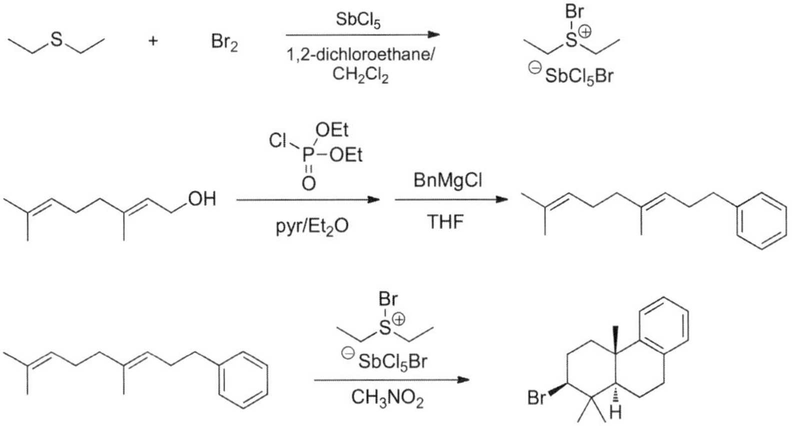

Discussion Addendum for:

ASYMMETRIC SYNTHESIS OF (M)-2- HYDROXYMETHYL-1-(2-HYDROXY-4,6-DIMETHYLPHENYL)NAPHTHALENE VIA A CONFIGURATIONALLY UNSTABLE BIARYL LACTONE

G. Bringmann,* T.A.M. Gulder, and T. Gulder

Original article published in *Org. Synth.* **2002**, *79*, 72-83

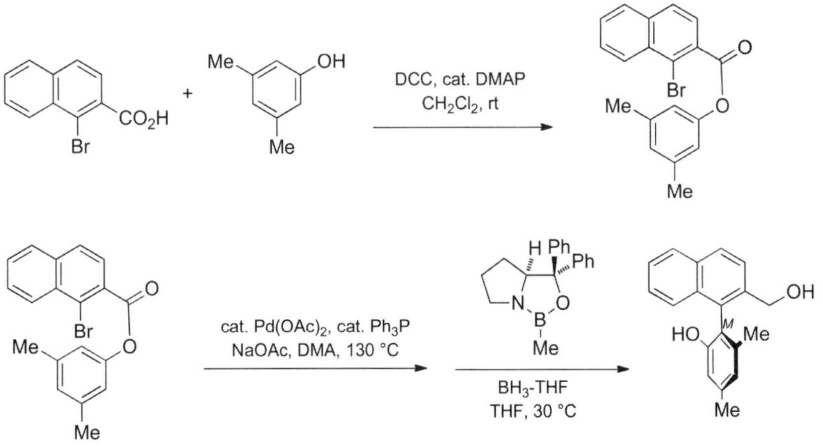

SYNTHESIS OF LITHIUM 2-PYRIDYLTRIOLBORATE AND ITS CROSS-COUPLING REACTION WITH ARYL HALIDES

Yasunori Yamamoto, Juugaku Sugai, Miho Takizawa, and Norio Miyaura*

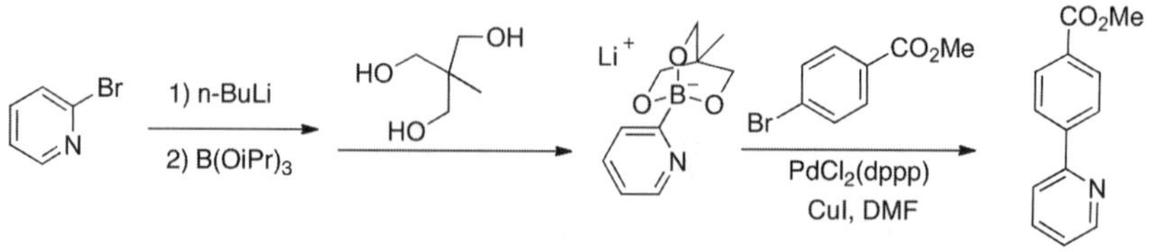

SYNTHESIS OF (+)-B-ALLYLDIISOPINOCAMPHEYLBORANE AND ITS REACTION WITH ALDEHYDES

Huikai Sun and William R. Roush*

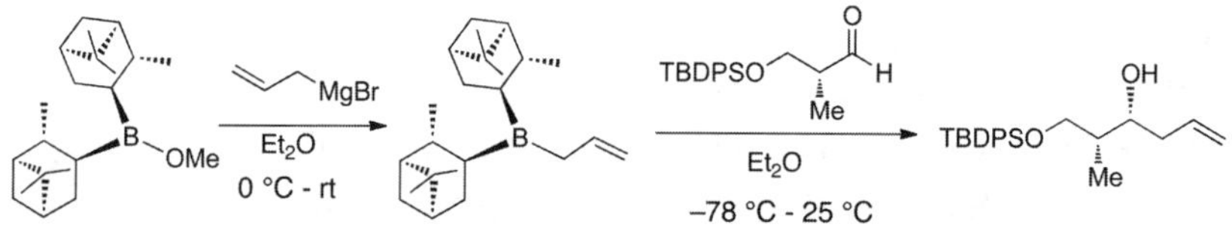

Discussion Addendum for:

PALLADIUM-CATALYZED CROSS-COUPLING OF (Z)-1-HEPTENYLDIMETHYLSILANOL WITH 4-IODOANISOLE: (Z)-(1-HEPTENYL)-4-METHOXYBENZENE

Scott E. Denmark* and Jack Hung-Chang Liu
Original article published in *Org. Synth.* **2005**, *81*, 42-53

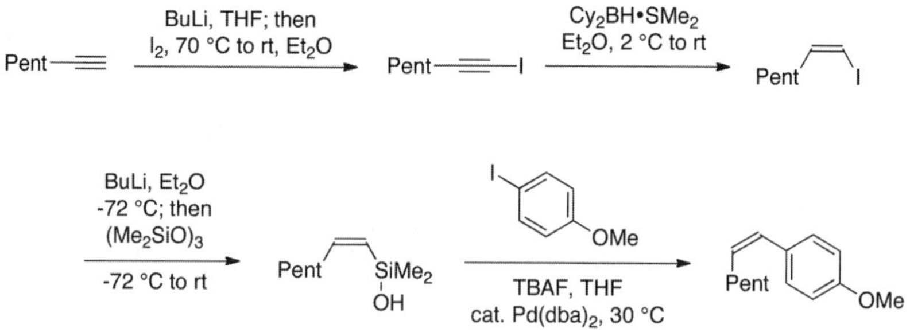

THE PREPARATION OF CYCLOHEPT-4-ENONES BY RHODIUM-CATALYZED INTERMOLECULAR [5+2] CYCLOADDITION

Paul A. Wender,* Adam B. Lesser, and Lauren E. Sirois

ORGANOCATALYTIC ENANTIOSELECTIVE SYNTHESIS OF BICYCLIC β-LACTONES FROM ALDEHYDE ACIDS VIA NUCLEOPHILE-CATALYZED ALDOL-LACTONIZATION (NCAL)

Henry Nguyen, Seongho Oh, Huda Henry-Riyad, Diana Sepulveda, and Daniel Romo*

PHOSPHINE-CATALYZED [3 + 2] ANNULATION: **138**
SYNTHESIS OF ETHYL 5-(*tert*-BUTYL)-2-PHENYL-
1-TOSYL-3-PYRROLINE-3-CARBOXYLATE

Ian P. Andrews and Ohyun Kwon*

PREPARATION OF HORNER-WADSWORTH-EMMONS   **152**
REAGENT: METHYL 2-BENZYLOXYCARBONYLAMINO-
2-(DIMETHOXY-PHOSPHINYL)ACETATE

Hiroki Azuma, Kentaro Okano, Tohru Fukuyama, and
Hidetoshi Tokuyama*

Discussion Addendum for: **162**

5-*ENDO-TRIG* CYCLIZATION OF 1,1-DIFLUORO-1-ALKENES: SYNTHESIS OF 3-BUTYL-2-FLUORO-1-TOSYLINDOLE (1*H*-INDOLE, 3-BUTYL-2-FLUORO-1-[(4-METHYLPHENYL)SULFONYL]-)

Junji Ichikawa*

Original article published in *Org. Synth.* **2006**, *83*, 111-120

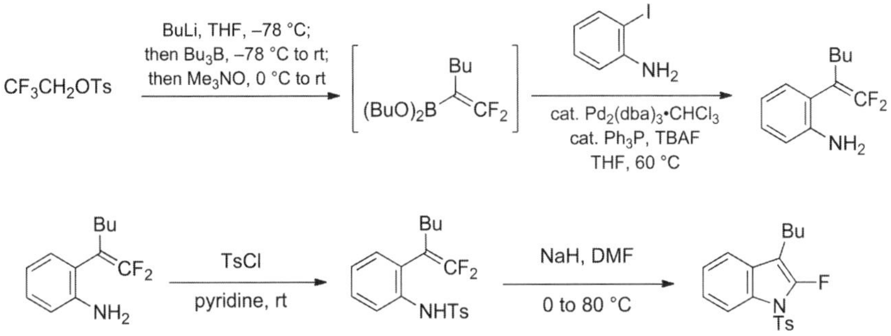

PREPARATION OF A TRIFLUOROMETHYL TRANSFER AGENT: 1-TRIFLUOROMETHYL-1,3-DIHYDRO-3,3-DIMETHYL-1,2-BENZIODOXOLE **168**

Patrick Eisenberger, Iris Kieltsch, Raffael Koller, Kyrill Stanek, and Antonio Togni*

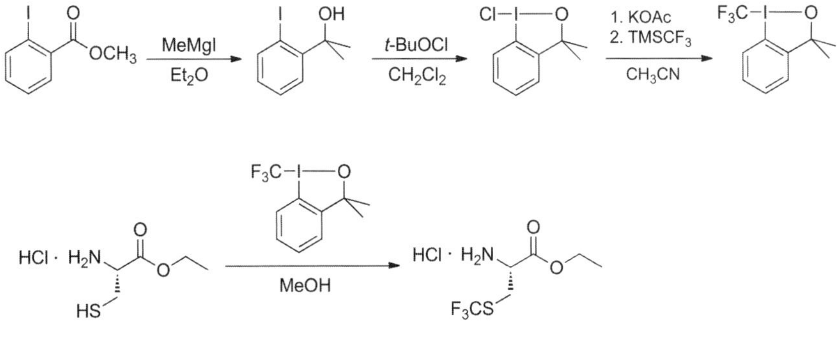

SYNTHESIS OF (*S,S*)-DIISOPROPYL TARTRATE (*E*)-CROTYLBORONATE AND ITS REACTION WITH ALDEHYDES: (2R,3R,4R)-1,2-DIDEOXY-2-ETHENYL-4,5-O-(1-METHYLETHYLIDENE)-XYLITOL

181

Huikai Sun and William R. Roush*

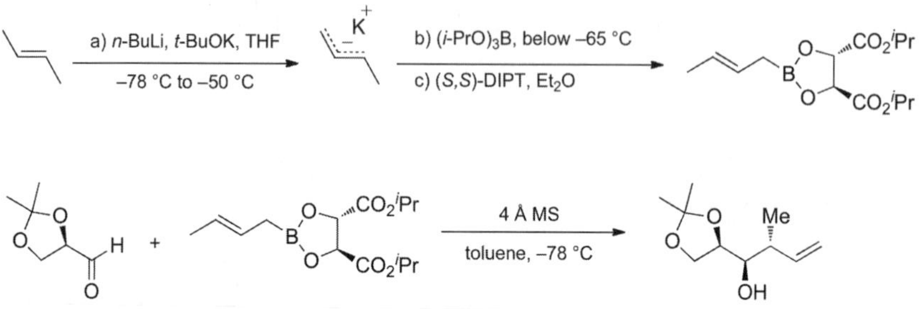

Discussion Addendum for:

197

4-METHOXY-4'-NITROPHENYL. RECENT ADVANCES IN THE STILLE BIARYL COUPLING REACTION AND APPLICATIONS IN COMPLEX NATURAL PRODUCTS SYNTHESIS

Robert M. Williams*

Original article published in *Org. Synth.* **1993**, *71*, 97-106

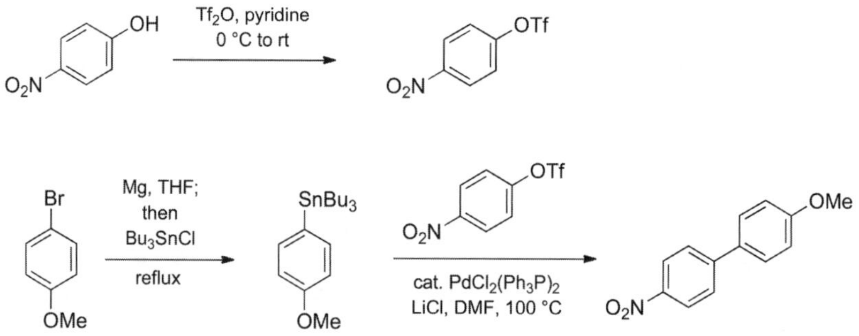

xliixlii

Discussion Addendum for:
PALLADIUM-CATALYZED REACTION OF
1-ALKENYLBORONATES WITH VINYLIC HALIDES:
(1*Z*,3*E*)-1-PHENYL-1,3-OCTADIENE
Norio Miyaura*
Original article published in *Org. Synth.* **1990**, *68*, 130-137

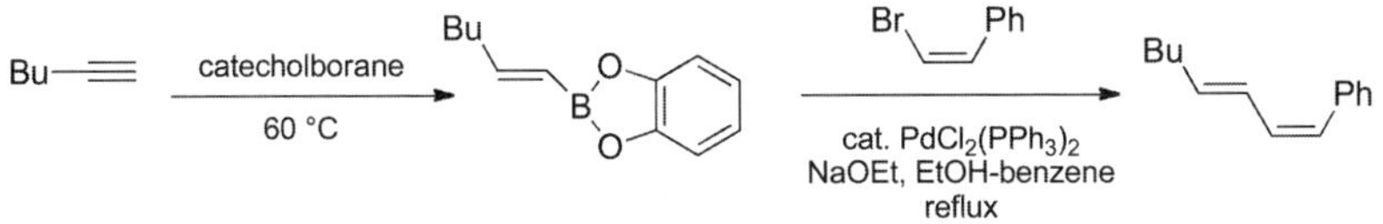

202

Bu—≡ →[catecholborane][60 °C] Bu⌒⌒B(O₂C₆H₄) →[cat. PdCl₂(PPh₃)₂][NaOEt, EtOH-benzene reflux] (Br / Ph) Bu⌒⌒⌒Ph

Discussion Addendum for:
PALLADIUM(0)-CATALYZED REACTION
OF 9-ALKYL-9-BORABICYCLO[3.3.1]NONANE
WITH 1-BROMO-1-PHENYLTHIOETHENE: 4-(3-
CYCLOHEXENYL)-2-PHENYLTHIO-1-BUTENE
Norio Miyaura*
Original article published in *Org. Synth.* **1993**, *71*, 89-96

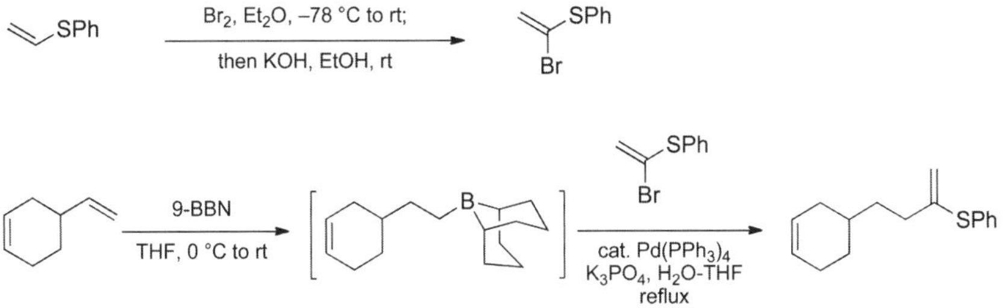

207

⌒SPh →[Br₂, Et₂O, –78 °C to rt;][then KOH, EtOH, rt] (SPh / Br)

cyclohexenyl-CH=CH₂ →[9-BBN][THF, 0 °C to rt] [B-alkyl-9-BBN] →[cat. Pd(PPh₃)₄][K₃PO₄, H₂O-THF reflux] product-SPh

PREPARATION OF ISOPROPYL 2-DIAZOACETYL(PHENYL)CARBAMATE

Hubert Muchalski, Amanda B. Doody, Timothy L. Troyer, and Jeffrey N. Johnston*

GRAM SCALE CATALYTIC ASYMMETRIC AZIRIDINATION: PREPARATION OF (2R,3R)-ETHYL 1-BENZHYDRYL-3-(4-BROMOPHENYL)AZIRIDINE 2-CARBOXYLATE

Aman A. Desai, Roberto Morán-Ramallal, and William D. Wulff*

Discussion Addendum for: **260**
**PALLADIUM-CATALYZED REDUCTION OF VINYL
TRIFLUOROMETHANESULFONATES TO ALKENES:
CHOLESTA-3,5-DIENE**
Sandro Cacchi,* Enrico Morera, and Giorgio Ortar
Original article published in *Org. Synth.* **1990**, *68*, 138-147

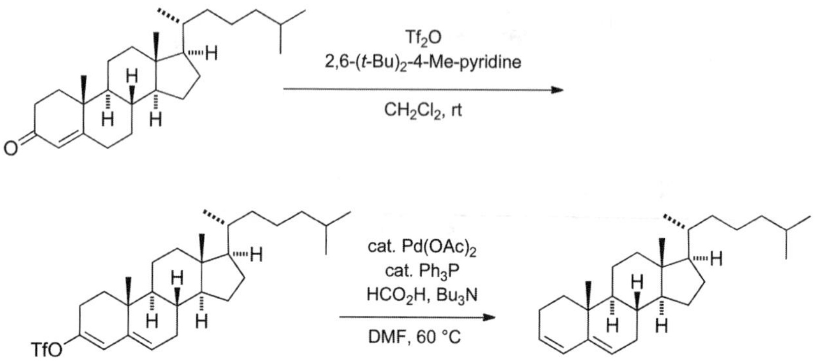

Tf₂O
2,6-(*t*-Bu)₂-4-Me-pyridine

CH₂Cl₂, rt

cat. Pd(OAc)₂
cat. Ph₃P
HCO₂H, Bu₃N

DMF, 60 °C

Discussion Addendum for: **291**
**SYNTHESIS OF INDOLES BY PALLADIUM-
CATALYZED REDUCTIVE *N*-HETEROANNULATION
OF 2-NITROSTYRENES: METHYL INDOLE-4-
CARBOXYLATE**
Björn C. Söderberg*
Original article published in *Org. Synth.* **2003**, *80*, 75-84

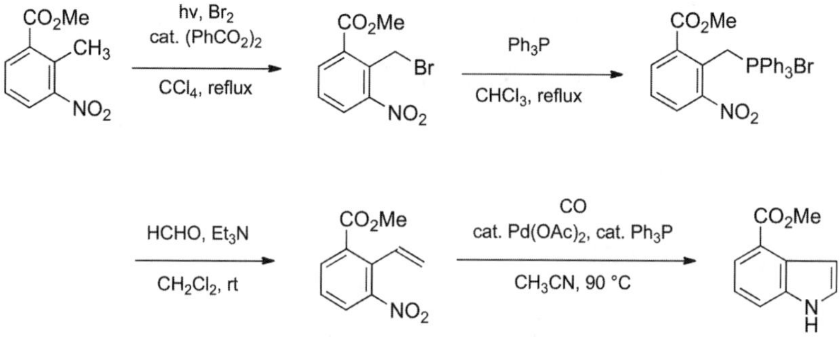

hv, Br₂
cat. (PhCO₂)₂

CCl₄, reflux

Ph₃P

CHCl₃, reflux

HCHO, Et₃N

CH₂Cl₂, rt

CO
cat. Pd(OAc)₂, cat. Ph₃P

CH₃CN, 90 °C

SYNTHESIS OF (*S*)-8a-METHYL-3,4,8,8a-TETRAHYDRO-   330
1,6-(2*H*,7*H*)-NAPHTHALENEDIONE VIA *N*-TOSYL-(*S*ₐ)-
BINAM-L-PROLINAMIDE ORGANOCATALYSIS

Ben Bradshaw, Gorka Etxebarria-Jardí, Josep Bonjoch,*
Santiago F. Viózquez, Gabriela Guillena, and Carmen Nájera

STEREOSELECTIVE NICKEL-CATALYZED   342
1,4-HYDROBORATION OF 1,3-DIENES

Robert J. Ely and James P. Morken*

SYNTHESIS OF 2,3-DISUBSTITUTED INDOLES VIA   377
PALLADIUM-CATALYZED ANNULATION OF INTERNAL
ALKYNES: 3-METHYL-2-(TRIMETHYLSILYL)INDOLE

Yu Chen, Nataliya A. Markina, Tuanli Yao, and Richard C. Larock*

AN INTRAMOLECULAR AMINATION OF ARYL HALIDES   388
WITH A COMBINATION OF COPPER (I) AND
CESIUM ACETATE: PREPARATION OF
5,6-DIMETHOXYINDOLE-1,2-DICARBOXYLIC
ACID 1-BENZYL ESTER 2-METHYL ESTER

Toshiharu Noji, Kentaro Okano, Tohru Fukuyama, and
Hidetoshi Tokuyama*

LIGAND-FREE COPPER(II) OXIDE NANOPARTICLES-   398
CATALYZED SYNTHESIS OF SUBSTITUTED
BENZOXAZOLES

Prasenjit Saha, Md Ashif Ali, and Tharmalingam Punniyamurthy*

1

(R)-3,3'-BIS(9-PHENANTHRYL)-1,1'-BINAPHTHALENE-   **406**
2,2'-DIYL HYDROGEN PHOSPHATE

Wenhao Hu,* Jing Zhou, Xinfang Xu, Weijun Liu, and Liuzhu Gong*

ENANTIOSELECTIVE THREE-COMPONENT REACTION   **418**
FOR THE PREPARATION OF β-AMINO-α-HYDROXY
ESTERS

Jing Zhou, Xinfang Xu, and Wenhao Hu*

MILD CONVERSION OF TERTIARY AMIDES TO ALDEHYDES USING Cp$_2$Zr(H)Cl (SCHWARTZ'S REAGENT)

Matthew W. Leighty, Jared T. Spletstoser, and Gunda I. Georg*

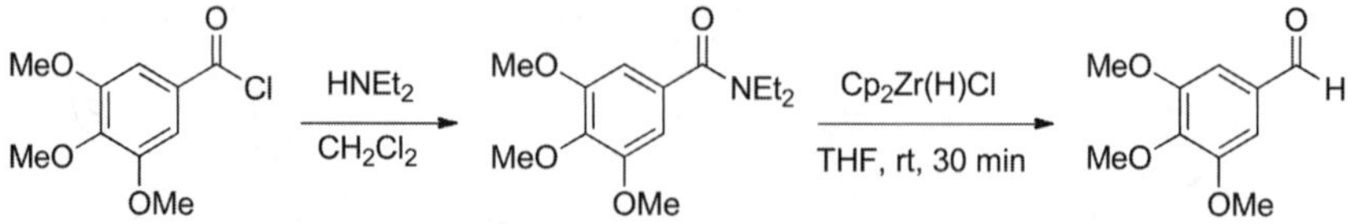

Editorial

Organic Syntheses:
The "Gold Standard" in Experimental Synthetic Organic Chemistry

Experimental Synthetic Chemistry: Art or Science?

It has often been suggested that the practice of synthetic organic chemistry is as much an art as a science. Commenting on the controversy over the reproducibility of the Rabe synthesis of quinine, the late William von Eggers Doering wrote

> *"It is almost never possible to reproduce published details. They assume an indefinable amount of experience and cannot be written for the first time cook who has never mastered the elementary techniques! The premise that the best, detailed descriptions suffice to guarantee reproducibility is contrary to universal experience. Try writing two sets of descriptions for playing a piano composition - the one to reproduce the performance according to Ogdon, the second Horowitz!"*[1]

Is this truly the case? Is it impossible to prepare a written experimental procedure that enables any chemist "skilled in the art" to reproduce the yields and selectivity reported for a synthetic reaction? This is the question that will be addressed in this editorial and its sequel.

Reproducibility: A Perennial Concern in Synthetic Chemistry

There is no denying that the reproducibility of reactions found in the chemical literature cannot be taken for granted. Who among experienced synthetic chemists has not at one time or another found themselves unable to obtain the reported yield for a reaction, even after repeated trials? Some in the community would even argue that the situation has grown worse in recent years. This may indeed be the case, and one can suggest several reasons why reproducibility may be more problematic now than in the past.

- **Increasing sensitivity of reactions to precise conditions**. Reactions today often call for reagents that are exceptionally sensitive to even traces of air and/or moisture. Many reactions are carried out at low temperature and are sensitive to exotherms, local heating effects, etc. The penalty for deviating from optimal conditions can be quite severe.

- **Catalytic reactions**. More and more often, reactions employ reagents in catalytic amounts. Catalytic processes can be especially sensitive to trace impurities and minor departures from the optimal protocol.
- **Reactions carried out on a very small scale**. It is not unusual in a paper reporting a new synthetic method to find that no case was run on a scale producing more than 10-25 mgs of product. This certainly increases the likelihood that problems will be encountered when the reaction is attempted on a "preparative" scale.
- **Carelessly prepared experimental procedures**. Unfortunately, it is all too common these days to find that experimental procedures lack adequate detail, are carelessly written, and contain unclear or ambiguous instructions that are easily misinterpreted. This may be a consequence of the fact that today experimental procedures are usually relegated to "supporting information" (SI) rather than appearing incorporated in the main text of an article. The frequency with which errors, ambiguous language, grammatical mistakes, etc. appear in SI procedures suggests that these procedures are not receiving the same scrutiny from the principal authors and from reviewers that would be afforded to procedures included in the main text.[2]

Organic Syntheses as the "Gold Standard" for Synthetic Chemistry

Since 1921, the mission of *Organic Syntheses* has been to provide the community with detailed and reliable procedures for the synthesis of organic compounds. Unique among chemistry publications, every reaction published in *Organic Syntheses* is "checked" for reproducibility in the laboratory of a member of the Board of Editors (BoE). The eleven academic and industrial scientists on the BoE serve 8-year terms (5 years in the case of editors from outside North America) and are elected by the current members of the Board. The chemists elected to the BoE are drawn from the most distinguished members of the synthetic organic chemistry community to ensure that checking is carried out by highly competent and experienced researchers under the supervision of the leading figures in the field.

The following system has been developed over the years in order to maximize the reproducibility of the initial versions of procedures submitted to checking, as well as the final versions published in *Organic Syntheses*.

1. Comprehensive and very explicit "Instructions to Authors" have been prepared that require submitters to provide much more detail in experimental procedures than is typical in other publications.
2. Every article submitted to *Organic Syntheses* is first reviewed by each member of the BoE who provides detailed comments to the Editor in Chief. If the consensus favors checking, the Editor in Chief assigns the procedure to the laboratory of one of the editors for checking.
3. In the initial review, BoE members carefully scrutinize each submitted experimental procedure to identify any aspects of the procedure that are unclear or ambiguous. Authors are contacted and asked to provide additional details clarifying such points before checking begins.
4. In order for an article to be accepted for publication, each experimental procedure must be reproduced at least twice in the checking editor's laboratory with yields and selectivity close to that reported by the submitters. If problems are encountered, the original authors are consulted for advice and assistance.
5. When results slightly differ, the checking editor's data are reported in the final article, with the submitting author's results generally mentioned in an accompanying note.

In spite of all of these measures, currently 3-5% of the articles submitted for publication eventually must be rejected due to the inability of the checkers to satisfactorily reproduce the results of the submitters. This figure actually represents an improvement in recent years, since during the period 1982-2005 the rejection rate was ca. 12%! Why do so many procedures prove not to be reproducible? Part 2 of this editorial will discuss the most common causes of problems based on the experiences of the Board of Editors.

Rick L. Danheiser
Editor in Chief,
Organic Syntheses

---

[1] W. E. Doering email to J. I. Seeman, Cambridge, MA, May 18, 2005 as reported in Seeman, J. I. *Angew. Chem. Int. Ed.* **2007**, *46*, 1378-1413.

[2] Beginning in 2011, the *Journal of Organic Chemistry* has published experimental procedures in the main text of articles (rather than in SI). It is hoped that other journals will follow suit!

α-ARYLATION OF ESTERS CATALYZED BY THE Pd(I) DIMER
[P(*t*-Bu)₃Pd(μ-Br)]₂

Submitted by David S. Huang, Ryan J. DeLuca, and John. F. Hartwig.[1]
Checked by David Hughes.

1. Procedure

A 500-mL, 3-necked, round-bottomed flask (Note 1) equipped with a 3-cm oval Teflon-coated magnetic stir bar is fitted with a gas inlet adapter connected to a nitrogen line and a gas bubbler. The other two necks are capped with rubber septa; a thermocouple probe is inserted through one of the septa. (Note 2) To the flask is added anhydrous toluene (100 mL, Note 3) and dicyclohexylamine (9.96 g, 54.9 mmol, 1.3 equiv). The flask is placed in an ice-water bath and cooled with stirring to +2 °C. *n*-Butyllithium (2.36 M in hexanes, 22.0 mL, 15.2 g, 51.9 mmol, 1.23 equiv) is added over 10 min to the cooled solution of dicyclohexylamine via a 50-mL disposable syringe (Notes 4 and 5). The reaction mixture is stirred for 20 min at 0–5 °C. To the resulting lithium dicyclohexylamide suspension is added methyl isobutyrate (5.40 mL, 4.80 g, 47.0 mol, 1.11 equiv) over 20 min via a disposable 10-mL syringe (Note 6). The reaction mixture is stirred for an additional 30 min at 0–5 °C. 3-Bromoanisole (5.40 mL, 7.90 g, 42.2 mol, 1.00 equiv) is then added over 1 min via a 10-mL disposable syringe. The mixture is degassed by two vacuum-nitrogen purge cycles (Note 7). A septum is removed, [P(*t*-Bu)₃Pd(μ-Br)]₂ (12.4 mg, 0.0160 mmol, 0.00038 equiv) is added under a flow of nitrogen, and then the septum is replaced (Note 8). The flask is removed from the ice-water bath, allowed to warm to ambient temperature, and the reaction mixture is stirred for one hour (Note 9). A septum is removed and additional [P(*t*-Bu)₃Pd(μ-Br)]₂ (13.8 mg, 0.0180 mmol, 0.00042 equiv) is added under a flow of nitrogen (Note 10). The reaction mixture is stirred at ambient temperature for 4 h. After

4

confirming reaction completion (Note 9), *t*-butyl methyl ether (100 mL) is added. One septum is removed and replaced with a 100-mL dropping funnel. Aqueous HCl (1N, 70 mL) is added to the reaction mixture over 10 min via the dropping funnel, resulting in a temperature rise to 30 °C and formation of a thick slurry (Note 11). The resulting suspension is stirred for 10 min and then is filtered through a 600-mL coarse-porosity sintered glass funnel. The precipitate is washed with *t*-butyl methyl ether (4 x 25 mL). The resulting mixture is transferred to a 500-mL separatory funnel, and the organic layer is separated, washed sequentially with saturated aqueous NaHCO$_3$ (50 mL) and brine (50 mL), and then is vacuum-filtered through a bed of Na$_2$SO$_4$ (50 g) in a 350-mL medium porosity sintered glass funnel. The cake is rinsed with *t*-butyl methyl ether (3 x 25 mL). The filtrate is concentrated by rotary evaporation (40 °C bath, 100 mmHg initial, lowered to 20 mmHg) to afford the crude product (10.2 g), which is purified by silica gel column chromatography (Note 12) to furnish methyl 2-(3-methoxyphenyl)-2-methylpropanoate (7.14–7.59 g, 81–86 % yield) as a clear yellow oil (Notes 13-15).

2. Notes

1.   All glassware was dried in an oven at 130 °C prior to use.
2.   The internal temperature is monitored using a J-Kem Gemini digital thermometer with a Teflon-coated T-Type thermocouple probe (12-inch length, 1/8 inch outer diameter, temperature range –200 to +250 °C).
3.   The following reagents and solvents were obtained from Sigma-Aldrich and used without further purification: toluene (ACS reagent, >99.5%, dried over 3A pelleted molecular sieves), 2.5 M BuLi in hexanes, dicyclohexylamine (99%), methyl isobutyrate (99%), 3-bromoanisole (98%), *t*-butyl methyl ether (ACS reagent, >99%), ethyl acetate (ACS reagent, >99.5%), and hexanes (ACS reagent, >98.5%). [P(*t*-Bu)$_3$Pd(μ-Br)]$_2$ was obtained from Strem and stored in a glove box freezer at –35 °C. Deionized tap water was used throughout.
4.   The mass of *n*-BuLi added was determined by weighing the syringe before and after addition. *n*-BuLi was titrated using diphenylacetic acid as described in Davies, S. G.; Fletcher, A. M.; Roberts, P. M. *Org. Synth*, **2010,** *87,* 143-160.

5. The reaction mixture warmed to 7 °C during the addition and became a yellow slurry as $LiNCy_2$ precipitated when the enolate was formed.

6. Adding the ester slowly is crucial to avoid the Claisen condensation product, which is difficult to remove from the product by flash chromatography. The mixture warmed to 6 °C during the addition.

7. The purge cycle was carried out by slowly drawing a vacuum in the flask, which results in bubbling as the mixture is degassed. After 2 minutes the bubbling nearly ceases and the flask is back-filled with nitrogen. The cycle is repeated to ensure all dissolved oxygen, which may be present, is removed.

8. The quality of the catalyst is vital to the reaction. The catalyst should be a dark metallic green. If there is concern about the quality of the catalyst, a ^{31}P NMR spectrum should be obtained. $^{31}P\{^{1}H\}NMR$ (500 MHz,C_6H_6, H_3PO_4) δ: 87.0 (s). Poorly performing catalyst is brown/black in color and contains species that appear in the ^{31}P NMR spectrum at δ: 107 (s). $[P(t\text{-}Bu)_3Pd(\mu\text{-}Br)]_2$ decomposes to $[Pd(P(t\text{-}Bu)_3)_2(C(CH_3)_2CH_2)(\mu\text{-}Br)]_2$ over time.[2]

9. The mixture warms to 20 °C over 20 minutes and changes from a brown mixture to yellow. The progress of the reaction is monitored by ^{1}H NMR (checker) or GC analyses (submitter). At the one-hour reaction point after the first catalyst addition, the reaction proceeds only 5-10 %. For the NMR analysis, a sample of the reaction mixture is quenched into a mixture of 1 mL of 1N HCl and 1 mL of $CDCl_3$. The bottom organic layer is filtered through a plug of sodium sulfate into an NMR tube. The methyl resonances of the methyl ester and methoxy group are diagnostic (OMe product resonance at 3.85 ppm, starting material at 3.84 ppm; CO_2Me product resonance at 3.70 ppm, starting material at 3.72 ppm). GC analyses were obtained on an Agilent 6890 GC equipped with an HP-5 column (25 m x 0.20 mm ID x 0.33 μm film) and an FID detector. The temperature program: hold at 80 °C for 1.5 min, ramp from 80 °C to 300 °C at 100 °C /min, hold at 300 °C for 3 min. t_R (3-bromoanisole) = 3.33 min, t_R (methyl 2-(3-methoxyphenyl)-2-methylpropanoate) = 3.89 min.

10. After the second charge of catalyst the mixture slowly warms from 23 °C to 33 °C over 30 min and then returns to room temperature over the next hour. The reaction is generally complete within an hour of the second charge. The checker found the double catalyst charge protocol provided optimum results, where the first charge is largely sacrificial. When added as

6

a single charge, the reaction times were variable (10-30 hours) and generally stalled at 90% completion. For stalled reactions, a second catalyst charge even after one-day reaction time will drive the reaction to completion.

11. The ^{1}H NMR spectrum of the precipitate matched the spectrum of dicylohexylammonium chloride.[3]

12. A 6-cm glass column is wet-packed (4% EtOAc/hexanes) with SiO_2 (250 g) topped with 0.5 cm sand. The crude reaction product is loaded neat on the column and eluted with 4% EtOAc/hexanes (2.5 L), collecting 100 mL fractions. TLC (UV visualization) is used to follow the chromatography. The Rf value of the title compound is 0.4 (10% EtOAc/hexanes). Fractions 15-22 are concentrated by rotary evaporation (40 °C bath, 20 mmHg), then held under vacuum (20 mmHg) at 22 °C for 20 h to constant weight (7.14–7.59 g).

13. Methyl 2-(3-methoxyphenyl)-2-methylpropanoate has the following physical and spectroscopic data: ^{1}H NMR (400 MHz, CDCl$_3$) δ: 1.59 (s, 6 H), 3.67 (s, 3 H), 3.81 (s, 3 H), 6.79 (ddd, J = 8.2, 2.5, 0.8 Hz, 1 H), 6.91–6.89 (m, 1 H), 6.93 (ddd, J = 7.8, 1.8, 0.8 Hz, 1 H), 7.26 (t, J = 8.0 Hz, 1 H). ^{13}C NMR (100 MHz, CDCl$_3$) δ: 26.7, 46.7, 52.3, 55.3, 111.7, 112.4, 118.3, 129.5, 146.6, 159.8, 177.3; IR (thin film): 2978, 2953, 2838, 1729, 1601, 1584, 1490, 1466, 1434, 1263, 1149, 1050 cm^{-1}. LC-MS calcd for [M + H]$^+$ 209.2; found 209.1; GC-MS (EI): 208 (M$^+$) (25 %), 149 ([M– CO$_2$Me]$^+$)(100 %). HPLC >99 area % purity at 215 nm detection (HPLC conditions, Zorbax extend C18 column (3 x 150 mm), 3.5 µM particle size; 0.75 mL/min flow; gradient eluent from 5/95 MeCN/ aq. pH 3.5 buffer to 100% MeCN over 9.5 min, hold for 3 min; 35 °C; product elutes at 8.5 min). An analytical sample was prepared by dissolving 100 mg of the product in 3 mL of hexanes, filtering through a 0.45 micron PTFE syringe filter, and concentrating to dryness under vacuum for 20 h. Anal. calcd. for C$_{12}$H$_{16}$O$_3$: C, 69.21; H, 7.74; found: C, 68.97; H, 7.73.

14. The product after chromatographic purification contains 0.5-1.0 % of the Claisen condensation product of methyl isobutyrate (methyl 2,2,4-trimethyl-3-oxopentanoate; NMR match with the literature, Mloston, G.; Romanski, J.; Linden, A.; Heimgartner, H. *Helv. Chim. Acta* **1999**, *82*, 1302-1310) as assessed by peak integration of the ^{13}C-^{1}H satellite resonances (0.55 %) corresponding to the OMe and CO$_2$Me protons of the product against the ^{1}H resonances corresponding to gem-dimethyl (δ 1.39) and Me$_2$CH (δ 1.09) protons of the Claisen impurity. For more details on using ^{13}C satellites for quantitative analysis of low level impurities, see Claridge, T. D. W.; Davies,

S. G.; Polywka, M. E. C.; Roberts, P. M.; Russell, A. J.; Savory, E. D.; Smith, A. D. *Org. Lett.* **2008**, *10*, 5433. This impurity was not detected by GC-MS or LC-MS.

15. The major by-product generated in the reaction (2-6 % yield) is *N*, *N*-dicyclohexyl-3-methoxyaniline arising from the C-N cross-coupling reaction between dicyclohexylamine and 3-bromoanisole. This impurity elutes prior to the main fraction in the column chromatography and is readily removed. A pure sample was obtained by combining early fractions from several reactions and re-chromatographing as follows. A 5-cm glass column is wet-packed (3% EtOAc/hexanes) with SiO$_2$ (150 g) topped with 0.5 cm sand. The crude amine (1.0 g) is loaded neat and eluted with 3% EtOAc/hexanes (700 mL), taking 50 mL fractions. Fractions 16-18 are concentrated by rotary evaporation (40 °C bath, 20 mmHg) to afford *N*, *N*-dicyclohexyl-3-methoxyaniline (0.68 g). ^{1}H NMR (500 MHz, CDCl$_3$) δ: 1.11-1.17 (m, 2 H), 1.28-1.36 (m, 4 H), 1.55-1.66 (m, 6 H), 1.75-1.82 (m, 8 H), 3.26 (tt, *J* = 3.3, 11.6 Hz, 2 H), 3.79 (s, 3 H), 6.36 (dd, *J* = 2.3, 8.2 Hz, 1 H), 6.50 (t, *J* = 2.3 Hz, 1 H), 6.57 (dd, *J* = 2.3, 8.3 Hz, 1 H), 7.09 (t, *J* = 8.2 Hz, 1 H); ^{13}C NMR (125 MHz, CDCl$_3$) δ: 26.2, 26.6, 32.2, 55.3, 58.0, 103.2, 106.7, 113.3, 128.9, 150.4, 160.1.

Safety and Waste Disposal Information

All hazardous materials should be handled and disposed of in accordance with "Prudent Practices in the Laboratory"; National Academy Press; Washington, DC, 1995.

3. Discussion

The palladium-catalyzed coupling of carbonyl compounds and aryl halides is a convenient method for the synthesis of the aryl C-C bond in α-aryl carboxylic acid derivatives.[4-9] The α-arylation of esters proceeds in high yields and tolerates a variety of functional groups on both the ester and the aryl halide.[5] The α-arylation of esters with aryl bromides is reported in the literature to proceed at ambient temperature for catalyst systems containing tri-*tert*-butylphosphine as ligand.[5d, 5f, 5i, 5j] The catalyst for these systems is either generated by treating a Pd^0 precursor with tri-*tert*-butylphosphine or from the palladium (I) dimer, [(P(*t*-Bu)$_3$Pd(μ-Br)]$_2$. [(P(*t*-Bu)$_3$Pd(μ-Br)]$_2$ is an effective catalyst for a number of different cross-coupling reactions.[5f, 5h-5j, 6c, 10-15] The α-arylation of esters with aryl bromides has been studied with the [(P(t-Bu)$_3$Pd(μ-Br)]$_2$ as the catalyst (Table 1). The coupling of esters with aryl bromides containing different functional groups and heteroatoms proceeds in moderate to high yield with low catalyst loadings.[5h] This catalyst is advantageous because it can be weighed in air, even though tri-*tert*-butylphosphine is pyrophoric. [(P(t-Bu)$_3$Pd(μ-Br)]$_2$ is a more active catalytic system for the coupling of esters and aryl bromides than other catalytic systems based on tri-*tert*-butylphosphine.[5d, 5h]

Table 1. α-Arylation of ester with aryl bromides catalyzed by [(P(t-Bu)$_3$Pd(μ-Br)]$_2$.[5h]

Entry	R^1	R^2	R^3	ArBr		Cat. loading	Yield^a
1	H	H	t-Bu	t-Bu—C$_6$H$_4$—Br		0.2%	83%
2				F—C$_6$H$_4$—Br		0.4%	82%
3				MeO—C$_6$H$_4$—Br		0.2%	86%
4				F$_3$C—C$_6$H$_4$—Br		0.4%	73%
5	Me	H	t-Bu	Cl—C$_6$H$_4$—Br		0.2%	83%
6					*p*-F	0.2%	88%
7					*m*-F	0.2%	90%
8				MeO-naphthyl-Br		0.2%	75%
9					*p*-OMe	0.25%	87%
10					*m*-OMe	0.25%	84%
11					*o*-OMe	0.25%	87%

Table 1. (continued)

Entry	R^1	R^2	R^3	ArBr		Cat. loading	Yield[a]
12				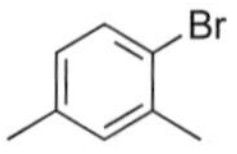		0.05%	72%
13	Me	Me	Me	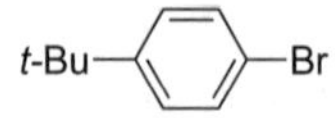 t-Bu—⟨ ⟩—Br		0.05%	72%
14				 Cl—⟨ ⟩—Br		0.5%	89%
15				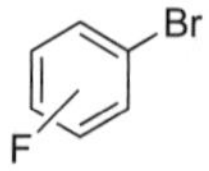	p-F	0.5%	85%
16					m-F	0.5%	72%
17				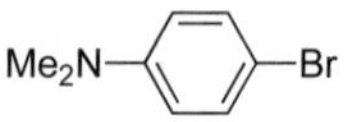 Me$_2$N—⟨ ⟩—Br		0.05%	88%
18				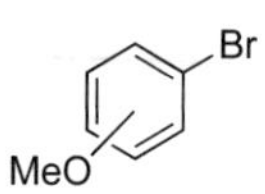	p-OMe	0.5%	85%
19					m-OMe	0.5%	88%
20				MeO	m-OMe	0.075%	77%[b]
21				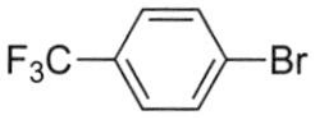 F$_3$C—⟨ ⟩—Br		0.5%	60%
22				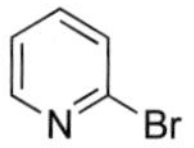		0.5%	71%
23				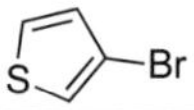		0.5%	75%

a) Isolated yields (average of two runs) for reaction of 1 mmol of bromoarene in 3 mL toluene. b) This work

1. Department of Chemistry, University of Illinois, A410 Chemical Life Science Lab, 600 S. Matthews Ave., Urbana, IL, 61801. E-mail: jhartwig@scs.illinois.edu
2. Barrios-Landeros, F.; Carrow, B. P.; Hartwig, J. F. *J. Am. Chem. Soc.* **2008,** *130*, 5842.
3. Gopalakrishnana, J.; Srinivas, J.; Srinivasamurthy, G.; Rao, M. N. S. *Indian J. Chem., Sect. B: Org. Chem. Incl. Med. Chem.* **1997,** *36B*, 47.
4. For a recent review of transition metal catalyzed α-arylation of carbonyl compounds see, Johansson, C. C. C.; Colacot, T. J. *Angew. Chem. Int. Ed.* **2010,** *49*, 676.
5. (a) Satoh, T.; Inoh, J.; Kawamura, Y.; Miura, M.; Nomura, M. *Bull. Chem. Soc. Jpn.* **1998,** *71*, 2239. (b) Moradi, W. A.; Buchwald, S. L. *J. Am. Chem. Soc.* **2001,** *123*, 7996. (c) Lee, S.; Beare, N. A.; Hartwig J.

10

F. *J. Am. Chem. Soc.* **2001**, *123*, 8410 (d) Jørgensen, M.; Lee, S.; Liu, X.; Wolkowski, J. P.; Hartwig, J. F. *J. Am. Chem. Soc.* **2002**, *124*, 12557. (e) Gaertzen, O.; Buchwald, S. L. *J. Org. Chem.* **2002**, *67*, 465. (f) Hama, T.; Liu, X.; Culkin, D. A.; Hartiwg, J. F. *J. Am. Chem. Soc.* **2003**, *125*, 11176. (g) Solé, D.; Serrano, O. *J. Org. Chem.* **2008**, *73*, 2476. (h) Hama, T.; Hartwig, J. F. *Org. Lett.* **2008**, *10*, 1545. (i) Hama, T.; Hartwig, J. F. *Org. Lett.* **2008**, *10*, 1549. (j) Bercot, B. A.; Caille, S.; Bostick, T. M.; Ranganathan, K.; Jensen, R.; Faul, M. M *Org. Lett.* **2008**, *10*, 5251 (k) Biscoe, M. R.; Buchwald, S. L. *Org. Lett.* **2009**, *11*, 1773.

6. (a) Shaughnessy, K. H.; Hamann, B. C.; Hartiwg J. F. *J. Am. Chem. Soc.* **1998**, *63* 6546. (b) Lee, S.; Hartwig, J. F. *J. Org. Chem.* **2001**, *66* 3402. (c) Hama, T.; Culkin, D. A.; Hartwig, J. F. *J. Am. Chem. Soc.* **2006**, *128*, 4976. (d) Arao, T.; Kondo, K.; Aoyama, T. *Chem. Pharm. Bull.* **2006**, *54*, 1743. (e) Kündig, E. P.; Seidel, T. M.; Jia, Y. X.; Bernardinelli, G. *Angew. Chem. Int. Ed.* **2007**, *46* 8484. (f) Hillgren, J. M.; Marsden, S. P. *J. Org. Chem.* **2008**, *73*, 6459. (g) Jia, Y. X.; Hillgren, M.; Watson, E. L.; Marsden, S. P.; Kündig, E. P. *Chem. Comm.* **2008**, 4040. (h) Altman, R. A. Hyde, A. M.; Huang, X.; Buchwald, S. L. *J. Am. Chem. Soc.* **2008**, *130*, 9613.

7. (a) Liu, X.; Hartwig, J. F. *Org. Lett.* **2003**, *5*, 1915. (b) Durbin, M. J.; Willis, M. C. *Org. Lett.* **2008**, *10*, 1413. (c) Altman, R. A. Hyde, A. M.; Huang, X.; Buchwald, S. L. *J. Am. Chem. Soc.* **2008**, *130*, 9613. (d) Jiang, L.; Weist, S.; Jansat, S. **2009**, *11*, 1543. (e) Taylor, A. M.; Altman, R. A. Buchwald, S. L. *J. Am. Chem. Soc.* **2009**, *131*, 9900.

8. (a) Kawatsusura, M. Hartwig, J. F.; *J. Am. Chem. Soc.* **1999**, *121*, 1473. (b) Stauffer, S. R.; Beare, N. A.; Stambuli, J. P., Hartwig, J. F. *J. Am. Chem. Soc.* **2001**, *123*, 4641. (c) Beare, N. A.; Hartwig, J. F. *J. Org. Chem.* **2002**, *67*, 541. (d) Millemaggi, A.; Perry, A.; Whitwood, A. C.; Taylor, R. J. K. *Eur. J. Org. Chem.* **2009**, 2947. (e) Storgaard, M.; Dörwald, Z.; Peschke, B.; Tanner, D. *J. Org. Chem.* **2009**, *74*, 5032.

9. Culkin, D. A.; Hartwig, J. F. *J. Am. Chem. Soc.* **2002**, *124*, 9330.

10. Dura-Vila, V.; Mingos, D. M. P.; Vilar, R.; White, A. J. P.; Williams, D. J. *J. Organomet. Chem.* **2000**, *600*, 198.

11. Stambuli, J.P.; Kuwano, R.; Hartwig, J. F. *Angew. Chem. Int. Ed. Engl.* **2002**, *41*, 4746.

12. Prashad, M.; Mak, X. Y.; Liu, Y.; Repic, O. *J. Org. Chem.* **2003**, *68*, 1163.

13. Hooper, M. W.; Utsunomiya, M.; Hartwig, J. F. *J. Org. Chem.* **2003**, *68*, 2861.
14. Huang, J.; Bunel E.; Faul, M. M. *Org. Lett.* **2007**, *9*, 4343
15. Ryberg, P. *Org. Process Res. Dev.* **2008**, *12*, 540.

Appendix
Chemical Abstracts Nomenclature; (Registry Number)

Dicyclohexylamine: Cyclohexanamine, *N*-cyclohexyl-; (101-83-7)
n-Butyllithium: Lithium, butyl-; (109-72-8)
Methyl isobutyrate: Propanoic acid, 2-methyl-, methyl ester; (547-63-7)
3-Bromoanisole: Benzene, 1-bromo-3-methoxy-; (2398-37-0)
Di-μ-bromobis(tri-*tert*-butylphosphine)dipalladium; (185812-86-6)
Methyl 2-(3-methoxyphenyl)-2-methylpropanoate: Benzeneacetic acid, 3-methoxy-α,α-dimethyl-, methyl ester; (32454-33-4)

John F. Hartwig received his A. B. degree from Princeton in 1986 and his Ph.D. from the University of California, Berkeley in 1990 before conducting postdoctoral research at the Massachusetts Institute of Technology. He began his independent career at Yale University in 1992 and joined the faculty at Illinois in July 2006. Professor Hartwig's research focuses on the discovery and mechanistic understanding of organic reactions catalyzed by organometallic complexes. He has developed palladium-catalyzed cross-coupling reactions to form carbon-heteroatom bonds, palladium-catalyzed couplings of enolates, the functionalization of C-H bonds with boron reagents, asymmetric iridium-catalyzed allylic substitution, and catalysts for olefin hydroamination.

12

Ryan DeLuca was born in 1984 in Salt Lake City, Utah. He graduated from Southern Utah University with a B.S. degree in chemistry in 2007. He spent a year at the University of Illinois Urbana-Champaign working under the guidance of Professor John F. Hartwig. He is currently pursuing his Ph.D. at the University of Utah with Prof. Matthew Sigman.

David Huang was born in 1983 in Ames, Iowa. He received his B.S. from UC Berkeley in 2006, where he preformed undergraduate research in the Toste lab. David joined the Hartwig group at the University of Illinois in the fall of 2006 as a graduate student. His research focuses on the palladium-catalyzed coupling with enolates.

THE PREPARATION OF AMIDES BY COPPER-MEDIATED OXIDATIVE COUPLING OF ALDEHYDES AND AMINE HYDROCHLORIDE SALTS

Submitted by Maxime Giguère-Bisson, Woo-Jin Yoo and Chao-Jun Li.[1]
Checked by Melissa J. Leyva and Jonathan A. Ellman.

1. Procedure

All the following manipulations are performed in air.

A mechanical stir paddle (60 mm in diameter) is fitted to an oven-dried 100-mL, 2-necked round-bottomed flask (24/40 joint). Copper (I) iodide (0.047 g, 0.25 mmol, 0.01 equiv) (Note 1), AgIO$_3$ (0.070 g, 0.25 mmol, 0.01 equiv) (Note 2), L-valine methyl ester hydrochloride (6.19 g, 36.9 mmol, 1.5 equiv) (Note 3) and CaCO$_3$ (2.10 g, 21.0 mmol, 0.9 equiv) (Note 4) are added sequentially to the reaction flask, and the mixture is stirred at the maximum rate (the submitters stirred at 230 rpm). Acetonitrile (5.5 mL) (Note 5) is added in one portion using a syringe under constant stirring. A white opaque mixture is obtained at the end of this step. Benzaldehyde (2.50 mL, 24.6 mmol, 1.0 equiv) (Note 6) is added over 15 seconds using a syringe under constant stirring, and upon addition the reaction mixture becomes an opaque yellow. A *tert*-butyl hydroperoxide solution (4.00 mL, 28.0 mmol, 1.1 equiv) (Note 7) is added using a syringe under constant stirring and the mixture turns green. The reaction flask is then placed in an oil bath at 40 °C for 6 h under constant stirring.

After placing the reaction flask in the oil bath, the reaction mixture gradually becomes a clear orange-gold solution. The reaction flask is taken out of the oil bath and is placed in an ice-bath. Upon cooling, the reaction mixture becomes a thick, opaque mustard-yellow. Hydrochloric acid (1.6 M, 7.5 mL) is added dropwise (ca. 4 drops per second) under constant stirring. Upon neutralization, the reaction mixture turns green. The reaction flask is taken out of the ice-bath and ethyl acetate (25 mL) (Note 8) and distilled water (20 mL) are added. The organic layer is light green and the aqueous layer white-blue and opaque. If left overnight at room temperature, the

Org. Synth. **2011**, *88*, 14-21
Published on the Web 9/8/2010

organic layer may turn orange. The mixture is transferred to a 250-mL separatory funnel. The 100-mL round-bottomed flask is rinsed with ethyl acetate (15 mL) and distilled water (15 mL). The rinsings are transferred to the separatory funnel. Brine (15 mL) is added to the separatory funnel to obtain a better phase separation. The two layers are separated and transferred into 250-mL Erlenmeyer flasks. The aqueous layer is then transferred back to the separatory funnel and extracted with ethyl acetate (3 x 25 mL). Brine (3 x 15 mL) is added to facilitate every extraction. The organic layers are combined and washed with a saturated solution of $NaHCO_3$ (25 mL) and poured into a 250-mL Erlenmeyer flask. Anhydrous magnesium sulfate (4.0 g) (Note 9) is added. The solution is filtered under vacuum using a 30-mL suction funnel equipped with a medium porosity fritted-disc into an oven-dried, 500-mL round-bottomed flask. The magnesium sulfate in the frit is washed with two portions of ethyl acetate (2 x 10 mL), and the combined filtrate is concentrated under vacuum (40 °C and 21 mmHg) to provide a white-beige solid. Dichloromethane (20 mL) (Note 10) is added to the round-bottomed flask. The flask is sonicated for 30 seconds resulting in a green brown solution. Silica gel (5.0 g) (Note 11) is added to the flask and the solvent is evaporated using a rotary evaporator at 55–60 °C (21 mmHg). Flash column chromatography is performed using a 5-cm-wide, 45-cm-high column packed with 230 g of silica gel. The column is packed with silica gel, flushed with 1000 mL of the eluent (hexanes: ethyl acetate = 3:2) (Note 12). The eluent level is adjusted to the upper level of the silica gel. The silica gel containing the compound is then loaded on the column and column chromatography is then performed (1.7 L of eluent is used). The collected fractions (15 mL each) are analysed using TLC (eluting with hexanes: ethyl acetate =3:2) and the spots are visualized using a UV lamp (254 nm) (Note 13). The fractions (41 to 92) containing the desired compound are combined and evaporated to dryness using a rotary evaporator. The product is further dried under reduced pressure (0.01 mmHg) for 2 h to give (S)-methyl-2-benzamido-3-methylbutanoate (4.91 g, 20.9 mmol, 85% yield) (Note 14).

2. Notes

1. Copper (I) iodide was obtained from Aldrich and used as received.
2. Silver iodate was obtained from GFS Chemicals and used as received.
3. L-Valine methyl ester hydrochloride 99% was obtained from

Aldrich and used as received.

4. Calcium carbonate A.C.S reagent was obtained from Aldrich and used as received.

5. Acetonitrile A.C.S. Grade was obtained from Fisher Scientific and used as received.

6. Benzaldehyde ReagentPlus ≥99% was obtained from Aldrich and used as received.

7. *tert*-Butyl hydroperoxide, T-hydro solution, 70 wt. % in water was obtained from Aldrich and used as received.

8. Ethyl acetate HPLC Grade was obtained from Fisher Chemicals and used as received.

9. Anhydrous magnesium sulfate was obtained from Fisher Chemicals and used as received.

10. Dichloromethane A.C.S. Reagent was obtained from ACP Chemicals and used as received.

11. Silica gel MP Silitech 32-63 D 60 Å was used as received.

12. Hexanes HPLC Grade was obtained from Fisher Chemicals and used as received.

13. The reaction and column fractions were monitored by TLC using Dynamic Adsorbents, Inc. glass plates coated with 250 mm F-254 silica gel: Hexanes:ethyl acetate (3:2). The benzaldehyde starting material has an R_f of 0.65, and the product (*S*)-methyl-2-benzamido-3-methylbutanoate has an R_f of 0.45.

14. The product is obtained as a white (slightly yellow) flaky solid and has the following physical and spectroscopic properties: Melting point: 108-109 °C, IR (neat): 3347, 2967, 1736, 1640, 1518, 1490, 1202, 1151, 994, 692 cm^{-1}. ^{1}H NMR (400 MHz, CDCl$_3$) δ: 0.96 (d, J = 7.0 Hz, 3 H), 0.98 (d, J= 7.0 Hz, 3 H), 2.25 (m, 1 H), 3.74 (s, 3 H), 4.74 (dd, J= 5.2, 8.4 Hz, 1 H), 6.64 (d, J= 8.4 Hz, 1 H), 7.39 (m, 2 H), 7.48 (m, 1 H), 7.77 (m, 2 H). ^{13}C NMR (100 MHz, CDCl$_3$) δ: 18.1, 19.2, 31.8, 52.4, 57.6, 127.2, 128.8, 131.9, 134.3, 167.5, 172.9. MS (ESI) m/z [M + H]$^+$ calcd for C$_{13}$H$_{18}$NO$_3$: 236.128. Found: 236.128. HPLC (Daicel Chiral AD-H, hexanes/isopropanol=95:5, flow rate 1.0 mL/min) t_R = 20.556 min (major), < 99.9 % ee. Submitters provided checkers with racemic product, t_R = 15.150 min (minor). Anal. calcd. for C$_{13}$H$_{17}$NO$_3$: C, 66.36; H, 7.28; N, 5.95. Found: C, 66.38; H, 7.44, N, 5.85. Specific rotation, $[\alpha]_D^{20}$ +34.0° (c = 0.1, CH$_2$Cl$_2$)

All hazardous materials should be handled and disposed of in accordance with "Prudent Practices in the Laboratory"; National Academy Press; Washington, DC, 1995.

3. Discussion

The amide functional group is ubiquitous in organic chemistry and is an important motif in polymers, natural products, and pharmaceuticals.[2] The most prevalent strategy for amide bond formation relies heavily upon the interconversion of activated carboxylic acid derivatives with an amine.[3] However, due to the lability of activated carboxylic acid derivatives, alternative strategies toward the synthesis of amides have been explored. Examples include the utilization of azides as amine equivalents in the modified Staudinger reaction,[4] hydrative amide syntheses with alkynes,[5] and thio acid/ester ligation methods.[6] Transition-metal-catalyzed carbonylation of alkenes,[7] alkynes,[8] and haloarenes[9] with amines has also been employed for amide synthesis. Finally, the direct coupling of aldehydes with amines under oxidative conditions can also serve as an attractive entry into amides.[10] This methodology has been hampered thus far by the need for expensive transition-metal catalysts and poor substrate scope. We previously reported a copper-catalyzed stereoselective oxidative esterification of aldehydes with β-dicarbonyl compounds using *tert*-butyl hydroperoxide (TBHP) as an oxidant.[11] In light of our success in that oxidative esterification reaction, we turned our attention to the much more challenging amidation reaction of simple aldehydes and amines. Subsequently, we developed a simple methodology[12] using CuI as catalyst, an aqueous solution of *tert*-butyl hydroperoxide as an oxidant, as well as $AgIO_3$ as an additive. This paper describes the scale up of this methodology. The specific reaction shown in this paper is the oxidative amidation of benzaldehyde with the hydrochloride salt of L-valine methyl ester. The $CuI/TBHP/AgIO_3$ methodology has also been used successfully on a smaller scale with other aldehydes and amine hydrochloride salts as shown in the following table.

Table 1. Copper Catalyzed Oxidative Amidation of Aldehydes with Amine Hydrochloride Salts[a,b]

Entry	R	R'	Isolated yield(%)
1	Ph	Et	91
2	Ph	Bn	71
3	Ph	CH_2Bn	89
4	Ph	$c\text{-}C_6H_{11}\text{-}$	73
5	Ph	t-Bu	39
6	Ph	CH_2CH_2Cl	89
7	Ph	CH_2CO_2Et	91
8	$4\text{-Me-}C_6H_4$	CH_2CO_2Et	91
9	$4\text{-MeO-}C_6H_4$	CH_2CO_2Et	78
10	$4\text{-Cl-}C_6H_4$	CH_2CO_2Et	81
11	$4\text{-NO}_2\text{-}C_6H_4$	CH_2CO_2Et	49
12	$c\text{-}C_6H_{11}$	CH_2CO_2Et	39

[a] See reference 12; [b] 0.9 mmol scale

1. Department of Chemistry, McGill University, 801 Sherbrooke Street West, Montreal, QC H3A 2K6, Canada. Email: cj.li@mcgill.ca.
2. Humphrey, J. M.; Chamberlin, A. R. *Chem. Rev.* **1997**, *97*, 2243-2266.
3. Larock, R. C. *Comprehensive Organic Transformation*; VCH: New York, 1999
4. (a) Saxon, E.; Bertozzi, C. R. *Science,* **2000**, *287*, 2007-2010. (b) Nilsson, B. L.; Kiessling, L. L.; Raines, R. T. *Org. Lett.* **2000**, *2*, 1939-1941. (C) Damkaci, F.; DeShong, P. *J. Am. Chem. Soc.* **2003**, *125*, 4408-4409.
5. (a) Cho, S.; Yoo, E.; Bae, I.; Chang, S. *J. Am. Chem. Soc.* **2005**, *127*, 16046-16047. (b) Cassidy, M. P.; Raushel, J.; Fokin, V. V. *Angew. Chem., Int. Ed.* **2006**, *45*, 3154-3157.
6. Dawson, P. E.; Muir, T. W.; Clark-Lewis, I.; Kent, S. B. *Science* **1994**, *266*, 776-779. (b) Shangguan, N.; Katukojvala, S.; Greenerg, R.; Williams, L. J. *J. Am. Chem. Soc.* **2003**, *125*, 7754-7755. (c) Merkx, R.; Brouwer, A. J.; Rijkers, D. T. S.; Liskamp, R. M. J. *Org. Lett.* **2005**, *7*, 1125-1128.
7. Beller, M.; Cornils, B.; Frohning, C. D. *J. Mol. Catal. A: Chem.* **1995**, *104*, 17-85.
8. (a) Ali, B. E.; Tijani, J. *Appl. Organomet. Chem.* **2003**, *17*, 921-931. (b) Knapton, D. J.; Meyer, T. Y. *Org. Lett.* **2004**, *6*, 687-689. (c) Uenoyama, Y.; Fukuyama, T.; Nobuta, O.; Matsubara, H.; Ryu, I. *Angew. Chem., Int. Ed.* **2005**, *44*, 1075-1078.
9. For recent examples, see: (a) Lin, Y.-S.; Alper, H. *Angew. Chem., Int. Ed.* **2001**, *40*, 779-781. (b) Uozumi, Y.; Arii, T.; Watanabe, T. *J. Org. Chem.* **2001**, *66*, 5272-5274. (c) Nanayakkara, P.; Alper, H. *Chem. Commun.* **2003**, 2384-2385.
10. (a) Tamaru, Y.; Yamada, Y.; Yoshida, Z. *Synthesis* **1983**, 474-476. (b) Naota, T.; Murahashi, S. *Synlett* **1991**, 693-694. (c) Tillack, A.; Rudloff, I.; Beller, M. *Eur. J. Org. Chem.* **2001**, 523-528.
11. Yoo, W.-J.; Li, C.-J. *J. Org. Chem.* **2006**, *71*, 6266-6268.
12. Yoo, W.-J.; Li, C.-J. *J. Am. Chem. Soc.* **2006**, 128, 13064-13065.

Appendix
Chemical Abstracts Nomenclature; (Registry Number)

L-Valine methyl ester hydrochloride: (*S*)-2-amino-3-methylbutanoate
   hydrochloride; (6306-52-1)
Silver iodate: iodic acid (HIO_3), silver(1+) salt (1:1); (7783-97-3)
Calcium carbonate: carbonic acid, calcium salt (1:1); (471-34-1)
tert-Butyl hydroperoxide: 1,1-dimethylethyl hydroperoxide; (75-91-2)
Benzaldehyde: benzenecarboxaldehyde; (100-52-7)
(*S*)-Methyl-2-benzamido-3-methylbutanoate: L-Valine, *N*-benzoyl-, methyl
   ester; (10512-91-1)

Chao-Jun Li (born in 1963) received his Ph.D at McGill University (1992). After a two year NSERC Postdoctoral position at Stanford University, he became Assistant Professor (1994), Associate Professor (1998) and Full Professor (2000) at Tulane University. In 2003, he became a Canada Research Chair (Tier I) in Organic/Green Chemistry and a Professor (E. B. Eddy Chair Professor since 2010) of Chemistry at McGill University in Canada. Currently, he serves as the Co-Chair of the Canadian Green Chemistry and Engineering Network, Director of CFI Facility for Green Chemistry and Green Chemicals, and Co-Director for FQRNT Center for Green Chemistry and Catalysis. His current research efforts are focused on developing innovative and fundamentally new organic reactions that will defy conventional reactivities and have high synthetic efficiency.

Maxime Giguère-Bisson was born in 1986 in Grand-Mère (Canada). He obtained his B. Sc. Degree in Chemistry Honours in 2008 from McGill University. He is currently doing his Masters in Chemistry under the supervision of Dr. Chao-Jun Li. His current research focuses on Asymmetric A^3-Coupling.

20

Woo-Jin Yoo (born in 1978) received his B.Sc. degree from the University of Guelph in 2003. In 2005, he received his M.Sc. degree from the University of Guelph under the supervision of William Tam, where he studied metal-catalyzed cross-coupling reactions and Diels-Alder cycloadditions. He then joined the research group of Chao-Jun Li at McGill University and studied copper-catalyzed oxidative coupling reactions and multicomponent coupling reactions. He obtained his Ph.D. degree in 2009 and is currently a postdoctoral fellow with Shū Kobayashi at The University of Tokyo. He has been the recipient of the Alexander Graham Bell Canada Graduate Scholarship (Ph.D.), an NSERC Postdoctoral Fellowship (declined), and a JSPS Postdoctoral Fellowship for Foreign Researchers.

Melissa J. Leyva was born in El Paso, Texas in 1982. She received her B.S. degree in Chemistry at the University of Texas, El Paso in 2005. She then began her doctoral studies at the University of California, Berkeley under the direction of Professor Jonathan A. Ellman. Her graduate research has focused on the identification of novel inhibitors for therapeutically important proteases.

SYNTHESIS OF 2-ARYL PYRIDINES BY PALLADIUM-CATALYZED DIRECT ARYLATION OF PYRIDINE N-OXIDES

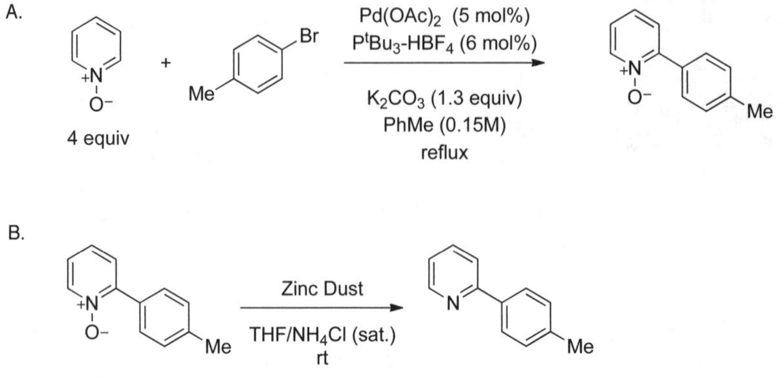

A.

Pd(OAc)$_2$ (5 mol%)
P^{t}Bu$_3$-HBF$_4$ (6 mol%)

K$_2$CO$_3$ (1.3 equiv)
PhMe (0.15M)
reflux

4 equiv

B.

Zinc Dust

THF/NH$_4$Cl (sat.)
rt

Submitted by Louis-Charles Campeau[1] and Keith Fagnou.[2]
Checked by Adnan Ganić and Andreas Pfaltz.

1. Procedure

A.   *2-(4'-Methylphenyl)-pyridine N-oxide.*   A 1-L, three-necked round-bottomed flask is equipped with a magnetic stir bar (cylindrical, 4 × 1 cm), two glass stoppers and a reflux condenser with a gas bubbler on the top and connection to the argon line (Note 1). Pd(OAc)$_2$ (0.560 g, 2.50 mmol, 5 mol%), P^{t}Bu$_3$·HBF$_4$ (0.870 g, 3.00 mmol, 6 mol%) (Notes 2, 3), potassium carbonate powder (K$_2$CO$_3$, 8.97 g, 65.0 mmol, 1.3 equiv) (Note 4), and pyridine *N*-oxide (19.0 g, 200 mmol, 4 equiv) (Note 5) are weighed in air and placed inside the flask. The whole setup is then evacuated (0.4 mmHg) and refilled with argon four times (Note 6).

A solution of 4-bromotoluene (8.55 g, 50 mmol, 1 equiv) (Note 7) in toluene (300 mL) (Note 8) is added under a steady flow of argon to the reaction mixture, and the glassware is rinsed with degassed toluene (30 mL) (Note 9). The obtained brown-orange suspension is immersed in the oil bath. Stirring (700 rpm) is commenced and the heating source is turned on (set to 125 °C). The mixture starts to reflux after approximately 45 min (Note 10).

Heating and stirring under a steady flow or argon is maintained for 16 h (overnight). The now black reaction mixture is allowed to cool to room temperature and transferred to 1-L round-bottomed flask. The reaction flask

22

is rinsed with toluene (50 mL) (Note 11), and the solvent is removed under reduced pressure (40 °C, 40 mmHg) on a rotary evaporator. A saturated aqueous solution of NH_4Cl (150 mL) and DCM (200 mL) are sequentially added to the dark slurry. The mixture is filtered through a Celite® pad (Note 12). The glassware is washed with water (50 mL) and DCM (2 × 50 mL). The resulting yellow biphasic solution is transferred into a 1-L separatory funnel and the layers are separated. The aqueous layer is washed with DCM (2 × 100 mL).the combined organic layers are dried over $MgSO_4$ (4 g) (Note 13), filtered and washed with DCM (2 × 50 mL). The solvent is removed under reduced pressure (40 °C, 500 mmHg). The resulting yellow solid (Note 14) is then dissolved in eluent (20 mL) (Note 15), charged on a column (7 × 8.5 cm) of silica gel (Note 16) and eluted with DCM/acetone/MeOH 90:8:2 mixture (3.5 L). After the first 300 mL of eluent, fraction collection (20-mL fractions) is started. The desired product is obtained from fractions 18-120 as beige solid (6.51 g, 35.1 mmol, 70%) (Note 17).

The solid obtained after column chromatography is placed in a 250-mL round-bottomed flask equipped with a magnetic stir bar (egg shaped, 2 × 1 cm) and a reflux condenser. Heptane (40 mL) and toluene (20 mL) are added, and the suspension is heated to 115 °C (700 rpm). Toluene is added in 5 mL portions through the reflux condenser over 20-30 30 min until a clear solution is obtained (the total volume of toluene used is 40 mL). A gas bubbler is placed on top of the condenser and connected to the argon line. The heating source is turned off and the clear (yellow solution) is allowed to cool down to rt overnight (14 h) under argon, while stirring is maintained. The obtained suspension is cooled down to 0 °C with the help of an ice bath and stirred for 1 h. The resulting solid is collected by filtration using a 50-mL Büchner funnel under a stream of nitrogen and washed with ice-cold heptane (3 × 20 mL). The obtained solid is transferred to a 100-mL round-bottomed flask and dried overnight (>12 h) at 0.1 mbar to provide the title compound as a white solid (5.98 g, 32.3 mmol, 65%) (Notes 18, 19).

B. 2-(4'-Methylphenyl)-pyridine. A three-necked, 500-mL round-bottomed flask equipped with a condenser, a thermometer, a glass stopper and a magnetic stirrer (egg shaped, 3 × 1.5 cm) (Note 20) is charged with 2-(4'-methylphenyl)-pyridine *N*-oxide (5.98 g, 32.3 mmol, 1 equiv). Tetrahydrofuran (120 mL) (Note 21) is added, followed by saturated aqueous NH_4Cl solution (120 mL). This addition caused a slight increase of the reaction temperature from 22 °C up to 28 °C, and a white solid is formed in the aqueous layer. Zinc dust (Note 22) is added in 4 portions (10.6 g, 161 mmol, 5 equiv) within 10 min under stirring (700 rpm) to this biphasic mixture. Again, an increase of the reaction temperature from 25 °C up to 38 °C over a period of 15 min is observed. The reaction mixture is stirred for 20 min until TLC analysis reveals complete consumption of the starting material (Note 23). The mixture is filtered through a *Celite*® pad (Note 25) and is washed with MTBE (250 mL). The filtrate is transferred to a 1-L separatory funnel, and the layers are separated. The aqueous layer is back-extracted with MTBE (125 mL) using a second 250-mL separatory funnel. The organic layers are consecutively washed with half-saturated aqueous NaHCO$_3$ solution (125 mL) and with brine (125 mL). The combined organic layers are then dried over MgSO$_4$ (3 g), filtered and concentrated (40 °C, 250 mmHg) to afford the title compound as yellow liquid (5.47 g, 32.3 mmol, 100%). Purification by Kugelrohr distillation (150 °C, 0.3 mmHg) gives the title compound as a colorless liquid (5.35 g, 31.6 mmol, 98%) (Notes 25, 26).

2. Notes

1.   Submitters setup: a 1-L pear-shaped flask is fitted with a Teflon sleeve and a cylindrical (4 × 1 cm) stir bar with a condenser.
2.   Palladium (II) acetate (46-1780) and phosphonium salt (15-6000) were purchased from *Strem* and used without any purification.
3.   By using 10 mol% of the phosphonium salt the checkers did not observe any increase in yield (70%).
4.   Purchased from *Aldrich* (347825) and used as is.
5.   Purchased from *Aldrich* (131652), particles greater than 1 cm^3 are crushed using a mortar. Pyridine *N*-oxide is a highly deliquescent solid,[3] and an undetermined amount of water is incorporated into the reaction mixture. The influence of water on the reaction was studied by the submitters. It is reported that the reaction occurs even when 5 equiv of water are added to the reaction mixture without affecting the yield.[4] An excess of this reagent is used to increase the yield of the reaction. The identical procedure can be performed with 1.5 equiv of pyridine *N*-oxide resulting in yields ranging from 65-70% (checkers: 58%). During the extraction protocol the excess *N*-oxide can be recovered from the aqueous phase by evaporation of water and chromatography of the residue with 10% MeOH in DCM.

6. The submitters used a rubber septum on the top of the condenser, which allows connection to the vacuum line. The submitters added argon from a balloon. The checkers used argon from an argon line.

7. Purchased from *Aldrich* (B82200). 4-Bromotoluene is a low melting solid (mp 30 °C). The submitters reported that it was stored in the freezer before use to facilitate its weighing as a solid. Alternatively, one could add it as a stock solution in toluene or as a liquid if heated. If it is added as a stock solution or a liquid, it should be added with the solvent. The checkers melted it using a water bath (40 °C) and weighed it as a liquid.

8. The submitters used certified A.C.S. Grade toluene purchased from *Fisher Scientific* (T324) and degassed it with argon (10 min) before use. The checkers used *J. T. Baker* (Baker analyzed) toluene and degassed the solution by bubbling argon for 15 min prior to its use.

9. Submitters reaction setup: Toluene (330 mL) is added in 50 mL portions under a steady flow of argon. After addition of the first 50 mL, the reaction is immersed in the oil bath. Stirring (400 rpm) is commenced and the heating source is turned on (set to 125 °C). After the remaining toluene has been added the reaction is kept under an argon atmosphere and the stirring speed is raised to 700 rpm.

10. The submitters report that the mixture starts to reflux after approximately 20 min and the color of the mixture pales to off-white after 30-40 min, which was not observed by the checkers. The reaction mixture remains heterogeneous, but slowly turns from brown-orange into a dark suspension.

11. Not all material is dissolved in toluene, and therefore the reaction flask is rinsed with DCM and sat. aq. NH_4Cl solution, which is added to the dark slurry.

12. *Celite*® (10 g) is weighed dry and packed using DCM (50 mL) in a 75-mL coarse fritted Büchner funnel. 20 g of sand is added to the top of the *Celite*® layer.

13. The submitters used 15 g of $MgSO_4$.

14. In some cases a yellow oil is obtained which can be reduced to a yellow solid under high vacuum or longer time on rotary evaporator.

15. The submitters used 20-35 mL of eluent for dissolving the material and report that sonication or heating with heat gun was necessary to fully dissolve the solid. The checkers dissolved the solid in 20 mL of eluent by stirring on a rotary evaporator without additional heating.

16. The submitters used 175 g of *Silicycle* (R10030B) silica gel on a 8.5 cm diameter column. 200 mL is collected in an Erlenmeyer flask followed by 15-mL fractions. The majority of the product was obtained from fractions 20-48. The product tailed off past fraction 55 but resulted only in a 2% increase in yield when collected. The checkers used silica gel (175 g) from *Fluka* (89943) and a sand (50 g) layer on the top. The fractions between 121-160 contained product that was contaminated with unreacted starting material.

17. The submitters report yields ranging from 79-83% (>97% purity by HPLC; contamination with <1% of tri-*tert*-butylphosphine oxide; mp 129–131 °C). This material was usually carried through to further transformations, although analytically pure material could be obtained by recrystallization. Submitters HPLC conditions: Zorbax SB C18, RRHT, 1.8 micron (50 × 4.6 mm) 0.1% H_3PO_4 in water / MeCN start 90:10, ramp to 50:50 over 8 min, ramp to 5:95 over 4 min. Total: 12 min, 2 mL/min, 35 °C, 215 nm, 5 uL. Retention times: Free base: 1.699 min, *N*-oxide: 3.250 min. The checkers obtained yields in the 68-70% range (99% purity by HPLC; mp 141–143 °C). Checkers HPLC conditions: Reprosil 100 C18, 3 micron (125 × 3 mm), 0.1% H_3PO_4 in water/MeCN start 80:20, ramp to 50:50 over 15 min, ramp to 15:85 over 15 min, total: 30 min, 0.5 mL/min, 35 °C, 254 nm, 1 mg/mL, 5 μL; retention time: 7.50 min.

18. The checkers obtained yields after recrystallization in the 62-65% range (>99% purity by HPLC). Analytical data: mp 144–145 °C (lit.[5] mp 145–146 °C; TLC (SiO$_2$, DCM/Acetone/MeOH 90:8:2): R$_f$ = 0.19; ^{1}H NMR (400 MHz, CDCl$_3$) δ: 2.41 (s, 3 H), 7.14–7.23 (m, 1 H), 7.24–7.32 (m, 3 H), 7.41 (dd, *J* = 7.8, 1.7 Hz, 1 H), 7.72 (d, *J* = 6.6 Hz, 2 H), 8.32 (d, *J* = 6.4 Hz, 1 H); ^{13}C NMR (101 MHz, CDCl$_3$) δ: 21.6, 124.3, 125.7, 127.3, 129.1, 129.3, 129.8, 139.9, 139.9, 140.7; IR (ATR) ṽ: 3064, 3043, 1478, 1430, 1403, 1328, 1237, 1186, 1144, 1110, 1028, 1011, 948, 842, 817, 798, 735, 713, 694, 573, 526 cm^{-1}; MS (EI, 70 eV, 150 °C) *m/z* (%): 185 (M$^+$, 63), 184 (M-H$^+$, 100), 169 (10), 156 (13), 117 (25), 78 (12); Anal. calcd. for $C_{12}H_{11}NO$: C, 77.81; H, 5.99; N, 7.56; found: C, 77.92; H, 6.22; N, 7.53.

19. Submitters' recrystallization conditions: *N*-oxide was suspended in 6 mL/g of heptane in a three-necked flask fitted with a reflux condenser and a mechanical stirrer. The suspension is heated to 95 °C in an oil bath and toluene is added (in 1 mL/g portions) until all solids dissolve (typically ~6 mL/g). The heat source on the oil bath is turned off and mixture is allowed to cool in the oil bath overnight. In the morning, the suspension is

immersed in an ice bath for 30 min and then filtered on a fritted Büchner funnel. The solid collected is then washed with 10 mL/g of heptane, and the solid is dried under a stream of N_2 and collected as a fluffy white solid which is >99% pure (Note 17). Recovery for this procedure is typically from 88-95%.

20. The Submitters used a mechanical stirrer.

21. The submitters used ACS Grade THF with BHT over sieves from A&C, which is used as is. The checkers use J. T. Baker (Baker analyzed) material.

22. Both submitters and checkers purchased zinc dust from Aldrich (20,998-8) and used as received.

23. The submitters stirred the reaction mixture for 40 min, and the reaction progress was checked by HPLC analysis. The checkers used TLC analysis (SiO_2, DCM/acetone/MeOH 90:8:2; R_f 0.87) to monitor the reaction progress.

24. The submitters used Solka floc®. They weighed Solka floc dry (15 g) and packed it using MTBE in a 60-mL coarse fritted Büchner funnel. The checkers used Celite® (10 g), which was weighed dry and packed using MTBE in a 75-mL coarse fritted Büchner funnel, adding 10 g of sand to the top of the Celite® layer.

25. HPLC analysis shows >99% purity (for conditions, see Note 18; retention time 4.19 min). Analytical data: ^{1}H NMR (400 MHz, CDCl$_3$) δ: 2.41 (s, 3 H), 7.20 (ddd, J = 6.6, 4.8, 2.1 Hz, 1 H), 7.29 (d, J = 7.9 Hz, 2 H), 7.64–7.78 (m, 2 H), 7.86–7.93 (m, 2 H), 8.65–8.71 (m, 1 H); ^{13}C NMR (101 MHz, CDCl$_3$) δ: 22.4, 120.4, 121.9, 126.9, 129.6, 136.7, 136.8, 139.1, 149.8, 157.6; IR (ATR) $\tilde{v}$: 3006, 2918, 1613, 1586, 1562, 1514, 1464, 1432, 1298, 1185, 1152, 1016, 829, 772, 741, 566, 529 cm^{-1}; MS (EI, 70 eV, rt) m/z (%): 169 (M$^+$, 100), 168 (49), 167 (16); Anal. calcd. for $C_{12}H_{11}N$: C, 85.17; H, 6.55; N, 8.28; found: C, 85.13; H, 6.84; N, 8.15.

26. In the submitters' procedure the product was not purified by distillation. It was stated that the isolated product was >98% pure (HPLC and exhibited identical spectral data as a commercial sample available from Aldrich (98870). Storing of the obtained material is recommended under inert atmosphere, because the obtained product turns yellow under air at rt.

Waste Disposal Information

All hazardous materials should be handled and disposed of in accordance with "Prudent Practices in the Laboratory"; National Academy Press; Washington, DC, 1995

3. Discussion

The limited number of successful metal-catalyzed cross-coupling protocols using 2-metallapyridines has fueled the investigations of methods that do not require their use.[6] One strategy is to use direct arylation methodology, where a simple arene replaces the organometallic component.[7] In 2005, we reported that pyridine N-oxide was a good substitute for 2-metallapyridine in cross-coupling reactions.[4,8] The 2-aryl pyridine N-oxide products can easily be converted to the corresponding 2-aryl pyridines under mild conditions and in high yield.[9] This protocol therefore allows for the rapid and easy synthesis of 2-aryl pyridines. More recently a full account describing optimized reaction conditions and reaction scope was reported.[10] The reaction tolerates a broad range of substitution patterns on both coupling partners. Substitution in the *ortho*, *meta* and *para* position is well tolerated for alkyl and both electron-donating and electron-withdrawing substituents. The pyridyl moiety can be substituted at the 2-, 3- or 4-position allowing for a broad range of potential products. The reaction also proceeds with other azines such as isoquinoline and pyrazine. The procedure is convenient and makes use of cheap, readily available and air stable reagents. No precautions in the storage or the purification of the starting materials were taken. The 2-aryl pyridines obtained via this procedure are found in a number of biologically active compounds and are of value as organic synthesis building blocks. Furthermore, the N-oxide moiety can serve as a useful handle in further elaboration of the pyridine core.

Table 1. Direct Arylation of Azine *N*-Oxides

Entry	N-Oxide	Aryl Halide	Product	Yield[b]
1		1		78[d]
2		2		80
3		2		74
4		1		97
5		1		74
6	R=CO₂Et			81[c]
7	R=OMe			81[c]
8		2		89
9		2		90[d]

[a]Conditions: Aryl halide (1 equiv), pyridine *N*-oxide (2-4 equiv), K₂CO₃ (2 equiv), Pd(OAc)₂ (0.05 equiv) and P[t]Bu₃ - HBF₄ (0.06 equiv) in toluene (0.15M) at 110 °C overnight. [b]Isolated yields. [c]With 4 equiv of *N*-oxide and 15 mol% P[t]Bu₃ - HBF₄. [d]With 1.1 equiv of *N*-oxide.

1. Current address: Department of Process Research and Development, Merck Frosst Canada, 16711 Trans Canada Hwy, Kirkland, Quebec, H9H 3L1.

2. Department of Chemistry, University of Ottawa, 10 Marie Curie, Ottawa, Ontario, K1N 5N6.

3. Mosher H. S.; Turner L. Carlsmith A. *Org. Synth.* **1953**, *33*, 79–81; *Org. Synth. Coll. Vol.* **1963**, *4*, 828–830.

4. Campeau, L.-C.; Rousseaux, S.; Fagnou, K. *J. Am. Chem. Soc.* **2005**, *127*, 18020–18021.

5. Butler D. E.; Bass P.; Nordin I. C.; Hauck, Jr. F. P.; L'Italien Y. J. *J. Med. Chem.* **1971**, *14*, 575–579.

6. Campeau, L.-C.; Fagnou, K. *Chem. Soc. Rev.* **2007**, *36*, 1058–1068.

7. For recent reviews, see: (a) Alberico, D.; Scott M. E.; Lautens, M. *Chem. Rev.* **2007**, *107*, 174–238. (b) Campeau, L.-C.; Fagnou, K. *Chem. Commun.* **2006**, 1253–1264. (c) Campeau, L.-C.; Stuart, D.R.; Fagnou, K. *Aldrich. Chim. Acta.* **2007**, *40*, 35–41. (d) Zhao, D.; Wang, W.; Yang, F.; Lan, J.; Yang, L.; Gao, G.; You, J. *Angew. Chem. Int. Ed.* **2009**, *48*, 3296–3300.

8. *!!Caution!!:* Pyridine *N*-oxides have been shown to exothermically decompose at very high temperature. Uncontrolled heating of the reaction media should be avoided. DSC analysis of this particular 2-arylpyridine *N*-oxide as well as others have revealed no exotherms at temperature up to 250 °C. Examples of exothermic onset temperatures (*T*o) for pyridine *N*-oxide, 288 °C; 2,6-lutidine *N*-oxide, 288 °C; nicotinic acid *N*-oxide, 302 °C; picoline *N*-oxide, 285 °C; picolinic acid *N*-oxide, 307 °C, from: Ando, T.; Fujimoto, Y.; Morisaki, S. *J. Haz. Mater.* **1991**, *28*, 251–280.

9. (a) Pd/C: Balicki, R. *Synthesis* **1989**, 645–646. (b) Zinc: Aoyagi, Y.; Abe, T.; Ohta, A. *Synthesis* **1997**, 891–894.

10. Campeau, L.-C; Stuart, D. R.; Leclerc, J.-P.; Bertrand-Laperle, M.; Villemeure, E.; Sun, H.-Y; Guimond, N.; Lasserre, S.; Lecavellier, M.; Fagnou, K. *J. Am. Chem. Soc.* **2009**, *131*, 3291–3306.

Appendix
Chemical Abstracts Nomenclature; (Registry Number)

Palladium acetate: Pd(OAc)$_2$; (3375-31-3)

P^{t}Bu$_3$HBF$_4$: Phosphine, tris(1,1-dimethylethyl)-, tetrafluoroborate(1-) (1:1); (155026-77-0)

Pyridine N-oxide: Pyridine, 1-oxide; (694-59-7)

4-Bromotoluene: Benzene, 1-bromo-4-methyl-: (106-38-7)

Potassium carbonate; (584-08-7)

Zinc; (7440-66-6)

2-(4'-Methylphenyl)-pyridine; (4467-06-5)

Keith Fagnou was born in 1971 in Saskatoon, Saskatchewan, Canada. He received a Bachelor of education (B.Ed.) degree with distinction from the University of Saskatchewan in 1995 and, after teaching at the high school level for a short period of time, he continued his studies in chemistry at the University of Toronto. In 2000, he received an M.Sc. degree and, in 2002, completed his Ph.D. requirements under the supervision of Mark Lautens. He then joined the chemistry faculty at the University of Ottawa, and had initiated research programs focusing on the development of new catalytic reactions for use in organic synthesis. Most recently, he was awarded the OMCOS Award, as well as the Rutherford Memorial Medal, for outstanding research contributions to the field. Keith Fagnou passed away suddenly after a short illness in November 2009.

Louis-Charles Campeau was born in 1980 in Cornwall, Ontario, Canada. In 2003, he received his bachelor's degree with distinction in biopharmaceutical sciences (medicinal chemistry option) from the University of Ottawa. He then joined the research group of Professor Keith Fagnou, where his Ph.D. research was directed towards the development of new transition-metal-catalyzed processes. Louis-Charles received the University's Pierre Laberge Thesis Prize for the Sciences. In the summer of 2007, he joined the process research group at Merck in Montreal. He then moved to Merck in Rahway NJ, in 2010.

Adnan Ganić was born in Travnik (Bosnia and Herzegovina) in 1980 and moved with his family to Switzerland in 1992. Fascinated by the world of chemistry during his apprenticeship in the F. Hoffmann-La Roche company he decided to study chemistry at University of Applied Science (Fachhochschule beider Basel) in Basel. After a half year industry internship again working for F. Hoffmann-La Roche Adnan continued his studies at the University of Basel. During this time he joined the group of Prof. Andreas Pfaltz, where he started his Ph.D. studies in March of 2008. His research interests include the development of new P,N-ligands for iridium catalyzed asymmetric hydrogenation and application of new substrates.

32

THE PREPARATION OF INDAZOLES VIA METAL FREE INTRAMOLECULAR ELECTROPHILIC AMINATION OF 2-AMINOPHENYL KETOXIMES

A.

B.

Submitted by Carla M. Counceller, Chad C. Eichman, Brenda C. Wray, Eric R. Welin, and James P. Stambuli.[1]

Checked by John L. Tucker and Margaret Faul.

1. Procedure

A.   (E)-1-(2-Aminophenyl)ethanone oxime (1).   A 250-mL, three-necked round-bottomed flask equipped with a Teflon-coated 3.2-cm egg-shaped magnetic stir bar, internal temperature probe and a reflux condenser with nitrogen inlet is charged with distilled water (17.0 mL) and ethanol (93.0 mL) (Notes 1, 2).   2'-Aminoacetophenone (9.00 mL, 10.0 g, 72.8 mmol, 1.00 equiv) (Note 3) is added *via* syringe, and the solution is stirred during the addition.   Hydroxylamine hydrochloride (15.4 g, 223 mmol, 2.97 equiv) (Note 4) is added in one portion, followed by sodium hydroxide (23.7 g, 594 mmol, 8.16 equiv) (Notes 5 and 6).   The reaction flask, which contains the resulting suspension of a white solid in a yellow liquid, is placed in an oil bath, and the internal temperature is held at 60 °C. After 1 h, TLC analysis (Note 7) revealed that the reaction is complete. The reaction flask is removed from the oil bath, and the white slurry is allowed to cool to ambient temperature.   The mixture is transferred to a 1-L, single-necked round-bottomed flask (Note 8) and concentrated by rotary evaporation (bath temperature increased from 28 to 46 °C, 30–60 mmHg).

Org. Synth. **2011**, *88*, 33-41
Published on the Web 9/15/2010

The solid residue is dissolved in distilled water (140 mL). The aqueous solution is transferred to a 250-mL separatory funnel and is extracted three times with ethyl acetate (3 x 100 mL) (Note 9). The combined organic layers are dried over MgSO₄ (6.00 g). The drying agent is removed by filtration before the organic layer is concentrated by rotary evaporation (the bath temperature is increased from 30 to 45 °C, 30–60 mmHg) yielding the crude product as a white solid. This solid is dissolved in dichloromethane (120 mL) (Notes 10, 11) at 25 °C, before hexanes (20.0 mL) (Note 12) is added over two min. Upon clouding, the solution is cooled in an ice/water bath to 0–5 °C, and hexanes (100.0 mL) is added over 10 minutes. Crystals are allowed to age for 1 h at 0–5 °C, before isolation on a Büchner funnel. The white cotton-like solid is washed with ambient temperature hexanes (30.0 mL). The mother liquor and wash are concentrated by rotary evaporation (bath temperature increased from 28 to 46 C, 30–60 mmHg), yielding an off-white solid that is purified *via* crystallization using dichloromethane (30 mL) and hexanes (30 mL) (Note 13). The first crop provided 7.50 g, and the second crop 0.91 g of product. The total yield of (*E*)-1-(2-aminophenyl)ethanone oxime is 8.41 g (56.0 mmol, 76.9%, 98.0% pure by GC) (Notes 14, 15).

B. *3-Methyl-1H-indazole (2)*. A 2-L single-necked round-bottomed flask equipped with a Teflon-coated 4-cm egg-shaped magnetic stir bar, a rubber septum, and nitrogen inlet is charged with 1-(2-aminophenyl) ethanone oxime (9.76 g, 65.0 mmol, 1.00 equiv) and dichloromethane (1 L) (Note 16). Triethylamine (18.1 mL, 130 mmol, 2.00 equiv) (Note 17) is added *via* syringe, and the reaction mixture is stirred for 15 min at ambient temperature before being cooled to 0–5 C in an ice/water bath. Methanesulfonyl chloride (6.00 mL, 78.0 mmol, 1.20 equiv) (Note 18) is added to 300 mL dichloromethane (Note 16), and the resulting solution is cooled to 0–5 C in an ice/water bath. The cold solution of methanesulfonyl chloride is added *via* cannula to the colorless oxime mixture over 30 min. The resulting yellow solution is stirred at 0–5 C for 1.5 h. Silica gel (Notes 19, 20) is added, and the green/brown solution is concentrated by rotary evaporation (bath temperature is 25 C, 30–60 mmHg). The resulting brown solid is loaded on a flash chromatography column (2.50 in. diameter glass column and *ca* 410 g of silica) (Note 21). The product is eluted using 3.00 L of 2:3 ethyl acetate-hexanes, followed by 2.00 L of 1:1 ethyl acetate-hexanes. The fractions (50 mL size) containing the desired product are combined and concentrated by rotary evaporation to afford 5.86 g (44.1 mmol, 68.2%) of **3** as a tan solid (Notes 22, 23, 24).

2. Notes

1. Ethanol (reagent grade) was purchased from Sigma Aldrich and used as received without further purification.
2. Distilled water (17.0 mL) and ethanol (93.0 mL) were used creating a 15% volume/volume solution of distilled water and ethanol, keeping the solution at 0.69 M in 2'-aminoacetophenone.
3. 2'-Aminoacetophenone (98%) was purchased from Aldrich Chemical Co., Inc. and used as received without further purification.
4. Hydroxylamine hydrochloride (97%) was purchased from Acros Organics and used as received without further purification.
5. Sodium hydroxide (97.0%) was purchased from Acros Organics.
6. The exothermic reaction warms the mixture to ~70 °C after addition of sodium hydroxide. The initial thick slurry will thin as the reaction progresses.
7. TLC analysis was done on silica gel using 20% ethyl acetate/hexanes as the eluent using UV lamps for visualization. The ketone starting material has an R_f = 0.64 and the oxime product has an R_f = 0.29.
8. Ethanol (50.0 mL) was used as a wash to aid in transfer.
9. Ethyl acetate (99.8%) was purchased from Sigma Aldrich and used as received without further purification.
10. Dichloromethane (99.5%) was purchased from Sigma Aldrich and used as received without further purification.
11. Brief warming to 30 °C may be required for full dissolution.
12. Hexanes (98.5%) was purchased from Sigma Aldrich and used as received without further purification.
13. Clouding is not observed prior to cooling to 0–5 °C.
14. mp 110–111 °C; ^{1}H NMR (400 MHz, *DMSO-d6*) δ: 2.17 (s, 3 H), 6.36 (br s, 2 H), 6.54 (ddd, J = 8.0, 7.0, 1.2 Hz, 1 H), 6.68 (dd, J = 8.0, 1.0 Hz, 1 H), 7.01 (ddd, J = 8.2, 7.0, 1.5 Hz, 1 H), 7.28 (dd, J = 7.8, 1.4 Hz, 1 H), 10.95 (s, 1 H); ^{13}C NMR (100 MHz, *DMSO-d6*) δ: 12.0, 114.9, 115.5, 117.5, 128.3, 128.6, 146.6, 155.8; IR (ATR): 3391(s), 3242(s), 3138(m), 3053(m), 2845(w), 2365(m), 1900(s), 1619(s), 1607(s), 1491(m), 1439(s), 1364(s), 1311(m), 1289(s), 1243(s), 1156(s), 1097(m), 996(s), 929(s),

859(m), 752(m), 709(m), 647(s), 544(m), 503(s) cm^{-1}; HRMS (ESI): calcd for $C_8H_{10}N_2O[M+H]^+$: 151.0866, found 151.0861.

15. By NMR, > 95 % E isomer.

16. Dichloromethane (99.5%) was purchased from Sigma-Aldrich and used without further purification.

17. Triethylamine (99%) was purchased from Alfa Aesar and used without further purification.

18. Methanesulfonyl chloride (98%) was purchased from Alfa Aesar and used without further purification.

19. Silica gel was EMD Silica Gel 60, 220–400 mesh.

20. 30 g of silica was used.

21. The column is wet-loaded using 2:3 ethyl acetate-hexanes. TLC analysis was done on silica gel using a 1:1 mixture of ethyl acetate-hexanes as the eluent using UV lamps for visualization. The oxime starting material has an $R_f = 0.67$ and the indazole product has an $R_f = 0.50$.

22. mp 110–111 °C; ^{1}H NMR (400 MHz, CDCl$_3$) δ: 2.61 (s, 3 H), 7.15 (t, J = 7.5 Hz, 1 H), 7.35 (t, J = 7.5 Hz, 1 H), 7.43 (d, J = 8.5 Hz, 1 H), 7.67 (d, J = 8 Hz, 1 H); ^{13}C NMR (125 MHz, CDCl$_3$) δ: 12.0, 109.8, 120.1, 122.7, 126.7, 141.1, 143.2; IR (ATR): 3304(w), 3062(s), 2919(m), 1959(s), 1615(m), 1497(m), 1440(m), 1387(m), 1364(m), 1335(s), 1271(s), 1254(s), 1177(s), 1157(s), 1115(m), 1067(w), 1006(s), 984(m), 940(s), 898(s), 763(m), 746(s), 676(w), 593(w), 530(w) cm^{-1}.; HRMS (ESI): calcd. for $C_8H_8N_2[M+H]^+$: 133.0760, found 133.0761. The crystalline material appears stable indefinitely at 0 C in a closed container.

23. Compound **3** was determined to be ≥98% pure by GC analysis using an Agilent 19091J-413 GC with a HP-5 5% Phenyl Methyl Siloxane column with the following parameters: Initial Oven Temp: 100 °C; Initial Time: 2.00 min; Rate: 25 °C/min to 225 °C over 5 min, then 25 °C/min to 250 °C, then 250 °C until 13 min. Initial Flow: 3.5 mL/min.

24. 3-Methyl-1-(methylsulfonyl)-1H-indazole (1.08 g, 5.16 mmol) was an isolated by-product of the reaction as a red/orange solid (92.7% pure by GC). $R_f = 0.58$ in 1:1 ethyl acetate-hexanes; mp 72–73 C; ^{1}H NMR (400 MHz, CDCl$_3$) δ: 2.62 (s, 3 H), 3.18 (s, 3 H), 7.34-7.41 (m, 1 H), 7.56 (d, J = 8.0 Hz, 1 H), 7.69 (d, J = 8.0 Hz, 1 H), 8.04 (d, J = 8.4 Hz, 1 H); ^{13}C NMR (125 MHz, CDCl$_3$) δ: 12.3, 40.5, 113.2, 120.7, 124.0, 125.8, 129.4, 141.0, 150.7; IR (ATR): 3059(m), 3022(s), 2925(s), 1608(s), 1444(s), 1356(m), 1247(s), 1169(s), 976(m), 754(m), 603(s), 541(s), 507(m), cm^{-1}.; HRMS (ESI): calcd. for $C_9H_{10}N_2O_2S[M+H]^+$: 211.0541, found 211.0534.

Waste Disposal Information

All hazardous materials should be handled and disposed of in accordance with "Prudent Practices in the Laboratory"; National Academic Press; Washington DC, 1995.

3. Discussion

The use of 1*H*-indazoles as anti-cancer, -inflammatory, and -microbial agents has been documented in recent patents and publications.[2-6] The diverse pharmacological properties exhibited by 1*H*-indazoles have sparked the emergence of novel methods toward their synthesis. Classical syntheses involve the use of harsh or inconvenient reaction conditions, such as diazotizations or nitrosation reactions.[7] Most of the modern approaches to prepare indazoles employ metal catalysts[8-9] or occur under impractical reaction conditions.[10-11] Although numerous methodologies had been reported to synthesize 1*H*-indazoles, a mild and general method remained a challenge.

Recent work by others has employed electrophilic amination chemistry to prepare tertiary amines providing evidence for a nucleophilic substitution mechanism at the sp^3-nitrogen.[12] Earlier work in electrophilic aminations demonstrated that nucleophilic substitution reactions at the sp^2-nitrogen center of a protected oxime can occur to form the corresponding substitution products,[13-16] yet this strategy to produce heterocycles has rarely been employed. We hypothesized that chemoselective activation of an oxime could occur with preference over arylamino group activation for *o*-aminobenzoximes. This would prompt an intramolecular attack onto the activated oxime by the adjacent aniline producing the desired 1*H*-indazole. Based on this hypothesis, we explored this reaction and have reported a simple, metal-free synthesis of substituted 1*H*-indazoles that occurs from readily available aminobenzoximes under mild conditions.[17] Many different indazoles may be prepared using this method, and examples are shown in Table 1. The mild conditions allowed for successful indazole formation in the presence of a wide range of functionality, and also enabled the use of aldoximes, which have been shown to readily undergo dehydration to the corresponding nitriles. The mechanism of the reaction requires the hydroxyl group of the oxime to be distal to the amino group. Therefore, only oximes containing the required geometry react under these conditions.

Table 1. Synthesis of 1*H*-Indazoles

Entry	Oxime	1*H*-Indazole	Yield(%)	Entry	Oxime	1*H*-Indazole	Yield(%)
1	CH₃, NHallyl	CH₃, Nallyl	70	7	Cl, Ph, NH₂	Cl, Ph	82
2	Et, NH₂	Et	81	8	Ph, NH₂	Ph	84
3	Bn, NH₂	Bn	86	9	C₆H₄-*p*-Br, NH₂	C₆H₄-*p*-Br	81
4	R, CH₃, NH₂	R, CH₃ (R = OCH₃)	75	10	C₆H₄-4-F, NH₂	C₆H₄-*p*-F	86
5	furan, NH₂	furan	86	11	H, NH₂	H, Ms	52[a]
6	CH₃, NH₂	CH₃	38	12	CH₃, NH, CH₃	CH₃, CH₃	87

[a] 2 equiv of MsCl used at -20 °C

1. Department of Chemistry, The Ohio State University, Columbus, OH 43210. stambuli@chemistry.ohio-state.edu

2. Cerecetto, H.; Gerpe, A.; Gonzalez, M.; Aran, V. J.; Ocariz, C. O. *Mini-Rev. Med. Chem.* **2005**, *5*, 869-878.

3. Feng, Y.; Cameron, M. D.; Frackowiak, B.; Griffin, E.; Lin, L.; Ruiz, C.; Schroeter, T.; LoGrasso, P. *Bioorg. Med. Chem. Lett.* **2007**, *17*, 2355-2360.

4. Goodman, K. B.; Cui, H.; Dowdell, S.E.; Gaitanopoulos, D. E.; Ivy, R. L.; Sehon, C. A.; Stavenger, R. A.; Wang, G. Z.; Viet, A. Q.; Xu, W.; Ye, G.; Semus, S. F.; Evans, C.; Fries, H. E.; Jolivette, L. J.; Kirkpatrick, R. B.; Dul, E.; Khandekar, S. S.; Yi, T.; Jung, D. K.; Wright, L. L.; Smith, G. K.; Behm, D. J.; Bentley, R.; Doe, C. P.; Hu, E.; Lee, D. *J. Med. Chem.* **2007**, *50*, 6-9.

Org. Synth. **2011**, *88*, 33-41

5. Zhu, G.-D.; Gandhi, V. B.; Gong, J.; Thomas, S.; Woods, K. W.; Song, X.; Li, T.; Diebold, R. B.; Luo, Y.; Liu, X.; Guan, R.; Klinghofer, V.; Johnson, E. F.; Bouska, J.; Olson, A.; Marsh, K. C.; Stoll, V. S.; Mamo, M.; Polakowski, J.; Campbell, T. J. *J. Med. Chem.* **2007**, *50*, 2990-3003.

6. Sun, J.-H.; Teleha, C. A.; Yan, J.-S.; Rodgers, J. D.; Nugiel, D. A. *J. Org. Chem.* **1997**, *62*, 5627-5629.

7. Eicher, T.; Hauptmann, S. *The Chemistry of Heterocycles.* Wiley-VCH: Weinheim, 2003.

8. Inamoto, K.; Katsuno, M.; Yoshino, T.; Arai, Y.; Hiroya, K.; Sakamoto, T. *Tetrahedron* **2007**, *63*, 2695-2711.

9. Vina, D.; Olmo, E. D.; Lopez-Perez, J. L.; Feliciano, A. S. *Org. Lett.* **2007**, *9*, 525-528.

10. O'Dell, D. K.; Nicholas, K. M. *Heterocycles* **2004**, *63*, 373-382.

11. Jin, T.; Yamamoto, Y. *Angew. Chem. Int. Ed.* **2007**, *46*, 3323-3325.

12. Campbell, M. J.; Johnson, J. S. *Org. Lett.* **2007**, *9*, 1521-1524.

13. Kemp, D. S.; Woodward, R. B. *Tetrahedron* **1965**, *21*, 3019-3035.

14. Hassner, A.; Patchornik, G.; Pradhan, T. K.; Kumareswaran, R. *J. Org. Chem.* **2007**, *72*, 658-661.

15. Tsutsui, H.; Ichikawa, T.; Narasaka, K. *Bull. Chem. Soc. Jpn* **1999**, *72*, 1869-1878.

16. Kitamura, M.; Suga, T.; Chiba, S.; Narasaka, K. *Org. Lett.* **2004**, *6*, 4619-4621.

17. Counceller, C. M.; Eichman, C. C.; Wray, B. C.; Stambuli, J. P. *Org. Lett.* **2008**, *10*, 1021-1023.

Appendix
Chemical Abstracts Nomenclature; (Registry Number)

2'-Aminoacetophenone; (551-93-9)
Hydroxylamine hydrochloride; (5470-11-1)
Triethylamine; (121-44-8)
Methanesulfonyl chloride; (124-63-0)
3-Methyl-1*H*-indazole; (3176-62-3)
(*E*)-1-(2-Aminophenyl)ethanone oxime; (4964-49-2)

 James Stambuli was born in Paterson, NJ and received his B.A. in Chemistry with Honors from Rutgers University in Newark, NJ in 1998. He obtained his Ph.D. from Yale University in the laboratory of Professor John Hartwig in 2003 and was an NIH Postdoctoral Fellow with Professor Barry Trost at Stanford University from 2003-2006. He began an independent research career at The Ohio State University in 2006.

 Carla Counceller obtained her B.S. in Chemistry from Indiana University while performing undergraduate research in Professor Jeffrey Johnston's research group. She obtained her Ph.D. in the laboratories of Professor James P. Stambuli at The Ohio State University in 2010. Her doctoral studies involved the development of metal-free syntheses of nitrogen containing heterocycles and the synthesis of oxazoles via nickel-catalyzed C-S activation. She is currently employed at Chemical Abstracts Service.

 Chad Eichman received his B.S. in Chemistry from the University of Wisconsin, where he performed undergraduate research with Professor Hans J. Reich. He obtained his Ph.D. in the laboratories of Professor James P. Stambuli at The Ohio State University in 2010. His doctoral studies involved transition-metal catalysis and the development of metal-free electrophilic amination processes. He is currently working as a postdoctoral fellow at Northwestern University in the laboratories of Professor Karl A. Scheidt.

Brenda Wray obtained her B.S. in Chemistry from the University of Missouri-Columbia. She is currently pursuing graduate studies in the laboratory of Professor James P. Stambuli at The Ohio State University. Her graduate work has included metal-free synthesis of 1-aryl-1*H*-indazoles and application towards the total synthesis of the Nigella natural products.

Eric R. Welin was born in Columbus, Ohio in 1987. He enrolled in the Honors Program at The Ohio State University in 2006. In early 2009, he began conducting research in the Stambuli group. His research focused on the synthesis of nitrogen containing heterocycles. He graduated in the spring of 2010 with a B.S. in chemistry, and is currently pursuing his Ph.D. in the laboratories of Professor David MacMillan at Princeton University.

John Tucker obtained his B.S. in Chemistry from the University of California at Irvine in 1995. Beginning his career at Pfizer in Groton CT, initially as a medicinal chemist, he moved to process chemistry where he helped to develop commercial synthetic routes for marketed medicines including Tarceva, Draxxin, and Convenia. In 2006 he became a founding scientist and director of Green Chemistry at BioVerdant, a biotechnology firm focused upon the use of enzymatic synthesis for chiral drug manufacture. He has recently assumed a leadership role in a GMP manufacturing facility as a scientist at Amgen in Thousand Oaks, CA, in the Small Molecule Process and Process Development Group.

THE PREPARATION OF (2*R*,5*S*)-2-*t*-BUTYL-3,5-DIMETHYLIMIDAZOLIDIN-4-ONE

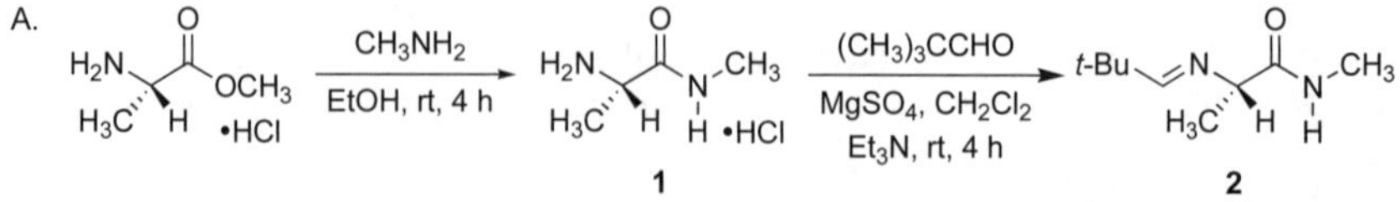

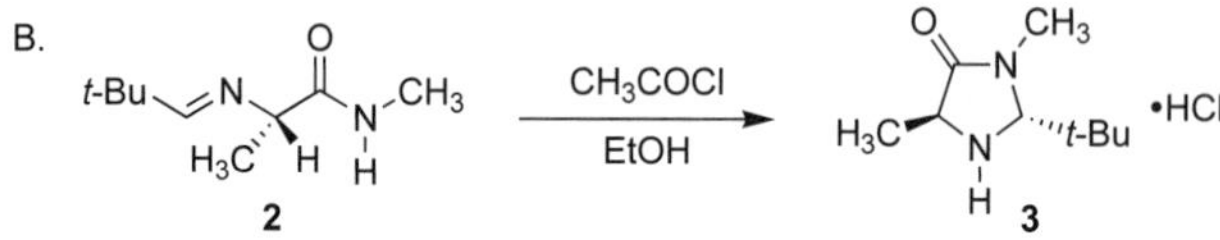

A.

B.

Submitted by Thomas H. Graham, Benjamin D. Horning and David W. C. MacMillan.[1]

Checked by David Hughes.[2]

1. Procedure

A. *(S,E)-2-(2,2-dimethylpropylidenamino)-N-methylpropanamide (2)*. A tared 1-L round-bottomed flask (Note 1) equipped with a 3-cm oval PTFE-coated magnetic stir bar is charged with a 31 wt% solution of methylamine in ethanol (112 mL, 85 g, 0.85 mol, 3.0 equiv) (Note 2) and placed in a room temperature water bath. To the stirred solution is added L-alanine methyl ester hydrochloride (40.0 g, 0.287 mol, 1.0 equiv) (Note 3) via a powder funnel followed by a rinse with ethanol (10 mL). The flask is fitted with a rubber septum through which is inserted both an 18-gauge needle connected to a nitrogen inlet with a gas bubbler and a thermocouple probe (Note 4). The mixture is stirred at 20–22 °C for 4 h (Notes 5 and 6). The stir bar is removed, and the mixture is concentrated by rotary evaporation (20 mm Hg, 45 °C bath temperature) to provide a wet solid (60 g). Toluene (100 mL) (Note 7) is added to the mixture, which is concentrated by rotary evaporation (20 mm Hg, 45 °C bath temperature) to provide a wet solid (65 g). The toluene (100 mL) flush is repeated, and the mixture is concentrated to 49 g of solids, which are dried in a vacuum oven (20 mmHg, 45 °C) for 4 h to afford the crude L-alanine-*N*-methylamide 1 as a pasty solid (45 g) (Notes 8 and 9).

The 1-L flask containing the crude L-alanine-*N*-methylamide 1 is equipped with a 3-cm oval PTFE-coated magnetic stirring bar. The solids

42

are scraped off the walls using a spatula (Note 10). The flask is immersed in a room temperature water bath and charged with anhydrous magnesium sulfate (30 g) and dichloromethane (140 mL) (Note 11). The mixture is stirred at ambient temperature and treated sequentially with triethylamine (60.0 mL, 43.6 g, 0.425 mol, 1.5 equiv) and pivaldehyde (95% purity, 35 mL, 28 g, 0.31 mmol, 1.07 equiv corrected for purity) (Notes 11 and 12). The flask is fitted with a rubber septum through which is inserted both an 18-gauge needle connected to a nitrogen inlet with a gas bubbler and a thermocouple probe (Note 4). The reaction mixture is stirred for 4 h at ambient temperature (Notes 13 and 14). Additional pivaldehyde (3 mL) and magnesium sulfate (5 g) are added, and the mixture is stirred for an additional 30 min at ambient temperature. The septum is removed and replaced with a 250-mL addition funnel. Toluene (200 mL) is added over 10 min, and the mixture is stirred for an additional 15 min. The mixture is then filtered through a 350-mL medium porosity sintered glass funnel to remove the triethylamine hydrochloride and magnesium sulfate. The filter cake is washed with toluene (3 x 50 mL). The combined filtrate is concentrated by rotary evaporation (100 mmHg initially to 20 mmHg, 45 °C bath temperature) to 70 g. Additional triethylamine hydrochloride precipitates during this concentration, so additional toluene (100 mL) is added and the mixture is filtered through a 60-mL medium porosity sintered glass funnel. The filtrate is concentrated by rotary evaporation (20 mmHg, 50 °C bath temperature), then vacuum dried (0.1 mmHg, 23 °C) for 4 h to afford (*S,E*)-2-(2,2-dimethylpropylidenamino)-*N*-methylpropanamide (**2**) (44.2–46.5 g, 95% purity, 86–90% yield) as a pale yellow oil. (Notes 15 and 16)

B.   *(2R,5S)-2-tert-butyl-3,5-dimethylimidazolidin-4-one (3).*   A 500-mL round-bottomed, three-necked flask, equipped with a 3-cm oval PTFE-coated magnetic stirring bar, is fitted with two septa and a 100-mL pressure-equalizing addition funnel connected to a nitrogen inlet with a gas bubbler. A thermocouple probe is inserted through one septum (Note 4). The flask is charged with ethanol (140 mL), placed in an ice-water bath, and cooled to 1–3 °C. The addition funnel is charged with acetyl chloride (22.3 mL, 24.5 g, 0.312 mol, 1.1 equiv) (Note 17). The acetyl chloride is added dropwise over 15 min to the stirred ethanol solution, resulting in a temperature rise to 18 °C. The solution is cooled to 5 °C with an ice bath. The addition funnel is replaced with a standard tapered glass funnel, and maintaining ice bath cooling, the crude imine (**2**) (45.7 g, 95% pure, 0.255 mol) is poured into the HCl/ethanol solution as one portion over 30 s. The flask that contained **2** is rinsed with ethanol (3 x 10 mL), and the rinses are added to the HCl/ethanol solution via the tapered glass funnel. The glass funnel is removed and replaced with an inlet adapter connected to a nitrogen line and gas bubbler. The reaction temperature rises from 5 °C to 30 °C over a 3 min period after addition of the imine, and crystallization occurs within 10 min (Note 18). The ice-bath is removed and replaced with a heating mantle, and the stirred mixture is warmed to 70 ± 2 °C over 30 min and held at this temperature for 20 min (Note 19). The mixture remains heterogeneous. The heating mantle is removed, and the stirred mixture is allowed to cool to 23 °C over 1.5 h, and stirring is continued for 2 h at ambient temperature. The resulting crystals are vacuum filtered using a 350-mL sintered glass funnel. The filter cake is washed with ethanol (2 x 30 mL) and air-dried to afford (2R,5S)-2-*tert*-butyl-3,5-dimethylimidazolidin-4-one (**3**) (43.2 g) as white crystalline material (Note 20). The mother liquors are concentrated to 80 mL by rotary evaporation (20 mmHg, 50 °C bath temperature) in a 500-mL round-bottomed flask, and the crystallization process is repeated to afford a second batch of **3** (3.9 g) as white crystalline material (Note 21). The first and second batches are combined to afford 47.1 g of **3** (89% yield for this step, 77–80% for 3 steps) (Notes 22-26).

2. Notes

1.   A 1-L flask was used to minimize bumping during concentration after completion of the reaction.

2.   Methylamine (33 wt% in ethanol) was purchased from Sigma-Aldrich and used as received. ^{1}H NMR analysis with a 5 second delay indicated the reagent contained 31 wt% methylamine and 69 wt% ethanol. The use of less than 3 equiv of methylamine led to incomplete conversion to the N-methyl amide.

3.   L-Alanine methyl ester hydrochloride (**1**) was purchased from Sigma-Aldrich (99%) and Alfa Aesar (99%) and used as received. The hydrochloride salt is hygroscopic and was weighed into a bottle that was capped after weighing to avoid exposure to air prior to addition to the reaction. ^{1}H NMR analysis of this starting material is recommended as one sample contained 15% of the diketopiperazine impurity. Water content was measured by Karl-Fischer titration and ranged from 0.2 to 1.0% for the various lots of material used.

44

4.   The internal temperature was monitored using a J-Kem Gemini digital thermometer with a Teflon-coated T-Type thermocouple probe (12-inch length, 1/8 inch outer diameter, temperature range –200 to +250 °C).

5.   The mixture warms from 21 °C to 29 °C within 5 min of solids addition, then cools to 20-22 °C over 15 min in the water bath. In the hands of the checker the reaction solution remained heterogeneous throughout. In one experiment, an aliquot of the reaction mixture was filtered at the end of the reaction. The solids were determined to be methylamine hydrochloride by ^{1}H NMR analysis (CD$_3$OD).

6.   The reaction progress is monitored by ^{1}H NMR analysis (CD$_3$OD) of aliquots of the reaction mixture. The starting material resonances at 3.74 (s, 3 H, OCH$_3$) and 1.34 (d, 3 H, CHCH$_3$) are monitored vs. product at 2.78 (s, 3 H, NHCH$_3$) and 1.29 (d, 3 H, CHCH$_3$). The reaction is complete after 1.5 h (<1% starting material based on a spiking experiment with starting material).

7.   Toluene (ACS reagent grade, >99.5%) was purchased from Sigma-Aldrich and used as received

8.   The solids are a mixture of methylamine and L-alanine-N-methylamide. The level of methylamine was typically ~40 mol% vs. product. The solids are vacuum dried until the level of ethanol and toluene are <3 mol% by ^{1}H NMR analysis. One experiment in which the toluene flushes were omitted afforded material that contained 15 mol% ethanol and 70 mol% methylamine relative to amide product. This material resulted in a 10 % lower yield in the imine formation.

9.   Spectroscopic data for crude L-alanine-N-methylamide hydrochloride: ^{1}H NMR (400 MHz, CD$_3$OD) δ: 1.49 (d, J = 7.1 Hz, 3 H, CHCH$_3$), 2.79 (s, 3 H, NHCH$_3$), 3.93 (q, J = 7.0 Hz, 1 H, CHCH$_3$); ^{13}C NMR (100 MHz, CD$_3$OD) δ: 17.8, 26.5, 50.4, 171.6. Methylamine: ^{1}H NMR (400 MHz, CD$_3$OD) δ: 2.55 (s); ^{13}C NMR (100 MHz, CD$_3$OD) δ: 25.5.

10.   Material dried on the walls of the flask tends not to react so optimum yields are obtained when this material is dislodged from the walls. L-Alanine-N-methylamide that is chunky or pasty in consistency performs well in the imine formation. The submitters report that thoroughly dried amide must be ground to a powder for optimum results.

11.   The following reagents and solvents were used as received for the imine formation: anhydrous magnesium sulfate powder (Fisher), toluene (Sigma-Aldrich, ACS reagent grade, >99.5%), triethylamine (Sigma-

Aldrich, 99%), and dichloromethane (Fisher Optima, >99.5%). Pivaldehyde (96%) was purchased from Sigma-Aldrich; ^{1}H NMR analysis revealed a number of low level impurities that collectively integrated to ~5%. The material was charged based on 95% purity.

12. The mixture warms from 24 °C to 30 °C over a 5 min period after addition of pivaldehyde.

13. The reaction progress is followed with ^{1}H NMR by adding an aliquot of the reaction mixture to CDCl$_3$ and filtering the sample. To determine the level of pivaldehyde remaining, the sample is analyzed directly, comparing the aldehyde proton (9.5 ppm) of pivaldehyde to the corresponding proton of the imine (7.6 ppm). To accurately measure unreacted amide, the solution is evaporated to remove dichloromethane, then taken up in CDCl$_3$ for analysis (imine **CH**CH$_3$ quartet at 3.7 ppm compared to amide at 3.5 ppm).

14. Complete consumption of pivaldehyde typically occurs within a two hour reaction time but depends on the consistency of the amide (chunky material takes longer to react). The reaction typically stalls at 91-94% conversion, requiring an additional charge of pivaldehyde and magnesium sulfate. Addition of 10% more pivaldehyde or 25% more magnesium sulfate at the beginning of the reaction does not lead to increased conversion.

15. By ^{1}H NMR analysis, the imine contained 1.5 wt% unreacted amide and 3.5 wt% toluene. Toluene levels up to 8% were used with no impact for the next step. Amide levels up to 8% were used for the next step with no impact for the first and second crop isolations but co-crystallized with product if a third crop was isolated. The submitters stored the imine under vacuum. The checker stored the imine in a flask sealed with a septum and observed approximately 2% hydrolysis to the amide in a week at room temperature.

16. Spectroscopic data for **2**: ^{1}H NMR (400 MHz, CDCl$_3$) δ: 1.08 (s, 9 H, C(**CH**$_3$)$_3$), 1.31 (d, J = 7.1 Hz, 3 H, CH**CH**$_3$), 2.84 (d, J = 5.0 Hz, 3 H, NC**H**$_3$), 3.68 (q, J = 7.0 Hz, 1 H, C**H**CH$_3$), 6.9 (bs, 1 H, **H**NCH$_3$), 7.52 (s, 1 H, (CH$_3$)$_3$C**H**N); ^{13}C NMR (100 MHz, CDCl$_3$) δ: 21.6, 26.0, 26.9, 36.6, 67.7, 173.2, 174.8.

17. Ethanol (ACS reagent, >99.5%, water content 0.37 mg/mL based on Karl Fischer titration) and acetyl chloride (reagent grade, 98%) were purchased from Sigma-Aldrich and used as received.

46

18. The reaction progress is monitored by ^{1}H NMR of a sample dissolved in CD_3OD. The reaction is complete within 10 min of imine addition with no resonances corresponding to the imine detectable.

19. The cyclization reaction forms a 3:1 mixture of trans:cis (**3:4**) diastereomers. Warming to 70 °C for 20 min results in equilibration to the thermodynamic ratio of 5:1 (this thermodynamic ratio in ethanol is also established starting with pure **3**). At the same time, decomposition of the diastereomers to amide **1** and the diethyl acetal of pivaldehyde occurs. At reflux (79 °C), the equilibration occurs within 5 – 10 min but degradation is also relatively rapid; therefore, the equilibration time and temperature were selected to afford complete equilibration with minimal decomposition (about 2%). The equilibration and decomposition of **3** is likely occurring via a reversible reaction with the imine as outlined below. Equilibration directly between **3** and **4** is unlikely as neither of the protons at positions 2 or 5 is exchanged in CD_3OD during equilibration, indicating that a deprotonation/ protonation process is not occurring. Starting with pure **3** in either methanol or ethanol solution, the equilibration of **3** and **4**, and their decomposition to amide **1** and the acetal of pivaldehyde, were followed by ^{1}H and ^{13}C NMR.

20. In the lab of the submitters, cooling the solution at a more rapid rate resulted in the entrainment of the minor diastereomer, (2*S*,5*S*)-2-*tert*-butyl-3,5-dimethylimidazolidin-4-one **4**. In addition, the submitters report cooling the solution below 20 °C, or allowing **3** to age with the mother liquors for >6 h, caused the minor diastereomer to begin crystallizing. In the lab of the checker, the minor diastereomer **4** was never detected in crystalline **3** (<0.2% by NMR) even with >12 h crystallization age times. It is important to adequately wash the filter cake to ensure that the mother liquors are not entrained.

21. ^{1}H NMR (CD$_3$OD) analysis of an evaporated sample of the mother liquors indicated a ratio of **3:4:1** of 49:37:12. Heating at 70 ± 2 °C for 20 min resulted in equilibration to a 5:1 ratio of **3:4**.

22. A third crop of **3** can be obtained by concentrating the remaining mother liquors to 40 mL and repeating the equilibration/crystallization procedure to afford 1.1 g of **3**.

23. (2R,5S)-2-*tert*-Butyl-3,5-dimethylimidazolidin-4-one (**3**) has the following physical and spectroscopic data: mp 211–216 °C with decomposition (ethanol); [α]$_D$ -43.4 (*c* 1.0, CH$_3$OH, 23 °C); IR (solid) 2873, 2641, 2514, 1719, 1584 cm^{-1}; ^{1}H NMR (400 MHz, CD$_3$OD) δ: 1.19 (s, 9 H, (CH$_3$)$_3$C), 1.59 (d, J = 7.0 Hz, 3 H, CHCH$_3$), 3.09 (s, 3 H, NCH$_3$), 4.28 (q, J = 7.0 Hz, 1 H, CHCH$_3$), 4.79 (s, 1 H, (CH$_3$)$_3$CCH); ^{13}C NMR (100 MHz, CD$_3$OD) δ: 14.9, 25.5, 32.5, 37.7, 54.9, 82.0, 171.3; HRMS (ESI-TOF) *m/z* calcd for C$_9$H$_{19}$N$_2$O ([M+H]$^+$) 171.1492, found 171.1492; Karl Fischer titration: 0.1% water; elemental analysis calcd. for C$_9$H$_{19}$ClN$_2$O: C, 52.29; H, 9.26; N, 13.55; found: C, 52.53; H, 9.51; N, 13.51; chloride titration (AgNO$_3$) calcd: Cl, 17.15; found Cl, 17.16.

24. The minor diastereomer, (2S,5S)-2-*tert*-butyl-3,5-dimethyl-imidazolidin-4-one (**4**), has the following spectroscopic data: ^{1}H NMR (400 MHz, CD$_3$OD) δ: 1.19 (s, 9 H, (CH$_3$)$_3$C), 1.59 (d, J = 7.2 Hz, 3 H, CHCH$_3$), 3.04 (s, 3 H, NCH$_3$), 4.13 (q, J = 7.1 Hz, 1 H, CHCH$_3$), 4.71 (s, 1 H, (CH$_3$)$_3$CCH); ^{13}C NMR (100 MHz, CD$_3$OD) δ: 14.9, 25.4, 31.5, 35.3, 54.8, 81.8, 171.6.

25. Imidazolidinone **3** is converted, as follows, to (2S,5S)-benzyl-2-*tert*-butyl-3,5-dimethyl-4-oxoimidazolidine-1-carboxylate for assessing the enantiopurity:

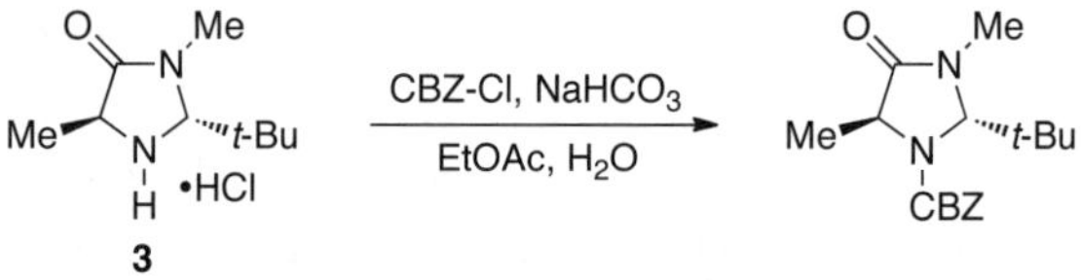

An 8 mL vial is charged with **3** (100 mg, 0.48 mmol, 1.0 equiv), solid NaHCO$_3$ (200 mg, 2.4 mmol, 5.0 equiv), ethyl acetate (1.0 mL) and water (1.0 mL). The mixture is treated with benzyl chloroformate (100 μL, 0.72 mmol, 1.5 equiv) and stirred at ambient temperature for 16 h. The layers are separated and the aqueous layer is extracted with ethyl acetate (2 mL). The combined organic layers are concentrated and purified on SiO$_2$ (10 g),

48

eluent: 25% ethyl acetate/ hexanes, 10 mL fractions. Fractions 8-13 are concentrated to afford (2S,5S)-benzyl-2-*tert*-butyl-3,5-dimethyl-4-oxoimidazolidine-1-carboxylate (130 mg, 88%) as a clear, colorless syrup: R$_f$ 0.4 (40% ethyl acetate/ hexanes); [α]$_D$ –15.3 (c 1.00, CH$_2$Cl$_2$, 22 °C); IR (neat) 2966, 1699, 1411, 1392, 1251, 1119, 698 cm^{-1}; ^{1}H NMR (500 MHz, CDCl$_3$) matches reported spectrum (Seebach, D.; Juaristi, E.; Miller, D. D.; Schickli, C.; Weber, T. *Helv. Chim. Acta* **1987**, *70*, 237-261) δ: 0.97 (s, 9 H, C(CH$_3$)$_3$), 1.55 (bs, 3 H, CHCH$_3$), 3.01 (s, 3 H, NCH$_3$), 4.04-4.06 (m, 1 H, CHCH$_3$), 5.05-5.21 (m, 3 H, CHC(CH$_3$)$_3$, PhCH$_2$O), 7.33-7.39 (m, 5 H, Ar**H**); ^{13}C NMR (125 MHz, CDCl$_3$) δ: 18 (br), 26.5, 32.2, 40.7, 56.0, 67.7, 81.2, 128.6, 128.80, 128.81, 136.0, 155 (br), 173.0; HRMS (ESI-TOF) m/z calcd for C$_{17}$H$_{25}$N$_2$O$_3$ ([M+H]$^+$) 305.1865, found 305.1865. Since both enantiomers were required to develop a chiral assay, the enantiomer of **3** was prepared from *D*-alanine methyl ester hydrochloride. Submitters chiral HPLC analysis: AD-H (250 x 4.6 mm, 5μm particle size), isocratic elution with 15% ethanol/hexanes, 1.0 mL/min, 254 nm), R$_t$(major) = 8.47 min, R$_t$(minor) = 6.87 min; checkers chiral SFC analysis: AD-H (250 x 4.6 mm, 5μm particle size), isocratic elution with 4% MeOH containing 25 mM isobutylamine, 3.0 mL/min, 200 bar, 35 °C, 215 nm, 6 min total run time; R$_t$(major) = 3.9 min, R$_t$(minor) = 4.5 min. None of the minor enantiomer was detected; spiking experiments indicated the ee was >99%.

26. Imidazolidinone **3** is non-hygroscopic at ambient humidity conditions and can be stored in a closed container at ambient temperature with no additional precautions.

Safety and Waste Disposal Information

All hazardous materials should be handled and disposed of in accordance with "Prudent Practices in the Laboratory"; National Academy Press; Washington, DC, 1995.

3. Discussion

Over the past decade, organocatalysis has emerged as a versatile method for the enantioselective preparation of organic molecules.[3] Among the numerous organocatalysts developed, the imidazolidinone family is one of the most versatile catalysts, mediating a variety of transformations including Diels-Alder,[4] 1,3-dipolar cycloadditions,[5] 1,4-conjugate additions,[6] α-oxidations,[7] α-chlorinations, α-fluorinations,[8] hydride reductions,[9] and epoxidations.[10] In addition, the mild reaction conditions and selectivity of the imidazolidinone organocatalysts allow for efficient cascade catalysis.[11] A new mode of activation, SOMO catalysis, involving the oxidative coupling of enamines and nucleophiles (SOMO-philes) allows for unprecedented enantioselective transformations including α-allylations,[12] α-enolations,[13] α-vinylation,[14] α-nitroalkylation,[15] α-arylations[16] and the carbo-oxidation of styrenes.[17]

Recently, (2R,5S)-2-*tert*-butyl-3,5-dimethylimidazolidin-4-one (**3**) has emerged as a privileged organocatalyst that mediates a variety of useful asymmetric transformations (Scheme 1). The α-chlorination of aldehydes occurs in 75-95% yield and 91-96% ee and the α-chloroaldehydes can be transformed into a variety of enantio-enriched epoxides, aziridines, α-chloro alcohols, α-cyano alcohols, α-hydroxy acids, and α-amino acids (Scheme 1, eq 1).[18] The α-trifluoromethylation of aldehydes occurs in 62-86% yield and 93-99% ee using visible light and 0.5 mol% of an iridium photoredox catalyst (Scheme 1, eq 2).[19] Similarly, bromoalkanes will add to the aldehyde with visible light and 0.5 mol% of a ruthenium photoredox catalyst (Scheme 1, eq 3).[20]

Scheme 1.

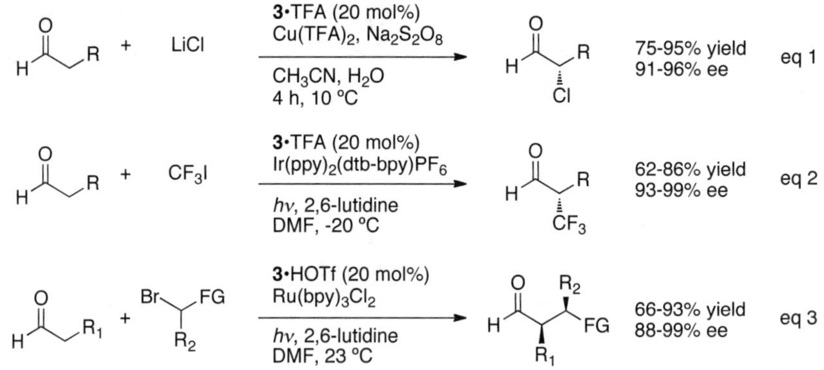

The procedure describes the preparation of (2R,5S)-2-*tert*-butyl-3,5-dimethylimidazolidin-4-one (**3**) from L-alanine methyl ester hydrochloride.[21] The present method consistently affords **3** as white crystals in good yield (77–80% for 3 steps) and high enantiomeric purity (>99% ee), without intermediate purifications or chromatography.

Org. Synth. **2011**, *88*, 42-53

1. Merck Center for Catalysis at Princeton University, Frick Laboratory, Washington Road, Princeton University, Princeton, NJ 08544; e-mail: dmacmill@princeton.edu

2. The checker thanks Zainab Pirzada for development of the chiral assay of the CBZ derivative of compound **3**, Mirlinda Biba for the rotation and chloride analyses, and Robert Reamer for NMR support.

3. (a) MacMillan, D. W. C. *Nature* **2008**, *455*, 304-308; (b) Lelais, G.; MacMillan, D. W. C. *Aldrichimica Acta* **2006**, *39*, 79-87.

4. (a) Ahrendt, K. A.; Borths, C. J.; MacMillan, D. W. C. *J. Am. Chem. Soc.* **2000**, *122*, 4243-4244; (b) Northrup, A. B.; MacMillan, D. W. C. *J. Am. Chem. Soc.* **2002**, *124*, 2458-2460; (c) Wilson, R. M.; Jen, W. S.; MacMillan, D. W. C. *J. Am. Chem. Soc.* **2005**, *127*, 11616-11617.

5. Jen, W. S.; Wiener, J. J. M.; MacMillan, D. W. C. *J. Am. Chem. Soc.* **2000**, *122*, 9874-9875.

6. (a) Paras, N. A.; MacMillan, D. W. C. *J. Am. Chem. Soc.* **2001**, *123*, 4370-4371; (b) Austin, J. F.; MacMillan, D. W. C. *J. Am. Chem. Soc.* **2002**, *124*, 1172-1173; (c) Paras, N. A.; MacMillan, D. W. C. *J. Am. Chem. Soc.* **2002**, *124*, 7894-7895; (d) Brown, S. P.; Goodwin, N. C.; MacMillan, D. W. C. *J. Am. Chem. Soc.* **2003**, *125*, 1192-1194; (e) Chen, Y. K.; Yoshida, M.; MacMillan, D. W. C. *J. Am. Chem. Soc.* **2006**, *128*, 9328-9329.

7. Brown, S. P.; Brochu, M. P.; Sinz, C. J.; MacMillan, D. W. C. *J. Am. Chem. Soc.* **2003**, *125*, 10808-10809.

8. Beeson, T. D.; MacMillan, D. W. C. *J. Am. Chem. Soc.* **2005**, *127*, 8826-8828.

9. (a) Ouellet, S. G.; Tuttle, J. B.; MacMillan, D. W. C. *J. Am. Chem. Soc.* **2005**, *127*, 32-33; (b) Tuttle, J. B.; Ouellet, S. G.; MacMillan, D. W. C. *J. Am. Chem. Soc.* **2006**, *128*, 12662-12663; (c) Lee, S.; MacMillan, D. W. C. *J. Am. Chem. Soc.* **2007**, *129*, 15438-15439; (d) Ouellet, S. G.; Walji, A.; MacMillan, D. W. C. *Acc. Chem. Res.* **2007**, *40*, 1327-1339.

10. Lee, S.; MacMillan, D. W. C. *Tetrahedron* **2006**, *62*, 11413-11424.

11. (a) Austin, J. F.; Kim, S. G.; Sinz, C. J.; Xiao, W. J.; MacMillan, D. W. C. *Proc. Nat. Acad. Sci.* **2004**, *101*, 5482-8487; (b) Huang, Y.; Walji, A. M.; Larsen, C. H.; MacMillan, D. W. C. *J. Am. Chem. Soc.* **2005**, *127*, 15051-15053; (c) Walji, A.; MacMillan, D. W. C. *Syn. Lett.* **2007**, *10*, 1477-1489; (d) Simmons, B.; Walji, A.; MacMillan, D. W. C. *Angew. Chem. Int. Ed.* **2009**, *48*, 4349-4353.

12. Beeson, T. D.; Mastracchio, A.; Hong, J.; Ashton, K.; MacMillan, D. W. C. *Science* **2007**, *316*, 582-585.
13. Jang, H.; Hong, J.; MacMillan, D. W. C. *J. Am. Chem. Soc.* **2007**, *129*, 7004-7005.
14. Kim, H.; MacMillan, D. W. C. *J. Am. Chem. Soc.* **2008**, *130*, 398-399.
15. Wilson, J. E.; Casarez, A. D.; MacMillan, D. W. C. *J. Am. Chem. Soc.* **2009**, *131*, 11332-11334.
16. Conrad, J. C.; Kong, J.; Laforteza, B. N.; MacMillan, D. W. C. *J. Am. Chem. Soc.* **2009**, *131*, 11640-11641.
17. Graham, T. H.; Jones, C. M.; Jui, N. T.; MacMillan, D. W. C. *J. Am. Chem. Soc.* **2008**, *130*, 16494-16495.
18. Amatore, M.; Beeson, T. D.; Brown, S. P.; MacMillan, D. W. C. *Angew. Chem. Int. Ed.* **2009**, *48*, 5121-5124.
19. Nagib, D. A.; Scott, M. E.; MacMillan, D. W. C. *J. Am. Chem. Soc.* **2009**, *131*, 10875-10877.
20. Nicewicz D. A.; MacMillan, D. W. C. *Science* **2008**, *322*, 77-80.
21. (a) Adamson, G. A.; Beckwith, A. L. J.; Chai, C. L. L. *Aust. J. Chem.* **2004**, *57*, 629-633; (b) Kazmierski, W. M.; Urbanczyk-Lipkowska, Z; Hruby, V. J. *J. Org. Chem.* **1994**, *59*, 1789-1795; (c) Naef, R.; Seebach, D. *Helv. Chim. Acta* **1985**, *68*, 135-143.

Appendix
Chemical Abstracts Nomenclature (Registry Number)

Acetyl chloride: (75-36-5)
L-Alanine methyl ester hydrochloride: (2491-20-5)
Benzyl chloroformate, carbobenzoxy chloride: (501-53-1)
Methylamine: (74-89-5)
Pivaldehyde, trimethylacetaldehyde: (630-19-3)
Triethylamine, *N,N*-diethyl-ethanamine: (121-44-8)

David W. C. MacMillan received his B.S. degree in chemistry in 1990 from the University of Glasgow, Scotland, and his Ph.D. degree in 1996 from the University of California, Irvine, where he worked under the direction of Professor Larry E. Overman. David then moved to Harvard University and completed his postdoctoral studies with Professor David A. Evans. In 1998, David joined the faculty at the University of California, Berkeley. In 2000, he moved to the California Institute of Technology, where, in 2003, he was promoted to the rank of full professor and the following year, he became the Earle C. Anthony Chair in Organic Chemistry. In 2006, David moved to Princeton University where he is the A. Barton Hepburn Professor of Chemistry, the Director of the Merck Center for Catalysis at Princeton and the Chairperson of the Department of Chemistry.

Thomas H. Graham received his B.S. in Chemical Engineering from Virginia Tech in 1995. He completed his undergraduate research in the laboratory of Professor Neal Castagnoli, Jr. He then moved to Research Triangle Park, North Carolina, where he worked for Eli Lilly and was involved in chemical synthesis, laboratory automation and technology development. From 2000 to 2006, he completed his graduate studies in organic chemistry with Professor Peter Wipf at the University of Pittsburgh. From 2006 to 2008, he was a postdoctoral fellow with Professor David MacMillan at Princeton University. He is currently employed at Merck Research Laboratories in Rahway, New Jersey.

Benjamin D. Horning received his B.S. in Biochemistry at the University of Oregon in Eugene, Oregon in 2007. He did undergraduate research with Professor Michael Haley studying metalla-benzenes. He is currently a graduate student at Princeton University with Professor David MacMillan studying natural product synthesis and cascade catalysis.

SYNTHESIS OF Et₂SBr•SbCl₅Br AND ITS USE IN BIOMIMETIC BROMINATIVE POLYENE CYCLIZATIONS

Submitted by Scott A. Snyder and Daniel S. Treitler.[1]
Checked by Nobuhiro Satoh and Tohru Fukuyama.

1. Procedure

> *Caution: Bromine, diethyl chlorophosphate, and antimony pentachloride solution are highly toxic, caustic liquids that may be fatal if inhaled, swallowed, or absorbed through skin. All manipulations should be carefully carried out in a well ventilated fume hood.*

A. *Bromodiethylsulfonium Bromopentachloroantimonate ("BDSB") (1).* An oven-dried, one-necked, 500-mL round-bottomed flask containing a magnetic stirring bar (egg-shaped, 45 mm length and 20 mm diameter) is purged with argon and sealed using a rubber septum and an argon inlet (Note 1). 1,2-Dichloroethane (200 mL) (Note 2) and bromine (4.27 mL, 13.3 g, 83.3 mmol, 1.0 equiv) (Note 3) are syringed into the flask, which is then cooled in a –30 °C dry ice-cooled acetone bath. To this vigorously stirring

Org. Synth. **2011**, *88*, 54-69
Published on the Web 10/14/2010

dark red solution is added diethyl sulfide (9.88 mL, 8.27 g, 91.7 mmol, 1.1 equiv) (Note 4) via syringe over the course of 1 min. The solution instantly lightens to a yellow-orange color. Maintaining the temperature of the bath between –30 and –35 °C by occasionally adding more dry ice, an antimony pentachloride solution (100 mL, 1 M in CH_2Cl_2, 100 mmol, 1.2 equiv) (Note 5) is cannulated slowly into the reaction flask over 10 min. The color of the solution first darkens to an orange-red, and as the addition proceeds, a yellow precipitate is observed. After the addition is complete, the reaction mixture is stirred at –30 °C for an additional 30 min and then the acetone bath is replaced with a cold water bath (~10 °C). The water bath is heated slowly (~2 °C per min) until all of the precipitate has dissolved (bath temperature = 26–32 °C), yielding a transparent homogeneous red solution. At this time, the rubber septum and argon inlet are quickly replaced with a tightly sealed cap and the flask is moved to a 4 °C refrigerator. After 8 h at 4 °C, the flask (now containing significant amounts of orange crystals) is moved to a –20 °C freezer, where it is kept for 14 h. The crystalline product (**1**) is isolated by decanting the majority of the solvent and then removing any residual solvent via pipette. The crystals are then washed with cold (–20 °C) CH_2Cl_2 (2 × 10 mL) (Note 6) and dried under high vacuum (23 °C, 1 mmHg, 2 h) to afford 39.2–39.8 g (86–87%) of pure BDSB (**1**) as light orange plates (Note 7).

B. *Geranyl Diethyl Phosphate (2).* To an argon-purged 500-mL, one-necked, round-bottomed flask equipped with a magnetic stirring bar (round, with removable pivot ring, 40 mm length and 8 mm diameter), a rubber septum, and argon inlet is added geraniol (14.0 mL, 12.3 g, 80.0 mmol, 1.0 equiv) (Note 8), pyridine (16.1 mL, 15.8 g, 200 mmol, 2.5 equiv) (Note 9) and diethyl ether (60 mL) (Note 10) via syringe. The clear solution is cooled in a –30 °C dry ice-cooled acetone bath and stirred vigorously while diethyl chlorophosphate (17.4 mL, 20.7 g, 120 mmol, 1.5 equiv) (Note 11) is added dropwise via syringe over the course of 10 min. Once the addition is complete, the reaction flask is removed from the cold bath (now –20 °C) and allowed to warm to 23 °C. Significant amounts of white precipitate appear as the reaction proceeds (Note 12). After stirring for 6.5 h at 23 °C, the reaction is quenched by the addition of ice-cold 1 M NaOH (150 mL) over ~1 min (a slight exotherm occurs). The biphasic mixture is stirred vigorously for 30 min, then transferred to a 1-L separatory funnel and extracted into EtOAc (3 × 150 mL) (Note 13). The combined organic layers are washed with 1 M HCl (2 × 150 mL), saturated aqueous $NaHCO_3$ (150 mL) and brine (150 mL), and dried over $MgSO_4$ (Note 14). The drying agent is removed by vacuum filtration (medium-pore frit) and the filtrate is concentrated by rotary evaporation (30 °C, 150→10 mmHg) and then under high vacuum (23 °C, 1 mmHg, 1 h). The crude yellow oil is pipetted onto a plug of packed silica gel (5 × 10 cm) (Note 15) and transferred quantitatively by adding CH_2Cl_2 washes (3 × 5 mL). The product is eluted with pressurized air directly into a 2-L round-bottomed flask with ice-cold hexanes (300 mL) (Note 16) followed by hexanes:EtOAc (2:3, 1.25 L). The bulk of the solvent is removed by rotary evaporation (30 °C, 150→10 mmHg) and the resultant geranyl diethyl phosphate (Note 17) is transferred to a 500-mL, one-necked round-bottom flask and co-evaporated with dry toluene (3 × 15 mL; 30 °C, 10 mmHg) (Note 18) to remove both residual EtOAc and trace water.

C. *Homogeranylbenzene (3)*. A magnetic stirring bar (round, with removable pivot ring, 40 mm length and 8 mm diameter) is added to the flask containing crude geranyl diethyl phosphate, which is dried under vacuum (20 min, 23 °C, 1 mmHg), back-filled with argon and sealed with a rubber septum and argon inlet. The viscous yellow oil is dissolved in dry THF (30 mL) (Note 19) and cooled in a –45 °C dry ice-cooled acetone bath. A solution of benzylmagnesium chloride in THF (0.5 M, 320 mL, 160 mmol, 2 equiv) (Note 20) is cannulated into the stirring geranyl diethyl phosphate solution over the course of 20 min (Note 21), with the continuous addition of dry ice to the cold bath to maintain its temperature between –40 °C and –45 °C. Upon completion of the addition of the Grignard reagent, the cold bath is allowed to warm slowly for 2.5 h (to 5 °C) and then is removed (Note 22). The light brown solution is stirred for an additional 2 h at 23 °C. Saturated aqueous NH_4Cl (75 mL) and water (75 mL) are added carefully to quench any remaining Grignard reagent and the biphasic mixture is stirred vigorously for 5 min (slight exotherm). This mixture is poured into a 1-L separatory funnel and extracted with hexanes:EtOAc (2:1, 3 × 150 mL). The combined organic layers are washed with saturated aqueous $NaHCO_3$ (150 mL) and brine (150 mL), dried over $MgSO_4$, filtered, then concentrated by rotary evaporation (30 °C, 150→10 mmHg) and dried under high vacuum (23 °C, 1 mmHg, 1 h). The crude yellow oil is then pipetted onto a column of packed silica gel (360 g, 80 mm diameter) (Note 23) and transferred quantitatively by adding hexanes rinses (3 × 5 mL). The column is eluted with hexanes (300 mL), then fraction collection is initiated (30 mL each) with further elution being performed with hexanes:CH_2Cl_2 (19:1, 900 mL) followed by hexanes:CH_2Cl_2 (9:1, 900 mL). The desired product is found in pure form in fractions 17–43 (fractions 44–50 are contaminated with

56

bibenzyl and are discarded), which are concentrated by rotary evaporation (30 °C, 300→10 mmHg) and dried under high vacuum (23 °C, 1 mmHg, 1 h) to afford 16.7 g (92% from geraniol) of homogeranylbenzene as a colorless oil (Notes 24 and 25).

D. *Cyclization of 3 to 4*. Homogeranylbenzene (1.14 g, 5.0 mmol, 1.0 equiv) is added to a 1-L, three-necked, round-bottomed flask containing a magnetic stirring bar (egg-shaped, 45 mm length and 20 mm diameter). The flask is purged with argon, and two of the necks are equipped with an argon inlet and a low-temperature thermometer, respectively. Dry nitromethane (485 mL) (Note 26) is poured through a funnel into the flask, and the third neck is capped. The colorless solution is cooled using a dry ice-cooled acetone bath (approx –30 °C) until the interior temperature reaches –25 °C. A solution of BDSB (3.02 g, 5.5 mmol, 1.1 equiv) in nitromethane (15 mL) that has been sealed and pre-cooled to –20 °C in a freezer is syringed in rapidly (Note 27) while stirring vigorously. The dark yellow solution is stirred for 5 min at –25 °C (Notes 28 and 29) at which time the flask is removed from the cold bath and the now light yellow solution is quenched by the addition of 2% aqueous Na$_2$SO$_3$ (Note 30; 300 mL) and saturated aqueous NaHCO$_3$ (100 mL). The biphasic mixture is stirred vigorously and allowed to warm to 23 °C over 60 min (Note 31). The quenched reaction mixture is then poured into a 2-L separatory funnel and extracted vigorously with hexanes (4 × 300 mL) (Note 32). The combined hexanes layers are washed with brine (300 mL), dried over MgSO$_4$, filtered, and concentrated by rotary evaporation (30 °C, 100 mmHg). The crude product is dried under high vacuum (23 °C, 1 mmHg, 1 h) to afford 1.45–1.62 g of a white solid that is approximately 85–90% pure product by ^{1}H NMR (Note 33). The crude product is dissolved in 100 mL of boiling methanol (Note 34), and the light yellow solution is allowed to cool slowly to 23 °C overnight. The crystallization is allowed to proceed for an additional 12 h at 4 °C and 24 h at –20 °C. The first crop is then isolated by vacuum filtration using a medium-pore fritted funnel, washed with cold (–20 °C) methanol (5 mL) and dried under vacuum (23 °C, 1 mmHg, 1 h) to afford 0.952–0.955 g of **4** as white plates. A second crop is obtained by concentrating the combined mother liquor and methanol wash to a volume of ~20 mL by rotary evaporation (35 °C, 80 mmHg) (Note 35), heating to reflux if necessary to render the solution homogenous, and cooling to 4 °C for 24 h followed by 24 h at –20 °C. Filtration and rinsing with cold methanol (2 mL) yields an additional 0.136–0.147 g of **4** as off-white needles, for a total of 1.09–1.10 g (71–72%) of pure **4** (Note 36).

2. Notes

1. For all reactions run under positive argon pressure: a balloon securely attached to the barrel of a 5-mL plastic syringe using Teflon tape and Parafilm is inflated with argon, capped with a needle and inserted through the rubber septum.

2. 1,2-Dichloroethane is purchased from Sigma-Aldrich (anhydrous, 99.8%) and used as received. A second experiment was also performed in which the 1,2-dichloroethane was distilled from calcium hydride onto 3 Å molecular sieves. The product yields and activities were identical, indicating that rigorous drying of the solvent is not necessary.

3. Bromine is purchased from Kanto Chemical (>99.0%) (checker), or Sigma-Aldrich (ACS Reagent Grade, 99.5+%) (submitter) and used as received.

4. Diethyl sulfide is purchased from Sigma-Aldrich (98%) and used as received.

5. Antimony(V) chloride is purchased from Sigma-Aldrich (1.0 M in dichloromethane). The entire 100 mL bottle is used for the reaction, and is thus assumed to contain 100 mL [previous experiments have established that slightly more or less than 1.2 equiv of antimony(V) chloride have negligible impact on the yield of BDSB obtained].

6. Dichloromethane (J.T. Baker, ACS Grade, 99.5% min) is purchased from VWR International and used as received.

7. Appearance: The color of BDSB is generally light orange, although different batches can vary from dark yellow to orange. The color of the crystals is also slightly temperature-dependent; they are noticeably more yellow in color when cold, and become darker orange if warmed. This color change appears to be reversible. Stability: BDSB has proven to be stable in a sealed container stored at –20 °C for at least one year, with no observed depreciation in reactivity. If left out in open air at 23 °C, BDSB will slowly hydrolyze to the sulfoxide over the course of minutes to hours depending on the size of the crystals as well as the humidity of the air. The short-term stability of BDSB in air allows it to be isolated, weighed, or otherwise manipulated in air. Solubility: BDSB is soluble in nitromethane, nitroethane, acetonitrile, dimethylsulfoxide, N,N-dimethylformamide, and ethyl acetate. It is slightly soluble in dichloromethane, dichloroethane, chloroform, and toluene, and insoluble in trifluoroethanol, hexafluoroisopropanol, benzene, hexanes, and pentane. The compound is not stable to ethereal or alcoholic solvents such as water, methanol, ethanol,

58

diethyl ether, or tetrahydrofuran. Characterization: mp = 102 – 105 °C (with decomposition); IR (KBr) v_{max} 2985, 2939, 1455, 1403, 1384, 1261, 932, 877 cm^{-1}; ^1H NMR (400 MHz, CD_3NO_2, solvent referenced at 4.33 ppm) δ: 1.67 (t, J = 7.3 Hz, 6 H), 3.92 (dq, J = 1.4, 7.3 Hz, 4 H); ^{13}C NMR (100 MHz, CD_3NO_2, solvent referenced at 63.8 ppm) δ: 11.3, 46.3. NMR spectra show trace amounts of diethyl sulfoxide (^1H NMR δ: 1.55 (t, J = 7.3 Hz, 6 H), 3.54 (q, J = 7.3 Hz, 4 H); ^{13}C NMR δ: 8.1, 44.3) if the sample is not prepared in a rigorously anhydrous manner.

8. Geraniol is purchased from Sigma-Aldrich (98%) and used as received.

9. Pyridine is purchased from Sigma-Aldrich (anhydrous, 99.8%), distilled from calcium hydride and stored over 3 Å molecular sieves.

10. Diethyl ether (Kanto Chemical, dehydrated, >99.5% (checker) or EMD Chemicals, OmniSolv, 99.9% min (submitter)) is dried using an anhydrous solvent delivery system equipped with activated alumina columns.

11. Diethyl chlorophosphate is purchased from Sigma-Aldrich (97%) and used as received.

12. The reaction can be followed by TLC (Merck 60 F_{254}, 0.25 mm thickness (checker) or EMD Chemicals 60 F_{254}, 0.25 mm thickness (submitter)). Solvent system: 1:1 Hex:EtOAc; stain: Cerium molybdate (prepared by dissolving 2.0 g ammonium cerium sulfate and 5.0 g ammonium heptamolybdate in 200 mL of 1 M aqueous sulfuric acid); R_f = 0.56 (S.M.), 0.31 (product).

13. Ethyl acetate (Kanto Chemical, 99.0% min (checker) or Malinckrodt Chemicals, ChromAR, 99.5% min (submitter)) is used as received.

14. Magnesium sulfate (anhydrous powder certified) is purchased from Wako Pure Chemical Industry (checker) or Fisher Chemical Company (submitter).

15. Silica gel (Kanto Chemical, spherical neutral, 40–100 μm particle size, 60 Å pore size (checker) or EMD Commercial grade, 40–63 μm particle size, 60 Å pore size (submitter)) is used as received.

16. Hexanes (Wako Pure Chemical Industry, 95.0% min (checker) or J.T. Baker, ACS Grade, 98.5% min (submitter)) is used as received. The exotherm that accompanies this first elution is enough to decompose some of the product (~3–5%) if room temperature hexanes is used.

17. The crude geranyl diethyl phosphate obtained by this procedure is ~90% pure by ^1H NMR analysis. The major impurities are ethyl acetate,

geraniol (~3%), linalool (~5%), and an unidentified diethylphosphate by-product (~2%). In order to obtain pure compound, the silica gel plug can be replaced by flash column chromatography (elution with a gradient of 20 to 70% EtOAc in hexanes provides the best results). The pure product exhibits the following spectral characteristics: IR (film) ν_{max} 2980, 2917, 1457, 1395, 1263, 1034, 820 cm^{-1}; ^{1}H NMR (400 MHz, CDCl$_3$) δ: 1.34 (dt, J = 0.9, 7.4 Hz, 6 H), 1.60 (s, 3 H), 1.68 (s, 3 H), 1.71 (s, 3 H), 2.03–2.13 (m, 4 H), 4.11 (quin, J = 7.3 Hz, 4 H), 4.57 (t, J = 7.8 Hz, 2 H), 5.08 (m, 1 H), 5.40 (m, 1 H); ^{13}C NMR (100 MHz, CDCl$_3$) δ: 16.8 (d, J = 7 Hz, 2 C), 17.1, 18.3, 26.3, 26.9, 40.2, 64.3 (d, J = 6 Hz, 2 C), 64.8 (d, J = 6 Hz), 119.7 (d, J = 6 Hz), 124.3, 132.6, 143.2; HRMS (FAB) calcd. for C$_{14}$H$_{27}$NaO$_4$P$^+$ [M+Na]$^+$ 313.1545, found 313.1534.

18. Toluene (Wako Pure Chemical Industry, 99.0% min) is used as received (checker). Toluene (BDH, ACS Grade, 99.5% min) is dried using an anhydrous solvent delivery system equipped with activated alumina columns (submitter).

19. Tetrahydrofuran (Kanto Chemical, dehydrated super, 99.5% min (checker) or EMD Chemicals, OmniSolv, 99.9% min (submitter)) is dried using an anhydrous solvent delivery system equipped with activated alumina columns.

20. A 0.5 M BnMgCl solution can be prepared by dilution of commercially available benzylmagnesium chloride solution (use of benzylmagnesium bromide must be avoided as it is seriously detrimental to the yield). Alternatively, 320 mL of 0.5 M benzylmagnesium chloride can be synthesized as follows: magnesium turnings (7.77 g, 320 mmol, 4 equiv) (Reagent Grade, 98%, purchased from Sigma-Aldrich) are activated by sequential rinsing in a medium-pore frit with 0.2 M HCl (40 mL), water (3 × 40 mL), acetone (2 × 40 mL), and diethyl ether (2 × 40 mL). The activated magnesium turnings are added to a 500 mL one-necked round-bottomed flask with a magnetic stirring bar and dried at 100 °C (1 mmHg) for 1 h; they are then allowed to cool to 23 °C and the flask is sealed under argon with a rubber septum and an argon inlet. Meanwhile, benzyl chloride (18.4 mL, 160 mmol, 2 equiv) (ReagentPlus, 99%, purchased from Sigma-Aldrich and used as received) and dry THF (300 mL) are syringed into an oven-dried 500-mL, one-necked, round-bottomed flask sealed under argon with a rubber septum and argon inlet. Approximately 15 mL of this solution is cannulated onto the magnesium turnings, which are stirred at 23 °C until a sudden exotherm indicates that the formation of the Grignard reagent is underway (initiation time varies from 20 s to ~5 min). Once the reaction has initiated,

60

the magnesium-containing flask is cooled to 0 °C using an ice-water bath and stirred at 0 °C while the remainder of the benzyl chloride solution is slowly cannulated down the inner wall of the flask over 20 min. An additional 60 min of stirring at 0 °C yields a gray-brown solution of benzylmagnesium chloride in THF (~0.5 M).

21. Due to the viscosity of the benzylmagnesium chloride solution, it may be necessary to pull a slight vacuum on the receiving flask to encourage the cannula transfer to proceed at a reasonable rate. This task is performed by attaching a needle to a hose secured to a vacuum line, insertion of this needle through the septum of the receiving flask, and very briefly opening the hose to vacuum every few minutes as necessary.

22. The reaction can be followed by TLC (silica gel plates using 10% CH_2Cl_2 in hexanes as eluent and visualization with cerium molybdate solution prepared as in Note 12. R_f = 0.02 (S.M.), 0.48 (product)).

23. The column is wet-packed (hexane) with Kanto Chemical, 40-100 μm particle size, 60 Å pore size silica gel (checker). The column is dry loaded with EMD Commercial grade, 40–63 μm particle size, 60 Å pore size silica gel and packed with three column volumes of hexanes using pressurized air (submitter). The desired product is visualized by TLC using hexanes:CH_2Cl_2 (9:1) to elute and a UV lamp (254 nm) to observe the product (R_f = 0.48). The major contaminant is bibenzyl (formed via Würtz-type coupling in the Grignard formation step; R_f = 0.41).

24. The product exhibits the following physical and spectral characteristics: bp = 84 °C (1 mmHg); IR (film) v_{max} 3085, 3062, 3027, 2966, 2923, 2855, 1496, 1454, 1376, 1108, 1030, 835 cm^{-1}; ^{1}H NMR (400 MHz, CDCl$_3$) δ: 1.55 (s, 3 H), 1.65 (s, 3 H), 1.69 (s, 3 H), 1.90–2.09 (m, 4 H), 2.31 (q, J = 7.4 Hz, 2 H), 2.64 (app t, J = 7.8 Hz, 2 H), 5.10 (m, 1 H), 5.19 (dt, J = 7.4, 0.9 Hz, 1 H), 7.18 (m, 3 H), 7.27 (m, 2 H); ^{13}C NMR (100 MHz, CDCl$_3$) δ: 16.6, 18.4, 26.4, 27.4, 30.6, 36.8, 40.4, 124.3, 125.0, 126.3, 128.9 (2 C), 129.2 (2 C), 132.0, 136.4, 143.1; HRMS (DART) calcd for $C_{17}H_{25}^+$ [M+H]$^+$ 229.1956, found 229.1945; Anal. calcd. for $C_{17}H_{24}$: C, 89.41; H, 10.59. Found: C, 89.66; H, 10.60.

25. 84% yield was obtained on half that scale. Submitter's yield is 83%.

26. Nitromethane (Aldrich, ACS reagent, >99.0%) is stored over 3 Å molecular sieves (checker). Nitromethane (Fisher Chemical, Certified ACS, 99.9%) is purchased from Fisher Scientific and used as received (submitter). In general, the nitromethane used as solvent for cyclization must be quite anhydrous. The submitters have found that most commercial nitromethane

can be used as received, but that once opened, nitromethane should be stored over activated 3 Å molecular sieves for best results (slight yellowing of the solvent is typical and is not detrimental to the reaction). Unfortunately, reducing the amount of this relatively expensive solvent by a factor of 10 results in an ~20% decrease in yield depending on the substrate. However, an alternative, biphasic reaction using hexanes:nitromethane (4:1) and 1.2 equivalents BDSB for 30 min at –25 °C enables a decrease in the total nitromethane volume without any decrease in product yield. It is not known whether this modification works for all substrates, or only very hydrophobic ones such as homogeranylbenzene.

27. A fast rate of addition (~5 s) is important for obtaining a reproducible yield. A slower rate of addition results in more side-products including significant amounts of the proton-cyclized product.

28. A slight exotherm of 2–3 °C is common immediately following the addition of the BDSB solution.

29. Although the reaction proceeds very quickly (usually complete within 1 min), TLC can be utilized to follow its progress (silica gel plates using 10% CH_2Cl_2 in hexanes as eluent and visualization with cerium molybdate solution prepared as in Note 12. R_f = 0.48 (SM), 0.40 (product)).

30. Sodium sulfite is purchased from Wako Pure Chemical Industry (>97.0%) (checker) or Sigma-Aldrich (98+%, ACS reagent) (submitter). The sodium sulfite solution is prepared freshly by dissolving 6 g in 300 mL de-ionized water (it has been found that stock solutions of sodium sulfite slowly lose their reducing potential over time).

31. Quenching the reaction leads to the formation of a white insoluble precipitate presumed to be made up of antimony salts. The sticky nature of this precipitate renders the extractions somewhat messy, but attempts to remove these salts prior to extraction by filtration were unsuccessful due to their propensity to clog fritted funnels. At the conclusion of the work-up, any residual precipitate can be cleaned from glassware by rinsing with 1 M HCl.

32. The separatory funnel contains three layers: hexanes, water, and nitromethane (from top to bottom). Because of this occurrence, thorough extraction is necessary to partition the desired product into the hexanes layer (each extraction consists of shaking the separatory funnel vigorously, with occasional venting, for approximately 2 min). At the conclusion of the extractions, the used nitromethane can be recovered, dried over $MgSO_4$, and filtered to yield approximately 95% of the original 500 mL. By ^{1}H NMR analysis, this recovered solvent is quite pure, containing ~3% water, ~2%

hexanes, and trace amounts of sulfide, thiol, and sulfoxide by-products (<1% each). After storing for 48 h over activated 3 Å molecular sieves (10% by weight), the water content is <0.5% by ^{1}H NMR analysis and this recycled nitromethane can be reused (without distillation) as the solvent for subsequent cyclizations without any decrease in reaction yield.

33. The major impurities appear to be a diastereomer of the product with opposite stereochemistry at the bromine position (~5%) and trace amounts of monocyclic products that failed to undergo the second, Friedel–Crafts-based cyclization step.

34. Methyl alcohol (Wako, 99.5% min (checker) or Malinckrodt Chemicals, ChromAR ACS Grade, 99.9% min (submitter)) is used as received.

35. The mother liquor may darken in color to red or brown when concentrated.

36. The product exhibits the following physical and spectral characteristics: mp = 104.1–105.9 °C; R_f = 0.49 (silica gel, hexanes:CH_2Cl_2, 4:1); IR (film) v_{max} 3059, 2969, 2947, 2838, 1488, 1475, 1448, 1392, 1377, 875, 763 cm^{-1}; ^{1}H NMR (400 MHz, CDCl$_3$) δ: 1.06 (s, 3 H), 1.16 (s, 3 H), 1.24 (s, 3 H), 1.47 (dd, J = 12.0, 2.3 Hz, 1 H), 1.59 (dt, J = 13.3, 3.7 Hz, 1 H), 1.81 (m, 1 H), 1.97 (m, 1 H), 2.21–2.43 (m, 3 H), 2.82–3.00 (m, 2 H), 4.05 (dd, J = 12.8, 4.1 Hz, 1 H), 7.02–7.22 (m, 4 H); ^{13}C NMR (100 MHz, CDCl$_3$) δ: 18.9, 21.3, 25.6, 31.2, 31.5, 32.2, 38.6, 40.6, 40.7, 51.9, 69.6, 125.1, 126.3, 126.6, 129.7, 135.4, 149.4; HRMS (DART) calcd for $C_{17}H_{23}^+$ [M-Br]$^+$ 227.1800, found 227.1806; Anal. calcd. for $C_{17}H_{23}Br$: C, 66.45; H, 7.55. Found: C, 66.37; H, 7.54.

Safety and Waste Disposal Information

All hazardous materials should be handled and disposed of in accordance with *Prudent Practices in the Laboratory*. National Academy Press: Washington, DC, 1995.

3. Discussion

Cation-π cyclizations are an important class of reactions for the rapid and stereoselective generation of molecular complexity.[2] As shown simplistically in Scheme 1, these reactions are initiated by electrophilic activation of an alkene, an event that then induces one or more ring constructions via the sequential attack of up to several suitably disposed olefins. To date, synthetic chemists have developed the means to effectively utilize proton,[3] oxygen,[4] iodine,[5] sulfur,[6] selenium,[6] mercury,[7] and transition metals such as gold,[8] palladium,[9] and platinum[10] as initiators for these cyclizations, forming up to 5 new rings at once.[11]

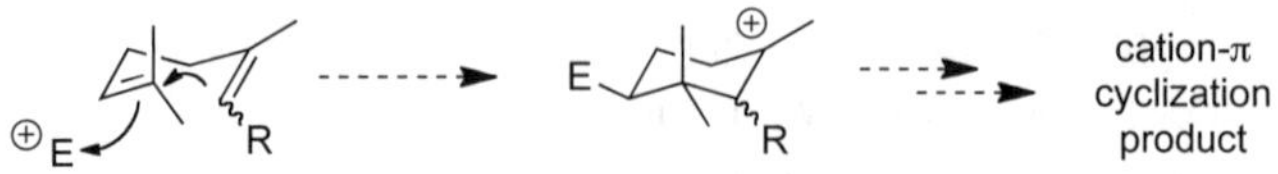

Scheme 1. Simplified form of a cation-π cyclization

Despite significant effort, however, challenges remain in initiating cation-π cyclizations with electrophilic bromine. Such a transformation would be of value given the existence of hundreds of brominated natural products that could be accessed from such an event as well as the potential ability to use the installed bromine atom as a handle to further elaborate the cyclized product. In nature, such reactions are achieved by a number of marine microorganisms that possess enzymes containing highly oxidized iron or vanadium metal centers that convert bromide into bromonium in a localized, highly controlled environment.[12] In the laboratory, by contrast, molecular bromine, N-bromosuccinimide, and 2,4,4,6-tetrabromocyclohexa-2,5-dienone (TBCO),[13] the most common sources of electrophilic bromine available to synthetic chemists, fail to broadly initiate cation-π cyclizations. Instead, numerous side-products and over-halogenated materials are typically formed in lieu of the desired material.[14] For example, none of these halogen sources can successfully effect the cyclization of electron-poor substrates, even if the R group in Scheme 1 is only moderately electron-withdrawing (such as CH₂OAc).

Attempts in our laboratory to use chiral bromodialkylsulfonium bromides for asymmetric bromination reactions (though unsuccessful) led to the discovery that reagents formed by sequestering the bromide anion of such species with the highly halophilic Lewis acid SbCl$_5$ could successfully initiate bromonium-induced cation π-cyclizations. An investigation of several dialkyl, diaryl, and alkyl-aryl bromosulfonium complexes of this type culminated in the observation that **1** (BDSB) was the best candidate given its solubility profile, stability, and ease of synthesis.[15] As illustrated in the procedure above, this reagent is readily prepared in large quantities from relatively inexpensive starting materials. Additionally, no discrete purification step is required, since the product crystallizes smoothly from the

Org. Synth. **2011**, *88*, 54-69

reaction solution. An additional benefit of **1** is its air-stability, as it can be easily weighed and otherwise manipulated on the bench.

As illustrated by several selected examples from our investigations in Table 1, BDSB (**1**) has significant substrate scope for cation-π cyclization, successfully cyclizing an array of polyenes derived from geraniol, farnesol, and nerol, whether electron-rich or deficient. Moreover, BDSB is capable of achieving cyclizations cleanly even when bromination of an electron-rich aromatic ring could be considered a competitive pathway. The reaction is presumed to proceed via a synchronous process given the observation that *E*-olefins yield *trans*-ring junctions while *Z*-olefins result in *cis*-ring junctions (for instance, entries 1 and 2), in line with the Stork-Eschenmoser hypothesis.[2] In terms of experimental execution, though reactions are easily set up and reaction times are only a few minutes, it is critical to use an appropriate reaction concentration to achieve optimal yield. On small scale (0.1 mmol), reaction yields are generally quite good at concentrations as high as 0.1 M. However, when scaling above such amounts, higher dilution is necessary to obtain equivalent yields (as is often the case for cation-π cyclizations).[16] For instance, on gram scale, reaction concentrations of 0.01 M are optimal; pleasingly, although more solvent is needed, it can be easily recovered and recycled as noted above.

In conclusion, we have developed a novel source of electrophilic bromine that is capable of initiating bromonium-induced cation-π cyclizations in good yields with a variety of terpene-derived substrates.

Table 1. Exploration of Reaction Scope

Entry	Starting Material	Product	Yield[b] (%) [Rxn Temp (°C)]
1[c]			80 [0]
2			71 [0]
3[d]			73 [23]
4[d]			56 [23]
5			79 [0]
6			75 [-25]
7			76 [-25]
8[e]			58 [-25]
9			67 [-25]

[a] All reactions performed on a 0.1 mmol scale at 0.05 M in CH_3NO_2 with 1.1 equivalents of **1** for 5 minutes. [b] Isolated yields. [c] Isolated as a 3.8:1 mixture of separable diastereomers at the C-4 position. [d] Isolated alkene mixtures. [e] CH_3SO_3H added to enhance yield of tetracyclic product.

1. Department of Chemistry, Havemeyer Hall, Columbia University, New York, NY 10027. E-mail: sas2197@columbia.edu

2. Seminal works: (a) Stork, G.; Burgstahler, A. W. *J. Am. Chem. Soc.* **1955**, *77*, 5068–5077. (b) Eschenmoser, A.; Ruzicka, L.; Jeger, O.; Arigoni, D. *Helv. Chim. Acta* **1955**, *38*, 1890–1904. Review: (c) Brunoldi, E.; Luparia, M.; Porta, A.; Zanoni, G.; Vidari, G. *Current Org. Chem.* **2006**, *10*, 2259–2282.

3. Review: Johnson, W. S. *Bioorg. Chem.* **1976**, *5*, 51–98.

4. Review: Yoder, R. A.; Johnston, J. N. *Chem. Rev.* **2005**, *105*, 4730–4756.

5. Sakakura, A.; Ukai, A.; Ishihara, K. *Nature* **2007**, *445*, 900–903.

6. Edstrom, E. D.; Livinghouse, T. *J. Org. Chem.* **1987**, *52*, 949–951.

7. (a) Hoye, T. R.; Caruso, A. J.; Kurth, M. J. *J. Org. Chem.* **1981**, *46*, 3550–3552. (b) Nishizawa, M.; Takenaka, H.; Nishide, H.; Hayashi, Y. *Tetrahedron Lett.* **1983**, *24*, 2581–2584. (c) Nishizawa, M.; Morikuni, E.; Asoh, K.; Kan, Y.; Uenoyama, K.; Imagawa, H. *Synlett* **1995**, 169-170. (d) Snyder, S. A.; Treitler, D. S.; Schall, A. *Tetrahedron* **2010**, *66*, 4796–4804.

8. Fürstner, A.; Morency, L. *Angew. Chem. Int. Ed.* **2008**, *47*, 5030–5033.

9. Overman, L. E.; Abelman, M. M.; Kucera, D. J.; Tran, V. D.; Ricca, D. J. *Pure & Appl. Chem.* **1992**, *64*, 1813–1819.

10. Koh, J. H.; Gagné, M. R. *Angew. Chem. Int. Ed.* **2004**, *43*, 3459–3461.

11. Fish, P. V.; Sudhakar, A. R.; Johnson, W. S. *Tetrahedron Lett.* **1993**, *34*, 7849–7852.

12. (a) Carter-Franklin, J. N.; Butler, A. *J. Am. Chem. Soc.* **2004**, *126*, 15060–15066. (b) Vaillancourt, F. H.; Yeh, E.; Vosburg, D. A.; Garneau-Tsodikova, S.; Walsh, C. T. *Chem. Rev.* **2006**, *106*, 3364–3378.

13. (a) van Tamelen, E. E.; Hessler, E. J. *Chem. Commun.* **1966**, 411–413. (b) Wolinsky, L. E.; Faulkner, D. J. *J. Org. Chem.* **1976**, *41*, 597–600. (c) Kato, T.; Ichinose, I.; Kumazawa, S.; Kitahara, Y. *Bioorg. Chem.* **1975**, *4*, 188–193. (d) Kato, T.; Mochizuki, M.; Hirano, T.; Fujiwara, S.; Uyehara, T. *J. Chem. Soc. Chem. Commun.* **1984**, 1077–1078.

14. (a) González, A. G.; Martín, J. D.; Pérez, C.; Ramírez, M. A. *Tetrahedron Lett.* **1976**, *17*, 137–138. (b) Kato, T.; Ichinose, I. *J. Chem. Soc. Perkin Trans. 1* **1980**, 1051–1056.

15. (a) Snyder, S. A.; Treitler, D. S. *Angew. Chem. Int. Ed.* **2009**, *48*, 7899–7903. (b) Snyder, S. A.; Treitler, D. S.; Brucks, A. P. *J. Am. Chem. Soc.* **2010**, *132*, 14303–14314.

16. (a) Demailly, G.; Solladie, G. *J. Org. Chem.* **1981**, *46*, 3102–3108. (b) Mente, N. R.; Neighbors, J. D.; Wiemer, D. F. *J. Org. Chem.* **2008**, *73*, 7963–7970. (c) Ungur, N.; Garcia, E. Z.; Gil, S.; Arques, J. S. *Synthesis* **2008**, 622–626. (d) Surendra, K.; Corey, E. J. *J. Am. Chem. Soc.* **2008**, *130*, 8865–8869.

Appendix
Chemical Abstracts Nomenclature; (Registry Number)

Bromine; (7726-95-6)

Diethyl sulfide: Ethane, 1,1'-thiobis-; (352-93-2)

Bromodiethylsulfonium Bromopentachloroantimonate: Sulfonium, bromodiethyl-, (OC-6-22)-bromopentachloroantimonate(1-) (1:1); (1198402-81-1)

Antimony pentachloride; (7647-18-9)

Geraniol: 2,6-Octadien-1-ol, 3,7-dimethyl-, (2E)-; (106-24-1)

Diethyl chlorophosphate: Phosphorochloridic acid, diethyl ester; (814-49-3)

Pyridine; (110-86-1)

Homogeranylbenzene: Benzene, [(3E)-4,8-dimethyl-3,7-nonadienyl]-; (22555-66-4)

Benzylmagnesium chloride: Magnesium, chloro(phenylmethyl)-; (6921-34-2)

Phenanthrene, 2-bromo-1,2,3,4,4a,9,10,10a-octahydro-1,1,4a-trimethyl-, (2R,4aR,10aS)-rel-: (1198206-88-0)

Scott A. Snyder pursued his undergraduate education at Williams College. He then obtained his Ph.D. with Professor K. C. Nicolaou at The Scripps Research Institute, during which time he co-authored *Classics in Total Synthesis II*. Scott then trained with Professor E. J. Corey at Harvard University. Since August of 2006, Scott has been an Assistant Professor of Chemistry at Columbia University where his group seeks to explore chemical space through natural product total synthesis. Recent honors include a Camille and Henry Dreyfus New Faculty Award, an Eli Lilly Grantee Award, an NSF CAREER Award, a Cottrell Scholar Award, a JSPS travel fellowship, and a Columbia University Presidential Teaching Award.

Daniel S. Treitler was born in Denville, New Jersey, in 1985. He earned a Bachelors Degree in Biology from Cornell University in 2007, with undergraduate research experience in polymer chemistry in the group of Professor Geoffrey Coates as well as with Novomer, LLC. He is currently in his third year of an organic chemistry Ph.D. at Columbia University as an NSF Predoctoral Fellow, where his research in the group of Professor Scott Snyder has focused broadly on halogenation reactions, particularly cation-π cyclizations.

Nobuhiro Satoh was born in 1984 in Sapporo, Japan. He received his M.S. degree under the direction of Professor Tohru Fukuyama at University of Tokyo in 2008, where he is currently pursuing a Ph.D. degree. His research interest is total synthesis of natural products.

ASYMMETRIC SYNTHESIS OF (M)-2-HYDROXYMETHYL-1-(2-HYDROXY-4,6-DIMETHYLPHENYL)NAPHTHALENE VIA A CONFIGURATIONALLY UNSTABLE BIARYL LACTONE

Prepared by G. Bringmann,*[1] T.A.M. Gulder, and T. Gulder.
Original article: Bringmann, G.; Breuning, M.; Henschel, P.; Hinrichs, J. *Org. Synth.* **2002**, *79*, 72.

The 'lactone concept' has in recent years been used for the atroposelective construction of a broad variety of axially chiral natural products,[2] especially for the synthesis of a large number of naphthylisoquinoline alkaloids with different coupling positions, substitution patterns, and oxidation states in the respective isoquinoline portions.[3]

An instructive example of the asymmetric synthesis of a representative of this group of alkaloids is the atropo-diastereodivergent construction of the two epimers korupensamine A [(P)-7] and B [(M)-7]

Org. Synth. **2011**, *88*, 70-78
Published on the Web 10/21/2010

from the same late precursor **4** (Scheme 1).[4] The two required building blocks, the readily available carboxylic acid **1**[5] and the optically pure tetrahydroisoquinoline **2**,[6] were linked together by esterification giving **3** in 74% yield. Pd-catalyzed intramolecular cross-coupling reaction of **3** using the 'Herrmann-Beller' catalyst[7] delivered the very rapidly interconverting biaryl lactones (P)- and (M)-**4** in 74% yield. The dynamic kinetic resolution by reductive ring cleavage of **4** proceeded highly atroposelectively when applying borane activated by **5**. Utilization of (S)-**5** afforded the ring-opened and therefore configurationally stable benzylic alcohol (M)-**6** in 58% yield and an excellent 92% de, while the analogous reaction with the enantiomeric reagent (R)-**5** diastereoselectively gave the epimer (P)-**6**, yet with a slightly lower asymmetric induction (88% de). The synthesis of both, (P)-**7** and (M)-**7**, was completed by using exactly the same reaction conditions starting from **6** in 10 steps and ca. 10% overall yield.[4]

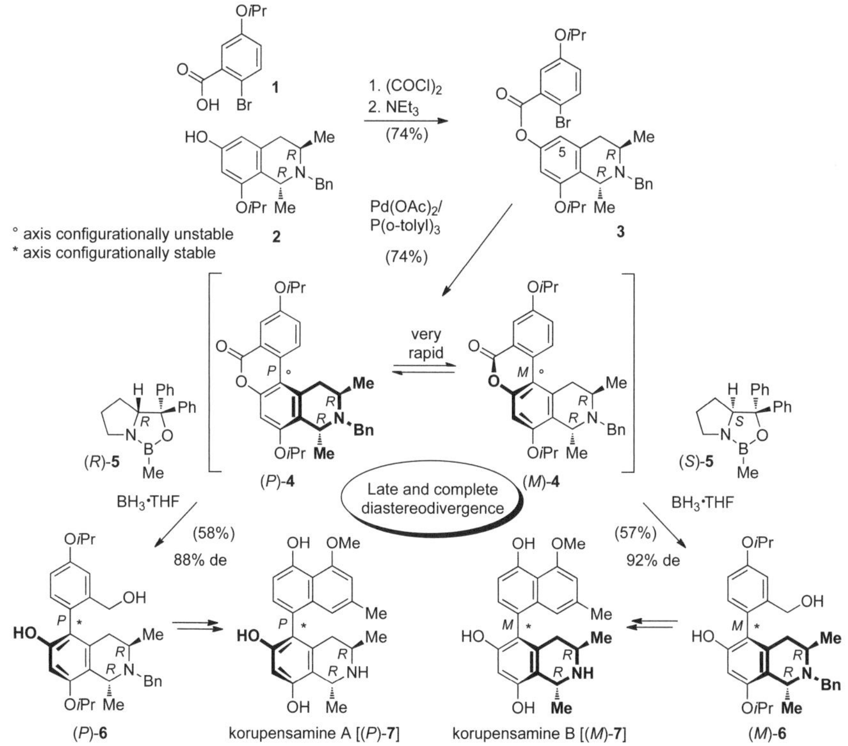

Scheme 1

Besides providing an atropo-diastereodivergent synthetic route to naphthylisoquinoline alkaloids, the 'lactone concept' paved the way for the synthesis of many other biaryl systems, like e.g. the phenylanthraquinone knipholone [(P)-**14**].[8] In this case, the key precursor **10** was obtained by esterification of the dibromide **8** with dimethoxyphenol **9** in 90% yield (Scheme 2). After pre-fixation of the two molecular portions, Pd-mediated *C-C* bond formation to give the configurationally unstable lactones **11** occurred in 68% yield, despite the large steric hindrance exerted by the presence of four substituents ortho to the biaryl axis. Dynamic kinetic resolution by atropo-enantioselective reduction utilizing (S)-**5** furnished (M)-**12**[9] in good yield and almost perfect enantioselectivity (96% ee), which was enhanced by a simple recrystallization to obtain enantiopure material (99% ee, 65% yield). The optically less pure mother liquor was recycled by a three-step oxidative procedure regenerating **11** in 71% yield. The target molecules, 6'-*O*-methylknipholone [(M)-**13**] and knipholone [(P)-**14**], were obtained from (M)-**12** by simple functional group conversions.[8]

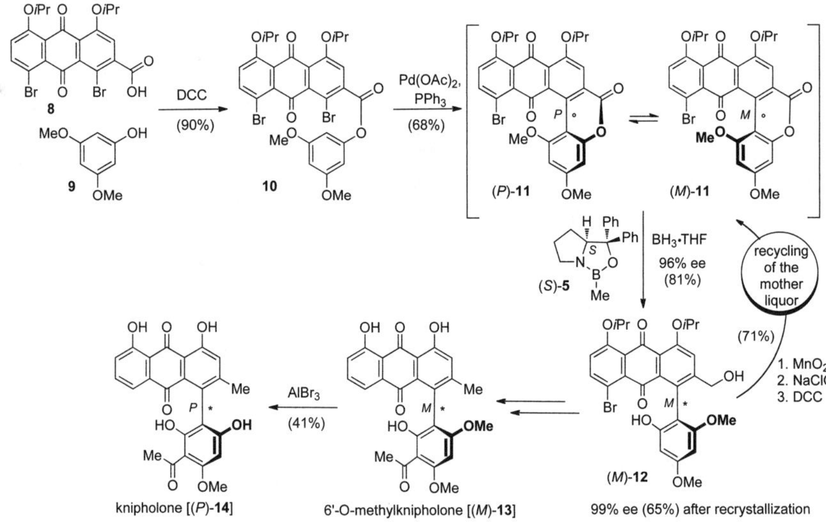

Scheme 2

Another example of the broad applicability of the 'lactone concept' is the atroposelective synthesis of the AB-fragment of vancomycin-type glycopeptides.[10] The key intermediate, ester **17**, was obtained by a simple esterification of carboxylic acid **15** with phenol **16** in good 81% yield (Scheme 3). Intramolecular Heck reaction gave the configurationally

72

unstable lactone **18** (64%) which after reductive cleavage of the heterocyclic ring by using borane activated (S)-**5**, yielded the biaryl (M)-**19** as a promising precursor for the vancomycin AB-fragment.

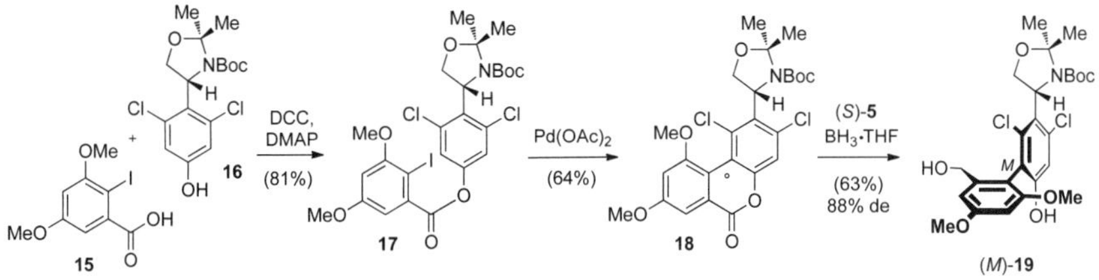

Scheme 3

The versatility of the 'lactone concept' has been demonstrated by our group in the total synthesis of several structurally diverse compounds, such as the bicoumarin isokotanin A [(M)-**20**],[11] the biscarbazole (M)-**21**,[12] which was attained by cleavage of the lactone bridge utilizing a chiral O-nucleophile [(R)-mentholate], and the biphenol mastigophorene A [(P)-**22**][13] (Scheme 4).

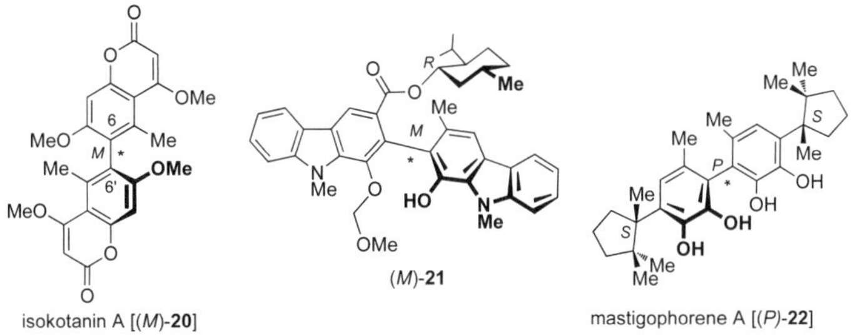

isokotanin A [(M)-**20**]   (M)-**21**   mastigophorene A [(P)-**22**]

Scheme 4

The 'lactone concept' for the construction of axially chiral natural products has also been used by other groups, i.a. by Abe, Harayama, et al.[14] Reductive cleavage of the configurationally unstable biaryl lactone **23** employing (S)-**5** gave (M)-**24**, a precursor for the formal total synthesis of (-)-steganone (M)-**25** (Scheme 6). The introduction of the chiral information into the biaryl system (M)-**24** was attained with excellent chemical (97%) and optical (83% ee) yields.

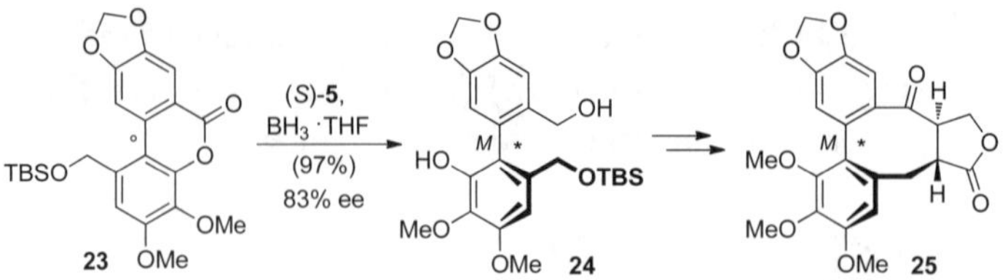

Scheme 5

The concept of using chiral *N*-nucleophiles in the atroposelectivity determining step, as previously elaborated in our group,[15,16] was also adopted by Suzuki et al. in the synthesis of benanomicin B (**28**).[17] Ring opening of the lactone system in **26** was thus achieved by using (S)-valinol to give (M)-**27** in good 90% yield and 82% de (Scheme 6). The asymmetric information of the biaryl axis was then transmitted to the stereoselective formation of the trans-diol in the final product by a semipinacol cyclization of an intermediate acetal-aldehyde with full stereocontrol. This example demonstrates that – by the use of the 'lactone concept' – axially chiral biaryls (like **27**) are now available so efficiently that they may even serve as a 'cheap' precursor for the construction of chiral target molecules with merely central chirality.

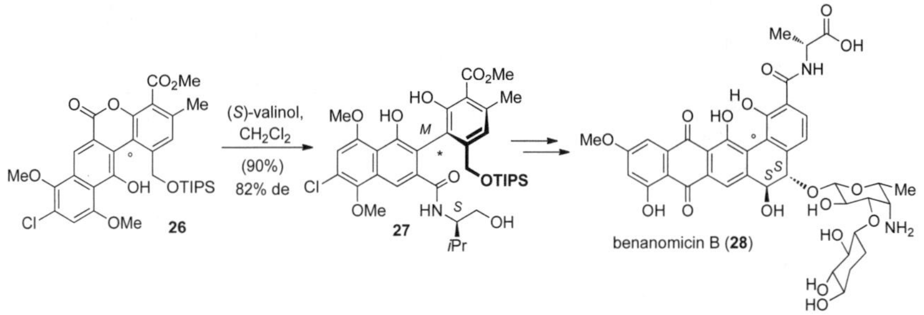

Scheme 6

The 'lactone concept' is not restricted to the dynamic kinetic resolution of biaryls that are configurationally labile due to the presence of a 6-membered lactone bridge, but can also be applied to 7-membered biaryl lactones, which are normally stable at the axis, thus just permitting a normal – non-dynamic – kinetic resolution.[3a] This alternative is well suited in particular for the preparation of constitutionally symmetric biaryls, since it avoids the need of building up two different aryl compounds (cf. Scheme 7). In the synthesis of the dimeric sesquiterpenes mastigophorene A [(P)-**22**]

74

and B [(M)-**22**][18] it thus provides the as yet shortest atroposelective approach (22 steps overall) to date by an atropo-diastereomer-differentiating reduction of **30** with borane and the CBS catalyst (R)-**5** in the key step, obtaining the diol (P)-**31** with a low 30% de, whilst the unreactive lactone (M)-**31** was recovered with a better diastereomeric excess of 62%.

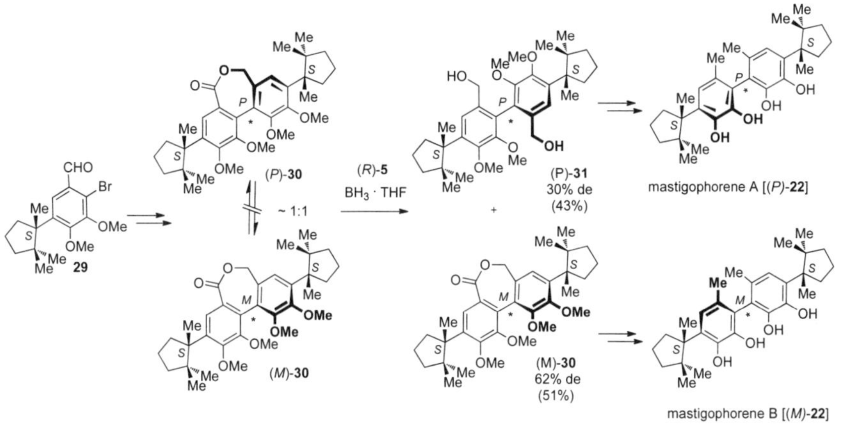

Scheme 7

A 7-membered lactone was also utilized by Molander et al. in the atroposelective construction of (M)-**32** in their synthesis of (+)-isoschizandrin [(M)-**35**][19] (Scheme 8). Starting from a racemic mixture of **32**, the undesired atropo-enantiomer was selectively consumed by enantiomer-differentiating reduction employing (R)-**5**, resulting in enantiopure unreacted lactone (M)-**33** (98% ee). The product of the reduction step, the P-configured diol (not shown), was re-oxidized and recyclized to again give **32**. Applying this recycling procedure, the overall yield of (M)-**33** was improved to 61%. DIBALH mediated reduction of (M)-**33** yielded aldehyde (M)-**34**, which was transformed to (+)-isoschizandrine [(M)-**35**] in six further steps.[19]

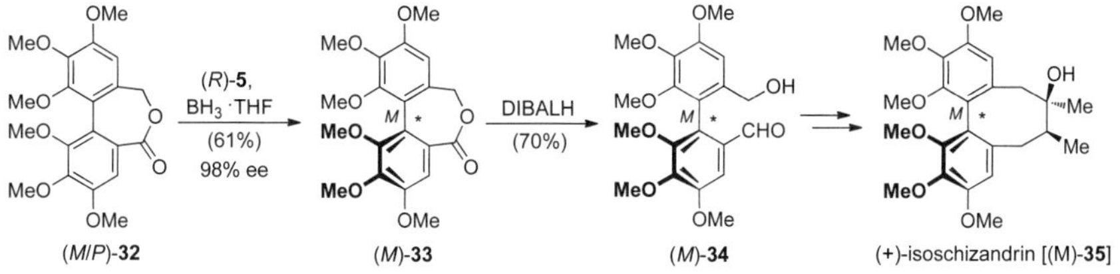

Scheme 8

More recently, Abe et al. applied the 'lactone concept' to the enantioselective construction of a valoneic acid derivative.[20,21] In addition, Yamada et al. elaborated catalytic atropo-enantioselective versions of this method using cobalt[22] and BINAP-based AgBF$_4$phosphine complexes.[23]

References

1. Institute of Organic Chemistry, Universty of Würzburg, Am Hubland, D-97074 Würzburg, Germany.
2. (a) Bringmann, G.; Menche, D. *Acc. Chem. Res.* **2001**, *34*, 615. (b) Bringmann, G.; Tasler, S.; Pfeifer, R.-M.; Breuning, M. *J. Organomet. Chem.* **2002**, *661*, 49. (c) Bringmann, G.; Price Mortimer, A. J.; Keller, P. A.; Gresser, M. J.; Garner, J.; Breuning, M. *Angew. Chem. Int. Ed.* **2005**, *44*, 5384.
3. (a) Bringmann, G.; Breuning, M.; Tasler, S. *Synthesis* **1999**, *4*, 525. (b) Bringmann, G.; Pokorny, F. In "The Alkaloids", Vol. 46; Cordell, G. A.; Ed.; Academic Press: New York, 1995; p 127.
4. Bringmann, G.; Ochse, M.; Götz, R. *J. Org. Chem.* **2000**, *65*, 2069.
5. Peters, K.; Peters, E.-M.; Ochse, M.; Bringmann, G. *Z. Kristallogr. – New Cryst. Struct.* **1998**, *213*, 559.
6. Bringmann, G.; Holenz, J.; Weirich, R.; Rübenacker, M.; Funke, C.; Boyd, M. R.; Gulakowski, R. J.; François, G. *Tetrahedron* **1998**, *54*, 497.
7. Herrmann, W. A.; Brossmer, C.; Öfele, K.; Reisinger, C.-P., Priermeier, T.; Beller, T.; Fischer, H. *Angew. Chem., Int. Ed. Engl.* **1995**, *34*, 1844.
8. (a) Bringmann, G.; Menche, D. *Angew. Chem. Int. Ed.* **2001**, *40*, 1687. (b) Bringmann, G.; Menche, D.; Kraus, J.; Mühlbacher, J.; Peters, K.; Peters, E.-M.; Brun, R.; B. Merhatibeb; Abegaz, B. M. *J. Org. Chem.* **2002**, *67*, 5595.
9. The configuration of the phenylanthraquinones **12-14** had been assigned incorrectly in the original publication
10. Bringmann, G.; Menche, D.; Mühlbacher, J.; Reichert, M.; Saito, N.; Pfeiffer, S. S.; Lipshutz, B. H. *Org. Lett.* **2002**, *4*, 2833.
11. Bringmann, G.; Hinrichs, J.; Henschel, P.; Kraus, J.; Peters, K.; Peters, E.-M. *Eur. J. Org. Chem.* **2002**, *6*, 1096.
12. Bringmann, G.; Tasler, S.; Endress, H.; Mühlbacher, J. *Chem. Commun.* **2001**, *8*, 761.

13. Bringmann, G.; Pabst, T.; Henschel, P.; Kraus, J.; Peters, K.; Peters, E.-V.; Rycroft, D. S.; Connolly, J. D. *J. Am. Chem. Soc.* **2000**, *122*, 9127.

14. (a) Abe, H.; Takeda, S.; Fujita, T.; Nishioka, K.; Takeuchi, Y.; Harayama, T. *Tetrahedron Lett.* **2004**, *45*, 2327. (b) Takeda, S.; Abe, H.; Takeuchi, Y.; Harayama, T. *Tetrahedron* **2007**, *63*, 396.

15. (a) Bringmann, G.; Walter, R.; Ewers, C. L. J. *Synlett* **1991**, 581. (b) Bringmann, G.; Güssregen, S.; Vitt, D.; Stowasser, R. *J. Mol. Model.* **1998**, *4*, 165. (c) Bringmann, G.; Breuning, M.; Tasler, S.; Endress, H.; Ewers, C. L. J.; Gobel, L.; Peters, K.; Peters, E. M. *Chem. Eur. J.* **1999**, *5*, 3029.

16. For a more recent example of a highly atropo-diastereoselective cleavage of configurationally unstable biaryl lactones with amino acid esters for the construction of new tweezer receptors for peptides, see: Bringmann, G.; Scharl, H.; Maksimenka, K.; Radacki, K.; Braunschweig, H.; Wich, P.; Schmuck, C. *Eur. J. Org. Chem.* **2006**, *19*, 4349.

17. Ohmori, K.; Tamiya, M.; Kitamura, M.; Kato, H.; Oorui, M.; Suzuki, K. *Angew. Chem. Int. Ed.* **2005**, *44*, 3871.

18. Bringmann, G.; Hinrichs, J.; Pabst, T.; Henschel, P.; Peters, K.; Peters, E.-M. *Synthesis* **2001**, 155.

19. Molander, G. A.; George, K. M.; Monovich, L. G. *J. Org. Chem.* **2003**, *68*, 9533.

20. Abe, H.; Harayama, T. *Heterocycles* **2008**, *75*, 1305.

21. Abe, H.; Sahara, Y.; Matsuzaki, Y.; Takeuchi, Y.; Harayama, T. *Tetrahedron Lett.* **2008**, *49*, 605.

22. Ashizawa, T.; Tanaka, S.; Yamada, T. *Org. Lett.* **2008**, *10*, 2521.

23. Ashizawa, T.; Yamada, T. *Chem. Lett.* **2009**, *38*, 246.

Gerhard Bringmann studied chemistry and biology in Gießen and Münster. After his Ph.D. with B. Franck in 1978, postdoctoral studies with D.H.R. Barton in Gif-sur-Yvette, and his habilitation in Münster, he received offers for full professorships at the Universities of Vienna and Würzburg, of which he accepted the latter in 1987. His research focuses on analytical, synthetic, and computational natural product chemistry, e.g., on axially chiral biaryls. He received, i.a., the Prize for Good Teaching (1999), the Adolf-Windaus Medal (2006), the Honorary Doctorate of the University of Kinshasa (2006), the Paul-J.-Scheuer Award (2007), and the Honorary Guest Professorship of Peking University (2008).

Tobias A. M. Gulder studied Chemistry at the University of Würzburg until 2004, followed by a Ph.D. with Prof. G. Bringmann in analytical and synthetic natural product chemistry, for which he received the 2009 DECHEMA PhD award. In 2008, he joined Prof. B.S. Moore at Scripps Institution of Oceanography (UCSD) as a DAAD postdoctoral fellow working on the elucidation and exploitation of marine natural product biosynthetic pathways. In the middle of 2010 he returned to Germany supported by a DAAD reintegration fellowship to start his independent research at the University of Bonn.

Tanja Gulder was born in 1978 in Weißenburg i. Bay. Germany, and received her diploma in chemistry from the University of Würzburg in 2004. After earning her Ph.D. with distinction under the supervision of Prof. G. Bringmann at University of Würzburg in 2008, she pursued postdoctoral studies with Prof. P. S. Baran at The Scripps Research Institute (La Jolla, CA) focusing on the synthesis of strained natural products. In summer 2010 she started her independent career at RWTH Aachen university supported by a Liebig fellowship.

SYNTHESIS OF LITHIUM 2-PYRIDYLTRIOLBORATE AND ITS CROSS-COUPLING REACTION WITH ARYL HALIDES

Submitted by Yasunori Yamamoto, Juugaku Sugai, Miho Takizawa, Norio Miyaura.[1]

Checked by Hyung Hoon Jung and Jonathan A. Ellman.

1. Procedure

A. Preparation of 2-Pyridyltriolborate. A three-necked, 500-mL round-bottomed flask is fitted with a mechanical stirrer (PTFE blade, 15 × 46 mm), a Claisen adapter with an internal thermometer and pressure-equalizing addition funnel, and a reflux condenser to which a nitrogen inlet and an oil bubbler are attached, and the apparatus is flushed with nitrogen (Note 1). The flask is charged with 300 mL of THF (Note 2) and 2-bromopyridine (**1**) (9.7 mL, 100 mmol) (Note 3). The mixture is cooled to –78 °C and 40.0 mL of *n*-butyllithium solution (2.5 M in hexane, 100 mmol) (Note 4) is added dropwise at a rate of 1 mL/min using the addition funnel (Note 5). The mixture is stirred for 45 min while maintaining the internal temperature at –78 °C. Triisopropylborate (23 mL, 100 mmol) (Note 3) is then added using the additional funnel (2 mL/min) (Note 6). The mixture is stirred for 1 h at –78 °C and then is allowed to warm to room temperature. The Claisen adapter is removed, and 1,1,1-tris(hydroxymethyl)ethane (12 g, 100 mmol) (Note 7) is added in one portion. The neck is tightly sealed with a glass stopper. The mixture is refluxed for 0.5 h and is cooled to room temperature. The mixture is poured into 1 L of hexanes. The solid product is isolated by filtration through a 17G3 fritted-glass funnel (Iwaki Glass, Co.), washed with hexanes (100 mL) and dried under vacuum to afford 19.8 g (93%) of 2-pyridyltriolborate (**2**) as a white solid, which is used in step B without purification (Note 8).

B. *Preparation of Methyl 4-(2-pyridyl)benzoate.* A three-necked, 250-mL, round-bottomed flask, is equipped with a rubber septum, a magnetic stirring bar (PTFE oval, 16 × 33 mm), an internal thermometer and a reflux condenser to which a nitrogen inlet and an oil bubbler are attached. The flask is charged with lithium 2-pyridyltriolborate (**2**) (6.39 g, 30.0 mmol, 3.0 equiv), methyl 4-bromobenzoate (**3**) (2.15 g, 10.0 mmol), dichloro (1,3-bis(diphenylphosphino)propane)palladium (118 mg, 0.200 mmol, 0.02 equiv) and copper iodide (0.190 g, 1.00 mmol, 0.1 equiv) and is flushed with nitrogen (Note 9, 10). Dry DMF (70 mL) (Note 11) is added, and the mixture is stirred for 21 h at 80 °C in an oil bath (internal temperature) and then is allowed to cool to room temperature. The mixture is transferred to a 500-mL separatory funnel charged with ethyl acetate (80 mL) and saturated aq. NH$_4$Cl solution (80 mL). The organic layer is separated. The aqueous layer is extracted with ethyl acetate (2 × 40 mL). The combined organic layer is washed with saturated aq. NaCl solution (5 × 80 mL). The organic layer is dried over anhydrous MgSO$_4$, filtered through a 17G3 fritted-glass funnel (Iwaki Glass, Co.) and concentrated on a rotary evaporator under reduced pressure to afford crude solid product. The crude product is dissolved in 5 mL of methanol at 40 °C and is allowed to cool to –30 °C for 1 h. The resulting crystals are collected by filtration on a 17G3 fritted-glass funnel (Iwaki Glass, Co.) and washed with 10 mL of cold methanol (1.65 g, 7.74 mmol, 77%). The mother liquor is concentrated, and the resulting solid is recrystallized from methanol (2 mL) to yield a second crop (0.29 g, 1.4 mmol, 14%). The process is repeated to yield a third crop (0.02 g, 0.01 mmol, 1.0%). The combined crystals are dried overnight at 0.1 mmHg to give 1.96 g (92%) of methyl 4-(2-pyridyl)benzoate (**4**) as a white crystalline solid (Note 12).

2. Note

1. All glassware was dried in an oven at 120 °C for 1 h, assembled while hot, and allowed to cool under a stream of nitrogen.
2. Tetrahydrofuran was passed through a column of activated alumina under nitrogen pressure immediately prior to use.

Org. Synth. 2011, *88*, 79-86

3. 2-Bromopyridine and triisopropylborate were purchased from Sigma-Aldrich Chemical Company and used without further purification.

4. *n*-Butyllithium (2.50 M in hexanes) was purchased from Sigma-Aldrich Chemical Company.

5. Adding the butyllithium solution at an approximate rate of 1 mL/min prevents the internal temperature from rising above −75 °C.

6. No significant increase in internal temperature was observed at an addition rate of 2.0 mL/min.

7. 1,1,1-Tris(hydroxymethyl)ethane was purchased from Acros Organics Company and was used without further purification.

8. The spectral data are as follows: mp >350 °C; IR 3449, 3230, 2950, 2870 cm^{-1}; ^{1}H NMR (500 MHz, DMSO-d_6) δ: 0.53 (s, 3 H), 3.67 (s, 6 H), 7.06 (td, J = 4.5, 1.5 Hz, 1 H), 7.33 (d, J = 8.5 Hz, 1 H), 7.51 (td, J = 7.5, 1.5 Hz, 1 H), 8.24 (d, J = 4.5 Hz, 1 H); ^{13}C NMR (125 MHz, DMSO-d_6) δ: 15.7, 34.4, 73.3, 120.5, 126.8, 134.1, 146.6; ^{11}B NMR (160 MHz, DMSO-d_6) δ: 1.10; MS (FAB$^+$): *m/z* (%): 213 (M$^+$, 6), 191 (53), 185 (100), 184 (24), 179 (6), 155 (6), 149 (18), 121 (11); HRMS (ESI): *m/z* calcd. for C$_{10}$H$_{14}$BLiNO$_3$ [M + H]$^+$ 214,1227, found 214.1228.

9. Methyl 4-bromobenzoate, triphenylphosphine and copper iodide were purchased from Sigma-Aldrich Chemical Company and used without further purification.

10. Dichloro(1,3-bis(diphenylphosphino)propane)palladium was purchased from Sigma-Aldrich Chemical Company and used without further purification.

11. DMF was purchased from EMD Chemicals Inc. and dried by distillation from calcium hydride under nitrogen before use.

12. Methyl 4-(2-pyridyl)benzoate: mp 96–97 °C; IR 3055, 2944, 1705, 1435, 1272, 1109 cm^{-1}; ^{1}H NMR (400 MHz, CDCl$_3$) δ: 3.95 (s, 3 H), 7.29 (dd, J = 8.8, 4.8 Hz, 1 H), 7.79 (d, J = 4.8 Hz, 2 H), 8.07 (d, J = 8.5 Hz, 2 H), 8.15 (d, J = 8.5 Hz, 2 H), 8.73 (d, J = 4.8 Hz, 1 H); ^{13}C NMR (100 MHz, CDCl$_3$) δ: 52.2, 121.0, 122.9, 126.8, 130.0, 130.3, 136.9, 143.5, 149.9, 156.2, 166.9; MS (EI) *m/z* (%): 213 (M$^+$, 78), 182 (100), 154 (56), 127 (25), 77 (19); HRMS (EI): calcd. for C$_{13}$H$_{11}$NO$_2$ [M$^+$] 213.0790, found 213.0788; Anal. calcd. for C$_{13}$H$_{11}$NO$_2$: C, 73.23; H, 5.20; N, 6.57. Found: C, 73.09; H, 5.39; N, 6.51.

Waste Disposal Information

All hazardous materials should be handled and disposed of in accordance with "Prudent Practices in the Laboratory"; National Academy Press; Washington, DC, 1995.

3. Discussion

Heteroaromatic biaryls are an important class of compounds due to the frequent occurrence of these fragments in natural products, pharmaceuticals, agrochemicals, and functional organic materials.[2] Cross-coupling reactions between arylboronic acids and aryl electrophiles provides simple access to such biaryls. Although cross coupling reactions are effective for a variety of organometallic reagents and electrophiles,[3] heteroaromatic boronic acids often fail to give biaryls due to the high sensitivity of the B-C bond of electron-deficient heteroaryl rings to hydrolytic B-C bond cleavage with water.[4] 2-Pyridylboronic acid is a typical example that undergoes very rapid cleavage with water. A recent advance in this field is the use of pinacol esters or diethanolamine esters of heteroaryl boronic acids.[5] The alkali metal triolborates have exceptionally high levels of stability in air and water thus allowing metal-catalyzed reactions in aqueous solvents. They also can be used in non-aqueous organic solvents such as DMF because of their reasonable solubility in organic solvents. Their high performance for bond-forming reactions have been demonstrated in palladium-catalyzed cross-coupling,[6] copper-catalyzed N-arylation (eq 1)[7] and rhodium-catalyzed asymmetric 1,4-addition (eq 2).[8]

82

$$(1)$$

$$(2)$$

(R)-Me-BIPAM

1. Division of Chemical Process Engineering, Graduate School of Engineering, Hokkaido University, Sapporo 060-8628, Japan.

2. (a) Stanforth, S. P. *Tetrahedron* **1998**, *54*, 263–303. (b) Anctil E. J. -G.; Snieckus, V. *Metal–Catalyzed Cross–Coupling Reactions, Second, Completely Revised and Enlarged Edition*; de Meijere A.; Diederich, F., Eds.; Wiley-VCH: Weinheim, 2005, pp. 761. (c) Campeau L. -C.; Fagnou, K. *Chem. Soc. Rev.* **2007**, *36*, 1058–1068. (d) Corbet J. -P.; Mignani, G. *Chem. Rev.* **2006**, *106*, 2651–2710.

3. For reviews, see: (a) Miyaura N.; Suzuki, A. *Chem. Rev.* **1995**, *95*, 2457–2483. (b) Suzuki, A. *Metal–Catalyzed Cross–Coupling Reaction*; Diederich, F.; Stang, P. J., Eds.; Wiley–VCH: Weinheim, 1998, pp. 49. (c) Miyaura, N. *Advances in Metal–Organic Chemistry*; Liebeskind, L. S., Ed.; JAI Press, Stamford, 1998, Vol. 6, p. 187. (d) Miyaura, N. *Top. Curr. Chem.* **2002**, *219*, 11–59. (e) Suzuki, A.; Brown, H. C. *Organic Synthesis via Boranes Vol. 3 Suzuki Coupling*; Aldrich Chemical Co., Milwaukee, 2003. (f) Miyaura, N. *Metal–Catalyzed Cross–Coupling*

Reactions, Second, Completely Revised and Enlarged Edition; de Meijere A.; Diederich, F., Eds.; Wiley-VCH: Weinheim, 2005, pp. 41–123.

4. Tyrrell, E.; Brookes, P. Synthesis 2003, 469–483.

5. (a) Ishikura, M.; Mano, T.; Oda, I.; Terashima, M. Heterocycles 1984, 22, 2417-2474. (b) Murafuji, T.; Mouri, R.; Sugihara, Y.; Takakura, K.; Mikata, Y.; Yano, S. Tetrahedron 1996, 52, 13933–13938. (c) Bouillon, A.; Lancelot, J. -C.; Sopkova de Oliveira Santos, J.; Collot, V.; Bovy, P. R.; Rault, S. Tetrahedron 2003, 59, 10043–10049. (d) Hodgson, P. B.; Salingue, F. H. Tetrahedron Lett. 2004, 45, 685–687. (e) Gros, P.; Doudouh, A.; Fort, Y. Tetrahedron Lett. 2004, 45, 6239–6241. (f) Jones, N. A.; Antoon, J. W.; Bowie, Jr., A. L.; Borak, J. B.; Stevens, E. P. J. Heterocycl. Chem. 2007, 44, 363–367. (g) Billingsley K. L.; Buchwald, S. L. Angew. Chem. Int. Ed. 2008, 47, 4695–4698. (h) Deng, J. Z.; Paone, D. V.; Ginnetti, A. T.; Kurihara, H.; Dreher, S. D.; Weissman, S. A.; Stauffer, S. R.; Burgey, C. S. Org. Lett. 2009, 11, 345-347. (i) Yang, D. S.; Colletti, S. L.; Wu, K.; Song, M.; Li, G. Y.; Shen, H. C. Org. Lett. 2009, 11, 381–384. (j) Knapp, D. M.; Gillis, E. P.; Burke, M. D. J. Am. Chem. Soc. 2009, 131, 6961–6963.

6. (a) Yamamoto, Y.; Takizawa, M.; Yu, X. -Q.; Miyaura, N. Angew. Chem. Int. Ed. 2008, 47, 928–931. (b) Yamamoto, Y.; Takizawa, M.; Yu, X. -Q.; Miyaura, N. Heterocycles 2010, 80, 359-368.

7. Yu, X. -Q.; Yamamoto, Y.; Miyaura, N. Chem. Asian J. 2008, 3, 1517–1522.

8. Yu, X. -Q.; Yamamoto, Y.; Miyaura, N. Synlett 2009, 994–998.

Appendix
Chemical Abstracts Nomenclature; (Registry Number)

2-Bromopyridine: Pyridine, 2-bromo-; (109-04-6)
Triisopropyl borate: Boric acid (H_3BO_3), tris(1-methylethyl)ester; (5419-55-6)
n-Butyllithium: Lithium, butyl-; (109-72-8)
1,1,1-Tris(hydroxymethyl)ethane: 1,3-Propanediol, 2-(hydroxymethyl)-2-methyl-; (77-85-0)
Lithium 2-pyridiltriolborate:Borate(1-), [2-[(hydroxy-κO)methyl]-2-methyl-1,3-propanediolato(3-)-κO1,κO3] -2-pyridinyl-, lithium

(1:1), (T-4)-; (1014717-10-2)

Methyl 4-bromobenzoate: Benzoic acid, 4-bromo-, methyl ester; (619-42-1)

Methyl 4-(2-pyridyl)benzoate: Benzoic acid, 4-(2-pyridinyl)-, methyl ester; (98061-21-3)

Dichloro[1,3-bis(diphenylphosphino)propane]palladium: Palladium, dichloro[1,1'-(1,3-propanediyl)bis[1,1-diphenylphosphine-κP]]-,

(SP-4-2)- (59831-02-6)

Triphenylphosphine: Phosphine, triphenyl-; (603-35-0)

Copper iodide; (1335-23-5)

Norio Miyaura was born in Hokkaido in 1946. He received his B. Eng. and Dr. Eng. from Hokkaido University. He became a Research Associate and an Associate Professor of the A. Suzuki research group, and then was promoted to Professor of the same group in 1994. In 1981, he joined the J. K. Kochi group at Indiana University as a postdoctoral fellow to study the epoxidation of alkenes catalyzed by salen-oxometal complexes. His current interests are mainly in the field of metal-catalyzed reactions of organoboron compounds, with an emphasis on applications to organic synthesis, for example, cross-coupling reactions, conjugate addition reactions, and addition and coupling reactions of diborons.

Yasunori Yamamoto was born in Hokkaido, Japan in 1968. He received his B. Eng. (1991) and M. Eng. (1993) from Hokkaido University under the direction of Professor Akira Suzuki. He worked at Mitsubishi Chemical Corporation as Researcher from 1993 to 1995. In 1995, he then moved to Hokkaido University as an Assistant Professor of the N. Miyaura research group and received his Ph. D degree in 2003 from Hokkaido University. His research interest is organic synthesis via metal-catalyzed reactions of organoboron compounds.

Juugaku Sugai was born in Sendai, Japan in 1985. He received his B. Eng. (2008) from Hokkaido University under the direction of Professor Norio Miyaura. Presently, he is pursuing his M.S. degree in the same group. The focus of his research is the cross-coupling reaction of triolborates.

Miho Takizawa was born in Niigata, Japan in 1984. She received her B. Eng. (2007) and M. Eng. (2009) from Hokkaido University under the direction of Professor Norio Miyaura. Currently she is working for Mitsubishi Chemical Corporation.

Hyung Hoon Jung was born in 1971 in Mokpo, Korea. After finishing military service, he studied organic chemistry at Hanyang University in Seoul, Korea where he completed his M.S. degree in Organic Chemistry under the supervision of Professor Chang Ho Oh. After working on palladium-catalyzed additions of organoboronic acids to alkynes as an assistant researcher at Hanyang University, he joined Professor Floreancig's group at the University of Pittsburgh in 2004. As a graduate student, he developed a gold-catalyzed synthesis of heterocycles and completed the syntheses of two natural products, (+)-andrachcinidine and (+)-leucascandrolide A. Currently, he is working as a postdoctoral fellow in Professor Ellman's group at Yale University on the development of rhodium-catalyzed additions of arylboronic acids to pharmaceutically interesting ketimines.

86

SYNTHESIS OF (+)-B-ALLYLDIISOPINOCAMPHEYLBORANE AND ITS REACTION WITH ALDEHYDES

Submitted by Huikai Sun and William R. Roush.[1]
Checked by David Hughes.[2]

1. Procedure

A. *(+)-B-allyldiisopinocampheylborane* *((+)-(Ipc)₂B(allyl)* or
(^{l}Ipc)₂B(allyl)) (1). A 500-mL, 3-necked oven-dried round-bottomed flask
equipped with a 3-cm oval Teflon-coated magnetic stir bar is fitted with a
gas inlet adapter connected to a nitrogen line and a gas bubbler. The other
two necks are capped with rubber septa; a thermocouple probe is inserted
through one of the septa. (Note 1)   The flask is charged with (+)-B-
methoxydiisopinocampheylborane ((+)-(Ipc)₂BOMe) (13.3 g, 42.1 mmol,
1.25 equiv) (Note 2) and diethyl ether (45 mL), which results in a clear and
colorless solution.   The solution is cooled to 3 °C in an ice/water bath and
vigorously stirred. Allylmagnesium bromide solution (1.0 M in diethyl ether,
40 mL, 40 mmol, 1.20 equiv) (Notes 2 and 3) is then added dropwise over
20 min via a 60 mL disposable syringe with 18-gauge needle, maintaining
the temperature below 6 °C. A large amount of white solids (presumably
MgBr(OMe)) precipitate during the addition. After the addition is complete,
the ice/water bath is removed, then the reaction mixture is vigorously stirred
(Note 4) for 1 h at room temperature. The resulting (^{l}Ipc)₂B(allyl) mixture in
ether is used immediately in the next step without further purification (Note
5).

B. *(2R,3R)-1-(tert-Butyldiphenylsilyloxy)-2-methylhex-5-en-3-ol* *(3)*. The heterogeneous mixture of (^{l}Ipc)$_2$B(allyl) generated in Step A is cooled to –75 °C with a dry ice/acetone bath under vigorous stirring, then a solution of (*R*)-3-(*tert*-butyldiphenylsilyloxy)-2-methylpropanal (**2**) (10.9 g, 33.4 mmol, 1.00 equiv) (Note 6) in diethyl ether (25 mL) is added dropwise over 20 min via syringe (Note 7), maintaining the temperature below –70 °C. The resulting mixture is vigorously stirred at –70 to –75 °C for 1.5 h (Note 8), then the dry ice/acetone bath is removed, and the reaction mixture is allowed to warm to room temperature (22 °C) over 1 h. The reaction mixture is cooled to 3 °C with an ice/water bath. A 125-mL gas equilibrating dropping funnel is attached to the flask, moving the gas adapter from one neck of the flask to the top of the dropping funnel. A premixed solution of 3M NaOH (64 mL) and 30 % H$_2$O$_2$ (26 mL) (Note 9) is carefully added via the dropping funnel over 10 min (exothermic), keeping the temperature below 15 °C, followed by addition of saturated aqueous NaHCO$_3$ (80 mL) over 3 min via the dropping funnel (Note 10). The resulting biphasic mixture is vigorously stirred for 10 h at room temperature (Note 11) to completely hydrolyze borinate ester products, then the organic phase is separated, and the aqueous phase is extracted with diethyl ether (2 x 80 mL). After the combined organic layers are washed with brine (2 x 50 mL), the ether solution is transferred to a 1-L round-bottomed flask, equipped with a 3-cm, egg-shaped Teflon-coated stir bar, then THF (150 mL), water (80 mL) and iron (II) sulfate heptahydrate salt (15.0 g) are added (Note 12). The flask is fitted with a gas inlet adapter connected to a nitrogen line and a gas bubbler. The resulting mixture is vigorously stirred for 14 h, then the organic phase is separated and the aqueous phase is extracted with diethyl ether (2 x 50 mL). The combined organic layers are washed with brine (2 x 50 mL), dried over 100 g of anhydrous Na$_2$SO$_4$ and vacuum filtered through a medium porosity sintered-glass funnel. The filtrate is concentrated (40 °C bath temperature, 100 mmHg initial to 20 mmHg) by rotary evaporation to give a light yellow residue (32 g), containing a 93 : 7 mixture of **3** and its anti diastereomer (**5**) (Note 13). This mixture is purified by column chromatography (Note 14) to provide 9.1 – 9.4 g (74 – 77%) of >98% pure (2*R*,3*R*)-1-(*tert*-butyldiphenylsilyloxy)-2-methylhex-5-en-3-ol **3** (Note 15) as a colorless oil.

2. Notes

1.   The internal temperature was monitored using a J-Kem Gemini digital thermometer with a Teflon-coated T-Type thermocouple probe (12-inch length, 1/8 inch outer diameter, temperature range –200 to +250 °C).

2.   (+)-B-Methoxydiisopinocampheylborane was purchased from Sigma-Aldrich and used as received. The submitters stored and transferred this material in the glove box. The checker stored it in the freezer and weighed and transferred (rapidly) in open air. Diethyl ether (anhydrous, ACS reagent) and allylmagnesium bromide (1.0 M in diethyl ether, stored in refrigerator) were purchased from Sigma-Aldrich and used as received.

3.   The specified amounts of (+)-(Ipc)$_2$BOMe (1.25 equiv) and allylmagnesium bromide   (1.20 equiv) are used in order to generate a sufficient amount of allylborane **1** to completely consume 1.0 equiv of aldehyde **2** in the allylboration step. Unreacted aldehyde (8-10%) remains at the end of the allylboration reaction when 1.0 equiv of both (+)-(Ipc)$_2$BOMe and allylmagnesium bromide are used. A slight excess of (+)-(Ipc)$_2$BOMe is used over allylmagnesium bromide in order to ensure complete consumption of the latter before addition of the aldehyde; the submitters observed that the allylboration diastereoselectivity decreased by 1-2% if a slight excess of (+)-(Ipc)$_2$BOMe is not used.

4.   The stirring of the thick mixture should be vigorous but balanced. Solids that stick on the walls above the mixture may contain entrapped reagent, perhaps in the form of the ate complex indicated in equation 1,[3] and cannot be washed back into the reaction mixture. In such cases, unreacted aldehyde is often observed after conclusion of the allylboration reaction.

equation 1

5.   (-)-B-allyldiisopinocampheylborane   ((-)-(Ipc)$_2$B(allyl)   or (^{d}Ipc)$_2$B(allyl)), can be prepared using the same procedure starting from (-)-B-methoxydiisopinocampheylborane ((-)-(Ipc)$_2$BOMe). The (Ipc)$_2$B(allyl) reagents are sensitive to air and moisture.

6.   (R)-3-(t-Butyldiphenylsilyloxy)-2-methylpropanal **2** was prepared (10.0 - 12.0 g scale) according to the procedure of Marshall et al.[4] and obtained in 75–80% yield as a white solid following column chromatography with 3% EtOAc/hexanes with $[\alpha]_D^{20}$ –26.4 (CHCl$_3$, c = 1.8).

Aldehyde **2** was stored at –20 °C for up to 2 weeks prior to use without racemization (as determined by measuring its optical rotation and ee determination of the allylation product **3**).

7. An oven-dried, 100-mL, single-necked, round-bottomed flask was charged with (*R*)-3-(*t*-butyldiphenylsilyloxy)-2-methylpropanal **2** (10.9 g), then 25 mL of diethyl ether was added, and the mixture was swirled to completely dissolve the aldehyde. The resulting solution was transferred to the reaction mixture via a 40 mL disposable syringe with a 15 cm needle (18 gauge) over 20 min. Additional diethyl ether (2 x 5 mL) was used to rinse the flask and then was added to the reaction mixture over 1 min. For the addition of the aldehyde solution to the cold reaction mixture, the tip of the syringe needle should be kept >5 cm above the surface and added at a steady rate to prevent crystallization of the aldehyde in the syringe. The submitters used a syringe pump for this addition.

8. The progress of the reaction was monitored by ^{1}H NMR spectroscopy monitoring the aldehyde proton at δ 9.8. (Typical procedure for ^{1}H NMR analysis: an aliquot of the reaction mixture was quickly transferred via a syringe to a small vial containing methanol (0.5 mL) at rt. The solvent was evaporated and the residue was dissolved in CDCl$_3$.)

9. 30% H$_2$O$_2$ was purchased from Fisher Chemical Company, stored at 5 °C and used as received.

10. The oxidative hydrolysis was not complete in 16 h at ambient temperature or under reflux without addition of the saturated NaHCO$_3$ solution. The pH of the hydrolysis reaction mixture after bicarbonate addition was 11.

11. The progress of the oxidative hydrolysis was monitored by ^{1}H NMR spectroscopy using CDCl$_3$ as solvent (An aliquot of the organic layer was evaporated to dryness and the residue dissolved in CDCl$_3$.) Completion of the oxidative hydrolysis was indicated by disappearance of the mutiplet resonances of the internal olefin hydrogen (δ 5.78 – 5.68) and terminal olefin hydrogens (δ 5.05 – 4.95) in the intermediate borinate ester (e.g., ROBIpc$_2$). In comparison, the chemical shift range of the multiplet of the corresponding olefinic hydrogens in the major product **3** are from δ 5.92 – 5.81 and 5.16 – 5.09. The hydrolysis was complete within 1.5 h in the hands of the submitters but typically required 8-10 h for the checkers. Agitation efficiency of the 3-phase mixture (aqueous, organic, solids) may affect the reaction rate.

90

12. Isopinocampheyl hydroperoxide **4** is produced in the reaction. This compound migrates with the *syn* product **3** on the TLC plate (EMD, silica gel, grade 60, F254) in this exemplified allylboration reaction, so the iron (II) sulfate reduction step is included to ensure that hydroperoxide **4** is completely consumed prior to product isolation. A reference sample of **4** was prepared by stirring a solution of Ipc$_2$BOMe (1.0 g) in THF under air for 1 h. This solution was cooled to 0 °C and a premixed solution of 3 N NaOH (2.3 mL) and 30% H$_2$O$_2$ (0.9 mL) was added followed by sat. NaHCO$_3$ (3.0 mL) solution. The resulting mixture was vigorously stirred for 1.5 h. The organic phase was separated and hydroperoxide **4** was separated from isopinocampheol (1 : 4 mixture, respectively) by column chromatography using 1:10 Et$_2$O-hexanes as eluent (R$_f$=0.5, 1:5 Et$_2$O-hexanes, staining with PMA).

13. The reaction diastereoselectivity was determined as follows. A small sample of the crude reaction product (80 mg) was purified (collecting all fractions containing product with R$_f$ = 0.4 to 0.5) by flash column chromatography (8 g silica gel) using ether/hexanes = 1/8 as the eluent to provide a mixture of three products. This mixture was analyzed by the submitters using normal phase HPLC (5% EtOAc in hexanes, 1.0 mL/min, 4.6 X 250 mm Varian column, UV detection at 254 nm; t$_R$(**3**) = 16.2 min; t$_R$(**5**) = 13.9 min; t$_R$(**6**) = 13.0 min). The checkers used a reverse phase assay to analyze diastereomeric ratios using an Agilent 1100 HPLC system; Ascentis Express C-18 fused-core column, 4.6 x 100 mm, 2.7 um particle size; 1.8 mL/min flow; temperature 40 °C; detection at 210 nm; gradient elution from 50/50 MeCN/water containing 0.1% H$_3$PO$_4$ to 95% MeCN/5% aq. over 14 min; t$_R$ (**3**) 9.6 min, t$_R$ (**5**) 9.8 min, t$_R$ (**6**) 11.6 min). The crude reaction product was determined to contain (2*R*,3*R*)-1-(*t*-butyldiphenylsilyloxy-2-methylhex-5-en-3-ol (**3**) (90%), (2*R*,3*S*)-1-(*t*-butyldiphenylsilyloxy)-2-methylhex-5-en-3-ol (**5**) (7%), and (2*R*, 3*R*)-1-(*t*-butyldiphenylsilyloxy)-2-methyl-5-methyleneoct-7-en-3-ol (**6**) (3%). The stereochemistry of the hydroxyl group of **6** is assumed, by analogy to the stereochemistry of the major product of the reaction (**3**). A use test (by direct reaction with aldehyde **2**) indicated that the solutions of allylmagnesium bromide in diethyl ether contain ca. 3% of 2-((bromomagnesium)methyl)-1,4-pentadiene.

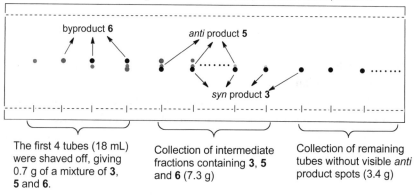

4   5   6

14. The crude product was purified by three flash column chromatography steps owing to the difficulty of separating the syn diastereomer **3** from the anti diastereomer **5** and the byproduct **6**. The first chromatography was performed using 300 g of silica gel (Fisher, 230-400 mesh, 60 Å) in a 5 cm diameter column using 10:1 hexanes-diethyl ether as the eluent (18-mL fractions). All fractions were analyzed by TLC (1:10 Et$_2$O-hexanes, 3 developments, staining with KMnO$_4$ solution), as depicted

Developed 3 times with Et$_2$O/hexanes = 1/10, stained with KMnO$_4$ solution.

byproduct **6**

anti product **5**

syn product **3**

| The first 4 tubes (18 mL) were shaved off, giving 0.7 g of a mixture of **3**, **5** and **6**. | Collection of intermediate fractions containing **3**, **5** and **6** (7.3 g) | Collection of remaining tubes without visible *anti* product spots (3.4 g) |

graphically below. Early fractions contained 0.7 g of a mixture of **5** and predominantly by-product **6**. Late eluting fractions without detectable *anti* diastereomer **5** were pooled, giving 3.7 g of *syn* diastereomer **3**. The intermediate fractions, containing 8.1 g of a mixture of **3**, **5** and **6** were pooled and subjected to a second chromatography as described above (250 g silica gel, 5 cm diameter column). Early eluting fractions containing predominantly **5**, **6** and a small amount of **3** were discarded. Late eluting fractions, without detectable *anti* diastereomer **5** according to TLC analysis, were combined to give an additional 3.4 g of *syn* diastereomer **3**. The intermediate, mixed fractions, consisting of 4.5 g of a mixture of **3**, **5** and **6** were subjected to a third column chromatography (200 g of silica gel in a

5 cm diameter column). This provided an additional 2.3 g of essentially pure *syn* diastereomer **3**, along with 1.5 g (12% yield) of mixed fractions that consisted of mixture **3** (ca. 80%), **4** (ca. 20%) and **6** (ca. 1 %). The latter fraction could be subjected to additional purification if desired. The three main fractions of *syn* diastereomer **3** were combined, giving 9.4 g (77% yield) which contained 1.3% of *anti* diastereomer **5** and 0.2% of by-product **6** according to HPLC analysis as described in note 13.

15. The submitters obtained pure samples of **3**, **5** and **6** by preparative HPLC for spectroscopic analysis (5% EtOAc in hexanes, 18.0 mL/min, 21.4 x 250 mm Varian Dynamax column, Microsorb 60-8; t_R(**3**) = 9.8 min; t_R(**5**) = 9.1 min; t_R(**6**) = 8.3 min). The enantiomers of both **3** and **5** have been synthesized and characterized previously.[5] The checker prepared a mixture of the (2*S*, 3*S*) and (2*S*, 3*R*) diastereomers by the same protocol starting with (*S*)-3-(*t*-butyldiphenylsilyloxy)-2-methylpropanal and (+)-B-methoxydiisopinocampheylborane, which produced a 78:22 mixture of (2*S*, 3*R*):(2*S*, 3*S*) diastereomers.

The *syn* product **3** exhibits the following physical and spectroscopic properties: colorless oil; $[\alpha]_D^{21}$ = +3.7 (c = 1.9, CHCl$_3$); ^{1}H NMR (400 MHz, CDCl$_3$) δ: 0.97 (d, *J* = 7.1 Hz, 3 H), 1.08 (s, 9 H), 1.77–1.82 (m, 1 H), 2.20–2.33 (m, 2 H), 2.73 (d, *J* = 3.4 Hz, 1 H), 3.70 and 3.77 (ABX, *J* = 10.1, 4.3 Hz, 2 H), 3.92–3.97 (m, 1 H), 5.09–5.16 (m, 2 H), 5.81–5.92 (m, 1 H), 7.30–7.48 (m, 6 H), 7.67–7.71 (m, 4 H); ^{13}C NMR (125 MHz, CDCl$_3$) δ: 10.5, 19.4, 27.1, 39.1, 39.3, 68.7, 73.5, 117.3, 128.0, 130.00, 130.04, 133.2, 133.4, 135.79, 135.81, 135.9; IR (KBr) 3468, 2955, 2858, 1606, 1515, 1471, 1427, 1112, 701 cm^{-1}; LC-MS calcd for [M+Na]$^+$ (C$_{23}$H$_{32}$NaO$_2$Si) 391.6, found, 391.7 m/z. Purity by reverse phase HPLC was >98% (see note 13 for method), t_R (**3**) 9.6 min (98.3%); t_R (**5**) 9.8 min (1.3%); t_R (**6**) 11.6 min (0.2%); **t$_R$** 9.1 min (unknown, 0.2%). A reverse phase chiral HPLC assay was developed to separate the (2*S*, 3*S*) and (2*R*, 3*R*) enantiomers: OJ-RH (150 x 4.6mm, 5um) isocratic 60% MeCN (pH 3.5, 2mM ammonium formate), 40% aqueous (pH 3.5, 2mM ammonium formate), 0.75mL/min, ambient temp, 215 nm, 20 min method time; t_R(**3**) (2*R*, 3*R*) 11.5 min; t_R (2*S*, 3*S*) 13.2 min; t_R (**5**) (2*R*, 3*S*) and (2*S*, 3*R*) co-elute 12.5 min. The enantiomeric purity of **3** was 99.0% indicating that aldehyde **2** did not racemize during its preparation and application in the exemplified procedure. An analytical sample of **3** was prepared by dissolving ~100 mg of the product from the pooled chromatographies in 5 mL of diethyl ether, filtering through a 0.45 micron PTFE syringe filter, and concentrating to dryness under vacuum for 16 h. Anal. calcd. for $C_{23}H_{32}O_2Si$: C, 74.95; H, 8.75; found: C, 74.85; H, 8.78.

The *anti* product **5** exhibits the following physical and spectroscopic properties: colorless oil; $[\alpha]_D^{21} = -2.6$ (c = 0.7, $CHCl_3$); ^1H NMR (400 MHz, $CDCl_3$) δ: 0.85 (d, J = 6.8 Hz, 3 H), 1.06 (s, 9 H), 1.80–1.84 (m, 1 H), 2.17–2.23 (m, 1 H), 2.34–2.39 (m, 1 H), 3.49 (d, J = 3.2 Hz, 1 H), 3.65 and 3.77 (ABX, J = 10.0, 4.4 Hz, 2 H), 3.65–3.73 (m, 1 H), 5.10–5.15 (m, 2 H), 5.88–5.98 (m, 1 H), 7.38–7.45 (m, 6 H), 7.67–7.69 (m, 4 H); ^{13}C NMR (100 MHz, $CDCl_3$) δ: 13.4, 19.1, 26.8, 39.4, 39.5, 68.6, 75.1, 117.2, 127.8, 129.8, 132.9, 135.3, 135.6, 135.7; IR (KBr) 3496, 2959, 2930, 2858, 1589, 1472, 1427, 1390, 1112, 701 cm^{-1}; LC-MS calcd for $[M+Na]^+$ ($C_{23}H_{32}NaO_2Si$) 391.6, found, 391.7 m/z.

The side product **6** exhibits the following physical and spectroscopic properties: colorless oil; ^1H NMR (400 MHz, $CDCl_3$) δ: 0.93 (d, J = 6.8 Hz, 3 H), 1.07 (s, 9 H), 1.76–1.80 (m, 1 H), 2.20–2.28 (m, 2 H), 2.49 (d, J = 3.6 Hz, 1 H), 2.82 (d, J = 6.8 Hz, 2 H), 3.68 and 3.73 (ABX, J = 10.4, 4.8 Hz, 2 H), 4.04–4.07 (m, 1 H), 4.90 (d, J = 0.8 Hz, 2 H), 5.05–5.10 (m, 2 H), 5.78-5.88 (m, 1 H), 7.38–7.46 (m, 6 H), 7.66–7.70 (m, 4 H); ^{13}C NMR (100 MHz, $CDCl_3$) δ: 10.4, 19.2, 26.9, 39.3, 40.6, 41.1, 68.0, 70.8, 112.9, 116.5, 127.7, 129.7, 129.8, 133.1, 133.3, 135.6, 135.7, 136.1, 145.3; IR (KBr) 3400, 2928, 2859, 1607, 1515, 1470, 1463, 1455, 1112, 822, 702, 505 cm^{-1}; LC-MS calcd for $[M+Na]^+$ ($C_{26}H_{36}NaO_2Si$) 431.6, found, 431.6 m/z.

Isopinocampheyl hydroperoxide **4** exhibits the following physical and spectroscopic properties: colorless oil; $[\alpha]_D^{21} = + 33.2$ (c = 0.7, $CHCl_3$); ^1H NMR (400 MHz, $CDCl_3$) δ: 0.92 (s, 3 H), 1.01 (d, J = 9.6 Hz, 1 H), 1.17 (d, J = 7.2 Hz, 3 H), 1.22 (s, 3 H), 1.79–1.86 (m, 2 H), 1.90–1.95 (m, 1 H), 2.00–2.03 (m, 1 H), 2.30–2.42 (m, 2 H), 4.27 (ddd, J = 3.6, 4.4 and 8.8 Hz, 1 H), 7.73 (s, 1 H); ^{13}C NMR (100 MHz, $CDCl_3$) δ: 1.41, 23.61, 27.30, 32.48, 33.56, 38.40, 40.87, 42.14, 47.24, 84.99; IR (KBr) 3391, 2908, 1453, 1367, 1158, 1035 cm^{-1}; LC-MS calcd for $[M-H_2O]^+$ ($C_{10}H_{18}O_2$) 152.1, found, 152 m/z. This compound is readily reduced by Fe_2SO_4 to give isopinocampheol.

Safety and Waste Disposal Information

All hazardous materials should be handled and disposed of in accordance with "Prudent Practices in the Laboratory"; National Academy Press; Washington, DC, 1995.

3. Discussion

Chiral allylmetal reagents are valuable intermediates in organic synthesis, and are especially useful for the synthesis of enantioenriched homoallylic alcohols.[6] Since the first chiral allylborane reagent was reported by Hoffman in 1978,[7] many more efficient and practical chiral allylmetal reagents or allylmetallation procedures have been reported,[6] including those by Brown,[8] Roush,[5,9] Corey,[10] Leighton,[11] Denmark,[12] Soderquist,[13] Hall[14] and Krische.[15] The present enantioselective aldehyde allylation procedure developed by Brown and his coworkers is one of the most widely adopted methods as it employs commercially available (+)-(Ipc)$_2$BOMe and commercially available allylmagnesium bromide for synthesis of the (^{l}Ipc)$_2$B(allyl) reagent;[8b] the enantiomeric (^{d}Ipc)$_2$B(allyl) species can be accessed from commercially available (-)-(Ipc)$_2$BOMe by using the same protocol.

The (Ipc)$_2$B(allyl) reagents were prepared in Brown's original procedure[8b] by treating (Ipc)$_2$BOMe with allylmagnesium bromide at –78 °C and then allowing the reaction mixture to warm to rt over 1 h. The procedure described herein follows an alternative protocol subsequently published by Brown, in which allylmagnesium bromide is added to a solution of (Ipc)$_2$BOMe in diethyl ether at 0 °C and then stirring the reaction mixture at rt for 1 h.[3] The allylboration of aldehydes can be performed after removal of the magnesium salts by filtration under an inert atmosphere (salt free procedure), which is reported to provide much improved enantioselectivites.[3] However, the procedure described here, by performing the allylboration in the presence of the magnesium salts, is more convenient as it avoids filtration and extra manipulations of the moisture and air sensitive allylborane species.

We found that use of slight excesses of both (Ipc)$_2$BOMe (1.25 equiv) and allylmagnesium bromide (1.2 equiv) were needed in order to achieve complete allylboration of aldehyde **2**. Generally, the secondary alcohol products can be isolated following oxidative hydrolytic workup by treating the intermediate borinate esters with a solution of 3 N sodium hydroxide and 30% hydrogen peroxide under reflux for several hours or at rt for 16 h. However, we found that the hydrolysis did not proceed to completion in the allylboration reaction reported here unless a solution of sat. sodium bicarbonate was also added. These optimized workup conditions not only resulted in complete oxidative hydrolysis in 10 h at room temperature, but also led to formation of two clear phases that facilitated the separation process. The submitters also found that isopinocampheyl hydroperoxide **4** is produced in 4-6% yield. While the origin of **4** has not been rigorously established, it seems likely that it arises via a radical process when the borinate ester intermediates are exposed to O_2 during reaction workup.[16] Hydroperoxide **4** is difficult to separate from *syn* product **3** in this exemplified reaction, so the submitters further modified the reaction workup by addition of an aqueous solution of iron (II) sulfate to reduce hydroperoxide **4** before product purification via column chromatography.[17]

The submitters also identified (2R, 3R)-1-(*tert*-butyldiphenylsilyloxy)-2-methyl-5-methyleneoct-7-en-3-ol (**6**, 3% according to HPLC analysis) as a by-product of this procedure. A use test (by direct reaction with aldehyde **2**) indicated that the solutions of allylmagnesium bromide in diethyl ether contains ca. 3% of 2-((bromomagnesium)methyl)-1,4-pentadiene. Production of analogous products deriving from 2-((bromomagnesium)methyl)-1,4-pentadiene as a contaminant of allylmagnesium bromide have not been described in the literature, to the best of the knowledge of the submitters. Whether the formation of this by-product can be avoided by using freshly prepared allylmagnesium bromide has not been determined.

Reactions of (Ipc)₂B(allyl) reagents with achiral aldehydes are reported to furnish the corresponding secondary homoallylic alcohols with 80 to 95% enantiomeric excess, as shown in Table 1. The substrate scope has been extended in the literature to a large number of aliphatic aldehydes, aromatic aldehydes and α, β-unsaturated aldehydes.[6f]

Table 1. Allylboration of achiral aldehydes with $(^d\text{Ipc})_2\text{B(allyl)}$ or $(^l\text{Ipc})_2\text{B(allyl)}$

Entry	aldehydes	(Ipc)₂B(allyl)	product	ee (%)	yield (%)[a]
1	(acetaldehyde)	$(^d\text{Ipc})_2\text{B(allyl)}$	(homoallylic alcohol)	93	74
2	(propanal)	$(^d\text{Ipc})_2\text{B(allyl)}$	(homoallylic alcohol)	86	71
3	(isobutyraldehyde)	$(^d\text{Ipc})_2\text{B(allyl)}$	(homoallylic alcohol)	90	86
4	(pivaldehyde)	$(^d\text{Ipc})_2\text{B(allyl)}$	(homoallylic alcohol)	83	88
5	(benzaldehyde)	$(^d\text{Ipc})_2\text{B(allyl)}$	(homoallylic alcohol)	96	81
6	(acrolein)	$(^d\text{Ipc})_2\text{B(allyl)}$	(homoallylic alcohol)	92	no reported yield
8[b]	(long chain aldehyde, 13)	$(^d\text{Ipc})_2\text{B(allyl)}$[c]	(homoallylic alcohol, 13)	80	78
9[b]	(long chain aldehyde, 13)	$(^l\text{Ipc})_2\text{B(allyl)}$[c]	(homoallylic alcohol, 13)	84	80

[a] Isolated yields; [b] see reference 18; [c] (Ipc)₂B(allyl) was prepared from the corresponding (Ipc)₂BCl

Allylboration of α-chiral aldehydes with (Ipc)₂B(allyl) reagents are also reported to produce secondary homoallylic alcohols in excellent diastereoselectivity and in good to excellent yields as shown in Table 2.[8c,19] In most cases (entries 1-3), with aldehyde substrates with very modest diastereofacial biases, the facial selectivity of the reaction is completely reversed upon switching the chirality of the chiral allylborane reagents. Numerous other applications of the use of the (Ipc)₂B(allyl) reagents in matched and mismatched double asymmetric reactions with chiral aldehydes are summarized in the cited review literature.[6a,6f]

Table 2. Allylboration of α-chiral aldehydes with $(^d\text{Ipc})_2\text{B(allyl)}$ or $(^l\text{Ipc})_2\text{B(allyl)}$

Entry	aldehyde	conditions	diastereomeric products/ratio		yield
1		$(^d\text{Ipc})_2\text{B(allyl)}$	96	4	81%
		$(^l\text{Ipc})_2\text{B(allyl)}$	5	95	83%
2	BzO	$(^d\text{Ipc})_2\text{B(allyl)}$	96	4	80%
		$(^l\text{Ipc})_2\text{B(allyl)}$	2	98	78%
3	OBz	$(^d\text{Ipc})_2\text{B(allyl)}$	94	6	80%
		$(^l\text{Ipc})_2\text{B(allyl)}$	4	96	75%
4	Ph	$(^d\text{Ipc})_2\text{B(allyl)}$	67	33	68%
		$(^l\text{Ipc})_2\text{B(allyl)}$	2	98	70%

1. Department of Chemistry, The Scripps Research Institute, Scripps Florida, 130 Scripps Way #3A2, Jupiter, FL 33458. E-mail: roush@scripps.edu. This research was supported by a grant from the National Institutes of Health (GM038436).

2. Checker is indebted for the analytical support from Kevin Maloney for the reverse-phase HPLC assay of diastereomers and Zainab Pirzada for the chiral HPLC assay.

3. Racherla, U. S.; Brown, H. C. *J. Org. Chem.* **1991**, *56*, 401-404.

4. Johns, B. A.; Grant, C. M.; Marshall, J. A. *Org. Synth.* **2002**, *79*, 59-71.

5. Roush, W. R.; Hoong, L. K.; Palmer, M. A. J.; Straub, J. A.; Palkowitz, A. D. *J. Org. Chem.* **1990**, *55*, 4117-4126.

6. (a) Roush, W. R., In *Comprehensive Organic Synthesis*, Trost, B. M., Ed. Pergamon Press: Oxford, **1991**; Vol. 2, p 1. (b) Yamamoto, Y.; Asao, N. *Chem. Rev.* **1993**, *93*, 2207. (c) Denmark, S. E.; Almstead, N. G., In *Modern Carbonyl Chemistry*, Otera, J., Ed. Wiley-VCH: Weinheim, **2000**; p 299. (d) Chemler, S. R.; Roush, W. R., In *Modern Carbonyl Chemistry*, Otera, J., Ed. Wiley-VCH: Weinheim, **2000**; p 403. (e) Denmark, S. E.; Fu, J. *Chem. Rev.* **2003**, *103*, 2763. (f) Lachance H.; Hall, D. G. *Org. React.* **2008**, *73*, 1.

7. (a) Herold, T.; Hoffmann, R. W. *Angew. Chem. Int. Ed.* **1978**, *17*, 768. (b) Hoffmann, R. W.; Herold, T. *Chem. Ber.* **1981**, *114*, 375.

8. (a) Brown, H. C.; Jadhav, P. K. *J. Am. Chem. Soc.* **1983**, *105*, 2092. (b) Jadhav, P. K.; Bhat, K. S.; Perumal, P. T.; Brown, H. C. *J. Org. Chem.* **1986**, *51*, 432. (c) Brown, H. C.; Bhat, K. S. *J. Am. Chem. Soc.* **1986**, *108*, 5919. (d) Brown, H. C.; Jadhav, P. K.; Bhat, K. S. *J. Am. Chem. Soc.* **1988**, *110*, 1535. (e) Brown, H. C.; Bhat, K. S.; Randad, R. S. *J. Org. Chem.* **1989**, *54*, 1570. (f) Brown, H. C.; Randad, R. S.; Bhat, K. S.; Zaidlewicz, M.; Racherla, U. S. *J. Am. Chem. Soc.* **1990**, *112*, 2389.

9. (a) Roush, W. R.; Walts, A. E.; Hoong, L. K. *J. Am. Chem. Soc.* **1985**, *107*, 8186. (b) Roush, W. R.; Banfi, L. *J. Am. Chem. Soc.* **1988**, *110*, 3979. (c) Roush, W. R.; Hoong, L. K.; Palmer, M. A. J.; Park, J. C. *J. Org. Chem.* **1990**, *55*, 4109. (d) Roush, W. R.; Ando, K.; Powers, D. B.; Palkowitz, A. D.; Halterman, R. L. *J. Am. Chem. Soc.* **1990**, *112*, 6339. (e) Roush, W. R.; Palkowitz, A. D.; Ando, K. *J. Am. Chem. Soc.* **1990**, *112*, 6348.

10. Corey, E. J.; Yu, C. M.; Kim, S. S. *J. Am. Chem. Soc.* **1989**, *111*, 5495.

11. (a) Kinnaird, J. W. A.; Ng, P. Y.; Kubota, K.; Wang, X.; Leighton, J. L. *J. Am. Chem. Soc.* **2002**, *124*, 7920. (b) Kubota, K.; Leighton, J. L. *Angew. Chem. Int. Ed.* **2003**, *42*, 946. (c) Hackman, B. M.; Lombardi, P. J.; Leighton J. L. *Org. Lett.* **2004**, *6*, 4375.

12. (a) Denmark, S. E.; Fu, J. *J. Am. Chem. Soc.* **2001**, *123*, 9488. (b) Denmark, S. E.; Fu, J.; Lawler, M. J. *J. Org. Chem.* **2006**, *71*, 1523,

13. (a) Burgos, C. H.; Canales, E.; Matos, K.; Soderquist, J. A. *J. Am. Chem. Soc.* **2005**, *127*, 8044. (b) Canales, E.; Prasad, K. G.; Soderquist, J. A. *J. Am. Chem. Soc.* **2005**, *127*, 11572,

14. (a) Lachance, H.; Lu, X.; Gravel, M.; Hall, D. G. *J. Am. Chem. Soc.* **2003**, *125*, 10160. (b) Kennedy, J. W. J.; Hall, D. G. *J. Org. Chem.*

2004, *69*, 4412. (c) Rauniyar, V.; Hall, D. G. *J. Am. Chem. Soc.* **2004**, *126*, 4518. (d) Rauniyar, V.; Hall, D. G. *Angew. Chem., Int. Ed.* **2006**, *45*, 2426. (e) Rauniyar, V.; Zhai, H.; Hall, D. G. *J. Am. Chem. Soc.* **2008**, *130*, 8481. (f) Rauniyar, V.; Hall, D. G. *J. Org. Chem.* **2009**, *74*, 4236.

15. (a) Kim, I. S.; Ngai, M.-Y.; Krische, M. J. *J. Am. Chem. Soc.* **2008**, *130*, 6340. (b) Kim, I. S.; Ngai, M.-Y.; Krische, M. J. *J. Am. Chem. Soc.* **2008**, *130*, 14891. (c) Bower, J. F.; Kim, I. S.; Patman, R. L.; Krische, M. J. *Angew. Chem. Int. Ed.* **2009**, *48*, 34. (d) Kim, I. S.; Han, S. B.; Krische, M. J. *J. Am. Chem. Soc.* **2008**, *130*, 6340. (e) Kim, I. S.; Han, S. B.; Krische, M. J. *Chem. Commun.* **2009**, 7278.

16. Davies, G. *J. Chem. Res.* **2008**, *7*, 361.

17. Gao, Y.; Hanson, R. M.; Klunder, J. M.; Ko, S. Y.; Masamune, H.; Sharpless, K. B. *J. Am. Chem. Soc.* **1987**, *109*, 5765.

18. Cardona, W.; Quinones, W.; Robledo, S.; Velez, I. D.; Murga, J.; Garcia-Fortanet, J.; Carda, M.; Cardona, D.; Echeverri, F. *Tetrahedron* **2006**, *62*, 4086.

19. Brown, H. C.; Bhat, K. S.; Randad, R. S. *J. Org. Chem.* **1987**, *52*, 319.

Appendix
Chemical Abstracts Nomenclature; (Registry Number)

(+)-B-Allyldiisopinocampheylborane ((+)-(Ipc)$_2$B(allyl) or (^{l}Ipc)$_2$B(allyl)); (106356-53-0)
(+)-B-Methoxydiisopinocampheylborane ((+)-(Ipc)$_2$BOMe); (99438-28-5)
Allylmagnesium bromide; (1730-25-2)
(*R*)-3-(*t*-Butyldiphenylsilyloxy)-2-methylpropanal: Propanal, 3-[[(1,1-dimethylethyl)diphenylsilyl] oxy]-2-methyl-, (2*R*)-; (112897-04-8)
Hydrogen peroxide; (7722-84-1)
(1*R*,2*R*,3*R*,5*S*)-2,6,6-Trimethylbicyclo[3.1.1]heptan-3-ol; (+)-isopinocampheol; (24041-60-9)
(2*R*,3*R*)-1-(*t*-Butyldiphenylsilyloxy)-2-methylhex-5-en-3-ol
(2*R*,3*S*)-1-(*t*-Butyldiphenylsilyloxy)-2-methylhex-5-en-3-ol
(2*R*,3*R*)-1-(*t*-Butyldiphenylsilyloxy)-2-methyl-5-methyleneoct-7-en-3-ol
(1*S*,2*S*,3*S*,5*R*)-3-hydroperoxy-2,6,6-trimethylbicyclo[3.1.1]heptane; (+)-isopinocampheylperoxide

William R. Roush is Professor of Chemistry, Executive Director of Medicinal Chemistry and Associate Dean of the Kellogg School of Science and Technology at The Scripps Research Institute, Florida. His research interests focus on the total synthesis of natural products and the development of new synthetic methodology. Since moving to Scripps Florida in 2005, his research program has expanded into new areas of chemical biology and medicinal chemistry. Dr. Roush was a member of the *Organic Syntheses* Board of Editors from 1993-2002 and was Editor of Volume 78. He currently serves on the *Organic Syntheses* Board of Directors (2003-present).

Huikai Sun received both his BS and MS degrees in Organic Chemistry from Nankai University in China. He joined the Department of Chemistry at Case Western Reserve University for his Ph. D. degree, which was completed in 2007 under the supervision of Professor Anthony J. Pearson. He is currently a postdoctoral research associate in the laboratory of Professor William R. Roush at Scripps Florida. His research interests focus on the total synthesis of natural products and the development of cysteine protease inhibitors.

Discussion Addendum for:
PALLADIUM CATALYZED CROSS-COUPLING OF (Z)-1-HEPTENYLDIMETHYLSILANOL WITH 4-IODOANISOLE: (Z)-(1-HEPTENYL)-4-METHOXYBENZENE

Prepared by Scott E. Denmark* and Jack Hung-Chang Liu.[1]
Original article: Denmark, S. E.; Wang, Z. *Org. Synth.* **2005**, *81*, 42.

The cross-coupling of organosilanols has emerged as a viable alternative to the classical methods of Suzuki (boronic acids), Stille (stannanes) and Negishi (organozincs).[2] The major developments over the past years have been the significant expansion of the scope of the organosilanol (or silanol precursor) and the introduction of non-fluoride activation of the silanols. Both advances will be summarized here.

Scope of Organosilanol Donor

Over the past ten years, a wide range of organosilanols have been prepared[3] and shown to be competent partners in the fluoride-activated cross coupling with aromatic and olefinic halides (including bromides[4]) and triflates.[5] Most notable has been the extension to pyranylsilanols,[6] cyclic siloxanes generated by intramolecular hydrosilylation,[7] intramolecular silylformylation,[8] and ring-closing metathesis[9] (Figure 1). The latter tandem process (RCM-cross-coupling) was featured in a total synthesis of (+)-brasilenyne (Figure 2).[10] In addition, silylcarbocyclization-formylation[11] allowed for the construction of the pyrrolidine core of isodomoic acids G and H (Figure 3).[12] The final cross-coupling step involved a fluoride mediated process that employed a buffered form of TBAF (octahydrate).

Org. Syn. **2011**, *88*, 102-108
Published on the Web 11/30/2010

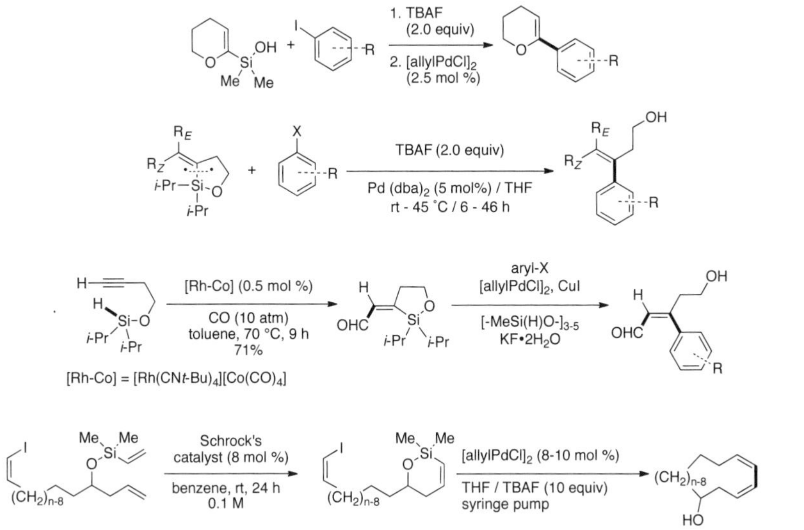

Figure 1. Newer variations of fluoride-promoted, silicon-based-cross coupling reactions.

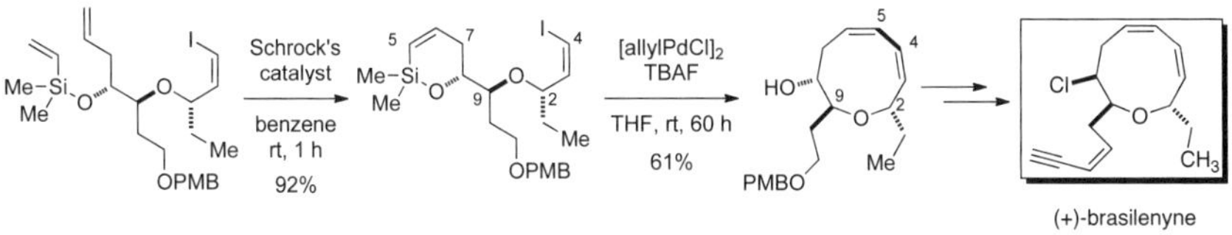

Figure 2. Fluoride-promoted intramolecular alkenyl-alkenyl cross-coupling for the syntheses of brasilenyne.

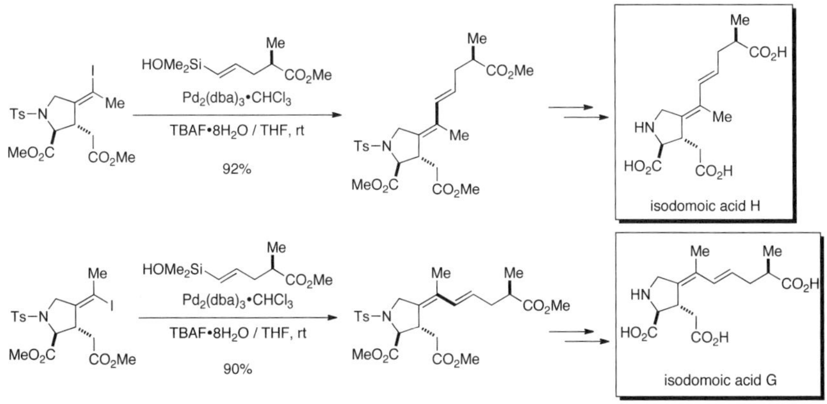

Figure 3. Fluoride-promoted intermolecular alkenyl-alkenyl cross-coupling for the syntheses of isodomoic acids G and H.

Fluoride-Free Cross-Coupling Reactions

By far the most important advance in the past five years has been the discovery of a preparatively useful and mechanistically distinct[13] pathway for cross-coupling of organosilanols that employs various Brønsted bases as activators.[2d] This discovery has allowed for a wider range of coupling partners to be incorporated and also for milder reaction conditions to be employed. For example is it now possible to effect the cross-coupling of simple alkenylsilanols and alkynylsilanols (with KOTMS),[14] arylsilanols (with Cs$_2$CO$_3$),[15] and 2-indolylsilanols (with KOt-Bu),[16] and isoxazolinylsilanols (with KOt-Bu)[17] (Figure 4).

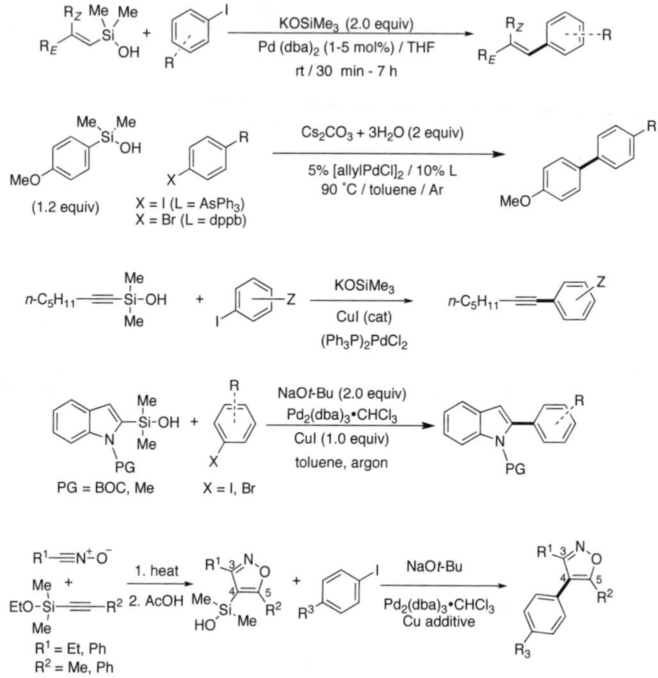

Figure 4. Variations of fluoride-free, cross-coupling reactions.

The ability to couple silanols under fluoride-free conditions has allowed the introduction of the preformed silanolate salts as viable coupling partners. The salts can be easily prepared by deprotonation with NaH or KH and are, in general, stable, free flowing powders. The silanolates couple directly without the need for added bases or activators. Accordingly, heterarylsilanolates derived from indoles, thiophenes, furans[18] as well as a

Org. Syn. **2011**, *88*, 102-108

wide range of aromatic silanolates[19] have been successfully employed. Finally, alkenylsilanolates also undergo high yielding and highly stereospecific cross-coupling with aryl chlorides[20] (Figure 5).

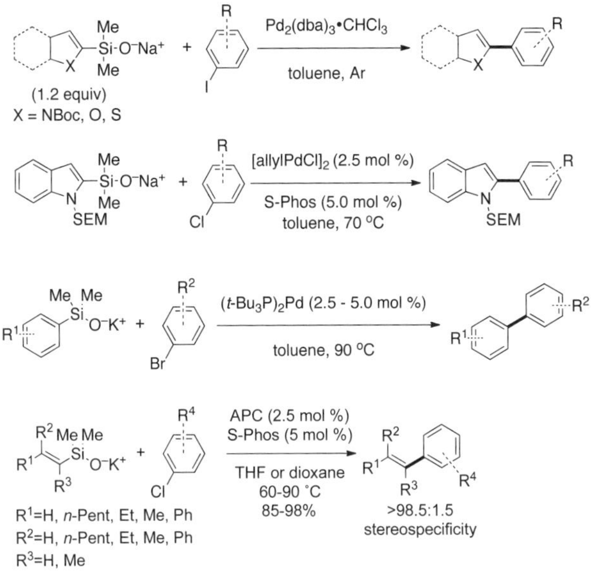

Figure 5. Cross-coupling reactions of preformed silanolate salts.

The demonstration of both fluoride and non-fluoride activation for silicon-based cross-coupling has led to the development of a conjunctive reagent that allows for sequential coupling at separate ends of a 1,4-butadiene unit.[21] This application was featured in the total synthesis of RK-397 (Figure 6).[22]

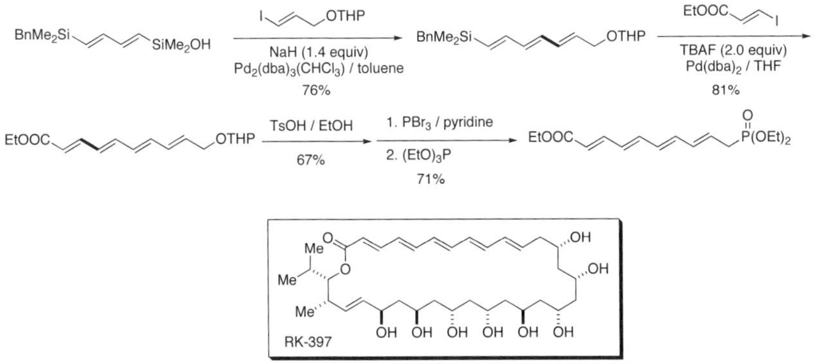

Figure 6. Sequential cross-coupling of a 1,4-bissilyl-1,3-butadiene.

The construction of the key aryl glycosidic bond en route to papulacandin D highlights the synthetic utility of the Brønsted base activation method (Figure 7).[23] The glycal silanol would not withstand activation by fluoride and the resorcinol coupling partner is highly deactivated. Nevertheless, the desired coupling could be achieved by the action of sodium *tert*-butoxide at 50 °C in the presence of $Pd_2(dba)_3 \cdot CHCl_3$. The coupled product contains the entire carbon framework of the sugar fragment of papulacandin D.

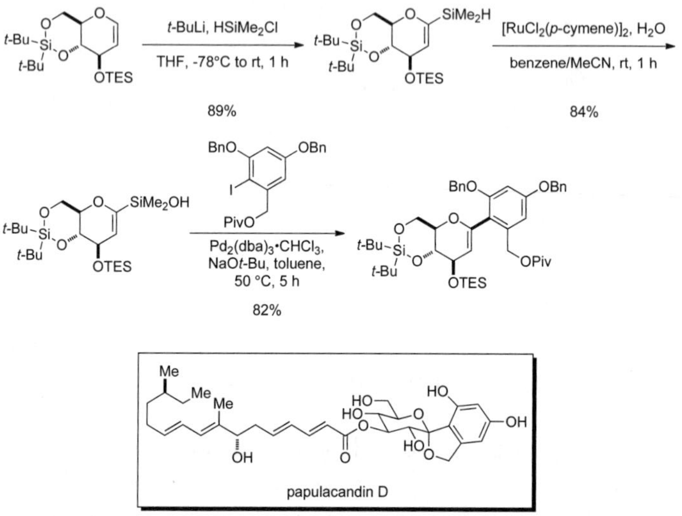

Figure 7. Fluoride-free alkenyl-aryl cross-coupling for the total synthesis of papulacandin D.

1. Department of Chemistry, University of Illinois at Urbana-Champaign, Urbana, IL, 61801.
2. For recent reviews see: (a) Denmark, S. E.: Sweis, R. F. *In Metal-Catalyzed Cross-Coupling Reactions*; De Meijere, A., Diederich, F., Eds.; Wiley-VCH Weinheim, Germany, 2004, 163-216. (b) Denmark, S. E.; Sweis, R. F. *Acc. Chem. Res.* **2002**, *35*, 835-846. (c) Denmark, S. E.; Ober, M. H. *Aldrichimica Acta* **2003**, *36*, 75-85. (d) Denmark, S. E.; Baird, J. D. *Chem. Eur. J.* **2006**, *12*, 4954-4963. (e) S. E. Denmark, C. S. Regens, *Acc. Chem. Res.* **2008**, *41*, 1486-1499. (f) S. E. Denmark, C. S. Regens, *Acc. Chem. Res.* **2008**, *41*, 1486-1499. (g) S. E. Denmark, *J. Org. Chem.* **2009**, *74*, 2915-2927. (h) S. E. Denmark, J. H.-C. Liu,

Angew. Chem. Int. Ed. **2010**, *49*, 2978-2986. (i) S. E. Denmark, J. H.-C. Liu, *Isr. J. Chem.* accepted.

3. Denmark, S. E.; Kallemeyn, J. M. *Org. Lett.* **2003**, *5,* 3483-3486.
4. Denmark, S. E.; Butler, C. R. *Org. Lett.* **2006**, *8*, 63-66.
5. Denmark, S. E.; Sweis, R. F. *Org. Lett.* **2002**, *4*, 3771-3774.
6. Denmark, S. E.; Neuville, L. *Org. Lett.* **2000**, *2*, 3221-3224.
7. (a) Denmark, S. E.; Pan, W. *Org. Lett.* **2001**, *3*, 61-64. (b) Denmark, S. E.; Pan, W. *Org. Lett.* **2002**, *4*, 4163-4166. (c) Denmark, S. E.; Pan, W. *Org. Lett.* **2003**, *5*, 1119-1122.
8. Denmark, S. E.; Kobayashi, T. *J. Org. Chem.* **2003**, *68*, 5153-5159.
9. (a) Denmark, S. E.; Yang, S. M. *Org. Lett.* **2001**, *3*, 1749-1752. (b) Denmark, S. E.; Yang, S. M. *J. Am. Chem. Soc.* **2002**, *124*, 2102-2103. (c) Denmark, S. E.; Yang, S. M. *Tetrahedron* **2004**, *60,* 9695-9708.
10. (a) Denmark, S. E.; Yang, S. M. *J. Am. Chem. Soc.* **2002**, *124*, 15196-15197. (b) Denmark, S. E.; Yang, S. M. *J. Am. Chem. Soc.* **2004**, *126*, 12432-12440. (c) Denmark, S. E.; Yang, S. M. In *Strategies and Tactics in Organic Synthesis*; Harmata, M. A., Ed.; Elsevier: Amsterdam, 2005, Vol. 6; Chapt. 4.
11. Denmark, S. E.; Liu, J. H.-C. *J. Am. Chem. Soc.* **2007**, *129*, 3737-3744.
12. (a) Denmark, S. E.; Liu, J. H.-C.; Muhuhi, J. M. *J. Am. Chem. Soc.* **2009**, *131,* 14188-14189. (b) Denmark, S. E.; Liu, J. H.-C.; Muhuhi, J. M. *J. Org. Chem.* **2010**, *75,* in press (doi: .
13. (a) Denmark, S. E; Sweis, R. F. *J. Am. Chem. Soc.* **2004**, *126*, 4876-4882. (b) Denmark, S. E.; Smith, R. C. *J. Am. Chem. Soc.* **2010**, *132*, 1243-1245.
14. (a) Denmark, S. E; Sweis, R. F. *J. Am. Chem. Soc.* **2001**, *123*, 6439-6440. (b) Denmark, S. E.; Tymonko, S. A. *J. Org. Chem.* **2003**. *68*, 9151-9154.
15. (a) Denmark, S. E.; Ober, M. H. *Org. Lett.* **2003**, *5*, 1357-1360. (b) Denmark, S. E.; Ober, M. H. *Adv. Synth. Catal.* **2004**, *346*, 1703-1714.
16. Denmark, S. E.; Baird, J. D. *Org. Lett.* **2004**, *6,* 3649-3652.
17. Denmark, S. E.; Kallemeyn, J. M. *J. Org. Chem.* **2005**, *70*, 2839-2842
18. (a) Denmark, S. E.; Baird, J. D. *Org. Lett.* **2006**, *8*, 793-795. (b) Denmark, S. E.; Baird, J. D.; Regens, C. S. *J. Org. Chem.* **2008**, *73*, 1440-1455. (c) Denmark, S. E.; Baird, J. D. *Tetrahedron* **2009**, *65*, 3120-3129.
19. Denmark, S. E.; Smith, R. C.; Chang, T. W.-t. Muhuhi, J. M. *J. Am. Chem. Soc.* **2009**, *131*, 3104-3118.
20. Denmark, S. E.; Kallemeyn, J. M. *J. Am. Chem. Soc.* **2006**, *128*, 15958-15959.
21. Denmark, S. E.; Tymonko, S. A. *J. Am. Chem. Soc.* **2005**, *127*, 8004-8005.

22. (a) Denmark, S. E.; Fujimori, S. *J. Am. Chem. Soc.* **2005**, *127*, 8971-8973. (b) Denmark, S. E.; Fujimori, S. In *Strategies and Tactics in Organic Synthesis Vol. 7*; Harmata, M. A., Ed. Elsevier: Amsterdam, 2007, Vol. 7; Chapt. 1.

23. (a) Denmark, S. E.; Regens, C. S.; Kobayashi, T. *J. Am. Chem. Soc.* **2007**, *129*, 2774-2776. (b) Denmark, S. E.; Kobayashi, T.; Regens, C. S. *Tetrahedron* **2010**, *66*, 4745-4759.

Scott E. Denmark was born in Lynbrook, New York on 17 June 1953. He obtained an S.B. degree from MIT in 1975 and his D.Sc.Tech. (with Albert Eschenmoser) from the ETH Zürich in 1980. That same year he began his career at the University of Illinois. He was promoted to associate professor in 1986, to full professor in 1987 and since 1991 he has been the Reynold C. Fuson Professor of Chemistry. His research interests include the invention of new synthetic reactions, exploratory organoelement chemistry and the origin of stereocontrol in fundamental carbon-carbon bond forming processes.

Jack Hung-Chang Liu was born in Taipei, Taiwan in 1979. He obtained a Hon. B. Sc. Degree at University of Toronto in 2002 (Working with Robert A. Batey and Mark Lautens). He then joined the research group of Scott E. Denmark at University of Illinois at Urbana-Champaign, focusing on the development and application of silicon-based cross-coupling. After obtaining his Ph. D. degree in 2009, he moved to University of California, Berkeley for post-doctoral research under F. Dean Toste.

THE PREPARATION OF CYCLOHEPT-4-ENONES BY RHODIUM-CATALYZED INTERMOLECULAR [5+2] CYCLOADDITION

Submitted by Paul A. Wender,[1] Adam B. Lesser,[1] and Lauren E. Sirois.[1]
Checked by Kay M. Brummond and Joshua M. Osbourn.

1. Procedure[2]

2-[(5-Oxo-1-cyclohepten-1-yl)methyl]-1H-isoindole-1,3(2H)-dione.
An oven-dried (>150 °C for 12 h), three-necked, 200-mL, round-bottomed flask, reflux condenser, thermometer adapter, thermometer, rubber septum, and Teflon-coated magnetic stir bar are assembled while hot, and the apparatus is cooled under a stream of nitrogen venting to a manifold (20 min). The septum is removed and the flask is charged with $[Rh(CO)_2Cl]_2$ (55.0 mg, 0.141 mmol, 0.005 equiv) (Note 1) and *N*-propargyl phthalimide (5.99 g, 32.3 mmol, 1.15 equiv) (Note 2). The rubber septum is replaced and the apparatus is gently re-flushed (5 min) with a stream of nitrogen and then kept under a positive nitrogen pressure using a mineral oil bubbler attached to the outlet. 1,2-Dichloroethane (52 mL) (Note 3) is added via syringe, and magnetic stirring is started. 1-(2-Methoxyethoxy)-1-vinylcyclopropane (VCP, 4.3 mL, 4.00 g, 28.1 mmol, 1.00 equiv) (Note 4) is added in a single portion via syringe (ca. 1 min later) (Note 5). At this time, the thermometer is immersed in the amber-colored heterogeneous reaction mixture and reads 25 °C. Under a positive pressure of nitrogen, the rubber septum is replaced by a Teflon stopper and the reaction vessel is lowered into a preheated oil bath (external bath temperature 90–100 °C). Within five min after the start of heating, the reaction mixture becomes golden yellow and homogeneous and reaches an internal reflux temperature of 82–84 °C. The progress of the reaction is followed by TLC analysis (Note 6) until consumption of the VCP is complete (25 min, during which time the homogeneous reaction solution

darkens in color from golden yellow to orange to a deep red-brown).

At this time the reaction vessel is removed from the oil bath, and stirring of the mixture is continued at ambient temperature for 90 min until the solution cools to an internal temperature of 25–30 °C. A 0.1 N HCl/EtOH solution (6.0 mL) (Note 7) is added in a single portion via syringe, and the red-brown homogeneous reaction solution is stirred at ambient temperature (30 min) until hydrolysis of the initial [5+2] cycloadduct **1** to ketone **2** is complete (as determined by TLC analysis) (Note 6). The solution is then transferred to a 1-L, round-bottomed flask. Two portions (75 mL each) of diethyl ether are used to rinse the reaction vessel and are combined with the transferred reaction solution (Note 8). Silica gel (30 g) (Note 9) is added, and the resulting slurry is concentrated in vacuo by rotary evaporation (Note 10) until the apricot-colored dry silica flows freely in the flask (Note 11).

The dry silica is loaded onto a chromatography column (pre-packed with a slurry of 260 g of silica gel in a 40% ethyl acetate/hexanes mixture), and the product is eluted using the same solvent mixture (Note 12). Collection of 25 mL fractions begins immediately, and product-containing fractions (as determined by TLC analysis) (Note 6) are combined in a 1-L round-bottomed flask and concentrated in vacuo using a rotary evaporator (37 °C bath, 10 mm Hg) to give a viscous, pale yellow oil. This oil is then transferred to a 500-mL, round-bottomed flask using successive rinses of dichloromethane (100 mL total volume). Evaporation of solvent (rotary evaporator followed by 12 h under high vacuum) (Notes 10 and 13) provides the purified [5+2] cycloadduct **2** as an off-white, powdery solid, mp = 71–73 °C (6.84 g, 90% yield) (Notes 14 and 15). The submitters report 95-96% yield (7.2-7.3 g) of **2** (mp = 71-72 °C) using freshly prepared 1-(2-methoxyethoxy)-1-vinylcyclopropane.

2. Notes

1. Chlorodicarbonylrhodium(I) dimer was purchased in 500 mg batches from Strem Chemicals, Inc. (product 45-0450), and stored at –20 °C in a vacuum sealed desiccator. The deep red crystalline complex was used as received.

2. *N*-Propargyl phthalimide (97%) was purchased from Sigma-Aldrich (product 696072) and used as received (white crystalline solid).

3. 1,2-Dichloroethane (99.8%) was purchased from Sigma-Aldrich

(product 34872) and distilled over calcium hydride under a nitrogen atmosphere prior to use.

4. 1-(2-Methoxyethoxy)-1-vinylcyclopropane was purchased in 1-gram ampules from Sigma-Aldrich (product 666246). For larger-scale applications, it can be prepared according to reference 2. Storage at $-20\ °C$ is recommended (colorless liquid).

5. For [5+2] reactions in general: when the alkyne is a liquid at ambient temperature, it is often preferable to add it to the reaction mixture last, in a single portion via syringe immediately after addition of the VCP. While not pertinent to the specific reaction conditions described above, when $[Rh(CO)_2Cl]_2$ is to be used for an intermolecular [5+2] cycloaddition at a higher catalyst loading (e.g., 5 mol %), it is advisable to purge the solution containing the catalyst and VCP (via bubbling of nitrogen from an immersed needle for ca. 10 min) prior to addition of the alkyne. This minimizes background rhodium-catalyzed formation of [5+2+1] cycloadducts[24] from any carbon monoxide liberated upon interaction of the VCP with $[Rh(CO)_2Cl]_2$ (the presumed liberation of CO gas is sometimes noticeable in the form of gentle bubbling).

6. Thin-layer chromatography analysis was performed on glass-backed, silica-coated plates purchased from EMD Chemicals, Inc. (silica gel 60 F_{254}, product 5715-7). Plates were eluted with a 50% EtOAc/hexanes mixture and visualized using short wave ultraviolet light (254 nm) and p-anisaldehyde stain (followed by gentle heating). 1-(2-Methoxyethoxy)-1-vinylcyclopropane has a $R_f = 0.64$ (purple), N-propargyl phthalimide has a $R_f = 0.53$ (UV-active), the [5+2] enol-ether intermediate 1 has a $R_f = 0.45$ (UV-active, blue), and the [5+2] ketone product 2 has $R_f = 0.36$ (UV-active, blue).

7. A 0.1 N hydrochloric acid solution was prepared by dilution of 12.1 N "concentrated" HCl (Fisher Scientific) with a pre-mixed solution of ethanol (95%, 190 proof, Fisher Scientific) and distilled water (98:2, EtOH:H$_2$O, vol:vol).

8. Unless otherwise noted, all solvents were purchased from Fisher Scientific and used as received.

9. Silica gel (40-63 μm, 60 Å) was purchased from Sorbent Technologies (product 30930M-25) and used as received.

10. The vacuum for the rotary evaporator was established using a water aspirator (ca. 10 mmHg). The water bath was slowly heated from 25 to 45 °C.

11. If desired for other [5+2] reactions, as an alternative to adsorption of the crude product on silica gel, a concentrated solution can be loaded onto the packed column for flash chromatography conducted as described in reference 3. When direct adsorption is not performed, it is preferable to filter the post-quench reaction solution through a short pad of silica gel (eluting with Et_2O or EtOAc) in order to prevent decomposition of potentially acid-sensitive products upon concentration for column loading.

12. The column is 70 mm wide and the height of the silica is 7 inches inside the column. After the silica containing the adsorbed crude product is poured over the pre-packed column bed, the 1-L vessel is rinsed with enough eluent to gently cover the additional silica. Sand and more eluent are added and forced-flow flash chromatography proceeds normally. The [5+2] product **2** is detected in fractions 56-82.

13. High vacuum is measured at <0.1 mm Hg via a digital manometer. The submitters state that if product **2** does not solidify upon initial exposure to high vacuum (up to 10 min), the viscous oil can be suspended in pentane (100 mL) and the flask immersed in a sonication bath for 5 min, followed by repeated rotary evaporation and static high vacuum to remove any trapped higher-boiling solvents.

14. Analytical data for product **2**: ^1H NMR (500 MHz, $CDCl_3$) δ: 2.39–2.34 (m, 4 H), 2.65–2.59 (m, 4 H), 4.23 (s, 2 H), 5.81 (t, $J = 5.5$ Hz, 1 H), 7.77–7.73 (m, 2 H), 7.88–7.85 (m, 2 H); ^{13}C NMR (125 MHz, $CDCl_3$) δ: 23.3, 25.8, 41.5, 42.0, 44.7, 123.3, 126.7, 131.8, 134.0, 135.1, 168.0, 212.6; IR (film, NaCl plate): 3467, 2915, 2848, 1770, 1708, 1612 cm^{-1}; MS (ESI+) m/z (relative intensity): 287 ($[M+NH_4]^+$, 100%), 270 ($[M+H]^+$, 58%); Exact mass (ESI+) $[M+H]^+$ calcd. for $C_{16}H_{16}NO_3$: 270.1130. Found: 270.1118; Anal. calcd. for $C_{16}H_{15}NO_3$: C, 71.36; H, 5.61; N, 5.20. Found: C, 71.24; H, 5.67; N, 5.16.

15. The submitters report >99% purity by gas chromatography: Agilent 7890A GC / 5975C inert MSD. Column: Agilent HP-5MS (part no: 19091S-433I), length: 30 m, I.D.: 0.25 mm, film: 0.25 μm, injector temperature: 280 °C, split ratio: 1/100. Temperature program: 35 °C for 3.75 min → 320 °C, 20 °C/min, then 320 °C for 7 min. The product **2** has a retention time of 16.65 min (injected as a solution in EtOAc).

Safety and Waste Disposal Information

All hazardous materials should be handled and disposed of in accordance with "Prudent Practices in the Laboratory"; National Academy Press; Washington, DC, 1995.

3. Discussion

The design and discovery of new reactions are of singular importance in efforts to synthesize molecules in a step economical and green, if not ideal, fashion.[4] New reactions provide new ways to think about bond construction, thereby creating new strategies and process options. While cycloadditions, such as the Diels-Alder [4+2] reaction,[5] have historically played a major role in organic synthesis, the introduction of new cycloadditions for medium-sized ring synthesis has only recently received attention. Metal-mediated and metal-catalyzed cycloadditions have figured prominently in such efforts.[6] The metal-catalyzed [5+2] cycloaddition of vinylcyclopropanes (VCPs) and π-systems was first introduced in 1995. It is a homolog of the Diels-Alder [4+2] reaction, in which a 5-carbon, 4-electron VCP conceptually replaces a 4-carbon, 4-electron diene.[7] Since the initial report involving tethered VCPs and alkynes,[7] the rhodium(I)-catalyzed intramolecular process has proven effective with other two-carbon π-components, such as alkenes[8] and allenes.[9] The rhodium-catalyzed intermolecular process has likewise been reported between VCPs and alkynes[10] or allenes.[11] While 1-(2-methoxyethoxy)-1-vinylcyclopropane is a convenient cycloaddition partner due to its synthetic accessibility and commercial availability, the intermolecular [5+2] reaction can also be conducted readily with 1-siloxy-VCPs or 1-alkyl-VCPs.[10]

Further development of the intramolecular [5+2] cycloaddition has included the introduction of new catalysts, some featuring alternative metals (namely, Ru, Fe, and Ni, although Rh remains the most effective metal for many cases).[12] Ligands have also been varied. A water-soluble catalyst has been introduced for conducting the cycloaddition in water or aqueous mixtures.[13] Chiral catalysts have also been reported that provide, in many cases, excellent enantioselectivity.[14] The [5+2] reaction has been reported to occur in the temperature range of −23 to 110 °C (depending on the substrates and catalyst employed); some of the most active catalysts effect the [5+2] reactions of VCPs and alkynes in minutes at room temperature and in high yields.[15] The functional group tolerance of the metal-catalyzed [5+2] cycloaddition is generally excellent, as evident from the representative substrate scope detailed in Table 1 (for 1-(2-methoxyethoxy)-1-vinylcyclopropane and various alkynes, $[Rh(CO)_2Cl]_2$ as the catalyst).[2] Rhodium-catalyzed [5+2] cycloadditions are generally most effective in dichloromethane (DCM), 1,2-dichloroethane (DCE), and 2,2,2-trifluoroethanol (TFE). However, the reaction can also be conducted in non-halogenated solvents, such as ethers (tetrahydrofuran and 2-methyl-THF, for example), acetone, and toluene, depending on the catalyst. Reaction concentrations of up to 0.5-1.0 M have been reported, with no or minimal loss in product yield in many cases.

The mechanism of the rhodium-catalyzed process has been investigated computationally, providing a theoretical foundation for understanding differences in rate for various VCPs and π-components.[16] The regioselectivity of both intra- and intermolecular [5+2] reactions has also been explored, along with some aspects of diastereoselectivity.[17] Partly as a result of the studies and improvements referenced herein, the [5+2] reaction has found increasing use in organic synthesis[18] as well as in the generation of small-molecule libraries.[19] The development of serial/tandem/cascade/domino transformations is one focus of ongoing efforts to achieve step economy in synthesis.[20] Examples of cascade catalysis or serialized processes based on the [5+2] reaction include single-flask [5+2]/Nazarov cyclizations, tandem [5+2]/[4+2] cycloadditions, and tandem allylic substitutions/[5+2] cycloadditions.[21] Significantly, the metal-catalyzed [5+2] reaction has also provided inspiration for the introduction of other new reactions, such as the [6+2] cycloaddition of vinylcyclobutanones and π-systems,[22] the [5+2] cycloaddition of allenylcyclopropanes and π-systems,[23] and higher-order, multicomponent cycloadditions, such as [5+2+1] and [5+2+1+1] reactions.[24]

114

Table 1.[2] Representative Intermolecular [5+2] Cycloadditions of 1-(2-Methoxyethoxy)-1-vinylcyclopropane and Alkynes using [Rh(CO)$_2$Cl]$_2$[a]

entry	alkyne	time	yield[b]	product
1		2 h	75%	
2	═─CO$_2$Me	10 min	84%	
3	OMe	15 min	92%	
4	OH	25 min	82%	
5	NHTs	15 min	87%	
6	CO$_2$H ₍₎₃	1.5 h	87%	
7	OH	12 min	89%	
8		7 h[c]	85%	
9	EtO$_2$C─═─CO$_2$Et	1 h	96%	
10	─═─CO$_2$Me	2 h	81%	

[a] Reaction conditions: 1-(2-methoxyethoxy)-1-vinylcyclopropane (1 mmol), alkyne (1.2-1.3 mmol), [Rh(CO)$_2$Cl]$_2$ (0.005 mmol), DCE (0.5 M concentration with respect to VCP), 80 °C; 1% HCl/MeOH, room temp. [b] Isolated yield of purified product. [c] Reaction conducted at room temp.

1. Department of Chemistry, Stanford University, Stanford, CA 94306-5080. E-mail: wenderp@stanford.edu. This research was supported by the NSF (Grant CHE-0450638 and a graduate fellowship to L.E.S.). A.B.L. thanks Amgen for financial support (graduate fellowship).
2. A related, less detailed procedure has been published: Wender, P. A.; Dyckman, A. J.; Husfeld, C. O.; Scanio, M. J. C. *Org. Lett.* **2000**, *2*, 1609-1611. The submitters would like to acknowledge these authors for preliminary exploration of large-scale [5+2] cycloadditions.
3. Still, W. C.; Kahn, M.; Mitra, A. *J. Org. Chem.* **1978**, *43*, 2923-2925.
4. (a) Wender, P. A.; Miller, B. L. In *Organic Synthesis: Theory and Applications*; Hudlicky, T., Ed.; JAI Press, Inc.: Greenwich, CT, 1993; Vol. 2, 27-66. (b) Wender, P. A.; Handy, S. T.; Wright, D. L. *Chem. Ind.* **1997**, 765-769. (c) Wender, P. A.; Verma, V. V.; Paxton, T. J.; Pillow, T. H. *Acc. Chem. Res.* **2008**, *41*, 40-49. (d) Wender, P. A.; Miller, B. L. *Nature* **2009**, *460*, 197-201.
5. Review of the Diels-Alder in synthesis: Nicolaou, K. C.; Snyder, S. A.; Montagnon, T.; Vassilikogiannakis, G. *Angew. Chem., Int. Ed.* **2002**, *41*, 1668-1698 and references therein.
6. Strategies for seven-membered ring synthesis: (a) Battiste, M. A.; Pelphrey, P. M.; Wright, D. L. *Chem. Eur. J.* **2006**, *12*, 3438-3447. (b) Butenschön, H. *Angew. Chem., Int. Ed.* **2008**, *47*, 5287-5290. Metal-catalyzed cycloadditions: (c) Lautens, M.; Klute, W.; Tam, W. *Chem. Rev.* **1996**, *96*, 49-92. (d) Yet, L. *Chem. Rev.* **2000**, *100*, 2963-3007. (e) Wender, P. A.; Croatt, M. P.; Deschamps, N. M. in *Comprehensive Organometallic Chemistry III*; Vol. 10 (Eds: R. H. Crabtree, D. M. P. Mingos), Elsevier, Oxford, **2007**, pp. 603-648.
7. First report: (a) Wender, P. A.; Takahashi, H.; Witulski, B. *J. Am. Chem. Soc.* **1995**, *117*, 4720-4721. Review of rhodium(I)-catalyzed [5+2] cycloadditions: (b) Wender, P. A.; Gamber, G. G.; Williams, T. J. In *Modern Rhodium-Catalyzed Organic Reactions*; Evans, P. A., Ed.; Wiley-VCH: Weinheim, 2005; pp 263.
8. Early report using alkenes: Wender, P. A.; Husfeld, C. O.; Langkopf, E.; Love, J. A. *J. Am. Chem. Soc.* **1998**, *120*, 1940-1941.
9. Wender, P. A.; Glorius, F.; Husfeld, C. O.; Langkopf, E.; Love, J. A. *J. Am. Chem. Soc.* **1999**, *121*, 5348-5349.
10. (a) Wender, P. A.; Rieck, H.; Fuji, M. *J. Am. Chem. Soc.* **1998**, *120*, 10976-10977. (b) Reference 2. (c) Wender, P. A.; Barzilay, C. M.; Dyckman, A. J. *J. Am. Chem. Soc.* **2001**, *123*, 179-180.

11. Wegner, H. A.; de Meijere, A.; Wender, P. A. *J. Am. Chem. Soc.* **2005**, *127*, 6530-6531.

12. Other representative catalysts. Rh: (a) Wender, P. A.; Sperandio, D. *J. Org. Chem.* **1998**, *63*, 4164-4165. (b) Gilbertson, S. R.; Hoge, G. S. *Tetrahedron Lett.* **1998**, *39*, 2075-2078. (c) Wang, B.; Cao, P.; Zhang, X. *Tetrahedron Lett.* **2000**, *41*, 8041-8044. (d) Wender, P. A.; Williams, T. J. *Angew. Chem., Int. Ed.* **2002**, *41*, 4550-4553. (e) Lee, S. I.; Park, S. Y.; Park, J. H.; Jung, I. G.; Choi, S. Y.; Chung, Y. K.; Lee, B. Y. *J. Org. Chem.* **2006**, *71*, 91-96. (f) Saito, A.; Ono, T.; Hanzawa, Y. *J. Org. Chem.* **2006**, *71*, 6437-6443. (g) Gómez, F. J.; Kamber, N. E.; Deschamps, N. M.; Cole, A. P.; Wender, P. A.; Waymouth, R. M. *Organometallics* **2007**, *26*, 4541-4545. Ru: (h) Trost, B. M.; Toste, F. D.; Shen, H. *J. Am. Chem. Soc.* **2000**, *122*, 2379-2380. Ni: (i) Zuo, G.; Louie, J. *J. Am. Chem. Soc.* **2005**, *127*, 5798-5799. Fe: (j) Fürstner, A.; Majima, K.; Martín, R.; Krause, H.; Kattnig, E.; Goddard, R.; Lehmann, C. W. *J. Am. Chem. Soc.* **2008**, *130*, 1992-2004.

13. Wender, P. A.; Love, J. A.; Williams, T. J. *Synlett* **2003**, 1295-1298.

14. (a) Wender, P. A.; Haustedt, L. O.; Lim, J.; Love, J. A.; Williams, T. J.; Yoon, J.-Y. *J. Am. Chem. Soc.* **2006**, *128*, 6302-6303. (b) Shintani, R.; Nakatsu, H.; Takatsu, K.; Hayashi, T. *Chem. Eur. J.* **2009**, *15*, 8692-8694.

15. For the most recent reports, see: Wender, P. A.; Sirois, L. E.; Stemmler, R. T.; Williams, T. J. *Org. Lett.* **2010**, *12*, 1604-1607, and references therein.

16. (a) Yu, Z.-X.; Wender, P. A.; Houk, K. N. *J. Am. Chem. Soc.* **2004**, *126*, 9154-9155. (b) Yu, Z.-X.; Cheong, P. H.; Liu, P.; Legault, C. Y.; Wender, P. A.; Houk, K. N. *J. Am. Chem. Soc.* **2008**, *130*, 2378-2379. (c) Liu, P.; Cheong, P. H.; Yu, Z.-X.; Wender, P. A.; Houk, K. N. *Angew. Chem., Int. Ed.* **2008**, *47*, 3939-3941.

17. Rh: (a) Wender, P. A.; Dyckman, A. J. *Org. Lett.* **1999**, *1*, 2089-2092. (b) Wender, P. A.; Dyckman, A. J.; Husfeld, C. O.; Kadereit, D.; Love, J. A.; Rieck, H. *J. Am. Chem. Soc.* **1999**, *121*, 10442-10443. (c) Liu, P.; Sirois, L. E.; Cheong, P. H.; Yu, Z.-X.; Hartung, I. V.; Rieck, H.; Wender, P. A.; Houk, K. N. *J. Am. Chem. Soc.* **2010**, *132*, 10127-10135. Ru: (d) Trost, B. M.; Shen, H. C. *Org. Lett.* **2000**, *2*, 2523-2525. (e) Trost, B. M.; Shen, H. C.; Schulz, T.; Koradin, C.; Schirok, H. *Org. Lett.* **2003**, *5*, 4149-4151. (f) Trost, B. M.; Shen, H. C.; Horne, D. B.;

Toste, F. D.; Steinmetz, B. G.; Koradin, C. *Chem. Eur. J.* **2005**, *11*, 2577-2590.

18. For uses of the intramolecular [5+2] reaction in synthesis and related applications, see: (a) Wender, P. A.; Fuji, M.; Husfeld, C. O.; Love, J. A. *Org. Lett.* **1999**, *1*, 137-139. (b) Wender, P. A.; Zhang, L. *Org. Lett.* **2000**, *2*, 2323-2326. (c) Wender, P. A.; Bi, F. C.; Brodney, M. A.; Gosselin, F. *Org. Lett.* **2001**, *3*, 2105-2108. (d) Trost, B. M.; Shen, H. C. *Angew. Chem., Int. Ed.* **2001**, *40*, 2313-2316. (e) Ashfeld, B. L.; Martin, S. F. *Tetrahedron* **2006**, *62*, 10497-10506. (f) Trost, B. M.; Hu, Y.; Horne, D. B. *J. Am. Chem. Soc.* **2007**, *129*, 11781-11790. (g) Trost, B. M.; Waser, J.; Meyer, A. *J. Am. Chem. Soc.* **2008**, *130*, 16424-16434.

19. Kumagai, N.; Muncipinto, G.; Schreiber, S. L. *Angew. Chem., Int. Ed.* **2006**, *45*, 3635-3638.

20. Discussions of serial/tandem/domino/cascade reactions and step economy: (a) Tietze, L. F.; Brasche, G.; Gericke, K. M. *Domino Reactions in Organic Synthesis*. Wiley-VCH: Weinheim, 2006. (b) Müller, T. J. J., Ed. *Top. Organomet. Chem.* **2006**, *19*. (c) Wasilke, J.-C.; Obrey, S. J.; Baker, R. T.; Bazan, G. C. *Chem. Rev.* **2005**, *105*, 1001-1020. (d) Nicolaou, K. C.; Edmonds, D. J.; Bulger, P. G. *Angew. Chem. Int. Ed.* **2006**, *45*, 7134-7186. (e) Reference 4.

21. Serial [5+2]/Nazarov: (a) Wender, P. A.; Stemmler, R. T.; Sirois, L. E. *J. Am. Chem. Soc.* **2010**, *132*, 2532-2533. Serial [5+2]/[4+2]: (b) Wender, P. A.; Gamber, G. G.; Scanio, M. J. C. *Angew. Chem., Int. Ed.* **2001**, *40*, 3895-3897. Tandem allylic substitution/[5+2]: (c) Ashfeld, B. L.; Miller, K. A.; Smith, A. J.; Tran, K.; Martin, S. F. *Org. Lett.* **2005**, *7*, 1661-1663.

22. Wender, P. A.; Correa, A. G.; Sato, Y.; Sun, R. *J. Am. Chem. Soc.* **2000**, *122*, 7815-7816.

23. Inagaki, F.; Sugikubo, K.; Miyashita, Y.; Mukai, C. *Angew. Chem., Int. Ed.* **2010**, *49*, 2206-2210.

24. (a) Wender, P. A.; Gamber, G. G.; Hubbard, R. D.; Zhang, L. *J. Am. Chem. Soc.* **2002**, *124*, 2876-2877. (b) Wender, P. A.; Gamber, G. G.; Hubbard, R. D.; Pham, S. M.; Zhang, L. *J. Am. Chem. Soc.* **2005**, *127*, 2836-2837. (c) Wang, Y.; Wang, J.; Su, J.; Huang, F.; Jiao, L.; Liang, Y.; Yang, D.; Zhang, S.; Wender, P. A.; Yu, Z.-X. *J. Am. Chem. Soc.* **2007**, *129*, 10060-10061. (d) Jiao, L.; Yu, Z.-X. *J. Am. Chem. Soc.* **2008**, *130*, 4421-4430. (e) Fan, X.; Tang, M.-X.; Zhuo, L.-G.; Tu, Y.

Org. Synth. **2011**, *88*, 109-120

Q.; Yu, Z.-X. *Tetrahedron Lett.* **2009**, *50*, 155-157. (f) Fan, X.; Zhuo, L.-G.; Tu, Y. Q.; Yu, Z.-X. *Tetrahedron* **2009**, *65*, 4709-4713.

Appendix
Chemical Abstracts Nomenclature; (Registry Number)

Chlorodicarbonylrhodium dimer: Di-μ-chloro-bis(dicarbonylrhodium):
  Tetracarbonyldi-μ-chlorodirhodium; (14523-22-9)
N-Propargylphthalimide: Phthalimide, *N*-2-propynyl-: 1*H*-Isoindole-
  1,3(2*H*)-dione, 2-(2-propynyl)-; (7223-50-9)
1,2-Dichloroethane: 1,2-Ethylene dichloride; (107-06-2)
1-(2-Methoxyethoxy)-1-vinylcyclopropane: Cyclopropane, 1-ethenyl-1-(2-
  methoxyethoxy)-; (278603-80-8)

Paul A. Wender was born in Pennsylvania. He completed his B.S. degree in Chemistry at Wilkes College in 1969 and his Ph.D. at Yale University in 1973 with Prof. Frederick E. Ziegler. After pursuing postdoctoral studies with Prof. Gilbert Stork at Columbia University as an NIH Fellow, he joined the faculty at Harvard University in 1974 and subsequently moved to Stanford University where he is currently the Bergstrom Professor of Chemistry and Professor (by courtesy) of Chemical and Systems Biology.

Adam Lesser was born in Massachusetts in 1983. He received his B.S. with honors in Chemistry from Trinity College (Hartford, CT) in 2006, where he performed undergraduate research on metallacyclic peptides in the lab of Prof. Timothy P. Curran. A recipient of Amgen and Eli Lilly graduate fellowships, Adam is currently a Ph.D. student in Prof. Wender's lab at Stanford University.

Lauren Sirois is a native of Massachusetts and New Hampshire. She obtained her A.B. in Chemistry in 2004 from Harvard College, where she conducted undergraduate research in C-H functionalization methods under the direction of Prof. M. Christina White. A recent National Science Foundation Fellow, Lauren is currently completing her Ph.D. studies in Prof. Wender's group at Stanford University.

Joshua Osbourn received his B.S. degree in chemistry from West Virginia University in 2007. He then joined the graduate program at the University of Pittsburgh and is working towards a Ph.D. under the direction of Professor Kay M. Brummond. His current research involves the thermal [2 + 2] cycloaddition reactions of allene-ynes.

120

ORGANOCATALYTIC ENANTIOSELECTIVE SYNTHESIS OF BICYCLIC β-LACTONES FROM ALDEHYDE ACIDS VIA NUCLEOPHILE-CATALYZED ALDOL-LACTONIZATION (NCAL)

Submitted by Henry Nguyen, Seongho Oh, Huda Henry-Riyad, Diana Sepulveda, and Daniel Romo.[1]

Checked by Gerri E. Hutson and Viresh H. Rawal.[2]

> *Caution: Care must be exercised in transferring and manipulating the ozonide reaction mixture due to the known instability of ozonides. It is best to perform the ozonide reduction step in the same fume hood and avoid transferring and manipulation of the reaction mixture.*

1. Procedure

A. (1,4-Dioxaspiro[4.5]dec-7-en-8-yloxy)trimethylsilane (1) (Note 1). An oven-dried 500-mL, two-necked, round-bottomed flask is equipped with a large stir bar (Note 2), and one neck is fitted with a 50-mL pressure-equalizing addition funnel, to which is attached a nitrogen inlet, and the second neck is sealed with a rubber septum. The reaction flask is charged with solid 1,4-cyclohexanedione mono-ethylene ketal (25.0 g, 160 mmol, 1.0 equiv), and then *N,N*-dimethylformamide (Note 3) (100 mL) is added to dissolve the solid. Triethylamine (52.0 mL, 368 mmol, 2.3 equiv) is added via syringe to the reaction vessel. Freshly distilled trimethylsilyl chloride (26.7 mL, 208 mmol, 1.3 equiv) is added to the addition funnel and then dispensed dropwise into the reaction vessel over a period of ~15 min. After the addition is complete, the addition funnel is replaced with a condenser, and the reaction mixture is then heated in an oil bath at 80 °C (bath temperature) for 12 h (Note 4). After cooling to ambient temperature (23 °C), the reaction is quenched by addition of cold H_2O (150 mL, cooled in an ice bath for 15 min), and the resulting mixture is diluted with hexanes (300 mL) and then transferred to a 1000-mL separatory funnel. The aqueous layer is removed and back-extracted with hexanes (2 x 100 mL). The combined hexane extracts are washed with saturated aqueous $NaHCO_3$ (1 x 100 mL), saturated aqueous NH_4Cl (2 x 150 mL), and then brine (1 x 150 mL). The organic layer is dried over anhydrous $MgSO_4$ (25 g) for 15 min and then concentrated on a rotary evaporator (25 mmHg, 30 °C). Further concentration under reduced pressure (0.5 mmHg, ambient temperature, 23 °C) delivers 32.6 g (89%) of the silyl enol ether **1** (Note 5) as a dark yellow oil, which is of sufficient purity for use in the subsequent step.

B. 3-(2-(2-Oxoethyl)-1,3-dioxolan-2-yl) propanoic acid (2). To a 1000-mL round-bottomed flask fitted with an egg-shaped magnetic stir bar (0.625 x 1.5 in) is added crude silyl enol ether **1** (32.0 g, 140 mmol, 1.0 equiv) and 500 mL of CH_2Cl_2. The reaction mixture is cooled to –78 °C (acetone/dry ice bath temperature) with stirring for 15 min. Stirring is stopped and ozone is bubbled through the solution with a fritted-glass gas dispersion tube for ~1-2 h until the blue color of ozone persists (Note 6). A stream of nitrogen is then bubbled through the solution until the blue color disappears (~ 1 h). Solid triphenylphosphine (37.1 g, 141 mmol, 1.01 equiv) is added with stirring, and the reaction mixture is allowed to

122

warm gradually to ambient temperature (23 °C) and stirred for ~18 h until the ozonide test provides a negative result (Note 7). The solvent is removed by means of a rotary evaporator (25 mmHg, 30 °C). Residual CH_2Cl_2 is removed by dissolving the crude mixture in diethyl ether (50 mL) and concentrating the solution on a rotary evaporator (25 mmHg, 30 °C). This step is repeated twice, and the resulting yellow slurry is then placed under high vacuum (0.5 mmHg) for 12 h. (Note 8). Diethyl ether (80 mL) is added, the solid/liquid mixture is stirred vigorously for 30 min, and the solid is then filtered off by vacuum filtration using a Büchner funnel fitted with filter paper (Whatmann #1, 70 mm). The solid is rinsed with ice cold ether (2 x15 mL). The ether filtrates are combined and transferred to a separatory funnel (250 mL) and extracted vigorously with water (4 x 80 mL) (Note 9). To a 500-mL Erlenmeyer flask containing the combined aqueous extracts (~250 mL) is added solid NaCl (70 g) with stirring, followed by additional NaCl until the solution is saturated. The aqueous solution is transferred to a 50-mL separatory funnel and extracted with hexanes (2 x 100 mL) to remove unreacted cyclohexane dione monoketal from step A (Note 4). The aqueous solution is then transferred to a clean 1000-mL separatory funnel and extracted vigorously with CH_2Cl_2 (3 x 150 mL). The aqueous solution is saturated again with additional sodium chloride (5 g) and extracted vigorously again with CH_2Cl_2 (3 x 150 mL). The combined CH_2Cl_2 extracts are dried over anhydrous $MgSO_4$ (50 g) for 15 min, filtered using a Büchner funnel equipped with filter paper (Whatman #1, 70 mm), and concentrated by rotary evaporation (25 mmHg, 30 °C) to deliver a light yellow oil. Further concentration under high vacuum (0.5 mmHg, 23 °C) provides 15.1 g (57%) of the aldehyde acid as a light yellow solid (Note 10), which is of sufficient purity for use in the subsequent step without further purification (Notes 11 and 12).

C. N-Methyl-2-chloropyridinium trifluoromethane sulfonate (3). A 1000-mL round-bottomed flask, equipped with an egg-shaped magnetic stir bar (0.625 x 1.5 in) and a rubber septum, is flame dried under a stream of nitrogen and cooled to ambient temperature. Following the addition of CH_2Cl_2 (250 mL) and 2-chloropyridine (13.1 mL, 139 mmol, 1.0 equiv), the flask is cooled in a dry ice/acetone bath (–78 °C, bath temperature) under nitrogen. Methyl trifluoromethanesulfonate (25.0 g, 153 mmol, 1.1 equiv) is then added slowly via syringe over 10 min through the septum down the side of the flask to allow cooling prior to mixing with the bulk solvent. The cold bath is removed and the reaction mixture is allowed to warm slowly to ambient temperature (23 °C) and stirred an additional 12 h, leading to a white precipitate. The reaction mixture is concentrated by rotary evaporation (25 mmHg, 30 °C) to deliver a white solid. Toluene (120 mL) is added to further induce precipitation, and after swirling under nitrogen for ~10 minutes, the toluene/CH_2Cl_2 mixture is removed from the solids via cannula under N_2 pressure. The white solid is dried under reduced pressure at 0.5 mmHg for 12 h at 23 °C to afford 35.9 g (93%) of N-methyl-2-chloropyridinium trifluoromethane sulfonate (3) as a white solid (mp 147–148 °C) (Note 13).

D. O-Trimethylsilylquinidine (4, O-TMS QND).[3] A 250-mL round-bottomed flask, equipped with a mechanical stirrer (PTFE stirrer blade 1.9 x 6 cm) is flame-dried with a stream of nitrogen and left to cool down to ambient temperature for 10 min. To the flask are added quinidine (4.00 g, 12.4 mmol, 1.0 equiv) and CH_2Cl_2 (100 mL) followed by distilled trimethylsilyl chloride (1.9 mL, 15 mmol, 1.2 equiv), which is added slowly via syringe pump over 15 min. The resulting solution is stirred at ambient temperature for 24 h and then transferred to a 500-mL separatory funnel, to which was added CH_2Cl_2 (100 mL) and $NaHCO_3$ (aq) (100 mL). After shaking, the layers are separated, and the aqueous layer is reextracted with CH_2Cl_2 (2 x 50 mL). The combined organic extracts are dried over $MgSO_4$, filtered, and concentrated by rotary evaporation (25 mmHg, 30 °C). The residue is then left to dry under reduced pressure at 0.5 mmHg for 12 h at 23 °C to afford 4.02 g (82%) of O-trimethylsilylquinidine (4) as a clear oil, which was of sufficient purity for use without purification (Note 14).

E. (1'S,5'R)-Spiro[1,3-dioxolane-2,3'-[6]oxabicyclo[3.2.0]heptan]-7'-one (5). N-Methyl-2-chloropyridinium trifluoromethane sulfonate (3) (Note 13) (33.2 g, 120 mmol, 1.5 equiv) is weighed out in a one-necked, oven-dried 1000-mL round-bottomed flask. A 4-cm stir bar is added, and the flask is purged with nitrogen and then charged with dry acetonitrile (240 mL). A solution of the catalyst, O-TMS QND (4) (3.16 g, 7.97 mmol, 0.1 equiv) dissolved in dry acetonitrile (100 mL) at 23 °C is transferred to the reaction flask via cannula with nitrogen pressure, which is followed by addition of freshly distilled diisopropylethylamine (Hünig's base) via syringe (35.4 mL, 199 mmol, 2.5 equiv). The reaction mixture is stirred for 10 min, and then a solution of 3-(2-(2-oxoethyl)-1,3-dioxolan-2-yl) propanoic acid (2) (15.0 g, 79.7 mmol, 1.0 equiv) in dry CH_3CN (60 mL) is added via syringe pump over ~1 h (30 mL syringe, ~1 mL/min, syringe is loaded twice). The reaction mixture changes from yellow to a dark red color

124

during this addition. The reaction mixture is stirred for an additional 18 h at ambient temperature, and the reaction progress is monitored by the disappearance of aldehyde acid **2** and the formation of β-lactone **5** by TLC analysis (R_f 0.29 and 0.52, respectively in 60% EtOAc/hexanes, KMnO$_4$ stain). When the reaction is judged complete, the reaction mixture is concentrated by rotary evaporation and then by high vacuum (0.5 mmHg, 30 °C) for 3 h to provide a dark brown, viscous oil (~ 75 g). The oil is diluted with 25 mL of CH$_2$Cl$_2$ and loaded onto a silica gel plug (188 g, 2.5 g silica gel per gram of crude product, 10-cm diameter flash column) pre-eluted with 50% EtOAc:hexanes, using a long stem funnel to carefully deliver the crude mixture onto the silica gel surface (washing with 5 mL of CH$_2$Cl$_2$ and then 10 mL of 50% EtOAc:hexanes). The product is quickly eluted with 50% EtOAc:hexanes (2000 mL). Forty fractions (~45 mL in 50 mL test tubes) are collected and fractions 7–35 containing the majority of the β-lactone (R_f = 0.52 (60% EtOAc:hexanes; KMnO$_4$)) are combined and concentrated by rotary evaporation (25 mmHg, 30 °C) and then under high vacuum (0.5 mmHg at 23 °C) to deliver 8.3 g of semi-crude β-lactone. Recrystallization (Note 15) from EtOAc:hexanes gives a total of 4.5 g (33%) of bicyclic-β-lactone (Notes 16 and 17). Enantiomeric excess was determined to be 90% ee by chiral HPLC analysis (Note 18).

2. Notes

1. Sources and purities of reagents used in the procedure are as indicated: 1,4-Cyclohexanedione mono-ethylene ketal (Aldrich, 97%), PPh$_3$ (Acros, 99%), 2-chloropyridine (Aldrich, 99%), methyl trifluoromethanesulfonate (Acros, 96%), quinidine (Acros, 99%), NEt$_3$ (Acros, 99%), TMSCl (Acros, 98%), *N,N*-dimethylformamide (Aldrich, 99.9%), acetonitrile (Aldrich, 99.8%), *N, N*-diisopropylethyl amine (Hünig's base, Acros, 98%), MgSO$_4$ (EMD, powder 98%), sodium bicarbonate (EMD, powder), dichloromethane (EMD, 99.8%), NaCl (EMD, crystals, 99%), ammonium chloride (EMD, 99.5%), hexanes (Fisher, 98.5%), diethylether (EMD, 99%), silica gel (Silicycle, 230-400 mesh), KI (EMD, 99%). Triethylamine, trimethylsilyl chloride, and Hünig's base were freshly distilled over CaH$_2$ prior to use. *N,N*-Dimethylformamide, dichloromethane and acetonitrile were dried through activated alumina using a converted MBraun System. Checkers used a solvent drying system manufactured by Innovative Technology, Inc.

2.   A 4-cm football-shaped stir bar is ideal since copious amounts of Et$_3$N•HCl formed during the course of the reaction makes stirring difficult.

3.   It is important to use very dry N,N-dimethylformamide (<150 ppm of H$_2$O) to avoid hydrolysis of TMSCl.

4.   While it is best to have all starting material converted to the silyl enol ether, reaction progress is not easily monitored by TLC analysis due to facile desilylation of the TMS enol ether **1**. Taking an aliquot for NMR analysis is appropriate. Generally, ensuring high quality and dryness of reagents/solvents and using specified reaction times ensured complete reaction. Any residual cyclohexanone is not deleterious to the subsequent ozonolysis and can be removed in the next step during extraction.

5.   The checkers obtained 17.03 g (93%) on a one-half scale run. The submitters reported obtaining 31–33 g (85–90%) of **1**. Data for silyl enol ether **1**: R_f = 0.65 (20% EtOAc/hexanes); IR (thin film) 1670 cm^{-1}; ^{1}H NMR (500 MHz, CDCl$_3$) δ: 0.17 (s, 9 H), 1.78 (t, J = 6.5 Hz, 2 H), 2.16–2.27 (m, 4 H), 3.93–3.98 (m, 4 H), 4.68–4.73 (m, 1 H); ^{13}C (125 MHz, CDCl$_3$) δ: 0.27, 28.5, 31.1, 33.9, 64.3, 100.6, 107.6, 149.8; [M + Na$^+$] calcd for C$_{11}$H$_{20}$O$_3$SiNa$^+$: 251.10739. Found: 251.10743.

6.   A Welbash Ozonizer (total pressure: 7–10 psi; O$_2$/ozone pressure: 3.0–4.5 psi) was used by the submitters and the checkers. The time required for the ozonolysis is dependent upon the O$_2$ pressure and ozone generation efficiency.

7.   The ozonide test is performed by adding 0.1 g of potassium iodide to 1.0 mL of glacial acetic acid and ~1.0 mL of the ozonolysis reaction mixture in a 5 mL round-bottomed flask. A brown mixture indicates the presence of ozonide. Blanks must always be prepared: 0.1 g of KI is added to 1.0 mL of CH$_2$Cl$_2$ and 1.0 mL of glacial acetic acid. The test solution has a very short shelf life and will naturally result in high blank values if stored for any length of time. *Caution: Care must be exercised in transferring and manipulating the ozonide reaction mixture due to the known instability of ozonides. It is preferable to perform the quenching step in the same fume hood and avoid transferring the reaction mixture.* A test for the presence of ozone resulting in a yellow solution, indicative of the absence of peroxides, is required before it is safe to proceed to work up.

8.   The CH$_2$Cl$_2$ must be removed at this step to minimize retention of aldehyde acid **2** in the organic layer during the first stage of the extraction procedure.

9. It is imperative that extractions of the aldehyde acid are done vigorously, with rapid end-to-end shaking, to ensure efficient transfer of the product to the aqueous phase. However, patience is then required to allow for separation of the layers due to emulsion formation. A copper wire loop was used by the submitters to assist in breakdown of the emulsion.

10. The submitters reported 57–65% yield of product **2** on full scale.

11. It is best to use the aldehyde acid soon after drying otherwise it should be stored in the freezer (–5 °C) under N_2 to avoid oxidation to the diacid. Data for aldehyde acid: R_f = 0.29 (60% EtOAc:hexanes); IR (thin film) 1723 cm^{-1}; ^{1}H NMR (500 MHz, CDCl$_3$) δ: 2.11 (t, J = 7.5 Hz, 2 H), 2.45 (t, J = 7.5 Hz, 2 H), 2.69 (d, J = 3.0 Hz, 2 H), 4.03 (s, 4 H), 9.73 (t, J = 3.0 Hz, 1 H); ^{13}C (125 MHz, CDCl$_3$) δ: 28.2, 33.0, 50.6, 65.2, 108.3, 178.9, 199.8; [M - H] calcd for $C_8H_{11}O_5$: 187.0612. Found: 187.0612.

12. A minor product formed during the ozonolysis was identified to be the corresponding diacid[4] formed by over oxidation and this is removed during the extraction procedure.

13. The pyridinium salt is somewhat hygroscopic and moisture sensitive and thus should be weighed out and transferred rapidly. IR (film) 3099, 1619, 1264 cm^{-1}; ^{1}H NMR (500 MHz, CD$_3$CN) δ: 4.29 (s, 3 H), 7.93–7.96 (m, 1 H), 8.12–8.14 (m, 1 H), 8.45–8.49 (m, 1 H), 8.78-8.80 (m, 1 H); ^{13}C NMR (125 MHz, CD$_3$CN) δ: 48.9, 122.3, 126.6, 127.6, 131.2, 148.6, 149.3.

14. Characterization data: IR (thin film) 2939, 1621, 1507 cm^{-1}; This compound is a mixture of two conformers at ambient temperature but low temperature ^{1}H NMR at –40 °C allowed assignment of one major conformer (500 MHz, –40 °C, CDCl$_3$) δ: 0.15 (s, 9 H), 0.75–0.89 (m, 1 H), 1.31–1.53 (m, 2 H), 1.68 (br s, 1 H), 2.11 (t, J = 11.5 Hz, 1 H), 2.15–2.25 (m, 1 H), 2.76–2.98 (m, 4 H), 3.38 (dd, J = 8.0, 12.0 Hz, 1 H), 3.92 (s, 3 H), 5.03 (d, J = 9.0 Hz, 1 H), 5.05 (d, J = 17.5 Hz, 1 H), 5.69 (br s, 1 H), 6.08 (apparent quint, J = 9.0 Hz, 1 H), 7.08 (br s, 1 H), 7.36 (dd, J = 2.5, 9.0 Hz, 1 H), 7.53 (d, J = 4.5 Hz, 1 H), 8.01 (d, J = 9.0 Hz, 1 H), 8.72 (d, J = 4.5 Hz, 1 H); ^{13}C NMR was taken at ambient temperature. ^{13}C NMR (125 MHz, CDCl$_3$) δ: 0.1, 19.8, 26.5, 28.2, 40.3, 49.7, 50.5, 55.6, 60.3, 73.3, 100.4, 114.3, 118.5, 121.5, 126.1, 131.8, 140.8, 144.3, 147.4, 147.6, 157.8.

15. Recrystallization is performed in a 50-mL Erlenmeyer flask by addition of hot EtOAc (20 mL) to the dried, semi-crude β-lactone. A stir bar is added, and the solution is then heated with stirring to ~50 °C on a hot plate to dissolve the solid. Hexanes (~16 mL) is added to the hot solution

with continued heating until the solution turns cloudy. On cooling to ambient temperature over ~1 h, crystals form and the flask is then placed in an ice-bath for an additional 30 min. The crystals are collected by rapid vacuum filtration and washed quickly with an ice-cold mixture of 50% EtOAc:hexanes (~10 mL) to give 4.0–4.2 g (30–31%) of the bicyclic-β-lactone as light yellow crystals. A second crop is collected by removal of ~1/3 of the mother liquor and placing the solution in a freezer (–10 °C) overnight to provide an additional 0.5–0.6 g (4–5%) as darker yellow crystals.

16. The submitters reported 34–35% yield of product 5 on full scale.

17. Higher yields of β-lactone could be achieved by using aldehyde acid purified by flash chromatography; however; a loss of material on the column led to further reduction of the overall yield. Characterization data for (1'S,5'R)-spiro[1,3-dioxolane-2,3'-[6]oxabicyclo[3.2.0]heptan]-7'-one (5): R_f = 0.52 (60% EtOAc/hexanes); mp 103–104 °C; $[\alpha]_D^{23}$ + 63.2 (c 0.95, CHCl$_3$); IR (thin film) 1825 cm^{-1}; ^{1}H NMR (500 MHz, CDCl$_3$) δ: 2.01 (dd, J = 8.5, 14.0 Hz, 1 H), 2.06 (dd, J = 5.0, 15.5 Hz, 1 H), 2.29 (dd, J = 1.0, 14.0 Hz, 1 H), 2.36 (dd, J = 1.5, 15.5 Hz, 1 H), 3.85–3.92 (m, 2 H), 3.97–4.07 (m, 3 H), 4.99 (t, J = 5.0 Hz, 1 H); ^{13}C NMR (125 MHz, CDCl$_3$) δ: 36.6, 39.5, 53.6, 64.8, 65.0, 74.0, 115.3, 170.6; [M + Na$^+$] calcd for C$_8$H$_{10}$O$_4$Na$^+$: 193.0471. Found: 193.0476.

18. Enantiomeric excess was determined to be 90% ee by chiral HPLC using the following conditions: Daicel IA column, 90% hexanes/10% isopropanol, 1 mL/min, 25 °C, Retention times were: 14.5 min (major), 16.6 min (minor). The submitters determined the enantiomeric excess of their product to be 86–88% using the following chiral HPLC conditions: Chiralcel OD, 250 x 4.6 mm (L x I.D.), solvent (isocratic) 85% hexanes, 15% isopropanol, flow rate 1.0 mL/min, wavelength λ = 220 nm. Retention times were: (1'R,5'S)-β-lactone, 18.06 min; (1'S,5'R)-β-lactone 22.96 min. An alternative method was developed previously[6a] using chiral GC with a non-commercially available chiral column: 2,3-di-OAc-6-TBS-CD and showed 92% ee with O-Ac quinidine as chiral promoter.

Safety and Waste Disposal Information

All hazardous materials should be handled and disposed of in accordance with "Prudent Practices in the Laboratory"; National Academy Press; Washington, DC, 1995.

128

3. Discussion

This procedure describes an organocatalytic, enantioselective method for the asymmetric synthesis of bicyclic β-lactones. The intramolecular nucleophile-catalyzed aldol-lactonization (NCAL) process effectively merges catalytic, asymmetric carbocycle synthesis with β-lactone synthesis leading to unique bicyclic β-lactones.[6] The demand for concise synthetic routes to optically active β-lactones continues to grow due to continued development of novel transformations of these heterocycles and a reappraisal of their utility as synthetic intermediates[7] including natural product synthesis,[8] their continued occurrence in natural products,[9] their potent activity as enzyme inhibitors[10] and their utility as activity based probes.[11] β-Lactones can be viewed as "activated aldol products" since they possess the structural features of aldol products, yet they also have inherent reactivity due to ring strain (β-lactones, 22.8 kcal/mole; epoxides, 27.2 kcal/mole, Figure 1).[12]

aldol adduct epoxide β-lactone

Figure 1. Comparison of structure and reactivity of β-lactones to aldol adducts and epoxides.

The aldol motif, the β-hydroxy carbonyl motif derived from polyketide biosynthesis, is common to many natural products; therefore asymmetric methods for their synthesis have been studied extensively. However, β-lactones arguably have greater utility over simple aldol products since they can undergo divergent nucleophilic cleavage processes occurring at either the acyl-oxygen (C_2-O_1) bond or the alkyl-oxygen (C_4-O_1) bond typically driven by release of ring strain. As a result, β-lactones undergo a number of interesting and useful stereospecific transformations making them versatile intermediates for organic synthesis.[7] Despite these facts, general methods for the direct synthesis of β-lactones in optically active form lag far behind those developed for epoxides and aldol adducts. Furthermore, there is much unexploited synthetic potential of β-lactones as chiral synthetic intermediates. Consequently, our group has been engaged in the development of efficient methods for the asymmetric synthesis of β-lactones and their subsequent transformations with applications to natural product synthesis.[13] Building on our initial reports of the NCAL process,[6] we recently developed optimized conditions that renders the process more practical with the recognition of the importance of the counterion of the Mukaiyama reagent.[6c] Several optically active cyclopentanes and cyclohexanes possessing fused β-lactones are now available in good yields and high enantioselectivities by this process (Table 1).

Table 1. Catalytic, Asymmetric Intramolecular NCAL Reactions Leading to Bicyclic β-Lactones

entry	β-lactone	Cmpd. no.	method[a]	% yield	% ee[b]	config.
1		9a	B	82	92	1R, 2S[c]
2[d]		5	C	74	92	1'S, 5'R[d]
3		9c	C	74	91	1R,2S[g]
4		9d	A	45	90	1R,2S[f]
5		ent-9a	A	51	86	1S, 2R[h]
6		10a	C	76	98	1R,2S[g]

[a]Method A: Pyridinium salt **11** was employed in CH_3CN for 108 h (original procedure, see ref. 6a). Method B: Pyridinium salt **3** in CH_3CN for 108 h. Method C: Pyridinium salt **12** was employed in CH_2Cl_2 for 48 h. [b]Enantiomeric excess was determined by chiral GC analysis. [c]Absolute configuration was assigned by reduction to the known diol[14] and comparison of optical rotations. [d]Note that the yield for the NCAL process leading to β–lactone **5** obtained previously (74%, entry 2) is higher than the yield reported in the procedure above. This discrepancy is primarily due to the fact that the aldehyde acid in the former case was purified by flash chromatography prior to the NCAL process (versus extraction) and 3.0 equiv of pyridinium salt **12** was employed rather than only 1.5 equiv as in the current procedure for reasons of practicality (see also Note 17). [e]Absolute configuration assigned by conversion to a known cyclopentene precursor of aristeromycin (ref. 6a). [f]Absolute configuration was assigned by x-ray analysis of a derived amide from ring opening (ref. 6a). [g]Predicted based on analogy to that determined for β-lactones **5** and **9a**. [h]O-Ac-Quinine was used as catalyst.

The utility of the dioxolane bicyclic β-lactone **5** described in this *Organic Syntheses* procedure was demonstrated in the synthesis of cyclopentane diol **13**, a useful intermediate in the synthesis of antiviral carbocyclic nucleosides including (–)-aristeromycin (Scheme 1).[6a]

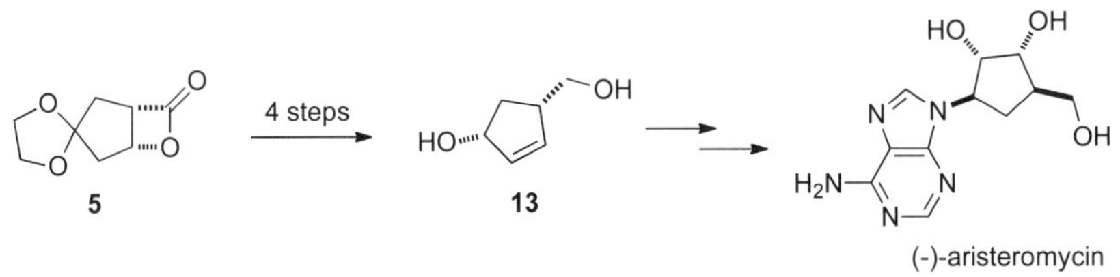

Scheme 1. Utility of β-lactone **5** toward a formal synthesis of (-)-aristeromycin.

Highly diastereoselective NCAL reactions utilizing chiral aldehyde acids **15** and **18** also provide expedient access to functionalized bicyclic-β-lactones **16** and **19**, respectively. Two examples are provided below, and the required aldehyde acid substrates are readily available from ozonolysis of (*R*)-citronellic acid (**14**, Scheme 2, eq 1) and from optically active β-hydroxy acid **18** readily obtained by Noyori reduction of ketoester **17** ultimately leading to the useful silyloxy substituted bicyclic-β-lactone **19** (Scheme 2, eq 2).

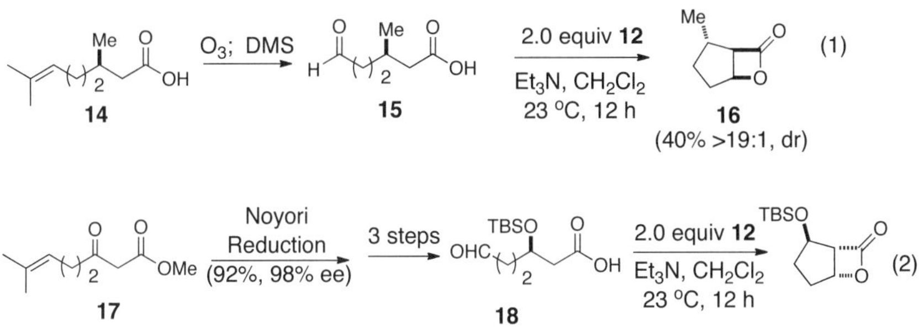

Scheme 2. β-Lactones via diastereoselective NCAL reactions of aldehyde acids.

The utility of the NCAL process was recently expanded to include the use of more tractable ketoacid substrates allowing access to bicyclic and tricyclic-β-lactones (Scheme 3).[15] This process was applied to an enantioselective synthesis of a reduced form of the DNA polymerase inhibitor, plakevulin A, and to a concise, 9-step bioinspired racemic and asymmetric syntheses of salinosporamide A and derivatives.[16] Furthermore,

131

novel transformations[7] including dyotropic rearrangments[17] provide further avenues for exploitation of β-lactones in chemical synthesis.

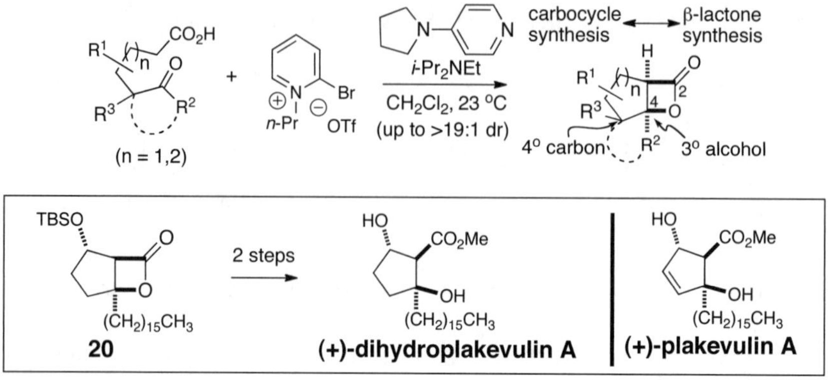

Scheme 3. Extension of the NCAL process to ketoacid substrates leading to bicyclic and tricyclic-β-lactones with application to the synthesis of (+)-dihydroplakevullin.

A number of natural products possess bicyclic-β-lactones within their structures including salinosporamide A, omuralide (proteasome inhibitors),[9b,c] vibralactone (pancreatic lipase inhibitor), and spongiolactone (unknown activity).[18] In addition, cyclopentyl-fused, bicyclic β-lactones can be envisioned as useful intermediates for the synthesis of several natural products including marine cembranoids such as verrillin and rameswaralide.[19]

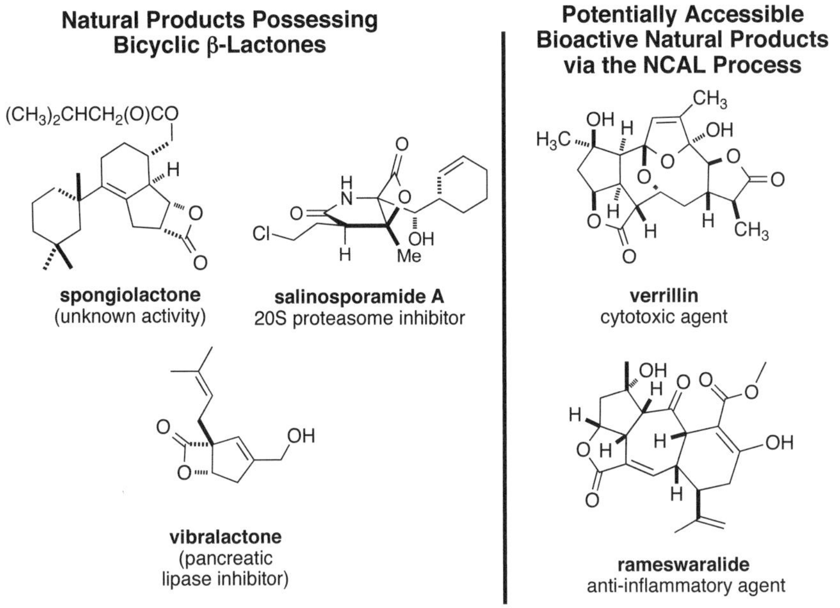

**Natural Products Possessing
Bicyclic β-Lactones**

spongiolactone
(unknown activity)

salinosporamide A
20S proteasome inhibitor

vibralactone
(pancreatic
lipase inhibitor)

**Potentially Accessible
Bioactive Natural Products
via the NCAL Process**

verrillin
cytotoxic agent

rameswaralide
anti-inflammatory agent

Figure 2. Some natural products possessing bicyclic β-lactones or structures potentially accessible from these intermediates.

1. Department of Chemistry, Texas A&M University, P. O. Box 30012, College Station, Texas, 77842-3012, USA. email: romo@tamu.edu. We gratefully acknowledge the NSF (CHE-0809747) and the Welch Foundation (A-1280) for support.

2. Department of Chemistry, The University of Chicago, 5735 South Ellis Avenue, Chicago, IL, 60637, USA.

3. This procedure mirrors that reported previously, see: Calter, M. A. *J. Org. Chem.* **1996**, *61*, 8006-8007.

4. Hon, Y.-S.; Lin, S.-W.; Chen, Y.-J. *Syn. Comm.* **1993**, *23*, 1543-1553.

5. Shitangkoon, A.; Vigh, G. *J. Chromatogr. A* **1996**, 31-42.

6. (a) Cortez, G. S.; Tennyson, R.; Romo, D. *J. Am. Chem. Soc.* **2001**, *123*, 7945-7946. (b) Cortez, G. S.; Oh, S.-H.; Romo, D. *Synthesis* **2001**, 1731-1736. (c) Oh, S.-H.; Cortez, G. S.; Romo, D. *J. Org. Chem.* **2005**, *70*, 2835-2838.

7. For reviews describing transformations of β-lactones to other functional arrays, see: (a) Pommier, A.; Pons, J.-M. *Synthesis* **1993**, 441-458. (b) Yang, H. W.; Romo, D. *Tetrahedron* **1999**, *51*, 6403-6434. For selected

more recent examples, see: (c) Donohoe, T. J.; Sintim, H. O.; Sisangia, L.; Harling, J. D. *Angew. Chem. Int. Ed.* **2004**, *43*, 2293. (d) Getzle, Y. D. Y. L.; Kundnani, V.; Lobkovsky, E. B.; Coates, G. W. *J. Am. Chem. Soc.* **2004**, *126*, 6842. (e) Mitchell, A. T.; Romo, D. *Heterocycles* **2005**, *66*, 627. (f) Shen, X.; Wasmuth, A. S.; Zhao, J.; Zhu, C.; Nelson, S. G. *J. Am. Chem. Soc.* **2006**, *128*, 7438. (g) Zhang, W.; Romo, D. *J. Org. Chem.* **2007**, *72*, 8939-8942. (h) Zhang, W.; Matla, A. S.; Romo, D. *Org. Lett.* **2007**, *9*, 2111-2114.

8. For a review on use of β-lactones in natural product total synthesis, see: Wang, Y.; Tennyson, R.; Romo, D. *Heterocycles* **2004**, *64*, 605-658.

9. For a review of naturally occurring β-lactones, see: (a) Lowe, C.; Vederas, J. C. *Org. Prep. Proced. Int.* **1995**, *27*, 305-346. (b) Reddy, L. R.; Saravanan, R.; Corey, E. J. *J. Am. Chem. Soc.* **2004**, *126*, 6230-6231. (c) Salinosporamide: R. H. Feling, G. O. Buchanan, T. J. Mincer, C. A. Kauffman, P. R. Jensen, W. Fenical, *Angew. Chem. Int. Ed.* **2003**, *42*, 355-357. Vibralactone: (d) Liu, D. E.; Wang, F.; Liao, T.-G.; Tang, J.-G.; Steglich, W.; Zhu, H.-J.; Liu, J.-K. *Org. Lett.* **2006**, *8*, 5749-5752.

10. (a) Romo, D.; Harrison, P. H. M.; Jenkins, S. I.; Riddoch, R. W.; Park, K. P.; Yang, H. W.; Zhao, C.; Wright, G. D. *Bioorg. Med. Chem.* **1998**, *6*, 1255-1272. (b) Rangan, V. S.; Joshi, A. K.; Smith J. W. *J. Bio. Chem.* **1998**, *273*, 34949-34953. (c) Ma, G.; Zancanella, M.; Oyola, Y.; Richardson, R. D.; Smith, J. W.; Romo, D. *Org. Lett.* **2006**, *8*, 4497-4500.

11. Böttcher, T.; Sieber, S. A. *Angew. Chem. Int. Ed.* **2008**, *47*, 4600-4603.

12. Greenberg, A.; Liebman, J. F. *"Strained Organic Molecules."* Academic Press: New York **1978**, Vol. *38*, Ch. 5.

13. (a) Wang, Y.; Romo, D. *Org. Lett.* **2002**, *4*, 3231-3234. (b) Tennyson, R. L.; Cortez, G. S.; Galicia, H. J.; Kreiman, C. R.; Thompson, C. M.; Romo, D. *Org. Lett.* **2002**, *4*, 533-536. (c) Schmitz, W. D.; Messerschmidt, N. B.; Romo, D. *J. Org. Chem.* **1998**, *63*, 2058-2059. (d) Yang, H. W.; Zhao, C.; Romo, D. *Tetrahedron* **1997**, *53*, 16471-16488. (e) Yang, H. W.; Romo, D. *J. Org. Chem.* **1997**, *62*, 4-5.

14. Inoguchi, K.; Fujie, N.; Yoshikawa, K.; Achiwa, K. *Chem. Pharm. Bull.* **1992**, *40*, 2921-2926.

15. (a) Henry-Riyad, H.; Lee, C. S.; Purohit, V.; Romo, D. *Org. Lett.* **2006**, *8*, 4363-4366. (b) Leverett, C. A.; Purohit, V. C.; Romo, D. *Angew. Chem. Int. Ed.* **2010**, *49*, 9479-9483.

16. (a) Ma, G.; Nguyen, H.; Romo, D. *Org. Lett.* **2007**, *9*, 2143-2146. (b) Nguyen, H.; Ma, G.; Romo, D. *Chem. Comm.* **2010**, *46*, 4803-4805. (c) Nguyen, H.; Ma, G.; Romo, D. *J. Org. Chem.* **2011**, *76*, 2-12.
17. Purohit, V. C.; Matla, A. S.; Romo, D. *J. Am. Chem. Soc.* **2008**, *130*, 10478-10479.
18. (a) Ooi, H.; Ishibashi, N.; Iwabuchi, Y.; Ishihara, J.; Hatakeyama, S. *J. Org. Chem.* **2004**, *69*, 7765-7768. (b) Reddy, L. R.; Saravanan, P.; Corey, E. J. *J. Am. Chem. Soc.* **2004**, *126*, 6230-6231. (c) Endo, A.; Danishefsky, S. *J. Am. Chem. Soc.* **2005**, *127*, 8298-8299.
19. (a) Tsuda, M.; Tadashi Endo,T.; Perpelescu M.; Yoshida, S.; Watanabe, K.; Fromont, J.; Mikami, Y. Kobayashi, J. *Tetrahedron* **2003**, *59*, 1137-1141. (b) Rodriguez, A.; Shi, Y-P. *J. Org. Synth.* **2000**, *65*, 5839-5842.

Appendix
Chemical Abstracts Nomenclature; (Registry Number)

(1,4-Dioxaspiro[4.5]dec-7-en-8-yloxy)trimethylsilane; (144810-01-5)

1,4-Cyclohexanedione mono-ethylene ketal: 1,4-Dioxaspiro[4.5]decan-8-one; (4746-97-8)

Trimethylsilyl chloride: Chlorotrimethylsilane; (75-777-4)

Triethylamine: Ethanamine, *N,N*-diethyl-; (121-44-8)

3-(2-(2-Oxoethyl)-1,3-dioxolan-2-yl) propanoic acid 360794-17-8

Triphenylphosphine; (603-35-0)

N-Methyl-2-chloropyridinium trifluoromethane sulfonate Pyridinium, 2-chloro-1-methyl-, 1,1,1-trifluoromethanesulfonate (1:1) 84030-18-2

2-Chloropyridine; (109-09-1)

Methyl trifluoromethanesulfonate; (333-27-7)

Quinidine: Cinchonan-9-ol, 6'-methoxy-; (56-54-2)

Diisopropylethylamine: 2-Propanamine, *N*-ethyl-*N*-(1-methylethyl)-; (7087-68-5)

Daniel Romo was born in San Antonio, Texas, USA in 1964. He received his B.A. in chemistry/biology from Texas A&M and a Ph.D. in Chemistry from Colorado State University as an NSF Minority Graduate Fellow under the tutelage of the late Professor Albert I. Meyers. Following postdoctoral studies at Harvard as an American Cancer Society Fellow, with Professor Stuart L. Schreiber he began his independent career at Texas A&M in 1993 and is currently Professor of Chemistry. In 2010 he became the Director of the Natural Products LINCHPIN Laboratory at Texas A&M University. Research interests in the Romo Group are at the interface of chemistry and biology focused on total synthesis and biomechanistic studies of natural products and the asymmetric synthesis and application of β-lactones in organic synthesis and their utility as both cellular probes and potential drug candidates.

Henry Nguyen was born in 1971 in Saigon, Vietnam. In 1993, he immigrated to America with his family and continued his undergraduate studies at Texas Woman's University. In 2000, he received his B.S. and then M.S. in chemistry in 2004 under the guidance of Professor James Johnson at Texas Woman's University. His research focused on mechanistic studies of methoxide ion substitution with Z and E isomers of methyl O-methylbenzothiohydroximates. In November 2004, he joined the research group of Prof. Romo at Texas A&M University and received his Ph.D. in 2010. His research focused on the enantioselective, biomimetic total synthesis and bioactivity of salinosporamide A and derivatives.

Seong Ho Ryan Oh was born in 1968 in Jeju, Korea and obtained his B.S. degree in 1991 and M.S. degree in 1993 under the direction of Kwan Soo Kim from Yonsei University in Seoul, Korea. In 1993, he joined the Chemical Research and Development group at LG Life Science, Taejon, Korea. In 2000, he continued as a Ph.D. student under the supervision of Prof. Romo at Texas A&M University. After completing his Ph.D. studies on the asymmetric synthesis of heterocycle-fused bicyclic β-lactones in 2005, he carried out postdoctoral research on the synthesis of orally active, antimalarial, anticancer, artemisinin-derived trioxane analogs in the laboratory of Gary Posner at Johns Hopkins University. In 2007, he joined the Biocatalyst and Chemical Development group at Codexis in Singapore.

136

Huda Henry-Riyad was born in 1974 in Khartoum, Sudan. In 2004 she obtained her Ph.D. from the University of Toronto under the supervision of Professor Thomas Tidwell. While at U of T, she examined ketenes and their reactivity with nitroxyl free radicals. In 2005 she joined the research group of Prof. Romo and worked on the synthesis of novel bicyclic β-Lactones via the nucleophile catalyzed aldol-lactonization (NCAL) process. She is currently the Director of The Key Research Center. The center's multi-disciplinary research idea is to bring different branches of knowledge together and to strengthen ties between social and the natural–physical sciences. Current research focuses on neuroscience and other related research areas that lead into the solution of mental, physical and social disorders.

Diana Sepulveda-Camarena was born in Guadalajara, Mexico. She attended the Instituto Tecnologico y de Estudios Superiores de Monterrey – Campus Monterrey where she received a B.Sc. in Chemistry. Prior to graduating in Fall 2006, she conducted her undergraduate research thesis project under the supervision of Prof. Romo at Texas A&M University. In February 2007, she joined BASF – Industrial Coatings division and worked in Apodaca, Mexico until she went back to Texas A&M in the Fall of 2008, where she is currently pursuing a Ph.D. in Chemistry under the supervision of Prof. Daniel Singleton.

Ms. Gerri E. Hutson obtained her B.S. in Chemistry in 2004 from Northern Illinois University, performing undergraduate research under Dr. Qingwei Yao. During her undergraduate studies she interned at Merck & Co. in Process Research in 2003 and 2004. She began her Ph.D. work under the supervision of Dr. Rawal in 2005, focusing her research on the utility of metal salens in enantioselective transformations. Her thesis title was "Modified Salen Catalyts in Atom-Economic Reactions: Enantioselective Carbonyl-ene and Nazarov Cyclization Reactions" and she was awarded a Ph.D. in Chemistry in 2010.

PHOSPHINE-CATALYZED [3 + 2] ANNULATION: SYNTHESIS OF ETHYL 5-(*tert*-BUTYL)-2-PHENYL-1-TOSYL-3-PYRROLINE-3-CARBOXYLATE

Submitted by Ian P. Andrews and Ohyun Kwon.[1]
Checked by John T. Colyer, Oliver R. Thiel, and Margaret Faul.

1. Procedure

A. Ethyl 5,5-dimethylhexa-2,3-dienoate (1). A flame-dried 2-L 3-necked, round-bottomed flask equipped with an overhead mechanical stirrer (Teflon paddle, 6 × 2 × 0.3 cm), a temperature probe, and a nitrogen inlet is charged with ethyl (triphenylphosphoranylidene)acetate (46.4 g, 133 mmol, 1.00 equiv) (Note 1), dichloromethane (665 mL) (Note 9), and triethylamine (20.4 mL, 146 mmol, 1.10 equiv) (Note 10). The mixture is stirred under nitrogen until a homogeneous solution is achieved. 3,3-Dimethylbutyryl chloride (18.5 mL, 133 mmol, 1.00 equiv) (Note 11) is added over 1 h using an addition funnel. The reaction mixture is stirred at room temperature for an additional 18 h during which time a precipitate forms, resulting in a cloudy heterogeneous mixture. The reaction mixture is added in three portions to a 1-L round-bottomed flask and concentrated to a volume of ca. 230 mL using a rotary evaporator (20 °C, 20–25 mmHg). Hexanes (300 mL) (Note 12) is added, and the flask is manually swirled for 2 min, resulting in a precipitate. The heterogeneous suspension is filtered through silica gel (100 g) (Note 13) in a 600-mL glass-fritted medium porosity Büchner funnel into a 2-L round-bottomed flask. The filter cake is rinsed with hexanes (600 mL). The resulting slightly yellow solution is concentrated under rotary evaporation (20 °C, 20–25 mmHg) until 750 mL of the solvent has been

138

removed (Note 14). The heterogeneous mixture is filtered through silica gel (100 g) (Note 13) in a 600-mL glass fritted medium porosity Büchner funnel into a 2-L round-bottomed flask. The filter cake is rinsed with 500 mL of hexanes/diethyl ether (19:1) (Note 15). The solution is concentrated (20 °C, 20–25 mmHg) and left on vacuum overnight to give ethyl 5,5-dimethylhexa-2,3-dienoate (1) (16.4–16.6 g, 73–74%) (Notes 16 and 17) as a pale yellow oil.

B. *Ethyl 5-(tert-butyl)-2-phenyl-1-tosyl-3-pyrroline-3-carboxylate (2).* A flame-dried, 1-L 3-necked, round-bottomed flask, equipped with an overhead mechanical stirrer (Teflon paddle, 6 × 2 × 0.3 cm), temperature probe, and a nitrogen inlet, is charged with (E)-N-benzylidene-4-methylbenzenesulfonamide (7.80 g, 30.0 mmol, 1.00 equiv) (Note 18) and purged with argon. Benzene (300 mL) (Note 19) is added to the flask via syringe, and the mixture is stirred until a homogeneous solution is achieved. Tributylphosphine (1.5 mL, 6.0 mmol, 0.20 equiv) (Note 20) is added via syringe and the reaction mixture is stirred for 5 min. Ethyl 5,5-dimethylhexa-2,3-dienoate (1) (6.06 g, 36.0 mmol, 1.20 equiv) is added over 5 min using a syringe pump (Note 21). The reaction is monitored using TLC (Note 22). Upon completion, the mixture is transferred into a 1-L round-bottomed flask and concentrated under rotary evaporation (30 °C, 20–25 mmHg) to a thick orange/brown oil and placed under high vacuum for 10 min (0.1–0.2 mmHg) to ensure the removal of residual solvent (Note 23). A sample of the crude oil (est. 0.5 g) is purified by column chromatography to afford seed crystals of 2 (Notes 24 and 25). A mixture of hexanes/ethyl acetate (9:1, 75 mL) is added to the crude oil in a 1-L round-bottomed flask and a reflux condenser is attached. A Teflon-coated stir bar is added and the crude residue is dissolved by heating in an oil bath with stirring until a homogeneous solution is achieved. After cooling to room temperature, the reflux condenser is removed and the flask is capped with a plastic stopper and left to stand overnight. The seed crystals are added as the reaction mixture is cooled. The following day the mixture is placed in a freezer at – 20 °C for 2 h. The crystals that precipitated are collected under vacuum filtration using a glass-fritted funnel (Notes 26 and 27). The crystals are rinsed with hexanes (200 mL) to give 11.4–11.8 g (88–92% yield) of 2 as a white to pale yellow solid (Note 28).

2. Notes

1. Ethyl (triphenylphosphoranylidene)acetate is commercially available from Aldrich or can be prepared readily from ethyl bromoacetate (Note 2) and triphenylphosphine (Note 3) using the following procedure provided by the submitters: A 1-L round-bottom flask equipped with a magnetic stir bar is charged with triphenylphosphine (39.4 g, 150 mmol) and toluene (200 mL) (Note 4). The mixture is stirred until a homogeneous solution is achieved. Ethyl bromoacetate (16.6 mL, 150 mmol) is added via syringe. The mixture is capped with a plastic stopper and stirred at room temperature for 24 h, during which time a white precipitate is formed, resulting in a thick slurry. The precipitate is collected under vacuum filtration using a glass-fritted Büchner funnel. The filtrate is washed with 75 mL of toluene (Note 5) and 75 mL of diethyl ether (Note 6). The white solid is transferred to a 1-L round-bottomed flask, placed under vacuum for 2 h, and then dissolved in 450 mL of deionized water (Note 7). A magnetic stir bar is added, followed by phenolphthalein indicator solution (200 µL) (Note 8). While stirring, 2 M potassium hydroxide is added dropwise until a pink color persists. The aqueous layer is extracted with dichloromethane (3 × 200 mL) (Note 9) and the combined organic extracts are washed with brine (400 mL), dried over sodium sulfate, filtered, and concentrated under rotary evaporation (30 °C, 20–25 mmHg). The resulting thick oil is redissolved in 20 mL of diethyl ether (Note 6) and concentrated under rotary evaporation (30 °C, 20–25 mmHg) to give an off-white solid. Drying under high vacuum (0.1–0.2 mmHg) yields 48.1 g (92%) of ethyl (triphenylphosphoranylidene)acetate as an off-white solid, mp 118–121 °C. The checkers used commercially available material from Aldrich (95%).

2. Ethyl bromoacetate (98%) was purchased from Alfa Aesar and used as received.

3. Triphenylphosphine (99%) was purchased from Alfa Aesar and used as received.

4. Toluene (99.8%) was purchased from Fisher Scientific and used as received.

5. Some of the precipitate may stick to the walls of the flask. If so, toluene can be used to rinse the round-bottomed flask.

6. Diethyl ether (99.9%) was purchased from Fisher Scientific and used without further purification.

7.   If the addition of water does not result in a homogeneous solution, heating is applied to aid the solubilization of the phosphonium salt.

8.   Phenolphthalein indicator solution is prepared by dissolving phenolphthalein powder (100 mg) (purchased from Mallinckrodt and used as received) in 2-propanol (10 mL) (99.9% purchased from Fisher Scientific and used as received). The addition of this indicator is used to help avoid saponification of the ethyl ester, which occurs upon the addition of excess potassium hydroxide.

9.   Dichloromethane (99.9%) as purchased from Fluka and used as received.

10.   Triethylamine ($\geq$ 99.5%) was purchased from Aldrich and used as received.

11.   3,3-Dimethylbutyryl chloride (99%) was purchased from Aldrich and used as received.

12.   Hexanes (98.5%) was purchased from Aldrich and used as received.

13.   Aldrich (200–400 mesh) silica gel was used as received. The silica gel is packed in the funnel from a slurry in hexanes. The surface of the packed silica is covered with filter paper to prevent disruption.

14.   Upon concentration of the solution, a white precipitate forms.

15.   After the rinse is complete, thin layer chromatography (TLC) is used to check the flow-through for the presence of compound **1**. The filter cake is rinsed with 6 x 100 mL portions of hexanes/diethyl ether (19:1) until TLC of the flow-through reveals the absence of **1**.

16.   The physical properties are as follows: R_f = 0.53 (hexanes/ethyl acetate, 19:1); IR: 2964, 2906, 2870, 1959, 1716, 1463, 1413, 1365, 1286, 1255, 1240, 1152, 1095, 1039, 874, 803 cm^{-1}. ^{1}H NMR (400 MHz, CDCl$_3$) δ: 1.12 (s, 9 H); 1.27 (t, J = 7.1 Hz, 3 H), 4.11–4.25 (m, 2 H), 5.59–5.63 (m, 2 H). ^{13}C NMR (100 MHz, CDCl$_3$) δ: 14.1, 29.9, 32.5, 60.5, 89.6, 106.6, 166.1, 210.4. HRMS [C$_{10}$H$_{16}$O$_2$ + H]$^+$: calcd. for 169.1223, found 169.1220. Anal. calcd. for C$_{10}$H$_{16}$O$_2$: C, 71.39; H, 9.59. Found: C, 70.79; H, 9.75.

17.   Differential scanning calorimetry of (**1**) was completed (DSC, 30 °C to 300 °C, ramp 10 °C/min). Exothermic decomposition was observed with an onset temperature of 90 °C. Heating of the product beyond 30 °C and purification by distillation is therefore not recommended without further safety testing.

18.   (*E*)-*N*-Benzylidene-4-methylbenzenesulfonamide was prepared as previously described.[2]

19. Benzene (≥99.5%) was purchased from Fluka and used as received.

20. Tributylphosphine (97%) was purchased from Aldrich and used without purification. ^{31}P NMR spectroscopy revealed that the reagent comprised 91% tributylphosphine (δ –30.65 ppm) and less than 5% tributylphosphine oxide (δ 48.49 ppm). Three other resonances (δ –17.35, <3%; δ 34.79, <1%; δ 132.33, <2%) were also observed.

21. The mixture slowly transforms from clear to yellow after the addition of ethyl 5,5-dimethylhexa-2,3-dienoate (1).

22. The reaction is monitored using TLC (hexanes/ethyl acetate, 3:1), observing the disappearance of (E)-N-benzylidene-4-methylbenzene-sulfonamide (R_f = 0.54) and the appearance of the product, ethyl 5-(tert-butyl)-2-phenyl-1-tosyl-3-pyrroline-3-carboxylate (2) (R_f = 0.66). Visualization is performed under UV light and through staining with potassium permanganate (prepared by dissolving 20 g of potassium carbonate in 300 mL of water followed by the sequential addition of 5 mL of 5% NaOH$_{(aq)}$ and 3 g of KMnO$_4$). Potassium permanganate was purchased from Fisher and used as received. The reaction typically reaches completion within 4–7 h. The reaction was checked for completion by TLC after 18 h prior to work-up.

23. If a vacuum of 0.1 mmHg is not reached, the oil can be dried overnight under vacuum (20–30 mmHg), then concentrated from diethyl ether (2 x 50 mL) and hexanes (50 mL) to provide a viscous brown oil.

24. The reproducibility of the crystallization is improved by generating a small amount of seed material through column chromatography.

25. Column chromatography was performed using a 40 mm-wide glass column packed with Aldrich (200–400 mesh) silica gel (90 g). The column is packed from a slurry of silica gel in hexanes/ethyl acetate (9:1). The crude residue is loaded onto the column in the oily state obtained immediately after rotary evaporation. The flask is rinsed twice with 2 mL portions of hexanes/ethyl acetate (9:1). The column is eluted with hexanes/ethyl acetate (9:1) and fractions are collected. Each collected fraction is analyzed using TLC, and those containing the product are collected, concentrated under rotary evaporation (30 °C, 20–25 mm Hg), and dried under vacuum to afford an oil that crystallized on standing.

26. The submitters obtained additional product from the filtrates using column chromatography (Note 27). The checkers did not isolate additional product from the filtrate, but determined by quantitative HPLC analysis that

142

an addition 960 mg of **2** was present in the combined mother liquors and wash.

27. Any small amounts of solid left on the filter funnel are dissolved in ethyl acetate and combined with the filtrate prior to concentrating. The filtrate is concentrated under rotary evaporation (35 °C, 20–25 mmHg) and purified through column chromatography to yield an additional 1.0 g (7.8%) of **2** as a white solid. Column chromatography was performed using a 4.75-cm-wide glass column packed with Aldrich (200–400 mesh) silica gel (65 g). The column is packed from a slurry of silica gel in hexanes/ethyl acetate (9:1). The crude residue is loaded onto the column in the oily state obtained immediately after rotary evaporation. The flask is rinsed twice with 2 mL portions of hexanes/ethyl acetate (9:1). The column is eluted with hexanes/ethyl acetate (9:1) and fractions of ca. 22 mL are collected. Each collected fraction is analyzed using TLC and those containing the product are collected, concentrated under rotary evaporation (30 °C, 20–25 mmHg), and dried under vacuum.

28. The physical properties are as follows: mp 93–98 °C; IR: 2957, 1714, 1335, 1232, 1164, 1102, 1089, 980 cm^{-1}. ^{1}H NMR (400 MHz, CDCl$_3$) δ: 0.80 (s, 9 H), 1.14 (t, J = 7.1 Hz, 3 H), 2.40 (s, 3 H), 4.12 (q, J = 7.1 Hz, 2 H), 4.37 (d, J = 2.3 Hz, 1 H), 5.89 (s, 1 H), 6.73 (dd, J = 1.2, 2.5 Hz, 1 H), 7.32–7.23 (comp, 5 H), 7.43 (d, J = 7.4 Hz, 2 H), 7.71 (d, J = 8.2 Hz, 2 H). ^{13}C NMR (100 MHz, CDCl$_3$) δ: 14.0, 21.5, 27.9, 35.9, 60.8, 68.4, 77.8, 127.5, 127.9, 128.0, 129.6, 134.1, 134.2, 139.7, 141.2, 143.8, 162.7. HRMS [C$_{24}$H$_{29}$NO$_4$S + H]$^+$: calcd. for 428.1890, found 428.1890. Anal. calcd. for C$_{24}$H$_{29}$NO$_4$S: C, 67.42; H, 6.84; N, 3.28. Found: C, 67.51; H, 7.14; N, 3.20.

Safety and Waste Disposal Information

All hazardous materials should be handled and disposed of in accordance with "Prudent Practices in the Laboratory"; National Academy Press; Washington, DC, 1995.

3. Discussion

The use of nucleophilic tertiary phosphines as catalysts in organic transformations has become a powerful tool for organic synthesis over the past two decades. The discovery that the zwitterions generated from the addition of a phosphine to an electron-deficient allene or alkyne can engage with various electrophiles to produce annulated products has generated much interest in the development of related reactions.[3] This flurry of activity is due, in part, to the complexity that can be generated from relatively simple starting materials, as well as the simplicity of the experimental procedures. For instance, workup typically requires no more that rotary evaporation and column chromatography. Most of the developed reactions have entailed the preparation of carbo- and heterocyclic products from simple starting materials, such as olefins, allenes, or acetylenes in combination with alkenes, imines, aldehydes, or dinucleophiles.[4] In the first examples of such reactions, reported by Xiyan Lu, activated allenes were reacted with electron-deficient alkenes and imines to produce cyclopentenes and pyrrolines, respectively.[5] Since Lu's original report of such [3 + 2] annulations, many enantioselective variants of the procedure have been described, as well as several applications in the realm of natural product synthesis.[6,7] Our laboratory discovered that α-alkyl allenoates behave as 1,4-dipole–like species in novel [4 + 2] annulations with imines to form tetrahydropyridines.[8] This broadly applicable reaction has since been applied to natural product synthesis[9] and has been developed into an asymmetric process.[10] The utility of the phosphine-catalyzed [4 + 2] annulation was further enhanced upon the successful application of activated alkenes to form functionalized cyclohexenes[11] and of trifluoromethyl ketones to form dihydropyrans.[12] Of particular importance to this discussion is the synthesis of 2,5-disubstituted pyrroline-3-carboxylates through the reactions of γ-substituted allenoates and imines.[13] 2,5-Disubstituted pyrroline-3-carboxylates are important scaffolds that are commonly used as intermediates en route to other important nitrogen atom–containing heterocycles.[14] Aside from their use as synthetic intermediates, some 2,5-disubstituted-3-pyrrolines themselves exhibit significant biological activity.[15] Recently, our laboratory synthesized a library of 2,5-disubstituted pyrroline-3-carboxylates through the phosphine-catalyzed [3 + 2] annulations of resin-bound allenoates and imines. Several of these compounds were identified as geranylgeranyltransferase type-I (GGTase-1) inhibitors (GGTIs) and one, P61A6, exhibits excellent antitumor effects when tested in mice.

Figure 1. Geranylgeranyltransferase type-I inhibitor P61A6.

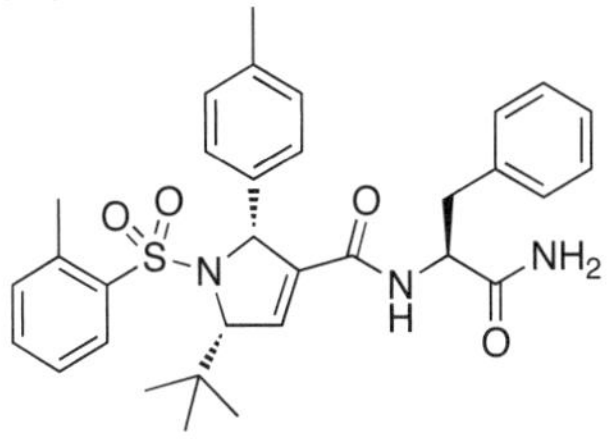

P61A6

Employing the procedure described above provides access to various 2,5-disubstituted pyrroline-3-carboxylates through the reactions of imines and 4-alkyl-2,3-butadienoates in the presence of a catalytic amount of tributylphosphine. Table 1 reveals that the reactions of various 4-alkyl- and 4-aryl-2,3-butadienoates all led to the desired 2,5-disubstituted pyrrolines in excellent yields (Table 1, entries 1–6). The process is highly diastereoselective, with preferential formation of the syn isomer ranging from 91:9 for the γ-methyl group (Table 1, entry 1) to exclusive for γ-isopropyl, *tert*-butyl, and phenyl groups (Table 1, entries 4–6). Note that product **2c** (Table 1, entry 3) was formed in a similar reaction employing the 2-butynoate, but in only 63% yield.[5b]

Table 1. Synthesis of Pyrrolines **2** From γ-Substituted Allenoates[a]

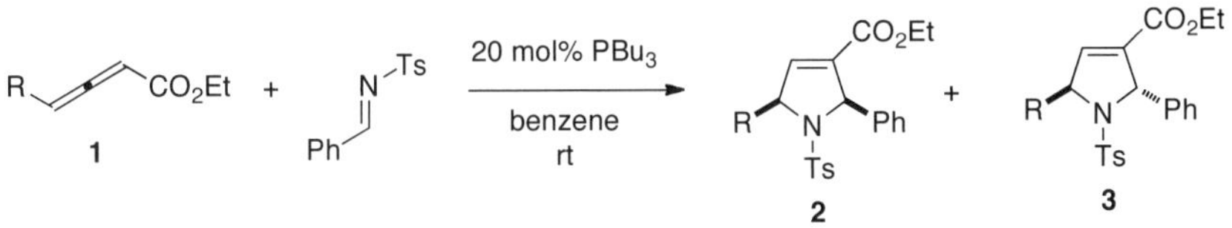

Entry	R	Time (hr)	Product	Yield (%)[b]	2/3[c]
1	Methyl (**1a**)	2	**2a/3a**	89	91:9
2	Ethyl (**1b**)	5	**2b/3b**	99	95:5
3	Propyl (**1c**)	5	**2c/3c**	98	96:4
4	Isopropyl (**1d**)	5	**2d**	99	100:0
5	*tert*-Butyl (**1**)	8	**2**	99	100:0
6	Phenyl (**1e**)	1.5	**2e**	99	100:0

[a] Reactions were performed on 1 mmol scale. [b] Isolated yield. [c] Determined using ^{1}H NMR spectroscopy.

The scope of the reaction was further established by reacting γ-substituted allenoates with various aryl imines (Table 2). The reactions of γ-isopropyl allenoate provided the corresponding 3-pyrrolines in excellent yields (Table 2, entries 1–3). The reactions of γ-tert-butyl allenoate with various aryl imines led to quantitative yields of all products (Table 2, entries 9–16).

Table 2. Reactions of γ-Substituted Allenoates With Aryl Imines[a]

R⤴CO₂Et (1) + Ar–CH=N–Ts → (20 mol% PBu₃, benzene, rt, 8 h) → 2 (pyrroline with CO₂Et, R, N–Ts, Ar)

Entry	R	Ar	Product	Yield (%)[b]
1	Isopropyl (1d)	2-FC₆H₄	2f	97
2	Isopropyl (1d)	3-BrC₆H₄	2g	95
3	Isopropyl (1d)	4-CF₃C₆H₄	2h	96
4	Isopropyl (1d)	4-(i-Pr)C₆H₄	2i	96
5	Phenyl (1e)	2-ClC₆H₆	2j	99
6	Phenyl (1e)	3-ClC₆H₆	2k	99
7	Phenyl (1e)	4-FC₆H₄	2l	99
8	Phenyl (1e)	4-MeOC₆H₄	2m	99
9	tert-Butyl (1)	2-ClC₆H₄	2n	>99
10	tert-Butyl (1)	3-ClC₆H₆	2o	>99
11	tert-Butyl (1)	3,4-Cl₂C₆H₃	2p	>99
12	tert-Butyl (1)	4-CNC₆H₄	2q	>99
13	tert-Butyl (1)	4-FC₆H₄	2r	>99
14	tert-Butyl (1)	4-MeC₆H₄	2s	>99
15	tert-Butyl (1e)	4-MeOC₆H₄	2t	>99
16	tert-Butyl (1e)	1-Naphthyl	2u	>99

[a] Reactions performed on 1 mmol scale. [b] Isolated yield.

Scheme 1 presents a proposed mechanism for the formation of 2,5-disubstituted-3-pyrrolines from γ-substituted allenoates and imines. Nucleophilic addition of the phosphine catalyst to the allenoate 1 results in

146

the intermediate zwitterion **4**, which then adds to the imine **5** to produce the phosphonium amide zwitterion **6**. A 5-endo-trig cyclization of the amide anion to the vinyl phosphonium cation results in the formation of the ylide **7**. After proton transfer to generate the β-phosphonium enolate intermediate **8**, liberation of the catalyst, through β-elimination of the phosphine, yields the final 2,5-disubstituted-pyrroline-3-carboxylate (**2**).

Scheme 1. Proposed Mechanism for the Phosphine-Catalyzed [3 + 2] Annulations of γ-Substituted Allenoates and Imines

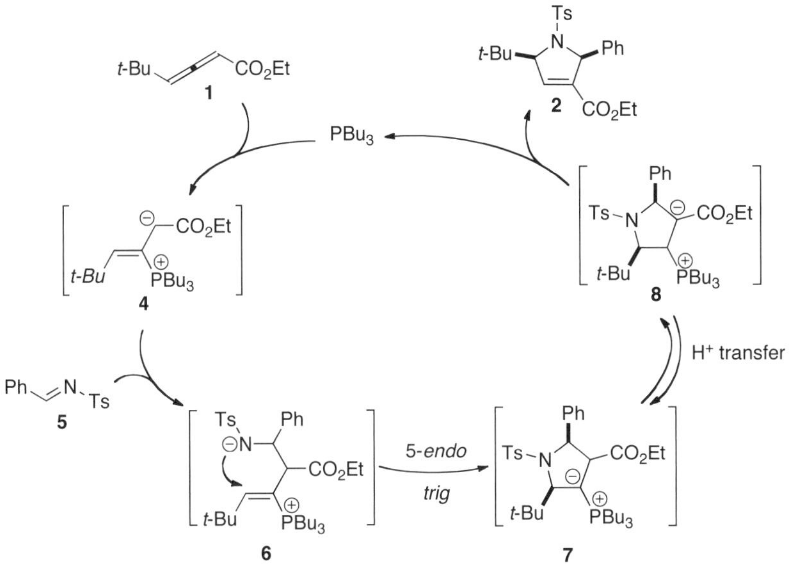

In summary, various 2,5-disubstituted 3-pyrrolines are readily available from the reactions of γ-substituted allenoates with aryl imines in the presence of catalytic amounts of a trialkylphosphine. These phosphine-catalyzed reactions are highly efficient, diastereoselective, operationally simple, and executed under mild conditions.

References

1. Department of Chemistry and Biochemistry, University of California, Los Angeles, 607 Charles E. Young Drive East, Los Angeles,

California 90095-1569. This research was supported by the U.S. National Institutes of Health (O.K.: R01GM071779, P41GM081282).

2. Lu, K.; Kwon, O. *Org. Synth.* **2009**, *86*, 212.

3. For reviews on phosphine-catalyzed reactions, see: (a) Lu, X.; Zhang, C.; Xu, Z. *Acc. Chem. Res.* **2001**, *34*, 535. (b) Valentine, D. H.; Hillhouse, J. H. *Synthesis* **2003**, *3*, 317. (c) Methot, J. L.; Roush, W. R. *Adv. Synth. Catal.* **2004**, *346*, 1035. (d) Lu, X.; Du, Y.; Lu, C. *Pure Appl. Chem.* **2005**, *77*, 1985. (e) Nair, V.; Menon, R. S.; Sreekanth, A. R.; Abhilash, N.; Biju, A. T. *Acc. Chem. Res.* **2006**, *39*, 520. (f) Denmark, S. E.; Beutner, G. L. *Angew. Chem., Int. Ed.* **2008**, *47*, 1560. (g) Ye, L.; Zhou, J.; Tang, Y. *Chem. Soc. Rev.* **2008**, *37*, 1140. (h) Kwong, C. K.-W.; Fu, M. Y.; Lam, C. S.-L.; Toy, P. H. *Synthesis* **2008**, 2307. (i) Aroyan, C. E.; Dermenci, A.; Miller, S. J. *Tetrahedron* **2009**, *65*, 4069. (j) Cowen, B. J.; Miller, S. J. *Chem. Soc. Rev.* **2009**, *38*, 3102. (k) Marinetti, A.; Voituriez, A. *Synlett* **2010**, 174. (l) Beata, K. *Cent. Eur. J. Chem.* **2010**, 1147. (m) Xu, S.; He, Z. *Sci. Sinca Chim.* **2010**, *40*, 856.

4. (a) Zhu, X.-F.; Henry, C. E.; Wang, J.; Dudding, T.; Kwon, O. *Org. Lett.* **2005**, *7*, 1387. (b) Zhu, X.-F.; Schaffner, A.-P.; Li, R. C.; Kwon, O. *Org. Lett.* **2005**, *7*, 2977. (c) Dudding, T.; Kwon, O.; Mercier, E. *Org. Lett.* **2006**, *8*, 3643. (d) Zhu, X.-F.; Henry, C. E.; Kwon, O. *J. Am. Chem. Soc.* **2007**, *129*, 6722. (e) Henry, C. E.; Kwon, O. *Org. Lett.* **2007**, *9*, 3069. (f) Creech, G. S.; Kwon, O. *Org. Lett.* **2008**, *10*, 429. (g) Creech, G. S.; Zhu, X.; Fonovic, B.; Dudding, T.; Kwon, O. *Tetrahedron* **2008**, *64*, 6935. (h) Sriramurthy, V.; Barcan, G. A.; Kwon, O. *J. Am. Chem. Soc.* **2007**, *129*, 12928. (i) Guo, H.; Xu, Q.; Kwon, O. *J. Am. Chem. Soc.* **2009**, *131*, 6318.

5. (a) Zhang, C.; Lu, X. *J. Org. Chem.* **1995**, *60*, 2906. (b) Xu, Z.; Lu, X. *Tetrahedron Lett.* **1999**, *40*, 549. (c) Xia, Y.; Liang, Y.; Chen, Y.; Wang, M.; Jiao, L.; Huang, F.; Liu, S.; Li, Y.; Yu, Z. *J. Am. Chem. Soc.* **2007**, *129*, 3470. (d) Mercier, E.; Fonovic, B.; Henry, C.; Kwon, O.; Dudding, T. *Tetrahedron Lett.* **2007**, *48*, 3617.

6. (a) Zhu, G.; Chen, Z.; Jiang, Q.; Xiao, D.; Cao, P.; Zhang, X. *J. Am. Chem. Soc.* **1997**, *119*, 3836. (b) Wilson, J. E.; Fu, G. C. *Angew. Chem., Int. Ed.* **2006**, *45*, 1426. (c) Cowen, B. J.; Miller, S. J. *J. Am. Chem. Soc.* **2007**, *129*, 10988. (d) Fang, Y.; Jacobsen, E. N. *J. Am. Chem. Soc.* **2008**, *130*, 5660. (e) Voituriez, A.; Panossian, A.; Fleury-

148

Bregeot, N.; Retailleau, P.; Marinetti, A. *J. Am. Chem. Soc.* **2008**, *130*, 14030.

7. (a) Du, Y.; Lu, X. *J. Org. Chem.* **2003**, *68*, 6463. (b) Wang, J.; Krische, M. J. *Angew. Chem., Int. Ed.* **2003**, *42*, 5855. (c) Pham, T. Q.; Pyne, S. G.; Skelton, B. W.; White, A. H. *J. Org. Chem.* **2005**, *70*, 6369.

8. (a) Zhu, X.; Lan, J.; Kwon, O. *J. Am. Chem. Soc.* **2003**, *125*, 4716. (b) Reference 2.

9. Tran, Y. S.; Kwon, O. *Org. Lett.* **2005**, *7*, 4289.

10. Wurz, R. P.; Fu, G. C. *J. Am. Chem. Soc.* **2005**, *127*, 12234.

11. Tran, Y. S.; Kwon, O. *J. Am. Chem. Soc.* **2007**, *129*, 12632.

12. Wang, T.; Ye, S. *Org. Lett.* **2010**, *12*, 4168.

13. Zhu, X.; Henry, C. E.; Kwon, O. *Tetrahedron* **2005**, *61*, 6276.

14. For examples of 3-pyrrolines used in natural product syntheses and in pharmaceutical analogues, see: (a) Burley, I.; Hewson, A. T. *Tetrahedron Lett.* **1994**, *35*, 7099. (b) Huwe, C. M.; Blechert, S. *Tetrahedron Lett.* **1995**, *36*, 1621. (c) Green, M. P.; Prodger, J. C.; Hayes, C. J. *Tetrahedron Lett.* **2002**, *43*, 6609.

15. (a) Castellano, S.; Fiji, H.D.G.; Kinderman, S.S.; Watanabe, M.; de Leon, P.; Tamanoi, F.; Kwon, O. *J. Am. Chem. Soc.* **2007**, *129*, 5843. (b) Watanabe, M.; Fiji, H.D.G.; Guo, L.; Chan, L.; Kinderman, S.S.; Slamin, D.J.; Kwon, O.; Tamanoi, F. *J. Biol. Chem.* **2008**, *283*, 9571. (c) Lu, J.; Chan, L.; Fiji, H.D.G.; Dahl, R.; Kwon, O.; Tamanoi, F. *Mol. Cancer Ther.* **2009**, *8*, 1218.

Appendix
Chemical Abstracts Nomenclature; (Registry Number)

Ethyl 5,5-dimethylhexa-2,3-dienoate: 2,3-Hexadienoic acid, 5,5-dimethyl-, ethyl ester; (35802-59-6)

Ethyl (triphenylphosphoranylidene)acetate: Acetic acid, 2-(triphenylphosphoranylidene)-, ethyl ester; (1099-45-2)

3,3-Dimethylbutyryl chloride: Butanoyl chloride, 3,3-dimethyl-; (7065-046-05)

Triethylamine: Ethanamine, *N,N*-diethyl-; (121-44-8)

Ethyl 5-(*tert*-butyl)-2-phenyl-1-tosyl-3-pyrroline-3-carboxylate: 1*H*-Pyrrole-3-carboxylic acid, 5-(1,1-dimethylethyl)-2,5-dihydro-1-[(4-

methylphenyl)sulfonyl]-2-phenyl-, ethyl ester, (2*R*,5*R*)-rel-; (861668-04-4)

(*E*)-*N*-Benzylidene-4-methylbenzenesulfonamide: Benzenesulfonamide, 4-methyl-*N*-(phenylmethylene)-, [N(*E*)]-; (51608-60-7)

Tributylphosphine; (998-40-3)

Ohyun Kwon was born in South Korea in 1968. She received her B.S. and M.S. in chemistry (with Eun Lee) from Seoul National University in 1991 and 1993, respectively. She came to the US in 1993 and obtained her PhD (with Samuel J. Danishefsky) from Columbia University in 1998. After a postdoctoral stint in the laboratory of Stuart L. Schreiber at Harvard University, she began her independent career in 2001 as an Assistant Professor at the University of California, Los Angeles (UCLA). In 2008, she was promoted to Associate Professor with tenure at UCLA. Her research involves the development of phosphine-catalyzed reactions and their application to natural product synthesis and chemical biology.

Ian P. Andrews was born in Monterey, California in 1983. He received his B.S. degree from the University of California, San Diego in 2005. He was a summer intern at Pfizer inc. working in medicinal chemistry on the discovery of new carbonic anhydrase inhibitors. He worked at Trinlink Biotechnologies in San Diego before pursuing a Ph.D. at the University of California, Los Angeles. He is currently conducting research in the Laboratory of Dr. Ohyun Kwon at UCLA.

John T. Colyer earned his B.S. in chemistry from Indiana University-Purdue University Indianapolis in 2000 working under the guidance of Professor William H. Moser. He completed his M.S. in 2004 at the University of Arizona under the guidance of Professor Michael P. Doyle, where he studied dirhodium (II) carboxamidate catalysis. In 2004 he joined the Chemical Process Research & Development department at Amgen in Thousand Oaks, CA.

Oliver R. Thiel studied chemistry at the Technical University Munich, Germany and completed a Diploma thesis on rhodium-catalyzed hydroaminations under supervision of Professor Matthias Beller. He then pursued a Ph.D. (1998-2001) at the Max-Planck-Institut für Kohlenforschung Mülheim, Germany under guidance of Professor Alois Fürstner, exploring RCM-reactions in natural product synthesis. After a postdoctoral appointment (2001-2003) at Stanford University with Professor Barry M. Trost, Oliver joined the Chemical Process Research & Development department at Amgen in Thousand Oaks, CA where he has been involved in the development of synthetic processes of numerous clinical candidates.

PREPARATION OF HORNER-WADSWORTH-EMMONS REAGENT: METHYL 2-BENZYLOXYCARBONYLAMINO-2-(DIMETHOXY- PHOSPHINYL)ACETATE

A.

B.

C.

Submitted by Hiroki Azuma,[1] Kentaro Okano,[1] Tohru Fukuyama,[2] and Hidetoshi Tokuyama.[1]
Checked by Alistair Boyer and Mark Lautens.

1. Procedure

A. α-*Hydroxy-N-benzyloxycarbonylglycine.* A 500-mL three-necked round-bottomed flask equipped with an overhead mechanical stirrer (teflon paddle, 75 x 20 mm), a glass stopper, and a reflux condenser fitted with an inert gas inlet (Note 1) is charged with benzyl carbamate (30.23 g, 200 mmol, 1.0 equiv) (Note 2) and glyoxylic acid monohydrate (20.25 g, 220 mmol, 1.1 equiv) (Note 3). The flask is evacuated and backfilled with inert gas, the glass stopper is removed under a stream of inert gas and the flask is charged with anhydrous Et$_2$O (200 mL) (Note 4). The resulting translucent solution is heated under reflux for 12 h (Note 5) with stirring at a rate of 200 rpm. Over this time, the product precipitates to give a white suspension. The white precipitate is collected by filtration, washed with hexanes-Et$_2$O (1:1) (6 x 10 mL) (Note 6), and dried *in vacuo* to yield α-hydroxy-*N*-benzyloxycarbonylglycine as fine white crystals

152

(32.80–35.37 g, 73–79%) (Note 7).

B. *Methyl α-methoxy-N-benzyloxycarbonylglycinate.* A 1-L one-necked, round-bottomed flask equipped with a Teflon-coated magnetic stir bar (octagonal, 38 mm) is charged with α-hydroxy-N-benzyloxycarbonylglycine (22.52 g, 100 mmol, 1.0 equiv) and anhydrous MeOH (200 mL) (Note 8). An inert gas inlet is attached (Note 1), and the flask is cooled to 1 °C (internal temperature). The gas inlet is temporarily removed and to the translucent solution is added conc. H_2SO_4 (5.0 mL) (Note 9) with a pipette over 5 min. The gas inlet is replaced, the cooling bath is then removed, and the mixture is stirred at room temperature for 15 h (Note 10). The reaction mixture is cooled to 1 °C (internal temperature) to which is added sat. aq. $NaHCO_3$ (120 mL) (Note 11) accompanied with vigorous gas evolution. The pH of the solution after addition is determined to be ca. 7. Methanol is removed under reduced pressure on a rotary evaporator (30 °C, 40 mmHg), and the residue is transferred into a 300-mL separatory funnel with the aid of water (50 mL) and EtOAc (200 mL). After partitioning, the aqueous layer is extracted with EtOAc (2 x 200 mL). The combined organic extracts are transferred into a 1-L separatory funnel and are washed with sat. aq. NaCl (1 x 200 mL), dried over $MgSO_4$ (20 g), and filtered (washing with 2 x 20 mL of EtOAc). The filtrate is concentrated under reduced pressure on a rotary evaporator (30 °C, 40 mmHg) and the residue is thoroughly dried *in vacuo* to give methyl α-methoxy-N-benzyloxycarbonylglycinate as a white solid (24.26–24.56 g, 96–97%) (Note 12).

C. *Methyl 2-benzyloxycarbonylamino-2-(dimethoxyphosphinyl)acetate.* A 500-mL flame-dried, two-necked, round-bottomed flask (Note 13) equipped with a Teflon-coated magnetic stir bar (octagonal, 38 mm), a rubber septum, and an inert gas inlet (Note 1) is charged with methyl α-methoxy-N-benzyloxycarbonylglycinate (20.27 g, 80.0 mmol, 1.0 equiv). The flask is evacuated and backfilled with inert gas. The flask is charged with anhydrous toluene (80 mL) (Note 14), and the resulting solution is heated at 70 °C. Phosphorus trichloride (6.98 mL, 11.0 g, 80.0 mmol, 1.0 equiv) (Note 15) is added to the solution over 5 min via a syringe. After stirring at 70 °C for 26 h (Note 16), trimethyl phosphite (9.44 mL, 9.93 g, 80.0 mmol, 1.0 equiv) (Note 17) is added to the mixture over 5 min via a syringe. The resulting mixture is stirred at 70 °C for an additional 2 h (Note 18). All volatile materials are removed under reduced pressure (40 then 10 mmHg, bath temperature: 70 °C) in a general distillation apparatus (a distillation head, a distillation adapter, and a receiver flask) to give a yellow viscous oil. The oil is transferred into a 300-mL separatory funnel with the aid of EtOAc (100 mL). The solution is washed with sat. aq. NaHCO₃ (3 x 50 mL). The combined aqueous washings are extracted with EtOAc (1 x 50 mL). The combined organic extracts are washed with sat. aq. NaCl (1 x 50 mL), dried over Na₂SO₄ (10 g), filtered (washing with 10 mL EtOAc), and concentrated on a rotary evaporator under reduced pressure (30 °C, 40 mmHg) to give a pale yellow oil (27.26 g). Upon addition of hexanes (40 mL) to the oil with vigorous stirring at room temperature, a white solid precipitates which is collected by filtration, washed with ice-cold hexanes (5 x 20 mL), and dried *in vacuo* to afford methyl 2-benzyloxycarbonylamino-2-(dimethoxyphosphinyl)acetate as a fine white powder (22.59–23.08 g, 85–87%) (Note 19).

2. Notes

1.   The use of either argon or nitrogen had no impact on the yield of the reaction.

2.   Benzyl carbamate (99%) was purchased from Sigma-Aldrich Co. and used as received without further purification.

3.   Glyoxylic acid monohydrate (98%) was purchased from Sigma-Aldrich Co. and used as received without further purification.

4.   Et₂O (puriss., dried over molecular sieves, ≤0.005% H₂O) was purchased from Sigma-Aldrich Co. and used as received without further purification (submitters used >99.5%, water content: <0.05% from Kanto Chemical Company, Inc).

5.   The submitters observed an incomplete reaction at 12 h and longer reaction time and/or higher reaction temperature did not improve the conversion.

6.   Concentration of the combined washings gave 13.35 g of the desired product: α–hydroxy-*N*-benzyloxycarbonylglycine, with a trace of glyoxylic acid monohydrate.

7.   The submitters report a yield of 58%. Data for product (without further purification): R_f = 0.60 (H₂O-MeOH-*n*-BuOH-EtOAc = 1:1:1:2; Merck silica gel 60F-254 aluminium-backed plates; visualized at 254-nm and with an ethanol solution of Ce₂(SO₄)₃ and phosphomolybdic acid followed by heating); mp = 198–200 °C (Et₂O); IR (film): 3333, 3039, 2946, 1732, 1694, 1542, 1536, 1454, 1340, 1266, 1246, 1085 cm⁻¹; ¹H NMR (400 MHz, DMSO-d_6) δ: 5.05 (s, 2 H), 5.22 (d, J = 8.8 Hz, 1 H), 7.27–7.41

(m, 5 H), 8.13 (d, J = 8.8 Hz, 1 H); The submitters report an additional ^{1}H NMR resonance at 6.26 (br s, 1 H); ^{13}C NMR (100 MHz, DMSO-d_6) δ: 66.2, 73.9, 128.5 (2 × C), 128.5, 129.0 (2 × C), 137.5, 156.2, 171.7; HRMS. [M + H]+ calcd. for $C_{10}H_{12}NO_5$: 226.0715. Found: 226.0716; Anal. calcd. for $C_{10}H_{11}NO_5$: C, 53.33; H, 4.92; N, 6.22. Found: C, 53.20; H, 5.27; N, 6.14.

8. Methanol (puriss., absolute, dried over molecular sieves, ≤0.01% H_2O) was purchased from Sigma-Aldrich Co. and used as received without further purification (submitters used 99.8%, water content: <50 ppm from Wako Pure Chemical Industries, Ltd.).

9. Conc. H_2SO_4 (95.0–98.0%) was purchased from Caledon Laboratories Ltd. and used as received without further purification (submitters used 95.0% from Wako Pure Chemical Industries, Ltd.).

10. The reaction typically requires 15 h to consume all the α-hydroxy-N-benzyloxycarbonylglycine and is monitored by TLC analysis. The R$_f$ values of the starting material and the product are baseline and 0.71, respectively (CH$_2$Cl$_2$-MeOH = 19:1; Merck silica gel 60F-254 aluminium-backed plates; visualized at 254-nm and with an ethanol solution of Ce$_2$(SO$_4$)$_3$ and phosphomolybdic acid followed by heating).

11. The submitters sometimes observed the solidification of the reaction mixture upon cooling to 0 °C. In this situation, the reaction mixture is warmed to room temperature and ice-cold sat. aq. NaHCO$_3$ (0 °C) is added at room temperature.

12. Data for product (without further purification): R$_f$ = 0.40 (hexanes-EtOAc = 2:1; Merck silica gel 60F-254 aluminium-backed plates; visualized at 254-nm and with an ethanol solution of Ce$_2$(SO$_4$)$_3$ and phosphomolybdic acid followed by heating); mp = 70–72 °C (EtOAc); IR (film): 3310, 3035, 2947, 1752, 1716, 1686, 1542, 1455, 1439, 1362, 1259, 1221, 1197, 1103 cm^{-1}; ^{1}H NMR (400 MHz, CDCl$_3$) δ: 3.47 (s, 3 H), 3.81 (s, 3 H), 5.14 (d, J = 12.3 Hz, 1 H), 5.17 (d, J = 12.3 Hz, 1 H), 5.36 (d, J = 9.4 Hz, 1 H), 5.84 (br s, 1 H), 7.31–7.39 (m, 5 H); ^{13}C NMR (100 MHz, CDCl$_3$) δ: 52.8, 56.2, 67.4, 80.6, 128.1, 128.3, 128.5, 135.7, 155.6, 167.9; HRMS. [M + Na]$^+$ calcd. for $C_{12}H_{15}NNaO_5$: 276.0848. Found: 276.0846; Anal. calcd. for $C_{12}H_{15}NO_5$: C, 56.91; H, 5.97; N, 5.53. Found: C, 56.83; H, 6.21; N, 5.56.

13. The use of an oversize flask allows for the removal of the reagents and solvent at the end of the reaction without the loss of product through bumping.

14. Toluene (dried max. 0.001% H_2O, puriss.) was purchased from Sigma-Aldrich Co. and used as received without further purification (submitters used 99.5%, water content: <30 ppm from Wako Pure Chemical Industries, Ltd.).

15. Phosphorus trichloride (ReagentPlus, 99%) was purchased from Sigma-Aldrich Co. and used as received without further purification (submitters used 99.0% from Wako Pure Chemical Industries, Ltd.).

16. The reaction typically requires 22–26 h to consume all the methyl α-methoxy-N-benzyloxycarbonylglycinate and is monitored by TLC analysis. The R_f values of the starting material and the intermediate are 0.71 and 0.50, respectively (CH_2Cl_2-MeOH = 19:1; Merck silica gel 60F-254 aluminium-backed plates; visualized at 254-nm and with an ethanol solution of $Ce_2(SO_4)_3$ and phosphomolybdic acid followed by heating). The intermediate of the reaction is *methyl N-benzyloxycarbonyl-α-chloroglycinate*. A small amount of the reaction mixture was concentrated *in vacuo* and analyzed by ^1H spectroscopy: ^1H NMR (400 MHz, $CDCl_3$) δ: 3.86 (s, 3 H), 5.10–5.24 (m, 2 H), 6.18 (br s, 1 H), 7.32–7.40 (m, 5 H).[3]

17. Trimethyl phosphite (>99%) was purchased from Sigma-Aldrich Co. and used as received without further purification.

18. The reaction typically requires 2 h to consume all the intermediate and is monitored by TLC analysis. The R_f values of the intermediate and the product are 0.50 and 0.45, respectively (CH_2Cl_2-MeOH = 19:1; Merck silica gel 60F-254 aluminium-backed plates; visualized at 254-nm and with an ethanol solution of $Ce_2(SO_4)_3$ and phosphomolybdic acid followed by heating).

19. Data for product (without further purification): R_f = 0.45 (EtOAc; Merck silica gel 60F-254 aluminium-backed plates; visualized at 254-nm and with an ethanol solution of $Ce_2(SO_4)_3$ and phosphomolybdic acid followed by heating); mp = 77–78 °C (hexanes); IR (film): 3229, 3034, 2963, 1749, 1716, 1535, 1427, 1332, 1277, 1240, 1213, 1030 cm^{-1}; ^1H NMR (400 MHz, $CDCl_3$) δ: 3.79 (d, $^3J_{H-P}$ = 11.0 Hz, 3 H), 3.82 (d, $^3J_{H-P}$ = 11.2 Hz, 3 H), 3.84 (s, 3 H), 4.93 (dd, $^2J_{H-P}$ = 22.3, J = 9.2 Hz, 1 H), 5.16 (d, J = 12.0 Hz, 1 H), 5.16 (d, J = 12.0 Hz, 1 H), 5.58 (d, J = 9.6 Hz, 1 H), 7.31–7.39 (m, 5 H); ^{13}C NMR (100 MHz, $CDCl_3$) δ: 52.0 (d, $^1J_{C-P}$ = 148 Hz), 53.3, 53.9 (d, $^2J_{C-P}$ = 6.8 Hz), 54.1 (d, $^2J_{C-P}$ = 6.5 Hz), 67.5, 128.1, 128.3, 128.5, 135.8, 155.5 (d, $^3J_{C-P}$ = 7.2 Hz), 167.1; HRMS. [M + H] calcd. for $C_{13}H_{19}NO_7P$: 332.0899. Found: 332.0905; Anal. calcd. for $C_{13}H_{18}NO_7P$: C, 47.14; H, 5.48; N, 4.23. Found: C, 47.51; H, 5.74; N, 4.34.

Waste Disposal Information

All hazardous materials should be handled and disposed of in accordance with "Prudent Practices in the Laboratory"; National Academic Press; Washington, DC, 1995.

3. Discussion

The Horner-Wadsworth-Emmons (HWE) olefination[4] has the following advantages over the Wittig reaction; 1) the phosphonate carbanions are more nucleophilic than phosphorus ylides, and even unreactive hindered ketones react readily in HWE olefinations; 2) water-soluble phosphonate byproducts facilitate the purification process; and 3) the product olefin geometry can be switched by the Corey-Kwiatkowski modification,[4a,5] the Still-Gennari modification,[4b,6] or the Ando modification.[7]

The present procedure describes a convenient and scalable preparation of the HWE reagent[8] that gives the (Z)-dehydroamino acid derivative, which allows facile access to Cbz-protected α-amino acid methyl esters via Rh-catalyzed enantioselective hydrogenation[9] (Scheme 1).

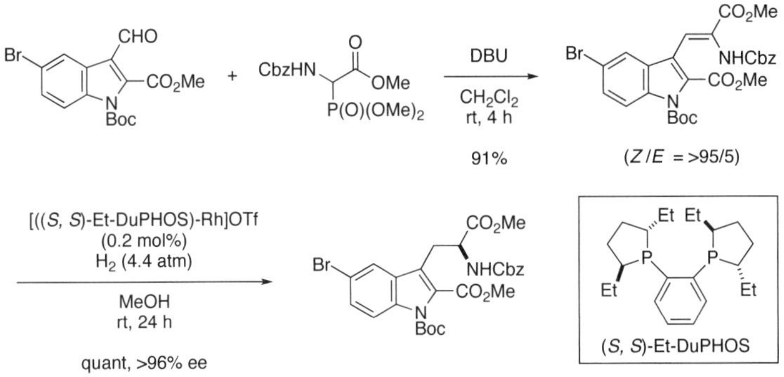

Scheme 1. Preparation of tryptophan derivative

The Cbz group in the HWE reagent is converted to a variety of protecting groups via hydrogenolysis followed by acylation[8b] (Table 1).

Table 1. Preparation of a series of related HWE reagents

hydrogenation	acylation	product
H$_2$ (3 atm), Pd/C MeOH, rt, 87%	Boc$_2$O CH$_2$Cl$_2$, rt, 80%	BocHN, OMe, P(O)(OMe)$_2$
H$_2$ (3 atm), Pd/C MeOH, rt, 87%	HCO$_2$H, DCC CH$_2$Cl$_2$, rt, 81%	H, N, O, OMe, P(O)(OMe)$_2$
H$_2$ (3 atm), Pd/C Ac$_2$O, MeOH, rt, 91%^{a}	–	AcHN, OMe, P(O)(OMe)$_2$

^{a} Debenzylation and acylation were carried out in one pot without isolation of the primary amine.

Recently, Ohfune and Shinada reported modified synthesis of a new Ando-type HWE reagent **1** providing an (*E*)-encarbamate.[10] The reagent was prepared from a hemiaminal, the product of step B in this procedure, by treatment with P(OPh)$_3$ and TMSOTf. Treatment of benzaldehyde with phosphonate **1** provided the corresponding (*E*)-dehydrophenylalanine derivative with excellent stereoselectivity (97:3). Their (–)-kaitocephalin synthesis features a substrate-controlled diastereoselective hydrogenation of (*E*)-enamide **2** prepared from the corresponding phosphonate **3** readily prepared from **1** (Scheme 2).

158

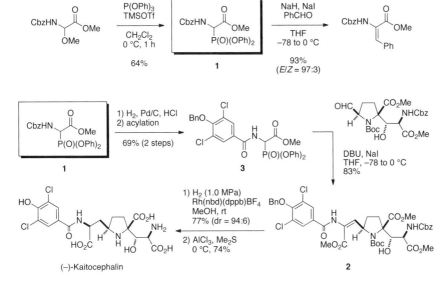

Scheme 2. Total synthesis of (−)-kaitocephalin featuring (E)-selective olefination and substrate-controlled diastereoselective hydrogenation

1. Graduate School of Pharmaceutical Sciences, Tohoku University, Aramaki, Aoba-ku, Sendai 980-8578, Japan. This work was supported by The Ministry of Education, Culture, Sports, Science and Technology, Japan.

2. Graduate School of Pharmaceutical Sciences, University of Tokyo, 7-3-1 Hongo, Bunkyo-ku, Tokyo 113-0033, Japan.

3. Williams, R. M.; Aldous, D. J.; Aldous, S. C. *J. Org. Chem.* **1990**, *55*, 4657.

4. (a) Boutagy, J.; Thomas, R. *Chem. Rev.* **1974**, *74*, 87. (b) Maryanoff, B. E.; Reitz, A. B. *Chem. Rev.* **1989**, *89*, 863.

5. Corey, E. J.; Kwiatkowski, G. T. *J. Am. Chem. Soc.* **1966**, *88*, 5654.

6. Still, W. C.; Gennari, C. *Tetrahedron Lett.* **1983**, *24*, 4405.

7. (a) Ando, K. *J. Org. Chem.* **1997**, *62*, 1934. (b) Ando, K. *J. Org. Chem.* **2000**, *65*, 4745. (c) Ando, K. *J. Synth. Org. Chem., Jpn.* **2000**, *58*, 869.

8. (a) Zoller, U.; Ben-Ishai, D. *Tetrahedron* **1975**, *31*, 863. (b) Schmidt, U.; Lieberknecht, A.; Wild, J. *Synthesis* **1984**, 53.

9. (a) Wang, W.; Xiong, C.; Zhang, J.; Hruby, V. J. *Tetrahedron* **2002**, *58*,

3101. (b) Crépy, K. V. L.; Imamoto, T. *Adv. Synth. Catal.* **2003**, *345*, 79. For a review on enantioselective hydrogenation, see: (c) Tang, W.; Zhang, X. *Chem. Rev.* **2003**, *103*, 3029.
10. Hamada, M.; Shinada, T.; Ohfune, Y. *Org. Lett.* **2009**, *11*, 4664.

Appendix
Chemical Abstracts Nomenclature (Registry Number)

Benzyl carbamate: Carbamic acid, Phenylmethyl ester; (621-84-1)
Glyoxylic acid monohydrate: Acetic acid, 2,2-dihydroxy-; (563-96-2)
Sulfuric acid: Sulfuric acid; (7664-93-9)
Phosphorous trichloride: Phosphorous trichloride; (123-08-0)
Trimethylphosphite: Phosphorous acid, trimethyl ester; (121-45-9)

Hidetoshi Tokuyama was born in Yokohama in 1967. He received his Ph.D. in 1994 from Tokyo Institute of Technology under the direction of Professor Ei-ichi Nakamura. He spent one year (1994-1995) at the University of Pennsylvania as a postdoctoral fellow with Professor Amos B. Smith, III. He joined the group of Professor Tohru Fukuyama at the University of Tokyo in 1995 and was appointed Associate Professor in 2003. In 2006, he moved to Tohoku University, where he is currently Professor of Pharmaceutical Sciences. His research interest is on the development of synthetic methodologies and total synthesis of natural products.

Hiroki Azuma was born in Fukushima in 1987. He received his B.S. in 2009 from the Faculty of Pharmaceutical Sciences, Tohoku University, where he carried out undergraduate research in the laboratories of Professor Hideo Takeuchi. In the same year, he then began his doctoral studies at the Graduate School of Pharmaceutical Sciences, Tohoku University under the supervision of Professor Hidetoshi Tokuyama. His research focuses on the total synthesis of azaspirocyclic natural products.

160

Kentaro Okano was born in Tokyo in 1979. He received his B.S. in 2003 from Kyoto University, where he carried out undergraduate research under the supervision of Professor Tamejiro Hiyama. He then moved to the laboratories of Professor Tohru Fukuyama at the University of Tokyo and started his Ph.D. research on synthetic studies toward antitumor antibiotic yatakemycin by means of the copper-mediated aryl amination strategy. In 2007, he started his academic carrier at Tohoku University, where he is currently an assistant professor in Professor Hidetoshi Tokuyama's group. His current research interest is natural product synthesis based on the development of new synthetic methodologies.

Tohru Fukuyama received his Ph.D. in 1977 from Harvard University with Yoshito Kishi. He remained in Kishi's group as a postdoctoral fellow until 1978 when he was appointed as Assistant Professor of Chemistry at Rice University. After seventeen years on the faculty at Rice, he returned to his home country and joined the faculty of the University of Tokyo in 1995, where he is currently Professor of Pharmaceutical Sciences. He has primarily been involved in the total synthesis of complex natural products of biological and medicinal importance. He often chooses target molecules that require development of new concepts in synthetic design and/or new methodology for their total synthesis.

Alistair Boyer was born in 1982 in Warrington, UK. He obtained his M.Sci. at the University of Cambridge in 2004, completing his masters project with Prof. Andrew B. Holmes. He stayed at Cambridge to perform his Ph.D. studies under the supervision of Prof. Steven V. Ley working on the synthesis of azadirachtin. In 2009, he moved to Toronto, Canada to become a postdoctoral research associate in the group of Prof. Mark Lautens, investigating novel rhodium-catalyzed reactions.

Discussion Addendum for:
5-*ENDO-TRIG* CYCLIZATION OF 1,1-DIFLUORO-1-ALKENES: SYNTHESIS OF 3-BUTYL-2-FLUORO-1-TOSYLINDOLE (1*H*-INDOLE, 3-BUTYL-2-FLUORO-1-[(4-METHYLPHENYL)SULFONYL]-)

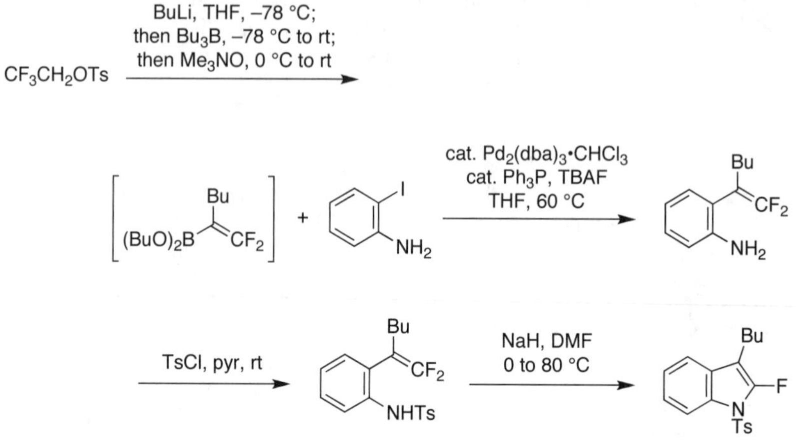

Prepared by Junji Ichikawa.*[1]

Original article: Ichikawa, J.; Nadano, R.; Mori, T.; Wada, Y. *Org. Synth.* **2006**, *83*, 111.

The precursors of 2-fluoroindoles, β,β-difluorostyrenes, are prepared by the palladium-catalyzed coupling reaction of 2,2-difluorovinylboron compounds with aryl iodides. The reaction of dialkyl(2,2-difluorovinyl)boranes **2** (Scheme 1) with an aryl iodide resulted in the coupling of not only the difluorovinyl but also the alkyl moieties on the boron. The contamination by alkyl-coupling product was suppressed by using fluoride salt as a base,[2a,3] and eventually overcome by the selective oxidation of the alkyl–boron bonds leading to the boronates prior to the coupling reaction as shown in the above scheme. Another way is a vinyl-selective transmetalation from boron to palladium via copper(I) species **6** (Scheme 1). On treatment of *in-situ* generated **2** with aryl iodides in the presence of cuprous iodide and a palladium catalyst, disubstituted difluoroalkenes **7** are obtained in excellent yield (Table 1).[2b,2c,4b]

Org. Synth. **2011**, *88*, 162-167
Published on the Web 1/19/2011

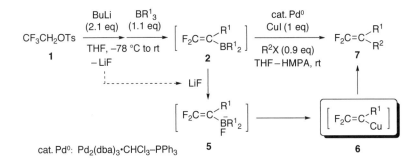

Scheme 1

cat. Pd⁰: Pd$_2$(dba)$_3$·CHCl$_3$–PPh$_3$

Alkenyl iodides and bromides,[5] and benzyl bromides[2c] are also successfully employed in the cross-coupling reaction. The configuration of 1-alkenyl halides is completely preserved during the reaction. The coupling reaction with alkynyl iodides[6] and allyl bromides[2c] proceeds without the palladium catalyst. These reactions provide important synthetic intermediates such as 1,1-difluoro-1,3- and -1,4-dienes and 1,1-difluoro-1,3-enynes (Scheme 1, Table 1).

Table 1. Synthesis of Disubstituted 1,1-Difluoro-1-alkenes

Product, Time (Halide); Yield from **1**		
F$_2$C=C, Bu, Ph	F$_2$C=C, Bu, CH=CHPh	F$_2$C=C, Bu, C≡C*c*-Hex
1 h (I); 90%	1 h (Br); 86%	1 h (I); 77%[a,b]
F$_2$C=C, Bu (cyclohexenyl)	F$_2$C=C, Bu, CH$_2$Ph	
1 h (Br); 66%[b]	15 h (Br); 61%	

[a] CuCl·SMe$_2$ (1 eq) was employed instead of CuI.
[b] The reaction was carried out in the absence of the Pd catalyst.

The sequence of reactions: (i) the 1,2-migration via borate complexes and (ii) the coupling reaction via difluorovinylcoppers **6** provides a general synthetic method for unsymmetrically disubstituted 1,1-difluoro-1-alkenes **7** by the introduction of two different carbon substituents (R^1 and R^2) onto *difluorovinylidene* (CF$_2$=C) unit in opposite polarities. That is, the carbon framework of difluoroalkenes can be constructed at will in this one-pot operation, where 2,2,2-trifluoroethyl *p*-toluenesulfonate (**1**) functions as a synthon of *difluorovinylidene* ambiphile (Scheme 2).

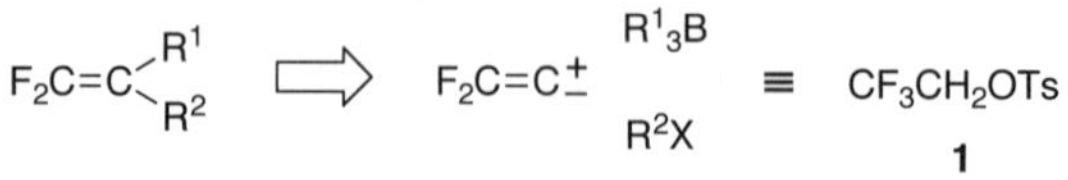

$$F_2C=C{\overset{R^1}{\underset{R^2}{\diagdown}}} \implies F_2C=C^{\pm} \quad \overset{R^1_3B}{\underset{R^2X}{}} \equiv \underset{\mathbf{1}}{CF_3CH_2OTs}$$

Scheme 2

Thus, procedure A in the original article, providing *o*-(1,1-difluorohex-1-en-2-yl)aniline, can be replaced with the following sequence (Scheme 3) on the same scale.[4a] The solution of 2,2-difluorovinylborane, generated from **1** (15.3 g, 60 mmol), is treated with hexamethylphosphoric triamide (HMPA, 20 mL), PPh₃ (1.26 g, 4.8 mmol), and Pd₂(dba)₃·CHCl₃ (1.24 g, 1.20 mmol). To the solution is added *N*-butylmagnesio-*o*-iodoaniline, which is generated *in situ* from *o*-iodoaniline (9.20 g, 42.0 mmol) and dibutylmagnesium (42 mL, 1.0 M in heptane, 42 mmol). MeMgI can be also used for deprotonation of the amino group. Copper(I) iodide (11.4 g, 60 mmol) is then added, and the reaction mixture is stirred at room temperature. After quenching the reaction with phosphate buffer, the mixture is treated with hydrogen peroxide (100 mL, 30% in water) at 0 °C and then at room temperature for 1 h. The mixture is filtered through a pad of Celite, and organic materials are extracted with ethyl acetate. After removing HMPA by short column chromatography on silica gel, the crude product is distilled under reduced pressure to give *o*-(1,1-difluorohex-1-en-2-yl)aniline (5.76 g, 65%). The coupling reaction of difluorovinylcoppers **6** can be also conducted with *o*-iodo-*p*-toluenesufonanilide without its deprotonation to give the indole precursors with an *N*-tosyl group directly, albeit in slightly lower yield.

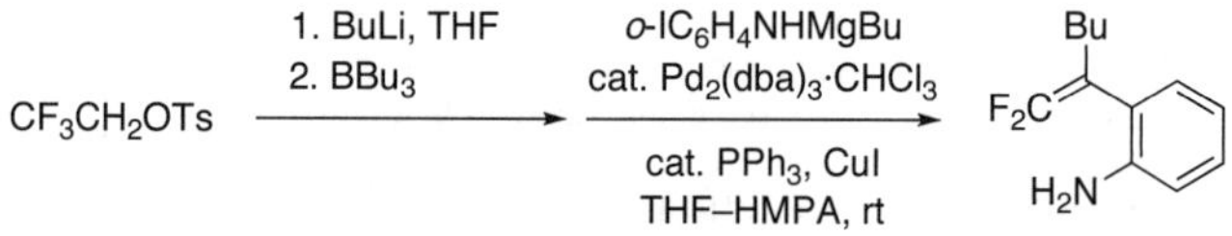

$$CF_3CH_2OTs \quad \xrightarrow[\substack{2.\ BBu_3}]{\substack{1.\ BuLi,\ THF}} \quad \xrightarrow[\substack{cat.\ PPh_3,\ CuI \\ THF-HMPA,\ rt}]{\substack{o\text{-}IC_6H_4NHMgBu \\ cat.\ Pd_2(dba)_3\cdot CHCl_3}}$$

Scheme 3

Moreover, difluorovinylcoppers **6** react as alkyl-substituted difluorovinyl anion with various electrophiles, such as acyl chlorides,[7] chlorodiphenylphosphine,[8] iodine,[9] NBS,[2c] and (methylene)ammonium iodides,[2c] which allows introduction of acyl, phosphine, iodine, bromine, and aminomethyl substituents to the difluorovinylic position. The directly-functionalized 1,1-difluoro-1-alkenes are readily supplied by this methodology.

While several methods for the activation of vinylboron compounds by transmetalation to copper have been reported,[10] all of them require a strong nucleophilic species, such as methylcopper, alkyllithium, or sodium methoxide, to induce borate-complex formation. In the above-mentioned transmetalation, the lithium fluoride formed *in situ* acts as the nucleophile, which allows the selective activation of the vinyl group on boron under mild conditions (Scheme 1).

Furthermore, this type of activation with fluoride ion and copper(I) salt can also be applied to fluorine-free alkenylboranes.[2b] The cross-coupling reaction of a *B*-(1-alkenyl)-9-borabicyclo[3.3.1]nonane (BBN), generated *in situ* via hydroboration of the corresponding 1-alkyene with 9-BBN, readily proceeds at room temperature within 1 h in the presence of cesium fluoride, cuprous iodide, and a palladium catalyst, while high temperature is normally required for efficient reaction rates in the Suzuki–Miyaura coupling (Scheme 4).[2b,2c,11] This is a useful activation method of alkenylboranes, which increases their reactivity as carbon nucleophiles.

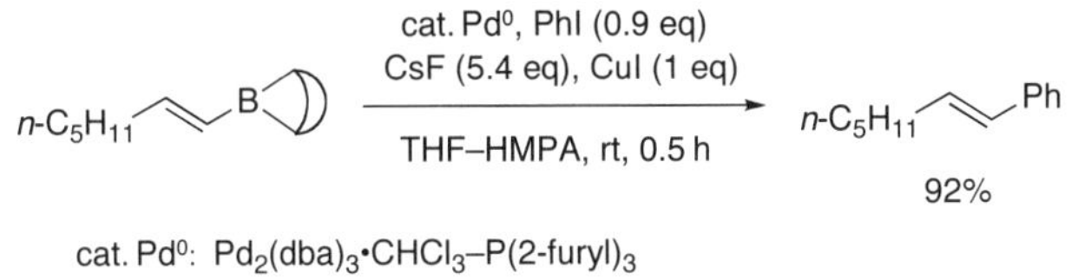

cat. Pd⁰, PhI (0.9 eq)
CsF (5.4 eq), CuI (1 eq)
THF–HMPA, rt, 0.5 h
n-C5H11
B
n-C5H11
Ph
92%
cat. Pd⁰: Pd2(dba)3•CHCl3–P(2-furyl)3

Scheme 4

1. Department of Chemistry, Graduate School of Pure and Applied Sciences, University of Tsukuba, Tsukuba, 305-8571, Japan. junji@chem.tsukuba.ac.jp

2. For the original reports on the synthesis of 1,1,-difluoro-1-alkenes, see: (a) Ichikawa, J.; Moriya, T.; Sonoda, T.; Kobayashi, H. *Chem. Lett.* **1991**, 961. (b) Ichikawa, J.; Minami, T.; Sonoda, T.; Kobayashi, H. *Tetrahedron Lett.* **1992**, *33*, 3779. (c) Ichikawa, J. *J. Fluorine Chem.* **2000**, *105*, 257 and references cited therein.
3. Since our report in 1991,[2a] *fluoride*-activation strategy has been widely

applied in the Suzuki–Miyaura coupling. See for example: (a) Wright, S. W.; Hageman, D. L.; McClure, L. D. *J. Org. Chem.* **1994**, *59*, 6095. (b) Desurmont, G.; Dalton, S.; Giolando, D. M.; Srebnik, S. *J. Org. Chem.* **1997**, *62*, 8907. (c) Wolfe, J. P.; Singer, R. A.; Yang, B. H.; Buchwald, S. L. *J. Am. Chem. Soc.* **1999**, *121*, 9550. (d) Littke, A. F.; Dai, C.; Fu, G. C. *J. Am. Chem. Soc.* **2000**, *122*, 4020. (e) Kotha, S.; Behera, M.; Shah, V. R. *Synlett* **2005**, 1877. (f) Korenaga, T.; Kosaki, T.; Fukumura, R.; Ema, T.; Sakai, T. *Org. Lett.* **2005**, *7*, 4915. (g) Ishikawa, S.; Manabe, K. *Chem. Lett.* **2006**, *35*, 164. (h) Molander, G. A.; Sandrock, D. L. *Org. Lett.* **2009**, *11*, 2369. (i) Butters, M.; Harvey, J. N.; Jover, J.; Lennox, A. J. J.; Lloyd-Jones, G. C.; Murray, P. M. *Angew. Chem. Int. Ed.* **2010**, *49*, 5156.

4. For the original reports on the synthesis of 2-fluoroindoles, see: (a) Ichikawa, J.; Wada, Y.; Fujiwara, M.; Sakoda, K. *Synthesis* **2002**, 1917. (b) Ichikawa, J.; Wada, Y.; Okauchi, T.; Minami, T. *Chem. Commun.* **1997**, 1537. See also: (c) Ichikawa, J.; Fujiwara, M.; Wada, Y.; Okauchi, T.; Minami, T. *Chem. Commun.* **2000**, 1887.

5. Ichikawa, J.; Ikeura, C.; Minami, T. *Synlett* **1992**, 739.

6. Ichikawa, J.; Ikeura, C.; Minami, T. *J. Fluorine Chem.* **1993**, *63*, 281.

7. Ichikawa, J.; Hamada, S.; Sonoda, T.; Kobayashi, H. *Tetrahedron Lett.* **1992**, *33*, 337.

8. Ichikawa, J.; Yonemaru, S.; Minami, T. *Synlett* **1992**, 833.

9. Ichikawa, J.; Sonoda, T.; Kobayashi, H. *Tetrahedron Lett.* **1989**, *30*, 6379.

10. For a review, see: Wipf, P. *Synthesis* **1993**, 537. For a recent example, see: Gerard, J.; Hevesi, L. *Tetrahedron* **2004**, *60*, 367.

11. For the recent reports on the room-temperature Suzuki–Miyaura coupling, see: (a) He, Y.; Cai, C. *Catal. Lett.* **2010**, *140*, 153. (b) Rahimi, A.; Schmidt, A. *Synlett* **2010**, 1327. (c) Navarro, O.; Marion, N.; Mei, J.; Nolan, S. P. *Chem. Eur. J.* **2006**, *12*, 5142. (d) Barder, T. E.; Walker, S. D.; Martinelli, J. R.; Buchwald, S. L. *J. Am. Chem. Soc.* **2005**, *127*, 4685 and references therein. (e) Savarin, C.; Liebeskind, L. S. *Org. Lett.* **2001**, *3*, 2149. For a review, see: (f) Lipshutz, B. H.; Ghorai, S. *Aldrichim. Acta*, **2008**, *41*, 59.

Junji Ichikawa was born in Tokyo in 1958. He received his B. Sc. and Dr. Sc. from the University of Tokyo under the supervision of Professor Teruaki Mukaiyama. He joined Kyushu University as an Assistant Professor in 1985. After working at Harvard University with Professor E. J. Corey (1989–1990), he moved to Kyushu Institute of Technology as a Lecturer. In 1999, he joined the University of Tokyo as an Associate Professor. He was then appointed Professor in Department of Chemistry, University of Tsukuba in 2007. His research interests lie in the area of synthetic methodology, specifically the development of novel reactions based on the properties of metals and fluorine.

PREPARATION OF A TRIFLUOROMETHYL TRANSFER AGENT: 1-TRIFLUOROMETHYL-1,3-DIHYDRO-3,3-DIMETHYL-1,2-BENZIODOXOLE

Submitted by Patrick Eisenberger, Iris Kieltsch, Raffael Koller, Kyrill Stanek, and Antonio Togni.[1]

Checked by Kay M. Brummond and Baptiste Manteau.

1. Procedure

A. 2-(2-Iodophenyl)propan-2-ol (**1**) A 250-mL three-necked, round-bottomed flask equipped with a reflux condenser, an argon inlet, a 50-mL dropping funnel with a rubber septum, a Teflon-coated magnetic stir bar and a rubber septum is charged with magnesium turnings (9.43 g, 388 mmol, 3.10 equiv) (Note 1). The vessel is flame dried under vacuum and maintained under an atmosphere of argon during the course of the reaction (Note 2). The flask is charged with diethyl ether (25 mL) (Note 3). The dropping funnel is charged with a solution of methyl iodide (17.3 mL, 278 mmol, 2.20 equiv) (Note 4) in diethyl ether (25 mL) by means of a syringe. The methyl iodide solution is added dropwise to the magnesium turnings. The reaction is initiated, as evidenced by reflux, upon the addition of 4 mL of methyl iodide solution. The reaction mixture is immediately diluted with additional diethyl ether (35 mL) through the septum of the flask using a syringe. Addition of methyl iodide is continued at a rate of 1 mL/min to maintain a gentle reflux. After the addition is complete the reaction mixture is allowed to cool to ambient temperature. The brownish reaction mixture is allowed to stand until the remaining magnesium turnings have settled, then the supernatant is transferred via cannula through the septa into

168

Org. Synth, **2011**, *88*, 168-180
Published on the Web 1/26/2011

a 500-mL three-necked, round-bottomed flask equipped with a reflux condenser with argon inlet, a 50-mL dropping funnel with a rubber septum, a large Teflon-coated magnetic stir bar and a rubber septum (Note 5). The magnesium turnings are rinsed with diethyl ether (25 mL), and the rinsings are transferred to the reaction flask by cannula, and the solution is cooled to 0 °C (ice/water bath). The dropping funnel is charged with a solution of methyl 2-iodobenzoate (19.1 mL, 126 mmol, 1.00 equiv) in diethyl ether (20 mL) by means of a syringe. The solution is added dropwise under vigorous stirring over 10 min (Note 6). Additional diethyl ether (13 mL) is used to rinse the dropping funnel. The reaction mixture is left in the cooling bath and allowed to warm to ambient temperature overnight (15 h). The brown suspension is heated to reflux for 1.5 h (Note 7), then cooled to 0 °C (ice water bath) and treated carefully with a saturated aqueous ammonium chloride solution (150 mL). A thick yellow precipitate forms and water (2 × 150 mL) is added until most of the solid material dissolves, and then the yellow suspension is filtered through a pad of celite. The organic phase is separated from the aqueous phase, and the aqueous phase is extracted with diethyl ether (4 x 200 mL). The combined ethereal phases are dried over potassium carbonate, filtered, the solvent is evaporated on a rotary evaporator (40 °C, 675 mmHg), and the residue dried under vacuum. The crude compound **1** (27.5 g, 105 mmol, 83% yield, 90% purity) is obtained as a brownish oil (Notes 8-10).

B. *1-Chloro-1,3-dihydro-3,3-dimethyl-1,2-benziodoxole* (**2**) A 100-mL two-necked, round-bottomed flask is equipped with an argon inlet, a Teflon-coated magnetic stir bar and a rubber septum and is protected from light by wrapping the flask in aluminum foil. The flask is charged with crude compound **1** (18.9 g, 64.9 mmol, 90% purity, 1.00 equiv) and purged with argon. Dichloromethane (60 mL) is added under a positive flow of argon (Note 11), and the solution is cooled to 0 °C. *tert*-Butyl hypochlorite (7.50 mL, 66.4 mmol, 1.02 equiv) is added over 20 seconds to this solution in the dark by means of a syringe (Note 12). Stirring overnight (17 h) results in a bright yellow orange solution that is concentrated using rotary evaporation (40 °C, 600 mmHg) and further dried under vacuum at room temperature for 30 min. The yellowish residue is dissolved in hot dichloromethane (50 mL) to give a bright yellow solution. Upon cooling in the freezer at -15 °C for 16 h large yellow crystals are formed and sub-sequently filtered (10.4 g, 35.1 mmol). The mother liquor is concentrated, dried under vacuum and treated with hot solutions of pentane (15 mL) and dichloromethane (15 mL) before being cooled in the freezer overnight (14 h) leading to the formation of another crop of yellow crystals, which are then filtered (3.50 g, 11.8 mmol). This sequence is repeated two more times to give 16.5 g of yellow crystals in total. The combined crystals are dried under vacuum to give compound **2** (16.5 g, 55.6 mmol, 86% based on 90% purity of compound **1**) as bright yellow crystals, mp = 148 – 150 °C (dec.) (Note 13).

> *Caution! During the preparation of 3, care should be taken not to heat the reaction mixture or crude product. DSC and TGA measurements reveal a good thermal stability below the melting point (78 °C) of the product 1-trifluoromethyl-1,3-dihydro-3,3-dimethyl-1,2-benziodoxole (3). However, above the melting point a rapid exothermic decomposition takes place (ca. 60 kcal mol^{-1}).*

C. 1-Trifluoromethyl-1,3-dihydro-3,3-dimethyl-1,2-benziodoxole (**3**) A 250-mL Schlenk flask equipped with a large Teflon-coated magnetic stir bar and a rubber septum is flame-dried under vacuum and maintained under an atmosphere of argon during the course of the reaction. The flask is charged with potassium acetate (6.66 g, 67.9 mmol, 1.68 equiv) and heated under vacuum using a heat gun (Note 14). After allowing the reaction flask to cool to room temperature, compound **2** (11.9 g, 40.3 mmol, 1.00 equiv) is added under a positive flow of argon followed by acetonitrile (100 mL) by means of a syringe. The yellow suspension is vigorously stirred for 1 h at ambient temperature giving a white suspension. Next, a 500-mL round-bottomed Schlenk-flask equipped with a Schlenk-frit with a rubber septum, and a Teflon-coated magnetic stir bar is flame-dried under vacuum and maintained under an atmosphere of argon. The white suspension from above is added to the Schlenk frit via cannula. The filtration is accomplished by creating a partial vacuum in the 500-mL Schlenk flask.

The 250-mL Schlenk flask is washed with additional acetonitrile (50 mL), and the acetonitrile wash is transferred to the Schlenk frit. Once all the liquid has been removed from the Schlenk-frit, it is replaced by a rubber septum under a positive flow of argon. The final solution in the 500-mL Schlenk flask is clear and almost colorless. Additional acetonitrile (50 mL) is added to the clear solution. The solution is cooled to -17 °C internal temperature (cryostat/isopropanol, -20 °C bath temperature) upon which the

170

acetoxy intermediate starts to precipitate giving a white suspension. (Trifluoromethyl)trimethylsilane (9.60 mL, 64.9 mmol, 1.61 equiv) is added by syringe, followed by dropwise addition of a solution of tetra-*n*-butyl-ammonium difluorotriphenylsilicate (0.065 g, 0.12 mmol, 0.3 mol%) in acetonitrile (2 mL) (Note 15). The reaction mixture is stirred for 16 h at -17 °C, and then warmed to -12 °C, at which time additional (trifluoro-methyl)trimethylsilane (1.30 mL, 8.80 mmol, 0.22 equiv) is added. The clear orange brown reaction mixture is warmed to ambient temperature over 3 h and then stirred at ambient temperature for an additional 3 h (Note 16). The volatile components of the mixture are removed using rotary evaporation (40 °C, 125 mmHg) and then under vacuum (0.8 mmHg) to give a slightly orangish-brown solid (Note 17). Dry *n*-pentane (150 mL) is added to the remaining brown solid (Note 18). A 250-mL Schlenk flask equipped with a Schlenk-frit (1.5 inch diameter), a 0.7 inch pad of aluminium oxide and a rubber septum is carefully flame-dried under vacuum and then maintained under argon. After allowing the apparatus to cool to room temperature, the solution is filtered through the pad of aluminium oxide via cannula into the Schlenk flask to give a clear, colorless solution (Note 19). Once all the liquid has been transferred from the Schlenk-frit, it is replaced by a rubber septum. The Schlenk flask is placed in a cool water bath (15 °C), and the solution is concentrated to dryness under vacuum to give compound **3** (11.3 g, 34.2 mmol, 85%) as a white solid, mp = 73 – 75 °C (Notes 20 and 21).

D. Ethyl (R)-2-amino-3-(trifluoromethylthio)propanoate hydrochloride (**4**) A 50-mL Schlenk tube equipped with a Teflon-coated magnetic stir bar and a rubber septum is charged with reagent **3** (730 mg, 2.21 mmol, 1.10 equiv) (Note 22). A second, 25-mL Schlenk tube equipped with a Teflon-coated magnetic stir bar and a rubber septum is charged with ethyl (*R*)-2-amino-3-mercaptopropanoate hydrochloride (373 mg, 2.01 mmol, 1.00 equiv) (Note 23). In both Schlenk tubes, an atmosphere of argon is established using three vacuum/argon cycles (Note 24). To both Schlenk tubes, methanol (4.5 mL each) is added via syringe (Note 25). Two colorless solutions are obtained and cooled to -78 °C (dry ice-acetone bath). The solution of the (*S*)-2-amino-3-mercaptopropanoic acid ethyl ester hydrochloride in methanol is added dropwise to the solution of reagent **3** using a cannula over 2 min. The solution turns yellow immediately. The 25-mL Schlenk tube is washed with methanol (2 x 2 mL) and the washing is also transferred to the 50-mL Schlenk tube via cannula. The solution is stirred at -78 °C for 30 min upon which it becomes colorless again. The cooling bath is replaced by a water bath (room temperature) and the solution is stirred for an additional 30 min, before it is concentrated under vacuum (Note 26). After complete removal of the solvent and further drying at room temperature under vacuum for 30 min, the oily colorless solid is washed with a mixture of hexanes and ethyl acetate (20:1; 3 x 20 mL) (Note 27). The suspension is filtered through a sintered glass filter and the Schlenk tube and the filter are washed with additional hexanes (2 x 20 mL). The remaining solid is dissolved in methanol (20 mL) and the clear colorless solution was taken to dryness using rotary evaporation (40 °C, 200 mmHg) and further dried under vacuum to give a white solid (495 mg, 1.95 mmol, 97%) with mp = 157-159 °C (Note 28).

2. Notes

1. Magnesium (purum) was purchased from Aldrich and used as received. Product **1** is obtained in higher purity when methyl magnesium iodide is freshly prepared. Performing the reaction using commercial methylmagnesium iodide results in the formation of significant quantities of 2-phenylpropan-2-ol.

2. The authors report that nitrogen can be used instead of argon.

3. Diethyl ether (puriss.) was purchased from Sigma-Aldrich and purified by passing through alumina using the Sol-Tek ST-002 solvent purification system.

4. Methyl iodide (99.5%) was purchased from Sigma-Aldrich and used as received.

5. The flask was flame-dried under vacuum prior to use.

6. Methyl 2-iodobenzoate (98%) was purchased from Alfa-Aesar and used as received.

7. The reaction can be monitored by treating an aliquot of the reaction mixture with a saturated solution of NH_4Cl and extracting it with Et_2O. The organic phase is then subjected to analytical thin layer chromatography (TLC) using precoated glass plates (TLC Silica Gel 60 F_{254}, EMD Chemicals), eluted with 20% ethyl acetate in hexane, and visualized with UV (254 nm) (compound **1** R_f = 0.50, 1-(2-iodophenyl)ethanone R_f = 0.59, methyl 2-iodobenzoate R_f = 0.66).

8. The authors report that analytically pure compound **1** can be obtained by bulb-to-bulb distillation at 110 °C (0.01 mmHg) which gives a slightly yellow, sticky oil that solidifies in a freezer (-18 °C).

Org. Synth, **2011**, *88*, 168-180

9. The submitters report that compound **1** can be prepared on a 200 g scale of product with slightly higher yields. On this larger scale it is necessary to use an overhead stirrer for the second step of the reaction and the extraction should be carried out using MTBE instead of Et_2O.

10. 2-(2-Iodophenyl)propan-2-ol can be stored under nitrogen at -18 °C without any decomposition, whereas at ambient temperature and when exposed to light it slowly decomposes. It has the following spectroscopic properties: ^1H NMR (300 MHz, CDCl$_3$) δ: 1.78 (s, 6 H), 2.62 (bs, 1 H), 6.91 (t, J = 7.8 Hz, 1 H), 7.34 (t, J = 7.8 Hz, 1 H), 7.65 (d, J = 7.8 Hz, 1 H), 7.98 (d, J = 7.8 Hz, 1 H); ^{13}C NMR (75 MHz, CDCl$_3$) δ: 29.8, 73.6, 93.2, 126.8, 128.2, 128.6, 142.7, 148.5; IR (neat): 3400, 3056, 2974, 2929, 1581, 1560, 1459, 1424, 1365, 1328, 1264, 1231, 1171, 1048, 1004, 952, 856, 757 cm^{-1}. The purity of the product was estimated by integrating the impurity (2-phenylpropan-2-ol) resonance at δ = 7.5 and the product resonance at δ = 7.6 in the ^1H NMR. From this the mol% of impurity is estimated to be 18% and the mass contribution calculated to be 10%.

11. Dichloromethane (99.5%) was purchased from Fisher Scientific and used as received.

12. *tert*-Butyl hypochlorite was purchased from TCI America and used as received. The authors synthesized *tert*-butyl hypochlorite according to *Org. Synth. Coll. Vol. 5*, 183. This procedure should be conducted in dim light and direct exposure to the hypochlorite should be avoided. The product should not be exposed to direct sunlight or rubber. Do not heat the product over its boiling point.

13. The authors report that the synthesis of compound **2** can be scaled up to 50 g of product without a significant drop in yield. *Caution:* The reaction is exothermic; on a larger scale it is important to have an adequate cooling bath and a reflux condenser. The reaction can be monitored by taking an aliquot of the reaction mixture and subjecting it to analytical thin layer chromatography (TLC) using precoated glass plates (TLC Silica Gel 60 F$_{254}$, EMD Chemicals), eluting with 20% ethyl acetate in hexane, and visualized with UV (254 nm) (compound **1** R$_f$ = 0.50, compound **2** R$_f$ = 0.05 – 0.40 smearing from the baseline). 1-Chloro-1,3-dihydro-3,3-dimethyl-1,2-benziodoxole is bench stable and can be stored in air. It has the following spectroscopic properties: ^1H NMR (300 MHz, CDCl$_3$) δ: 1.57 (s, 6 H), 7.17–7.20 (m, 1 H), 7.52–7.61 (m, 2 H), 8.03–8.06 (m, 1 H); ^{13}C NMR (75 MHz, CDCl$_3$) δ: 29.3, 85.1, 114.6, 126.1, 128.4, 130.4, 131.0, 149.5; IR (neat): 3081, 2970, 2924, 1590, 1562, 1460, 1438, 1378, 1364, 1275, 1255, 1178,

1154, 1110, 1030, 1001, 942, 864, 760, 720 cm^{-1}. Anal. Calcd. for C$_9$H$_{10}$ClIO: C, 36.45; H, 3.50; I, 42.80. Found: C, 36.41; H, 3.36; I, 42.52.

14. Potassium acetate (puriss) was purchased from Sigma-Aldrich and used as received, acetonitrile (puriss) was purchased from Sigma-Aldrich and purified by distillation over CaH$_2$ under argon.

15. (Trifluoromethyl)trimethylsilane and tetra-n-butylammonium difluorotriphenylsilicate were purchased from Sigma-Aldrich, tetra-n-butylammonium difluorotriphenylsilicate is dried under vacuum at room temperature for 1 h prior to use.

16. At -12 °C the reaction mixture should be a clear solution, if it is a suspension it should be cooled to -17 °C and tetra-n-butylammonium difluorotriphenylsilicate (0.065 g, 0.120 mmol) in acetonitrile (2 mL) is added to reinitiate the reaction. Then the reaction mixture is warmed to -12 °C over 4 h. For the checkers, a clear solution was always obtained at -12 °C and additional aliquots of tetra-n-butylammonium difluorotriphenyl-silicate were not necessary.

17. The product sublimes very easily. To prevent its loss, the Schlenk-flask should be cooled in an ice/water bath while drying the product under vacuum.

18. Pentane was purchased from EMD Chemicals and purified by distillation from sodium benzophenone ketyl.

19. Neutral aluminum oxide, activity I, was purchased from Sigma-Aldrich and flame dried under vacuum prior to use.

20. The yields of the synthesis may vary slightly. It is imperative to control and maintain the given temperatures accurately. Temperatures refer to the reaction mixture, not the cooling bath. The reaction can be monitored by taking an aliquot of the reaction mixture and subjecting it to analytical thin layer chromatography (TLC) using precoated glass plates (TLC Silica Gel 60 F$_{254}$, EMD Chemicals), eluted with 50% ethyl acetate in hexanes and visualized with UV (254 nm) (compound **2** R$_f$ = 0.05 – 0.48 smearing from the baseline, 1-acetoxy-1,3-dihydro-3,3-dimethyl-1,2-benziodoxole R$_f$ = 0.05 – 0.43 smearing from the baseline, compound **3** R$_f$ = 0.65). 1-Trifluoromethyl-1,3-dihydro-3,3-dimethyl-1,2-benziodoxole is moisture sensitive and should be stored under nitrogen or argon at −18 °C. Under these conditions the authors did not observe any decomposition over prolonged periods of time. DSC and TGA measurements reveal a good thermal stability below the melting point of the substance (78 °C). However, above the melting point a rapid exothermic decomposition takes place

174

(ca. 60 kcal mol^{-1}). High-purity samples of **3** may be obtained by sublimation at 40 °C (0.02 mmHg) during 4 h. Compound **3** is suspected to be toxic and should be handled with appropriate protection.

21. Trifluoromethyl-1,3-dihydro-3,3-dimethyl-1,2-benziodoxole has the following spectroscopic properties: ^{1}H NMR (400 MHz, CDCl$_3$) δ: 1.50 (s, 6 H), 7.40–7.45 (m, 2 H), 7.53–7.56 (m, 2 H); ^{13}C NMR (100 MHz, CDCl$_3$) δ: 30.8, 76.5, 110.6 (q, J = 3.0 Hz), 110.7 (q, J = 396.1 Hz), 127.3, 127.8 (q, J = 2.7 Hz), 129.8, 130.6, 149.2; ^{19}F NMR (376.6 MHz, CDCl$_3$) δ: –40.1; IR (neat): 2969, 2925, 1565, 1461, 1439, 1374, 1357, 1273, 1248, 1164, 1087, 999, 959, 871, 748 cm^{-1}; HRMS (MS ES+) calcd for C$_{10}$H$_{11}$F$_3$IO: 330.9807 (M + H). Found: 330.9793 (M + H). Anal. Calcd. for C$_{10}$H$_{10}$F$_3$IO: C, 36.39; H, 3.05; F, 17.27; I, 38.45. Found: C, 36.46; H, 3.04; F, 17.46; I, 38.29.

22. A Schlenk tube was chosen for the sake of convenience, a two-necked, round-bottomed flask with an argon inlet can be used instead.

23. Ethyl (*R*)-2-amino-3-mercaptopropanoate hydrochloride (98%) was purchased from Sigma Aldrich and used as received.

24. Nitrogen can also be used.

25. Methanol (anhydrous) was purchased from Sigma-Aldrich and used as received.

26. A rotary evaporator can be used instead.

27. *n*-Hexanes (96 %) and ethyl acetate (puriss) were purchased from Fisher Scientific and both were used as received. This step removes trace quantities of 2-(2-iodophenyl)propan-2-ol, and the checkers found that it was necessary to do additional washings to remove this compound.

28. Ethyl (*R*)-2-amino-3-(trifluoromethylthio)propanoate hydrochloride is stable to air and has the following spectroscopic properties ^{1}H NMR (400 MHz, MeOH-*d4*) δ: 1.36 (t, J = 7.1 Hz, 3 H), 3.62 (dd, J = 6.1, 15.2 Hz, 1 H), 3.68 (dd, J = 5.7, 15.2 Hz, 1 H), 4.35 (q, J = 7.1 Hz, 2 H), 4.48 (t, J = 5.8 Hz, 1 H), 5.19 (bs, 3 H); ^{13}C NMR (100 MHz, MeOH-*d4*) δ: 13.0, 28.9, 52.3, 63.0, 130.4 (q, J = 306.0 Hz), 166.7. ^{19}F NMR (376.6 MHz, MeOH-*d4*) δ: –42.9; IR (KBr pellet): 2923, 1740, 1598, 1571, 1487, 1390, 1352, 1424, 1118, 1013, 982, 855, 758, 740 cm^{-1}; HRMS (MS ES+) calcd for C$_6$H$_{11}$NO$_2$F$_3$S: 218.0463 (M + H). Found: 218.0450 (M + H); Anal. Calcd. for C$_6$H$_{11}$NO$_2$F$_3$SCl: C, 28.41; H, 4.37; F, 5.52; F, 22.47. Found: C, 28.52; H, 4.26; N, 5.55; F, 22.34.

Safety and Waste Disposal Information

All hazardous materials should be handled and disposed of in accordance with "Prudent Practices in the Laboratory"; National Academy Press; Washington, DC, 1995.

3. Discussion

Compounds **3** and **6** represent a new generation of electrophilic trifluoromethylation reagents, originally reported by Togni and co-workers in 2006.[2] The preparation of these compounds is conveniently carried out on a multi-gram-scale from easily available starting materials. Both reagents offer the advantage of being potentially recyclable. In fact, the byproducts resulting from a trifluoromethylation reaction either with reagent **3** or **6** are alcohol **1** or 2-iodobenzoic acid, respectively, which are the starting materials for the preparation of the reagents. Thus, compound **1** and 2-iodobenzoic acid may be readily separated from the main products by column chromatography.

Scheme 1. Possible transformations of various substrates using reagents **3** and **6**.

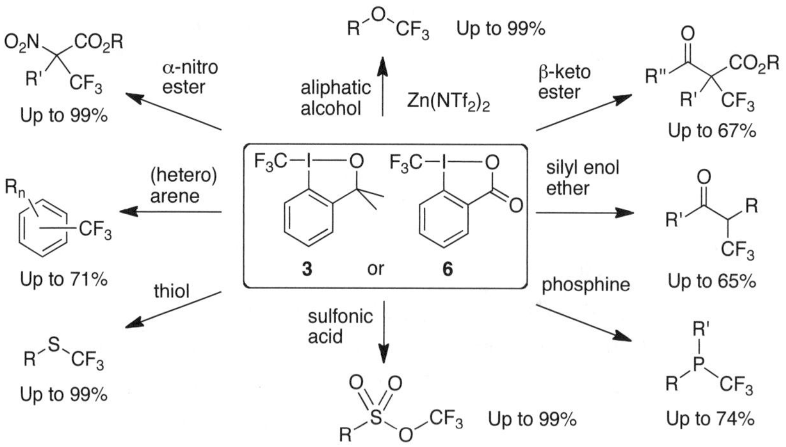

Reagents **3** and **6** are suited for the trifluoromethylation of a variety of carbon-,[3] sulfur-,[3,4] phosphorus-,[5] and oxygen-centered nucleophiles (see summary in Scheme 1).[6-8] Carbon nucleophiles such as β-keto esters or

176

silylenol ethers do afford the corresponding α-trifluoromethyl carbonyl derivatives in yields up to 60-70%, but the reaction is somewhat sluggish. α-Nitro esters give better yields but require the presence of a Cu(I) salt as a catalyst, typically 15 mol% CuBr·SMe$_2$. Electron-rich arenes and heterocycles react in terms of an electrophilic aromatic substitution. For nitrogen-containing heterocyclic compounds, a pronounced regioselectivity is observed in favor of the position adjacent to the nitrogen atom.

Among heteroatom nucleophiles, thiols are the best substrates, cleanly reacting preferentially with reagent **3** to the corresponding trifluoromethyl thioethers in excellent yields (up to quantitative).[3] This particular reaction shows an exceptional functional-group tolerance.[4] Primary phosphines react stepwise with reagent **6** to give the products of mono- or bis(trifluoromethylation), whereby the second step requires the presence of a base, typically DBU.[5]

Phenols (or phenolates) only undergo *O*-trifluoromethylation in low yields (up to 15%) when both the ortho and para positions already bear a substituent. However, also in this case, the major products are quinoid derivatives containing the CF$_3$ group at a quaternary center.[6]

Finally, when reagent **6** is activated by Zn(NTf$_2$)$_2$ primary and secondary aliphatic alcohols are converted to the corresponding trifluoromethyl ethers. However, this reaction requires an excess of the alcohols to ensure quantitative conversions of the reagent and to avoid decomposition side reactions. Simple alcohols that are liquid at room temperature may be used as solvents for this transformation.[7]

1. Department of Chemistry and Applied Biosciences, Swiss Federal Institute of Technology, ETH Zurich, Wolfgang-Pauli-Strasse 10, CH-8093 Zurich, Switzerland. atogni@ethz.ch. The work reported here has been supported by ETH Zurich and the Swiss National Science Foundation.

2. Eisenberger, P.; Gischig, S.; Togni, A. *Chem. Eur. J.* **2006**, *12*, 2579.

3. Kieltsch, I.; Eisenberger, P.; Togni, A. *Angew. Chem. Int. Ed.* **2007**, *46*, 754.

4. Capone, S.; Kieltsch, I.; Flögel, O.; Lelais, G.; Togni, A.; Seebach, D. *Helv. Chim. Acta* **2008**, *91*, 2035.

5. Eisenberger, P.; Kieltsch, I.; Armanino, N.; Togni, A. *Chem. Commun.* **2008**, 1575.

6. Stanek, K.; Koller, R.; Togni, A. *J. Org. Chem.* **2008**, *73*, 7678.
7. Koller, R.; Stanek, K.; Stolz, D.; Aardoom, R.; Niedermann, K.; Togni, A. *Angew. Chem. Int. Ed.* **2009**, *48*, 4332.
8. Koller, R.; Huchet, Q.; Battaglia, P.; Welch, J. M.; Togni, A. *Chem. Commun.* **2009**, 5993.

Appendix
Chemical Abstracts Nomenclature; (Registry Number)

2-(2-Iodophenyl)propan-2-ol (69352-05-2)

1-Chloro-1,3-dihydro-3,3-dimethyl-1,2-benziodoxole (69352-04-1)

1-Trifluoromethyl-1,3-dihydro-3,3-dimethyl-1,2-benziodoxole
(887144-97-0)

Iodomethane (74-88-4)

Methyl 2-iodobenzoate (610-97-9)

Tert-butyl hypochlorite (507-40-4)

(Trifluoromethyl)trimethylsilane (81290-20-2)

Tetrabutylammonium difluorotriphenylsilicate(IV) (163931-61-1)

2-Iodobenzoic acid (88-67-5)

Sodium (meta)periodate (7790-28-5)

1-Hydroxy-1,2-benziodoxol-3-(1*H*)-one (131-62-4)

Acetic anhydride (108-24-7)

1-Acetoxy-1,2-benziodoxol-3-(1*H*)-one (1829-26-1)

Cesium fluoride (13400-13-0)

1-(Trifluoromethyl)-1,2-benziodoxol-3(1*H*)-one (887144-94-7)

Ethyl (*R*)-2-amino-3-mercaptopropanoate hydrochloride (868-59-7)

Antonio Togni was born in Switzerland in 1956. He did his undergraduate and graduate studies (with L. M. Venanzi) at the ETH Zurich from 1975 to 1983. After a postdoctoral stay at Caltech with John E. Bercaw he joined in 1985 the Central Research Laboratories of Ciba-Geigy Ltd. in Basel, Switzerland, where he started working in the field of enantioselective catalysis. In 1992 he moved back to ETH becoming a full professor of organometallic chemistry in 1999. His research interests include asymmetric catalysis and organofluorine chemistry.

Patrick Eisenberger was born in Wettingen (Switzerland) in 1978. He studied chemistry at ETH, Zurich and obtained his Diploma degree in 2003. He then joined the group of Prof. Antonio Togni working on the synthesis and application of hypervalent iodine-based trifluoromethylating reagents and received his Ph.D. in 2007. In 2008 he accepted a postdoctoral position in the group of Prof. Laurel L. Schafer at UBC, Vancouver (Canada) where he was working on early transition-metal catalyzed syntheses of small N-containing molecules by hydroamination and hydroaminoalkylation. He is currently a postdoctoral researcher with Prof. Cathleen M. Crudden at Queen's University, Kingston (Canada).

Iris Kieltsch was born in Agnetheln (Romania) in 1979. She obtained her diploma in 2004 from the University of Marburg, Germany, and her Ph.D. in 2008 under the supervision of Prof. Antonio Togni at the ETH Zurich. After completing her doctoral work she moved to the University of Hawaii for postdoctoral studies with Prof. David Vicic.

Raffael Koller was born in Baden (Switzerland) in 1982. He studied chemistry at ETH, Zurich and completed his masters thesis in the research group of Prof. Sue Gibson at Imperial College, London. In 2006 he obtained his Masters degree at ETH Zurich, and then joined the group of Prof. Antonio Togni working on the application of hypervalent iodine reagents for the trifluoromethylation of oxygen-, carbon-, and phosphorus-centered nucleophiles. In 2010, after completing his Ph.D., he moved to Stanford University for postdoctoral studies with Prof. Barry M. Trost to work on total synthesis of natural products.

Kyrill Stanek was born in Weiningen (Switzerland) in 1980. He studied chemistry at ETH Zurich and completed his masters thesis in the research group of Prof. Peter H. Seeberger in 2005. In 2006 he joined the group of Prof. Antonio Togni working on the synthesis and application of electrophilic trifluoromethylating agents to improve the trifluoromethylation of oxygen centered nucleophiles. In addition, he investigated remote fluorine-metal interactions in late transition-metal complexes. He is currently working for Bachem AG.

Baptiste Manteau was born in 1982 in Poitiers, France. In 2006, he obtained his engineering degree in chemistry from ESCOM in Paris. In 2009, he completed his Ph.D. in chemistry with Dr F. Leroux from the University of Strasbourg and in collaboration with Bayer CropScience. He worked on the development of a general method to access trifluoromethoxy-heterocyclic building-blocks. He is currently pursuing post doctoral studies at the University of Pittsburgh under the guidance of Prof. Kay Brummond. His research is currently focusing on SAR and protein binding studies of a synthetic Chk1-phosphorylation inhibitor, which has been recently discovered from a diversity oriented synthesis library founded in his current laboratory.

SYNTHESIS OF (S,S)-DIISOPROPYL TARTRATE (E)-CROTYLBORONATE AND ITS REACTION WITH ALDEHYDES: (2R,3R,4R)-1,2-DIDEOXY-2-ETHENYL-4,5-O-(1-METHYLETHYLIDENE)-XYLITOL

Submitted by Huikai Sun and William R. Roush.[1]
Checked by David A. Candito, Mathieu Blanchot, and Mark Lautens.

1. Procedure

A. (S,S)-Diisopropyl tartrate (E)-crotylboronate **2**. An oven-dried 500-mL, three-necked, round-bottomed flask equipped with a 5-cm, egg-shaped stir bar, a rubber septum, and a digital and traceable thermocouple (Note 1) is connected to an argon line and is charged with *t*-BuOK (20.0 g, 95% purity, 170 mmol, 1.0 equiv) (Note 2) and anhydrous THF (85 mL) (Note 3) added by a syringe. This mixture is cooled to –78 °C with a dry ice/acetone bath, and then *trans*-2-butene (19.2 mL, 204 mmol, 1.2 equiv) (Note 4), condensed from a gas lecture cylinder into a 25-mL graduated cylinder immersed in a dry ice/acetone bath, is added via cannula. *n*-BuLi (2.5 M in hexane, 68.0 mL, 170.0 mmol) (Note 5) is added dropwise over 50 min using a syringe pump to make sure that the internal temperature does not rise above –65 °C. After the addition is complete, the cooling bath is removed, and the resultant yellow mixture is allowed to warm until the internal temperature reaches –50 °C (about 5 min). Then the reaction flask is quickly moved to an acetone bath precooled with dry ice to –50 °C. The internal temperature of the reaction solution is maintained between –50.0 °C to –50.5 °C for 25 min (Note 6) and then immediately recooled to –78 °C by moving the reaction flask to a dry ice/acetone bath.

Triisopropylborate (39.9 mL, 98% purity, 170 mmol) (Note 7) is added dropwise over 50 min via a syringe pump to the above orange solution of (*E*)-crotylpotassium **1** to make sure that the internal temperature does not rise above –65 °C. After completion of the addition, the reaction mixture is allowed to cool until the internal temperature remains constant around – 72.5 °C, maintained at this temperature for 10 min, and then rapidly poured into a 1-L separatory funnel containing 1N HCl solution (320 mL) saturated with NaCl and shaken vigorously. The aqueous layer is adjusted to pH 1 by adding 1N HCl solution (110 mL), and then a solution of (*S,S*)-diisopropyl tartrate ((*S,S*)-DIPT) (40.6 g, 98% purity, 170 mmol) (Note 8) in Et$_2$O (60 mL) is added and the mixture is shaken vigorously. The organic phase is separated, and the aqueous layer is extracted with Et$_2$O (4 x 80 mL). The combined organic layers are dried with MgSO$_4$ (120 g) over 2.5 h and then vacuum filtered through a fritted glass funnel under an argon blanket (Note 9) into a 1-L, oven-dried and pre-tared round-bottomed flask. The filter cake is washed with anhydrous Et$_2$O (2 x 70 mL) (Note 10), and then the combined filtrate is concentrated on the rotary evaporator (from 40 mmHg to 10 mmHg at rt) to a colorless thick liquid, which is further concentrated on a vacuum line (0.5–1 mmHg) with stirring by a 5-cm, egg-shaped and pre-tared stir bar to constant weight (57 g). The crude product is dissolved in anhydrous toluene (120 mL) (Note 11) to form a clear solution (173 mL) of (*S,S*)-diisopropyl tartrate (*E*)-crotylboronate **2** (Note 12). The material so obtained has ≥98% isomeric purity (Note 13), the resulting toluene solution has a reagent concentration of 0.65 mol/L, and the yield of reagent is 66% according to titration analysis (Note 14).

  B. *(2R,3R,4R)-1,2-Dideoxy-2-ethenyl-4,5-O-(1-methylethylidene)-xylitol* **4**. An oven-dried, 500-mL, single-necked, round-bottomed flask equipped with a 5-cm, egg-shaped stir bar, a rubber septum, and an argon balloon, is charged with powdered 4 Å molecular sieves (12.0 g) (Note 15) and anhydrous toluene (160 mL) added by syringe. After a solution of (*E*)-crotylboronate **2** in toluene (72.0 mL, 0.69 mol/L, 49.7 mmol, 1.2 equiv) is added by syringe, the resulting mixture is stirred at rt for 30 min and then cooled to –78 °C with a dry ice/acetone bath. A solution of D-(*R*)-glyceraldehyde acetonide **3** (6.0 g, 90% purity, 41.5 mmol) (Note 16) in anhydrous toluene (15.0 mL) is added dropwise over 50 min (Note 17) via a syringe pump, and after completion of the addition, the reaction mixture is stirred at this temperature for 2.0 h (Note 18). The reaction is quenched by slowly adding aqueous 2 N NaOH (130 mL) over 5 min by syringe, then the

182

cooling bath is removed. After ambient temperature is reached (about 1.5 h), the mixture is vigorously stirred for an additional 20 min. The toluene layer is separated, and then the reaction flask containing some solid residue is rinsed with Et_2O (120 mL), which is then used to extract the aqueous solution. The same rinse and extraction operation is repeated two more times with Et_2O (2 x 120 mL). The combined organic layers are washed with sat. $NaHCO_3$ (100 mL), brine (2 x 100 mL), dried with Na_2SO_4 (120 g) and vacuum filtered through a Büchner funnel. The filtrate is concentrated (from 35 to 8 mmHg) by rotary evaporation at ambient temperature to give a colorless liquid, containing a >98% of **4** (Note 19). Purification of the product by flash column chromatography (Note 20) using ether/hexanes (1/4 to 1/3) as the eluent provides 5.1 g (66% yield) of (2*R*,3*R*,4*R*)-1,2-dideoxy-2-ethenyl-4,5-*O*-(1-methylethylidene)-xylitol **4** as a colorless oil (Note 21).

2. Notes

1.   The digital and traceable thermocouple (Fisherbrand, –200 °C ~ +1370 °C) was purchased from Fisher Scientific Company.

2.   Potassium *tert*-butoxide (95% purity) was purchased from Sigma-Aldrich chemical Company, Inc., and stored and transferred in a glove box.

3.   Tetrahydrofuran was purchased from Fisher Chemical Company and dried by fresh distillation from sodium/benzophenone ketyl under an atmosphere of dry argon.

4.   The checkers purchased *trans*-2-butene from TCI (min. 99.0%, GC). The submitters purchased *trans*-2-butene (99+% purity) from Aldrich Chemical Company, Inc. and used as stored. It is necessary to use 1.1–1.2 equiv of *trans*-2-butene in order to compensate for the loss of the material during transfer.

5.   *n*-Butyllithium (2.5 M in hexane) was purchased from Aldrich Chemical Company, Inc., stored at 5 °C and used as received.

6.   The temperature of the central part of the reaction solution was monitored. Placing the probe of the thermocouple against reaction flask wall resulted in the isomeric purity of the crotylboronate reagent being reduced by 1-2%. The internal temperature of the solution was maintained between –55 °C and –50 °C by removing the flask from the cooling bath (dry ice/acetone) and allowing it to approach –50 °C, at which time the flask was resubmerged to lower the temperature. No deterioration of selectivity was observed.

7. Triisopropylborate (≥98% purity) was purchased from Sigma-Aldrich Chemical Company, Inc. and used as stored.

8. (*S,S*)-Diisopropyl tartrate ((*S,S*)-DIPT) (98% purity) was purchased from Sigma-Aldrich Chemical Company, Inc. and used as received.

9. The fritted glass funnel was covered with a Fisherbrand long-stem analytical funnel, to which an argon flow was applied through the long-stem end in order to prevent the solution from being exposed to moisture in air.

10. Anhydrous diethyl ether was purchased from Fisher Chemical Company and purified by passage through activated alumina using a GlassContour solvent purification system under argon.

11. Toluene was purchased from Fisher Chemical Company and purified by passage through activated alumina using a GlassContour solvent purification system under argon.

12. (*S,S*)-DIPT (*E*)-crotylboronate **2** is moisture sensitive, so it is handled as a solution in toluene. The total volume was measured when the toluene solution was transferred to an oven-dried, 250-mL, single-necked, round bottom flask by a syringe. No apparent decomposition was observed after this solution was stored at –20 °C under Ar for several months. It was previously reported that the reagent could be purified by distillation,[2] but the submitters found the procedure not to be reproducible owing to ease of decomposition. Therefore, for the past two decades the reagent has been consistently used in the submitter's laboratory as a solution in toluene without further purification.

13. The checkers determined isomeric purity by GC/MS (Elmer Autosystem XL GC coupled to Perkin Elmer turbomass MS; Column: 0.25 mm x 30 m, 0.25 μm film thickness, ZB5 (5% diphenyl/95%dimethyl-polysiloxane); Temperature program 70 °C to 170 °C over 10 min, then hold at 170 °C for 10 min; t_R for **2** is 11.85 min). The submitters determined isomeric purity of the product by GC analysis (30 m x 0.32 mm Agilent capillary metal column packed with silicon polymers; temperature program: 70 °C to 170 °C over 10 min, then hold at 170 °C for 10 min; t_R for **2** is 11.9 min; t_R for (*Z*)-crotylboronate **5** is 12.0 min). 11% of tartrate *n*-butylboronate **6** (t_R is 11.6 min) was present, reflecting incomplete metallation of *trans*-2-butene in this experiment, along with 8% of DIPT (t_R is 8.2 min) according to GC analysis.

$CO_2{}^iPr$ ... $CO_2{}^iPr$ ... Me ... Me

5   6   7   8

14. The checkers determined the yield and concentration of tartrate (*E*)-crotylboronate **2** by NMR. A 1-mL aliquot of the solution of (*E*)-crotylboronate **2** was withdrawn and transferred to a 25-mL round-bottomed flask and the solvent was removed under high vacuum. Then 30 μL of mesitylene (Aldrich, >99% purity) was measured accurately and added using a syringe. Then CDCl₃ was added (approximately 5 mL) and the flask was swirled a few times to ensure a homogenous concentration. Some of the solution was transferred to an NMR tube and the spectra was recorded using a 10 second relaxation delay. The concentration was then determine by the relative integration of the internal standard peak at 6.78 ppm (s, 3 H) and the signal of **2** centered on 5.47 ppm (m, 2 H). The submitters determined the yield and concentration of the solution of tartrate (*E*)-crotylboronate **2** by a titration procedure as described below. An oven-dried, 10-mL, single-necked, round-bottomed flask equipped with a stir bar, a rubber septum, and an argon balloon, is charged with cyclohexanecarboxaldehyde (112 mg, 1.0 mmol) and toluene (1.5 mL). A toluene solution (1.0 mL) of reagent **2** was added and stirred at rt for 1.0 h. The reaction solution was cooled to 0 °C by an ice bath, and methanol (1.0 mL) was added followed by NaBH₄ (113 mg, 3.0 mmol, 3.0 equiv). The resulting reaction mixture was vigorously stirred at rt for 1.5 h, and then 2 N NaOH (2 mL) was added carefully and the mixture vigorously stirred at rt for 30 min to hydrolyze the tartrate ester. The organic phase was separated and the aqueous layer was extracted with ether (3 x 5 mL). The combined organic layers were dried over K₂CO₃ and concentrated in vacuo (> 25 mmHg) to remove Et₂O, and then more toluene was added to make a 10 mL solution for GC analysis (the same temperature program was used as described in Note 13). Under these conditions cyclohexylmethanol (from reduction of unconsumed aldehyde) elutes at 3.0 min and homoallylic alcohol products (anti and syn) elute at 6.2 min. From the measured ratio of homoallylic alcohols over cyclohexylmethanol, the submitters calculate the concentration of reagent **2** solution and its yield accordingly. A standard curve was made by using cyclohexylmethanol (4.0 mg/mL) as internal standard and samples with ratios of homoallylic alcohols/cyclohexylmethanol of 0.5, 0.75, 1.0, 2.0 and 4.0.

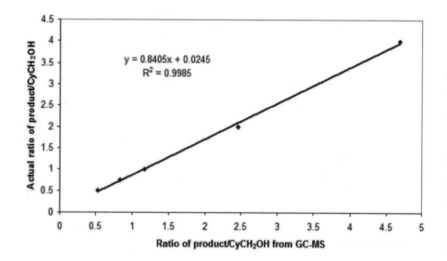

y = 0.8405x + 0.0245
$R^2 = 0.9985$

15. Powdered 4 Å molecular sieves (activated, 2.5 μm) were purchased from Sigma-Aldrich Chemical Company, Inc., further activated by flame heating under vacuum (0.5–1 mmHg) for 15 min and used a hour after cooling to rt under dry Ar atmosphere.

16. D-(R)-glyceraldehyde acetonide **3** was prepared according to the procedure of Schmid and Bryant.[3] The product (roughly 90% purity by ^{1}H NMR analysis) was obtained as a colorless liquid by distillation under vacuum (19–20 mmHg); fractions distilling at 46–48 °C were collected with an oven-dried flask immersed in an ice/water bath. The aldehyde **3** so obtained was used in the crotylboration reaction 30–45 min after distillation.

17. Aldehyde **3** (6.0 g) was dissolved in anhydrous toluene (12 mL) in an oven-dried, 50-mL, single-necked, round-bottomed flask. Then the resultant solution was added to the reaction mixture via a syringe pump over 45 min. An additional 3 mL of anhydrous toluene was used to rinse the flask and then was added to the reaction mixture via a syringe pump over 5 min.

18. The progress of the reaction cannot be monitored by TLC analysis because aldehyde **3** overlaps with impurities on the TLC plate. Submitters performed a small-scale reaction under the same conditions for 3.0 h and observed no increase in yield after product purification. Therefore, it was concluded that the reaction is complete after 2.0 h.

19. The checkers determined the isomeric purity of the crude reaction mixture to be >98% by GC/MS (Elmer Autosystem XL GC coupled to

186

Perkin Elmer turbomass MS; Column: 0.25 mm x 30 m, 0.25 μm film thickness, ZB5 (5% diphenyl/95%dimethylpolysiloxane); Temperature program 60 °C for 40 min, then increased to 170 °C over 5.5 min and hold at 170 °C for 10 min; t_R for **4** is 32.19 min. The submitters report that the crude product contains >98.0% of **4**, <1.5% of the (2R,3S,4S)-isomer **7** and <0.5% of the (2R,3S,4R)-diastereomer **8** (which derives from the minor (Z)-crotylboronate present in the reagent). Accordingly, the diastereoselectivity of this reaction was determined to be > 98.5:1.5. These data were obtained from GC-MS analysis (temperature program: 60 °C for 40 min, then increased to 170 °C over 5.5 min and hold at 170 °C for 10 min; t_R for **4** is 25.9 min; t_R for **7** is 27.6 min; t_R for **8** is 26.1 min).

20. Flash column chromatography was performed on a silica gel (EMD, grade 60, 40-63 μm) column. 10% Et$_2$O in hexanes (10 mL) was used to load the crude product onto a glass column (6.5 cm diameter) packed with 410 g of silica gel. The column was eluted with Et$_2$O/hexanes = 1/4 (1000 mL) followed by Et$_2$O/hexanes = 1/3 (ca 2000 mL) at 40 mL/min flow rate. Individual fractions (ca. 25 mL) were analyzed by TLC, and fractions were combined based on their composition as determined by TLC analysis. The first set of pooled fractions (1-2% yield, R_f = 0.57, Et$_2$O/hexanes = 1/3, 3 developments, UV and KMnO$_4$ active) contains a mixture of several unidentified compounds. The second set of pooled fractions (ca 15 to 20 test tubes) contain the desired product **4** (R_f = 0.52, Et$_2$O/hexanes = 1/3, 3 developments, UV and KMnO$_4$ active). The purity of **4** obtained in this way was determined to be 99+% by both ^{1}H NMR and GC-MS (Note 19). Mixed, trailing fractions containing ca. 80 mg of **4** (85-90% purity by ^{1}H NMR) mixed together with **8** (R_f = 0.42, Et$_2$O/hexanes = 1/3, 3 developments, UV and KMnO$_4$ active) and some unidentified compounds (R_f = 0.40, Et$_2$O/hexanes = 1/3, 3 developments, UV and KMnO$_4$ active) were recovered. A small sample of adduct **8** (from a separate run) was purified by a second column purification with Et$_2$O/hexanes = 1/3, which gave **8** with 92% purity according to GC-MS analysis. Finally, a small sample of diastereomer **7** (R_f = 0.34, Et$_2$O/hexanes = 1/3, 3 developments, UV and KMnO$_4$ active) was obtained in 96% purity from the first column as determined by GC-MS analysis. Diastereomers **4**, **7** and **8** displayed physical properties in agreement with data previously reported.[4]

21. Product **4** exhibits the following properties: colorless oil; ^{1}H NMR (400 MHz, CDCl$_3$) δ: 1.11 (d, J = 6.9 Hz, 3 H), 1.36 (s, 3 H), 1.42 (s, 3 H), 2.20 – 2.30 (m, 2 H), 3.39 (dd, J = 10.6, 5.0 Hz, 1 H), 3.73 (dd, J = 8.0, 7.0

Hz, 1 H), 4.00 (dd, J = 8.0, 6.5 Hz, 1 H), 4.10 (dd, J = 12.9, 6.5 Hz, 1 H), 5.00–5.11 (m, 2 H), 5.87 (ddd, J = 17.2, 10.4, 8.1 Hz, 1 H). ^{13}C NMR (101 MHz, $CDCl_3$) δ: 16.90, 25.64, 26.77, 41.51, 66.34, 75.44, 77.22, 109.41, 115.72, 139.71. $[\alpha]^D_{26}$ 15.7 (c 0.93, CH_2Cl_2). IR (film): 3476 (br), 2985, 2877, 1458, 1373, 1215, 1157, 1072, 1045, 914, 860 cm^{-1}. The material so obtained was identical in all respects with the compound described in the literature.[4]

Safety and Waste Disposal Information

All hazardous materials should be handed and disposed of in accordance with "Prudent Practices in the Laboratory"; National Academy Press; Washington, DC, 1995.

3. Discussion

Chiral crotylmetal reagents are important and useful reagents that react with aldehydes to give α-methyl-homoallylic alcohols with excellent stereoselectivity.[5] The reactions of (E)-crotylboron reagents with aldehydes is particularly useful, in that these reagents provide 2,3-*anti* diastereomers typically with greater selectivity than can be obtained by using propionate enolates via aldol reactions. To date, many efficient and practical chiral crotylmetal reagents or crotylmetallation procedures have been reported,[5] including those by Roush,[2,6] Brown,[7] Corey,[8] Leighton,[9] Denmark,[10] Soderquist,[11] Hall[12] and Krische,[13] among others. The DIPT modified (E)-crotylboronate reagent developed by Roush and his coworkers, and which is illustrated in the present procedure, is one of the most widely adopted reagents due to the ease of preparation, storage and handling, and the high selectivity obtained in crotylboration reactions of chiral aldehydes— especially in the matched double asymmetric mode.[2,6c-f]

The current procedure for preparation of (S,S)-diisopropyl tartrate (E)-crotylboronate (2) follows the protocol previously reported by the Roush group.[2] This procedure involves the metallation of *trans*-2-butene with *n*-BuLi and *t*-BuOK in THF at −50 °C for 25 min to generate (E)-crotylpotassium 1 with high isomeric purity (>98%). Treatment of 1 with (*i*-PrO)$_3$B followed by aqueous hydrolysis and esterification with (S,S)-diisopropyl tartrate (DIPT) furnished the (S,S)-DIPT (E)-crotylboronate reagent, 2. Care must be taken in this procedure to minimize isomerization

188

of (E)-crotylpotassium **1** to the thermodynamically more stable (Z)-isomer. The conditions for metallation of *trans*-2-butene reported here are those that maximize conversion while minimizing production of the (Z)-crotylpotassium (and subsequently the (Z)-crotylboronate) isomer. Thus, the reagent obtained as described here has >98% isomeric purity, but contains 10–12% of butylboronate **6** from borylation of residual BuLi that remains owing to incomplete metallation of 2-butene under these conditions. Previous studies in the submitter's laboratory indicate that amount of **6** can be suppressed by increasing reaction temperature to –45 °C for 15 min, or by increasing the reaction time at –50 °C to up to 45 min, but with some erosion of isomeric purity. With these modifications, the isomeric purity of **2** is reduced by 1–2% (e.g., 96–97% isomeric purity for **2**).[2]

(R,R)-DIPT (E)-crotylboronate **9** can be prepared from *trans*-2-butene and (R,R)-DIPT according to the same procedure as described here.[2] The preparation of (Z)-crotylboronates **5** and **10** by using *cis*-2-butene and the appropriate enantiomer of DIPT is analogous to the procedure described for (E)-crotylboronates with the exception that the metallation of (Z)-2-butene is performed at –20 to –25 °C for 30–45 min, which ensures high conversion in the metallation step to form (Z)-crotylpotassium. In comparison to preparation of (E)-crotylpotassium, the temperature control is less critical since (Z)-crotylpotassium is highly favored at equilibrium (>99:1). The remaining procedure is the same as that described for the synthesis of (E)-crotylboronate **2**.

The crotylboration reactions of aldehydes are performed in toluene solution at –78 °C for aliphatic aldehydes and in THF for aromatic aldehydes.[2, 6c, 6h] Best results are obtained by performing the reactions in the presence of powdered 4 Å molecular sieves to maintain an anhydrous reaction environment. Otherwise, crotylboronic acid resulting from hydrolysis of the moisture-sensitive chiral crotylboronate can function as a competitive, but achiral crotyl transfer reagent.

Reactions of crotylboronate reagents (R,R)-(E)-**9** and (R,R)-(Z)-**10** with achiral aliphatic aldehydes are reported to furnish the corresponding secondary homoallylic alcohols with 70 to 88% enantiomeric excess, as summarized in Table 1.[2] In some cases, the enantioselectivity can be increased to 91% ee when crotylboration is performed at –95 °C (entry 4). The reactions with pivalaldehyde, α,β-unsaturated aldehydes and aromatic aldehydes are less enantioselective (entries 8-13). *Anti* products **12** are usually obtained with >99% diastereoselectivity when 98% isomeric pure (*E*)-crotylboronate reagents are used. However, the reactions with 99% isomeric pure (*Z*)-crotylboronates afford *syn* products **13** with 97–98% diastereoselectivity. In addition, *anti* products **12** also exhibit higher enantiomeric purity than *syn* products **13**. These different results between **12** and **13** indicate that (*E*)-crotylboronates are more reactive and more enantioselective than their *Z*-crotyl counterparts

Double asymmetric reactions of the tartrate ester modified crotylboronates with chiral, non-racemic α-branched aldehydes also provide high diastereoselectivity in matched and in many mismatched double asymmetric reactions, as shown in Table 2.[6c-g] The facial selectivity of the crotylboration reaction is reversed upon switching the chirality of the chiral crotylboronate reagents, especially in reactions with chiral aldehydes with very modest intrinsic diastereofacial preferences. Numerous applications of these tartrate ester modified crotylboronate reagents have been reported in the total syntheses of natural products in the past two decades.[5a, 5f, 15]

Table 1. Crotylboration of Achiral Aldehydes with (R,R)-(E)-**9** and (R,R)-(Z)-**10**.

Entry	aldehydes (11)	reagent	isomeric purity	reaction time (yield)	anti:syn (12 : 13)	ee of major isomer
1	n-C$_9$H$_{19}$CHO	(R,R)-E-**9**	99.4	3 h (87%)	>99 : 1	88%
2		(R,R)-Z-**10**	98	6 h (80%)	3 : 99	82%
3	cyclohexyl-CHO	(R,R)-E-**9**	99.3	3 h (94%)	>99 : 1	85%
4		(R,R)-E-**9**	98	4 h (100%) (−95 °C)	>99 : 1	91%
5		(R,R)-Z-**10**	99	6 h (90%)	2 : 98	83%
6	TBSO-CH$_2$CH$_2$-CHO	(R,R)-E-**9**	98	4 h (71%)	>98 : 2	85%
7		(R,R)-Z-**10**	98	4 h (68%)	2 : >98	72%
8	t-Bu-CHO	(R,R)-E-**9**	99	6 d (41%)	95 : 5	73%
9		(R,R)-Z-**10**	99.5	6 d (66%)	1 : >99	70%
10	n-C$_7$H$_{15}$-CH=CH-CHO	(R,R)-E-**9**	99	4 h (91%)	>99 : 1	74%
11		(R,R)-Z-**10**	98	6 h (83%)	3 : 97	62%
12	Ph-CHO	(R,R)-E-**9**	99.3	3 h (91%)	>99 : 1	66%
13		(R,R)-Z-**10**	99	6 h (94%)	2 : 98	55%

Table 2. Double Asymmetric Crotylboration of Chiral Aldehydes with Tartrate Ester Modified Crotylboronate Reagents

Entry	aldehydes (14)	reagent[a]	yield[b]	15 (%)	16 (%)	17 (%)	18 (%)
1		(R,R)-(E)-9	80%	[97]	3	—	—
2	TBSO~~~CHO (Me) **14a**	(S,S)-(E)-2	--	16	[81]	3	—
3		(S,S)-(Z)-5	71%	—	4	[95]	1
4[c]		(R,R)-(Z)-10	--	12	2	45	41
5		(R,R)-(E)-9	--	[93]	5	1	1
6	BzO~~~CHO (Me) **14b**	(S,S)-(E)-2	--	15	[85]	—	—
7		(S,S)-(Z)-5	--	—	3	[88]	9
8		(R,R)-(Z)-10	--	2	—	46	[52]
9		(R,R)-(E)-9	56%	[98]	2	1	1
10	TBDPSO~~~CHO (Me Me) **14c**	(S,S)-(E)-2	55%	16	[84]	—	—
11		(S,S)-(Z)-5	51%	—	—	[94]	6
12		(R,R)-(Z)-10	52%	6	—	16	[78]
13[d]		(R,R)-(E)-9	87%	[87]	9	—	4
14[e]	(dioxolane aldehyde)	(S,S)-(E)-2	70%	2	[97]	1	—
15		(S,S)-(Z)-5	--	—	—	16	84
16		(R,R)-(Z)-10	80%	—	—	1	[99]
17[f]	PMB / OMOM aldehyde	(S,S)-(E)-2	86%	6	[94]	—	—

[a] 98-99% isomeric pure reagents were used. [b] Yields are for two steps including preparation of aldehydes. [c] In retrospect, the crotylboronate had pure isomeric purity. [d] 1.3-1.5 equiv. of aldehyde was used. [e] The reaction was performed under the conditions as described in the present procedure. [f] See reference 14.

Org. Synth. **2011**, *88*, 181-196

1. Department of Chemistry, The Scripps Research Institute, Scripps Florida, 130 Scripps Way #3A2, Jupiter, FL 33458. E-mail: roush@scripps.edu. This research was support by NIH Grant GM038436.
2. Roush, W. R.; Ando, K.; Powers, D. B.; Palkowitz, A. D.; Halterman, R. L. *J. Am. Chem. Soc.* **1990**, *112*, 6339
3. Schmid, C. R.; Bryant, J. D. *Org. Synth.* **1995**, *72*, 6.
4. Roush, W. R.; Adam, M. A.; Walts, A. E.; Harris, D. J. *J. Am. Chem. Soc.* **1986**, *108*, 3422.
5. (a) Roush, W. R., In *Comprehensive Organic Synthesis*, Trost, B. M., Ed. Pergamon Press: Oxford, **1991**; Vol. 2, p 1. (b) Yamamoto, Y.; Asao, N., *Chem. Rev.* **1993**, *93*, 2207. (c) Denmark, S. E.; Almstead, N. G., In *Modern Carbonyl Chemistry*, Otera, J., Ed. Wiley-VCH: Weinheim, **2000**; p 299. (d) Chemler, S. R.; Roush, W. R., In *Modern Carbonyl Chemistry*, Otera, J., Ed. Wiley-VCH: Weinheim, **2000**; p 403. (e) Denmark, S. E.; Fu, J., *Chem. Rev.* **2003**, *103*, 2763. (f) Lachance H.; Hall, D. G. *Org. React.* **2008**, *73*, 1.
6. (a) Roush, W. R.; Walts, A. E.; Hoong, L. K. *J. Am. Chem. Soc.* **1985**, *107*, 8186. (b) Roush, W. R.; Halterman, R. L. *J. Am. Chem. Soc.* **1986**, *108*, 294. (c). Roush, W. R.; Palkowitz, A. D.; Ando, K. *J. Am. Chem. Soc.* **1990**, *112*, 6348. (d) Coe, J. W.; Roush, W. R. *J. Org. Chem.* **1989**, *54*, 915. (e) Roush, W. R.; Hoong, L. K.; Palmer, M. A. J.; Straub, J. A.; Palkowitz, A. D. *J. Org. Chem.* **1990**, *55*, 4117. (f) Roush, W. R.; Palkowitz, A. D.; Palmer, M. J. *J. Org. Chem.* **1987**, *52*, 316. (g) Roush, W. R.; Grover, P. T. *J. Org. Chem.* **1995**, *60*, 3806. (h) Roush, W. R.; Hoong, L. K.; Palmer, M. A. J.; Park, J. C. *J. Org. Chem.* **1990**, *55*, 4109. (i) Roush, W. R.; Grover, P. T. *Tetrahedron* **1992**, *48*, 1981. (j) Roush, W. R.; Coe, J. W. *Tetrahedron Lett.* **1987**, *28*, 931.
7. (a) Brown, H. C.; Jadhav, P. K. *J. Am. Chem. Soc.* **1983**, *105*, 2092. (b) Brown, H. C.; Bhat, K. S. *J. Am. Chem. Soc.* **1986**, *108*, 5919. (c) Brown, H. C.; Jadhav, P. K.; Bhat, K. S. *J. Am. Chem. Soc.* **1988**, *110*, 1535. (d) Brown, H. C.; Bhat, K. S.; Randad, R. S. *J. Org. Chem.* **1989**, *54*, 1570.
8. Corey, E. J.; Yu, C. M.; Kim, S. S. *J. Am. Chem. Soc.* **1989**, *111*, 5495.
9. (a) Kinnaird, J. W. A.; Ng, P. Y.; Kubota, K.; Wang, X.; Leighton, J. L. *J. Am. Chem. Soc.* **2002**, *124*, 7920. (b) Hackman, B. M.; Lombardi, P. J.; Leighton, J. L. *Org. Lett.* **2004**, *6*, 4375.

10. (a) Denmark, S. E.; Fu, J. *J. Am. Chem. Soc.* **2001**, *123*, 9488. (b) Denmark, S. E.; Fu, J.; Lawler, M. J. *J. Org. Chem.* **2006**, *71*, 1523.

11. (a) Burgos, C. H.; Canales, E.; Matos, K.; Soderquist, J. A. *J. Am. Chem. Soc.* **2005**, *127*, 8044. (b) Canales, E.; Prasad, K. G.; Soderquist, J. A. *J. Am. Chem. Soc.* **2005**, *127*, 11572.

12. (a) Lachance, H.; Lu, X.; Gravel, M.; Hall, D. G. *J. Am. Chem. Soc.* **2003**, *125*, 10160. (b) Kennedy, J. W. J.; Hall, D. G. *J. Org. Chem.* **2004**, *69*, 4412. (c) Rauniyar, V.; Hall, D. G. *J. Am. Chem. Soc.* **2004**, *126*, 4518. (d) Rauniyar, V.; Hall, D. G. *Angew. Chem., Int. Ed.* **2006**, *45*, 2426. (e) Rauniyar, V.; Zhai, H.; Hall, D. G. *J. Am. Chem. Soc.* **2008**, *130*, 8481. (f) Rauniyar, V.; Hall, D. G. *J. Org. Chem.* **2009**, *74*, 4236.

13. (a) Kim, I. S.; Ngai, M.-Y.; Krische, M. J. *J. Am. Chem. Soc.* **2008**, *130*, 6340. (b) Kim, I. S.; Ngai, M.-Y.; Krische, M. J. *J. Am. Chem. Soc.* **2008**, *130*, 14891. (c) Bower, J. F.; Kim, I. S.; Patman, R. L.; Krische, M. J. *Angew. Chem. Int. Ed.* **2009**, *48*, 34. (d) Kim, I. S.; Han, S. B.; Krische, M. J. *J. Am. Chem. Soc.* **2008**, *130*, 6340. (e) Kim, I. S.; Han, S. B.; Krische, M. J. *Chem. Commun.* **2009**, 7278.

14. Kim, C. H.; An, H. J.; Shin, W. K.; Yu, W.; Woo, S. K.; Jung, S. K.; Lee, E. *Angew. Chem., Int. Ed.* **2006**, *45*, 8019.

15. For recent applications, see: (a) Prantz, K.; Mulzer, J. *Angew. Chem., Int. Ed.* **2009**, *48*, 5030. (b) Dunetz, J. R.; Julian, L. D.; Newcom, J. S.; Roush, W. R. *J. Am. Chem. Soc.* **2008**, *130*, 16407. (c) Canova, S.; Bellosta, V.; Bigot, A.; Mailliet, P.; Mignani, S.; Cossy, J. *Org. Lett.* **2007**, *9*, 145. (d) Wrona, I. E.; Garbada, A. E.; Evano, G.; Panek, J. S. *J. Am. Chem. Soc.* **2005**, *127*, 15026. (e) Smith, A. B., III; Adams, C. M.; Barbosa Lodise, S. A.; Degnan, A. P. *Proc. Natl. Acad. Sci.* **2004**, *101*, 12042. (f) Chemler, S. R.; Roush, W. R. *J. Org. Chem.* **2003**, *68*, 1319. (g) Francavilla, C.; Chen, W.; Kinder, F. R., Jr. *Org. Lett.* **2003**, *5*, 1233. (h) Yakelis, N. A.; Roush, W. R. *J. Org. Chem.* **2003**, *68*, 3838. (i) Roush, W. R.; Bannister, T. D.; Wendt, M. D.; Jablonowski, J. A.; Scheidt, K. A. *J. Org. Chem.* **2002**, *67*, 4275.

Appendix
Chemical Abstracts Nomenclature; (Registry Number)

Potassium *tert*-butoxide; (865-47-4)

Trans-butene; (624-64-6)

Triisopropylborane; (5419-55-6)

(*S,S*)-Diisopropyl tartrate ((*S,S*)-DIPT); Diisopropyl *D*-tartrate (*D*-DIPT); (62961-64-2)

(*S,S*)-Diisopropyl tartrate (*E*)-crotylboronate; (99687-40-8)

(*R,R*)-Diisopropyl tartrate (*E*)-crotylboronate; (99745-86-5)

(*S,S*)-Diisopropyl tartrate (*Z*)-crotylboronate; (106357-33-9)

(*R,R*)-Diisopropyl tartrate (*Z*)-crotylboronate; (106357-20-4)

D-(*R*)-glyceraldehyde acetonide; (15186-48-8)

(*3R,4R,5R*)-3-Methyl-5,6-O-isopropylidene-hex-1-ene-4-ol ((2R,3R,4R)-1,2-dideoxy-2-ethenyl-4,5-O-(1-methylethylidene)-xylitol); (88424-94-6)

(*3S,4S,5R*)-3-Methyl-5,6-O-isopropylidene-hex-1-ene-4-ol; (88424-95-7)

(*3S,4R,5R*)-3-Methyl-5,6-O-isopropylidene-hex-1-ene-4-ol; (96094-43-8)

William R. Roush is Professor of Chemistry, Executive Director of Medicinal Chemistry and Associate Dean of the Kellogg School of Science and Technology at The Scripps Research Institute, Florida. His research interests focus on the total synthesis of natural products and the development of new synthetic methodology. Since moving to Scripps Florida in 2005, his research program has expanded into new areas of chemical biology and medicinal chemistry. Dr. Roush was a member of the *Organic Syntheses* Board of Editors from 1993-2002 and was Editor of Volume 78. He currently serves on the *Organic Syntheses* Board of Directors (2003-present).

Huikai Sun received both his BS and MS degrees in Organic Chemistry from Nankai University in China. He joined the Department of Chemistry at Case Western Reserve University for his Ph. D. degree, which was completed in 2007 under the supervision of Professor Anthony J. Pearson. Then he moved to his current position as a postdoctoral research associate in Professor William R. Roush's group. His current research interests focus on the total synthesis of natural products and the development of cysteine protease inhibitors.

David A. Candito completed his undergraduate degree at York University in 2007 where he had the opportunity to work with Professor Michael Organ on palladium-catalyzed cross-coupling reactions employing the Pd-PEPPSI-IPr catalyst system. He then joined the group of Professor Mark Lautens at the University of Toronto where he is currently pursuing a Ph.D. His doctoral research has focused upon palladium-catalyzed domino reactions involving the use of norbornene to functionalize aryl C–H bonds and he is also engaged in exploring the use of aryne intermediates in transition metal catalyzed processes.

Mathieu Blanchot was born in Autun (France) in 1982 and studied chemistry in Dijon and Lyon (France). After completing his Diploma-Thesis under the direction of Prof. Genevieve Balme, he joined the group of Prof. Lukas Gooßen in Kaiserslautern (Germany) where he received his Ph.D. for his work on the Hydroamidation of Alkynes. He is currently working as a Postdoctoral researcher with Prof. Mark Lautens in Toronto, ON (Canada) where his research is focused upon palladium-catalyzed domino direct arylation/N-arylation.

4-METHOXY-4'-NITROPHENYL. RECENT ADVANCES IN THE STILLE BIARYL COUPLING REACTION AND APPLICATIONS IN COMPLEX NATURAL PRODUCTS SYNTHESIS

Prepared by Robert M. Williams.*[1]
Original article: Stille, J.[2]; Echavarren, A.; Hendrix, J.; Albrecht, B.[3]; Williams, R.[1] *Org. Synth.* **1993**, *71*, 97.

The use of the palladium-catalyzed Stille cross-coupling reaction[4] for the synthesis of biaryls has become a popular and practical method[5] mainly due to the air and moisture stability of organostannanes, the wide functional group compatibility under the reaction conditions and the generally readily available starting materials. Since the initial *Organic Syntheses* report of the preparation of 4-methoxy-4'-nitrophenyl,[6] numerous advances have been made regarding substrate scope, improved reaction conditions and catalyst design. Herein we report on some of these advances[7] as well as recent applications in total syntheses.[8]

The majority of the recent advances of the Stille coupling have come concomitantly with the advancement of ligand design for palladium catalysis.[9] The advent of electron rich, bulky phosphine and carbene ligands allowed for much milder reaction conditions and increased substrate scope. Initially, the Stille reaction was limited to aryl iodides, bromides and triflates, typically at elevated temperature. In 1999 Fu and co-workers reported the first general method for the Stille cross-coupling of aryl chlorides using tri-*t*-butyl phosphine as a ligand for palladium and cesium fluoride to activate the tin reagent.[10] More recently, Fu and co-workers have shown that similar reaction conditions can be utilized to cross-couple arylbromides at room temperature. In an analogous manner, Verkade and co-workers have shown that proazaphosphatrane ligands behave in a similar manner to the tri-*t*-butyl phosphine system utilized by Fu.[11] Finally, Baldwin and co-workers have shown that the inclusion of copper(I) salts[12] to the conditions reported by the Fu group significantly enhanced the reactivity of a variety of aryl bromides.

The Stille reaction has contributed significantly to numerous applications in the total synthesis of complex natural products because of the chemical stability of the coupling partners, the mild reaction conditions, and the functional group compatibility. In 2001, we reported a Stille coupling of 7-iodoisatin **1** and a suitably protected stannyl tyrosine derivative **2** (Scheme 1).[13] Treatment of these coupling partners under routine conditions afforded biaryl **3** as a model system towards the total synthesis of TMC-95A/B (**4**).

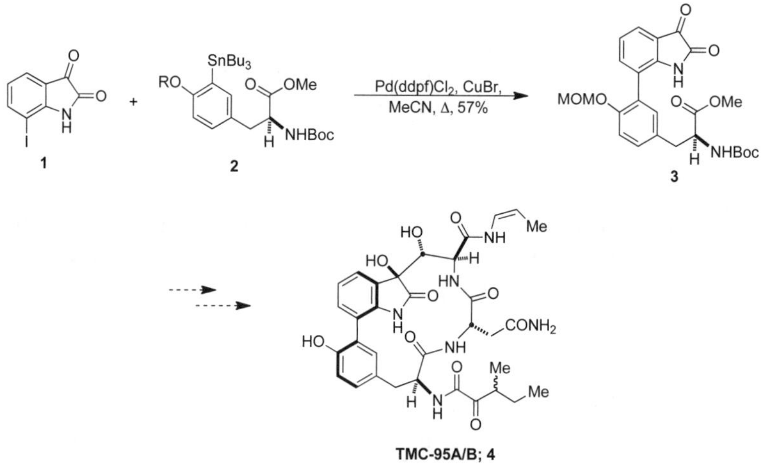

TMC-95A/B; 4

Scheme 1. Albrecht and Williams' Studies toward TMC-95A/B

In 2003, the Hoveyda and Snapper labs utilized a Stille biaryl coupling as a key macrocycle-forming reaction to provide a 17-membered ring in the total synthesis of chloropeptin **5**.[14] Macrocyclic precursor **6** was treated under the conditions reported by Fu, in the presence of collidine forming macrocycle **7** (Scheme 2). It is thought that the presence of collidine in the reaction stabilizes the active palladium complex thus preventing palladium black precipitation. The completion of chloropeptin was accomplished in an additional two steps.

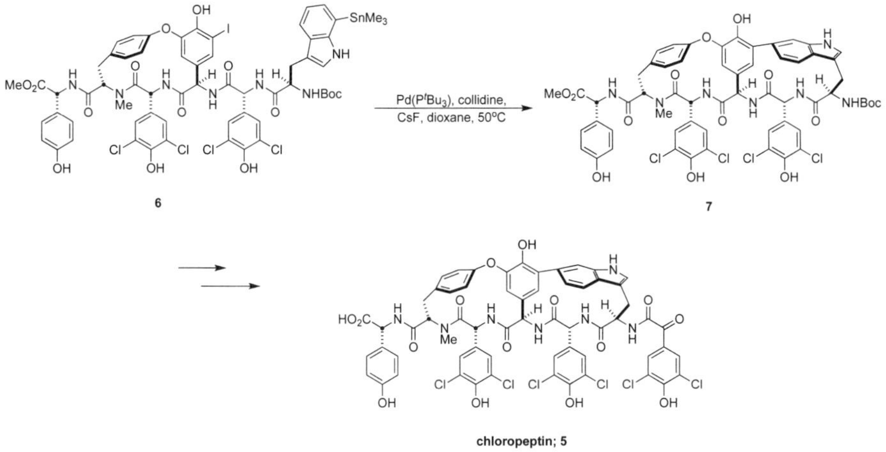

Scheme 2. Hoveyda and Snapper's total synthesis of chloropeptin.

Danishefsky and co-workers utilized a Stille biaryl coupling in the total synthesis and stereochemical revision of himastatin.[15] Stannylation of aryliodide **8** afforded stannane **9** which was resubjected to aryliodide **8** under standard Stille conditions to afford dimer **10** (Scheme 3). This advanced biaryl intermediate was ultimately converted to himastatin thus allowing for the correct stereochemical assignment.

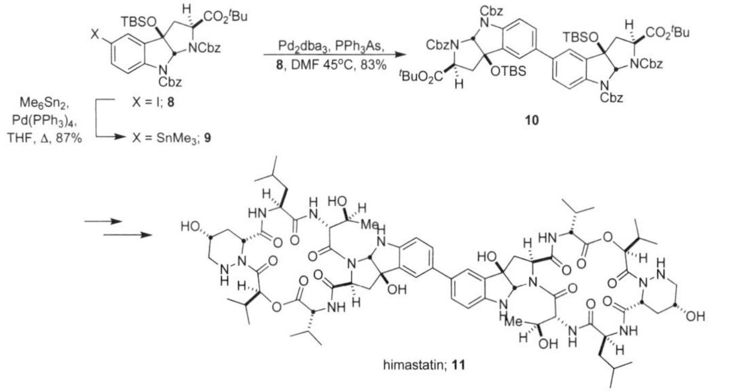

Scheme 3. Danishefsky's total synthesis of himastatin.

199

Since its first report, the Stille coupling has been a widely utilized carbon-carbon bond forming reaction. Recent advancements in homogenous palladium catalysis have rendered the Stille coupling reaction a much more attractive approach to the synthesis of complex biaryl systems. The examples above fully support the versatility and utility of this simple biaryl coupling reaction to highly functionalized and sterically demanding substrates. Efforts towards optimizing the Stille[16] protocol are continuously being pursued allowing for milder reaction conditions and increased substrate scope. These on-going efforts demonstrate that the Stille reaction is one of the most synthetically useful biaryl coupling reactions in the synthesis of complex molecules.[17]

1. Department of Chemistry, Colorado State University, Fort Collins, CO 80523 and the University of Colorado Cancer Center, Aurora, Colorado.

2. Deceased.

3. Department of Medicinal Chemistry, Amgen, Inc., One Kendall Square, Bldg. 1000, Cambridge, MA 02139.

4. Stille, J.K. *Angew. Chem. Int. Ed. Engl.* **1986**, *25*, 508-524.

5. For an excellent review of the Stille reaction see: Farina, V.; Krishnamurthy, V.; Scott, W.J. *Org. React.* **1997**, *50*, 1-652.

6. Stille, J.K.; Echavarren, A.M.; Williams, R.M.; Hendrix, J.A. *Org. Synth.* **1993**, *71*, 97-106.

7. For some recent reviews and highlights on the Stille reaction see: (a) Espinet, P.; Echavarren, A.M. *Angew. Chem. Int. Ed.* **2004**, *43*, 4704-4734. (b) Echavarren, A.M. *Angew. Chem. Int. Ed.* **2005**, *44*, 3962-3965.

8. For a recent review of the Stille Reaction in natural product synthesis see: Nicolaou, K.C.; Bulger, P.G.; Sarlah, D. *Angew. Chem. Int. Ed.* **2005**, *44*, 4442-4489.

9. Bedford, R. B.; Cazin, C. S. J.; Holder, D. *Coord. Chem. Rev.* **2005**, *248*, 2283-2321.

10. Littke, A.F.; Fu, G.C. *Angew. Chem. Int. Ed.* **1999**, *38*, 2411-2413.

11. Su, W. Urgaonkar, S.; McLaughlin, P.A.; Verkade, J.G. *J. Am. Chem. Soc.* **2004**, *126*, 16433-16439.

12. (a) Mee, S.P.H.; Lee, V.; Baldwin. *Angew. Chem. Int. Ed.* **2004**, *43*, 1132-1136. (b) Mee, S.P.H.; Lee, V.; Baldwin. *Chem. Eur. J.* **2005**,

11, 3294-3308. In addition to the inclusion of copper(I) salts a solvent change from dioxane to DMF was made.

13. Albrecht, B.K.; Williams, R.M. *Tetrahedron Lett.* **2002**, *42*, 2755-2757.

14. Deng, H.; Jung, J-K.; Liu, T.; Kuntz, K.W.; Snapper, M.L.; Hoveyda, A.H. *J. Am. Chem. Soc.* **2003**, *125*, 9032-9034.

15. Kamenecka, T.M.; Danishefsky, S.J. *Chem. Eur.J*.**2001**, *7*, 41-63.

16. Although not relevant to biaryl couplings the following are some noteworthy reports of Stille couplings catalytic in tin and of alkyl halides. For cross-couplings catalytic in tin see Gallagher, W.P.; Maleczka, R.E. *J. Org. Chem.* **2005**, *70*, 841-846, and references cited therein. For cross-coupling of alkyl halides see ref. 1b and references cited therein.

17. The work performed at Colorado State University was financially supported by the National Institutes of Health and the National Science Foundation.

Robert M. Williams recieved his B.A. degree in Chemistry in 1975 from Syracuse University. He obtained the Ph.D. degree in 1979 at MIT (W.H. Rastetter) and was a post-doctoral fellow at Harvard (1979-1980; R.B. Woodward/Yoshito Kishi). He joined Colorado State University in 1980 and was named a University Distinguished Professor in 2002. His interdisciplinary research program at the chemistry-biology interface is focused on the total synthesis of biomedically significant natural products, biosynthesis of secondary metabolites, studies on antitumor drug-DNA interactions, HDAC inhibitors, amino acids and peptides.

Discussion Addendum for:

PALLADIUM-CATALYZED REACTION OF 1-ALKENYLBORONATES WITH VINYLIC HALIDES: (1*Z*,3*E*)-1-PHENYL-1,3-OCTADIENE

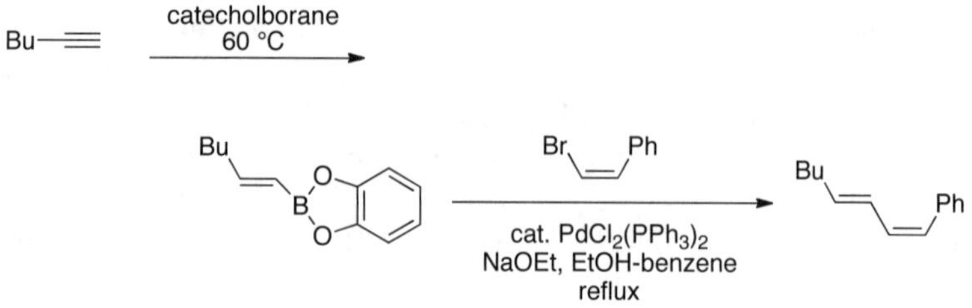

Prepared by Norio Miyaura.[†]
Original article: Miyaura, N.; Suzuki, A. *Org. Synth.* **1990**, *68*, 130.

The cross-coupling reactions of organoboronic acids have proved to be a general reaction for a wide range of selective carbon-carbon bond forming reactions in laboratories and in industry since they involve convenient reagents that are generally thermally stable and inert to water and oxygen, thus allowing handling without special precautions. These reactions have been reviewed.[1]

Synthesis of 1-Alkenylboron Compounds

Hydroboration of alkynes is especially valuable in the synthesis of stereodefined 1-alkenylboron compounds. Disiamylborane ($HB(Sia)_2$), dicyclohexylborane, and 9-BBN are very mild and selective hydroboration reagents to obtain 1-alkenylboranes. The addition of catecholborane (HBcat)[2] or dihaloborane ($HBCl_2{\cdot}SM_2$, $HBBr_2{\cdot}SMe_2$)[3] to alkynes followed by hydrolysis with water is a method for the synthesis of air-stable 1-alkenylborinic acids (**3**). Since hydroboration yields (*E*)-adducts through the *anti*-Markovnikov and *syn*-addition of an H-B bond to terminal alkynes, (*Z*)-1-alkenylboronates have been synthesized by a two-step method based on intramolecular S_N2-type substitution of 1-halo-1-alkenylboronates with metal hydrides[4] or *cis*-hydrogenation of 1-alkynylboronates.[5] Rhodium(I)/^{i}Pr$_3$P-catalyzed hydroboration is a new variant for the one-step

Org. Synth. **88**, *2011*, 202-206
Published on the Web 2/4/2011

synthesis of (Z)-1-alkenylboron compounds (**6**) from terminal alkynes.[6] On the other hand, the palladium-catalyzed borylation of 1-alkenyl halides or triflates with bis(pinacolato)diboron provides (Z)-1-alkenylboronic pinacol esters.[7] The pinacol esters (**5**, **6**) are advantageous over the boronic acids with regard to the preparation and handling of pure and stable materials since they are stable to air and moisture, GC analysis, and chromatographic isolation on silica gel. Treatment of boronic acids with KHF_2 results in spontaneous precipitation of stable and highly insoluble [1-alkenylBF_3]K (**4**).[8,18] All of those derivatives have been successfully used for various cross-coupling reactions.

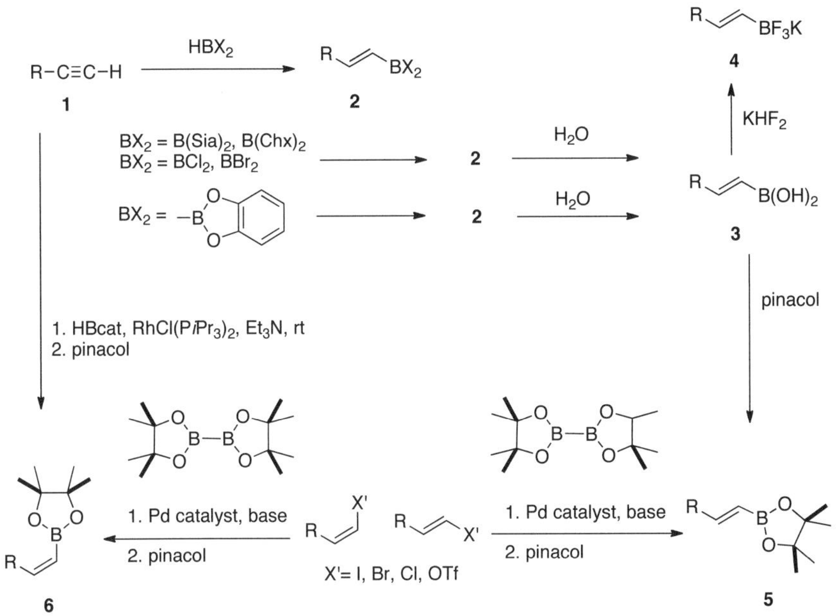

Scheme 1

Cross-Coupling Conditions (Table 1)

Cross-coupling reactions of 1-alkenylboron compounds with 1-alkenyl halides require a relatively strong base in the presence of a palladium/phosphine catalyst. The relative rate is in the order of their basic strength and affinity of the counter cations for halide anions (TlOH > KOH > K_3PO_4 > Na_2CO_3 > NaOAc). Aqueous NaOH has been used for 1-alkenylboronic acids or esters in refluxing THF-H_2O, DME-H_2O (entry 3), or benzene-H_2O (entry 6) and aqueous LiOH (entry 2) for disiamylborane derivatives. In spite of its toxicity, TlOH is an excellent base that enables completion of the coupling within one hour at room temperature (entry 4). Since an aqueous solution of TlOH precipitates brown-black solids under careful storage conditions, addition of TlOEt to aqueous THF was recently recommended as a suitable replacement for air-sensitive TlOH (entry 5).

Table 1. Conditions for Alkenyl-Alkenyl (sp^2-sp^2) Coupling

$$R^1CH=CHX + R^2CH=CHB< \xrightarrow{\text{Pd catalyst/base}} R^1CH=CHCH=CHR^2$$

entry	X=	B<	catalyst	base/solvent	temp/°C	ref.
1	Br	B(OH)$_2$	Pd(PPh$_3$)$_4$	NaOEt, benzene-EtOH	reflux	[9]
2	I	B(Sia)$_2$	Pd(PPh$_3$)$_4$	LiOH, THF-H$_2$O	reflux	[10]
3	I, Br	B(OR)$_2$	Pd(PPh$_3$)$_4$	NaOH, THF or DME -H$_2$O	reflux	[11]
4	I	B(OH)$_2$	Pd(PPh$_3$)$_4$	TlOH, THF-H$_2$O	rt	[12]
5	I	B(OH)$_2$	Pd(PPh$_3$)$_4$	TlOEt, THF-H$_2$O	rt	[13]
6	I, Br	B(OR)$_2$	Pd(PPh$_3$)$_4$	NaOH, benzene-H$_2$O	70	[14]
7	OTf	B(OR)$_2$	Pd(PPh$_3$)$_4$	K$_3$PO$_4$ nH$_2$O, dioxane	80	[15]
8	OTf	B(OR)$_2$	PdCl$_2$(PPh$_3$)$_2$	Na$_2$CO$_3$, THF-H$_2$O	40	[16]
9	OTs	B(OH)$_2$	PdCl$_2$(PPh$_3$)$_2$	KF, THF-H$_2$O	60	[17]
10	I, Br	BF$_3$K	PdCl$_2$(dppf)	Et$_3$N, BuNH$_2$ or Cs$_2$CO$_3$, PrOH-H$_2$O	reflux	[8, 18]

Sia=CHMeCHMe$_2$; (OR)$_2$= diol esters of boronic acids

Synthetic Applications

Alkenyl-alkenyl cross-coupling affords various stereodefined dienes, trienes, and further conjugated polyenes for the synthesis of biologically active natural products,[1] including palytoxine,[19] (-)-bafilomycin A$_1$,[20] a combinatorial synthesis of vitamin D$_3$ derivatives,[21] a macrolide antibiotic, rutamycin B,[22] and 5,6-DiHETE Methyl Esters.[23]

†   Graduate School of Engineering, Hokkaido University, Sapporo 060-8628, Japan.

1.   Reviews, (a) *Metal-Catalyzed Cross-Coupling Reactions - Second, Completely Revised and Enlarged Edition*, A. de Meijere, F. Diederich, Eds.; Wiley-VCH (2004); pp 41-123. (b) Suzuki, A.; Brown, H. C. *Organic Syntheses Via Boranes Vol. 3: Suzuki Coupling*, Aldrich

Org. Synth. **88**, *2011*, 202-206

(2003). (c) *Topics in Current Chemistry* Vol. 219, Miyaura, N. Ed.; Springer-Verlag (2002); pp 11-59. (d) *Metal-Catalyzed Cross-Coupling Reactions*, Diederich, F.; Stang, P. J. Eds.; Wiley-VCH (1998); 49-97. (e) Miyaura, N.; Suzuki, A. *Chem. Rev.* **1995**, *95*, 2457.

2. Brown, H. C.; Gupta, S. K. *J. Am. Chem. Soc.* **1972**, *94*, 4370.
3. (a) Brown, H. C.; Campbel, J. B. *J. Org. Chem.* **1980**, *45*, 389. (b) Brown, H. C.; Bhat, N. J.; Sommayaji, V. *Organometallics*, **1983**, *2*, 1311.
4. Brown, H. C.; Imai, T. *Organometallics* **1984**, *3*, 1392.
5. Srebnik, M.; Bhat, N. G.; Brown, H. C. *Tetrahedron Lett.* **1988**, *29*, 2635.
6. Ohmura, T.; Yamamoto, Y.; Miyaura, N. *J. Am. Chem. Soc.* **2000**, *122*, 4990.
7. (a) Takagi, J.; Takahashi, K.; Ishiyama, T.; Miyaura, N. *J. Am. Chem. Soc.* **2002**, *124*, 8001. (b) Takagi, J.; Kamon, A.; Ishiyama, T.; Miyaura, N. *Synlett* **2002**, 1880.
8. Batey, R. A.; Quach, T. D. *Tetrahedron Lett.* **2001**, *42*, 9099.
9. Miyaura, N.; Yamada, K.; Suginome, H.; Suzuki, A. *J. Am. Chem. Soc.* **1985**, *107*, 972.
10. Kobayashi, Y.; Shimazaki, T.; Taguchi, H.; Sato, F. *J. Org. Chem.* **1990**, *55*, 5324.
11. (a) White, J. D.; Kim, T-S.; Nambu, M. *J. Am. Chem. Soc.* **1997**, *119*, 103. (b) Sugai, T.; Yokoyama, M.; Yamazaki, T.; Ohta, H. *Chem Lett.* **1997**, 797.
12. Uenishi, J-I.; Beau, J-M.; Amstrong, R. W.; Kishi, Y. *J. Am. Chem. Soc.* **1987**, 109, 4756.
13. (a) Frank, S. A.; Chen, H.; Kunz, R. K.; Schnaderbeck, M. J.; Roush, W. R. *Org. Lett.* **2000**, *2*, 2691. (b) Kobayashi, S.; Mori, K.; Wakabayashi, T.; Yasuda, S.; Hanada, K. *J. Org. Chem.* **2001**, *66*, 5580.
14. Hanisch, I.; Brückner, R. *Synlett*, **2000**, 374.
15. Oh-e, T.; Miyaura, N.; Suzuki, A. *J. Org. Chem.* **1993**, *58*, 2201.
16. (a) Occhiato, E. G.; Trabocchi, A.; Guarna, A. *Org Lett.* **2000**, *2*, 1241. (b) Alvarez, R.; Iglesias, B.; Lera, A. R. *Tetrahedron* **1999**, *55*, 13779.
17. Wu, J.; Zhu, Q.; Wang, L.; Fathi, R.; Yang, Z. *J. Org. Chem.* **2003**, *68*, 670.
18. (a) Molander, G. A.; Bernardi, C. R. *J. Org. Chem.* **2002**, *67*, 8424. (b) Molander, G. A.; Rivero, M. R. *Org. Lett.* **2002**, *4*, 107.

19. Armstrong, R. W.; Beau, J-M.; Cheon, S. H.; Christ, W. J.; Fujioka, H.; Ham, W-H.; Hawkins, L. D.; Jin, H. L. D.; Kang, S. H.; Kishi, Y.; Martinelli, M. J.; McWhorter, W. W.; Mizuno, M.; Nakata, M.; Stutz, A. E.; Talamas, F. X.; Taniguchi, M.; Tino, J. A.; Ueda, K.; Uenishi, J-I.; White, J. B.; Yonaga, M. *J. Am. Chem. Soc.* **1989**, 111, 7525.
20. Scheidt, K. A.; Bannister, T. D.; Tasaka, A.; Wendt, M. D.; Savall, B. M.; Fegley, G. J.; Roush, W. R. *J. Am. Chem. Soc.* **2002**, *124*, 6981.
21. Hanazawa, T.; Wada, T.; Masuda, T.; Okamoto, S.; Sato, F. *Org. Lett.* **2001**, *3*, 3975.
22. Evans, D. A.; Ng, H. P.; Rieger, D. L. *J. Am. Chem. Soc.* **1993**, *115*, 11446.
23. Nicolaou, K. C.; Ramphal, J. Y.; Palazon, J. M.; Spanevello, R. A. *Angew. Chem. Int. Ed.* **1989**, *28*, 587.

Norio Miyaura was born in Hokkaido in Japan in 1946. He received his B. Eng. and Dr. Eng. from Hokkaido University. He became a Research Associate and an Associate Professor in the A. Suzuki research group, and then was promoted to the rank of Professor in the same group in 1994. He is now emeritus and a specially appointed Professor after his retirement from Hokkaido University in 2010. In 1981, he joined the J. K. Kochi group at Indiana University as a postdoctoral fellow to study the epoxidation of alkenes catalyzed by metal-salen complexes. His current interests are mainly in the field of metal-catalyzed reactions of organoboron compounds, with emphasis of applications to organic synthesis such as catalyzed hydroboration, palladium-catalyzed cross-coupling reactions of organoboronic acids, rhodium- or palladium-catalyzed conjugate addition reactions of arylboronic acids, and addition and coupling reactions of diborons and pinacolborane for the synthesis of organoboronic esters.

PALLADIUM(0)-CATALYZED REACTION OF 9-ALKYL-9-BORABICYCLO[3.3.1]NONANE WITH 1-BROMO-1-PHENYLTHIOETHENE: 4-(3-CYCLOHEXENYL)-2-PHENYLTHIO-1-BUTENE

Prepared by Norio Miyaura.[†]
Original article: Miyaura, N.; Ishiyama, T.; Suzuki, A. *Org. Synth.* **1993**, *71*, 89.

The cross-coupling reactions of organoboron compounds have proved to be a general method for a wide range of selective carbon-carbon bond forming reactions.[1] In 1989, the cross-coupling reaction of 9-alkyl-9-BBN with 1-alkenyl and aryl halides or triflates was found to proceed smoothly in the presence of PdCl$_2$(dppf) and K$_3$PO$_4$·nH$_2$O.[2] This coupling reaction of B-alkyl compounds has been reviewed.[3,4] The reaction is limited to *primary* alkylboranes; hydroboration of terminal alkenes with 9-BBN is the most convenient way to furnish the desired boron reagents. The reaction is catalyzed by PdCl$_2$(dppf),[2] PdCl$_2$(dppf)/2Ph$_3$As,[5] or other palladium-phosphine complexes in the presence of a base (Table 1). Since the presence of water greatly accelerates the reaction, the use of the hydrate of inorganic bases such as K$_3$PO$_4$·nH$_2$O (entry 1) or aqueous bases (entries 2 and 4) is generally recommended.[5] On the other hand, solid sodium methoxide added to 9-alkyl-9-BBN dissolves in THF by forming the corresponding ate-complex, which enables room temperature coupling under non-aqueous conditions (entries 3 and 5).[2] Treatment of 9-methoxy-9-BBN with *primary*-alkyllithiums is an alternative for *in situ* preparation of analogous boron ate-complexes.[6] The presence of KBr (1 equiv) is often critical to prevent decomposition of the catalyst for reactions of aryl and 1-alkenyl triflates (entry 6).[7]

Sp^3-sp^3 bond formation between two alkyl derivatives has been much less successful among the possible combinations of different-type nucleophiles and electrophiles. Difficulties arise from the oxidative addition of haloalkanes (RCH_2CH_2X) to a palladium(0) complex due to accompanying formation of $RCH=CH_2$ and RCH_2CH_3 and from the susceptibility of alkylpalladium(II) intermediates to β-hydride elimination.[8] In spite of these difficulties, sp^3-sp^3 bond formation occurs smoothly between *primary*-alkyl halides and *primary*-alkylboron compounds where each reactant possesses β-hydrogen (entries 8 and 9).[9] The coupling with secondary alkyl halides has been limited to cyclopropyl iodides.[10,11]

The reactions of the corresponding alkylboronic acids and [alkylBF$_3$]K[12] are significantly slower than that of trialkylboranes, but methylboroxine (MeBO)$_3$ or methylboronic acid alkylates bromoarenes with a common palladium/triphenylphosphine catalyst (entry 10).[13] Analogous reactions of alkylboronic acids possessing β-hydrogen are achieved by the use of Qphos (2) for aryl or 1-alkenyl bromides, triflates and chlorides (entry 11),[14] a dppf complex for iodides, bromides and triflates,[12,15] and N-cyclic carbene (1)[16] for arene diazonium salts. These reactions are limited to use for *primary*-alkylboronic acids; however, cyclopropylboronic acid derivatives alkylate aryl and 1-alkenyl halides or triflates[17] and acyl chlorides[18] without loss of stereochemistry of the cyclopropane ring (eq 1).

$$\text{(1)}$$

Table 1. Reaction Conditions for Coupling of primary-Alkylboron Derivatives

$$R^1CH=CH_2 \xrightarrow{HBX_2} R^1CH_2CH_2BX_2 \xrightarrow[\text{Pd catalyst, base}]{R^2X} R^1CH_2CH_2\text{-}R^2$$

entry	BX₂=	R²-X	catalyst/base/solvent	temp/°C	ref.
1	9-BBN	alkenyl, aryl I, Br	PdCl₂(dppf), K₃PO₄·nH₂O, DMF	rt-50	[4,6,7]
2	9-BBN	alkenyl, aryl I, Br	PdCl₂(dppf), NaOH, THF-H₂O	rt-reflux	[4]
3	9-BBN	alkenyl, aryl I, Br	PdCl₂(dppf), NaOMe, THF	rt-reflux	[4]
4	9-BBN	alkenyl, aryl I	PdCl₂(dppf)/2AsPh₃, Cs₂CO₃, DMF-H₂O	rt	[5]
5	9-BBN	aryl Cl	Pd(OAc)₂/NHC (1), KOMe, THF	reflux	[19]
6	9-BBN	alkenyl OTf	PdCl₂(dppf)/AsPh₃, Cs₂CO₃, KBr, DMF-H₂O	rt	[20]
7	9-BBN	alkenyl OP(O)(OPh)₂	Pd(PPh₃)₄, NaHCO₃, DMF-H₂O	50	[21]
8	9-BBN	prim-alkyl Br	Pd(OAc)₂/PCy₃, K₃PO₄, THF	rt	[9a]
9	9-BBN	prim-alkyl Cl	Pd₂(dba)₃/PCy₃, CsOH, dioxane	90	[9b]
10	(MeBO)₃	aryl I, Br	Pd(PPh₃)₄, K₂CO₃, dioxane-H₂O	reflux	[13]
11	B(OH)₂	prim-alkyl Cl	Pd₂(dba)₃/Qphos (2), K₃PO₄, toluene	100	[14]
12	BF₃K	aryl I, Br	PdCl₂(dppf), Cs₂CO₃, THF-H₂O	reflux	[12]

dppf   NHC (1)   Q-phos (2)

The connection of two fragments *via* the hydroboration-cross coupling sequence has found a wide range of applications in the synthesis of natural products and functional molecules,[1,3,4] including bacterial metabolites epothilone A and B,[22] ciguatoxin,[23] clinically useful 2-alkylcarbapenems,[24] and a novel class of glycomimetic compounds, aza-C-disaccharides.[25]

† Graduate School of Engineering, Hokkaido University, Sapporo 060-8628, Japan.

1. Reviews, (a) *Metal-Catalyzed Cross-Coupling Reactions - Second, Completely Revised and Enlarged Edition*, A. de Meijere, F. Diederich, Eds.; Wiley-VCH (2004); pp 41-123. (b) Suzuki, A.; Brown, H. C. *Organic Syntheses Via Boranes Vol. 3: Suzuki Coupling*, Aldrich (2003). (c) *Topics in Current Chemistry* Vol. 219, Miyaura, N. Ed.; Springer-Verlag (2002); pp 11-59. (d) *Metal-Catalyzed Cross-Coupling Reactions*, Diederich, F.; Stang, P. J. Eds.; Wiley-VCH (1998); 49-97.

(e) Miyaura, N.; Suzuki, A. *Chem. Rev.* **1995**, *95*, 2457.

2. Miyaura, N.; Ishiyama, T.; Sasaki, H.; Ishikawa, M.; Satoh, M.; Suzuki, A. *J. Am. Chem. Soc.* **1989**, *111*, 314. This report describes the use of K_3PO_4; however, the chemical company purchased this reagent later changed the label to its hydrate, $K_3PO_4 \cdot nH_2O$ whereby n is 2 to 3.

3. Chemler, S. R.; Trauner, D.; Danishefsky, S. J. *Angew .Chem. Int. Ed.* **2001**, *40*, 4545.

4. Netherton, M. W.; Fu, G. C. *Adv. Synth. Catal.* **2004**, *346*, 1525.

5. Johnson, C. R.; Braun, M. P. *J. Am. Chem. Soc.* **1993**, *115*, 11014.

6. Marshall, J. A.; Johns, B. A. *J. Org. Chem.* **1998**, *63*, 7885.

7. Oh-e, T.; Miyaura, N.; Suzuki, A. *J. Org. Chem.* **1993**, *58*, 2201.

8. (a) Ishiyama, T.; Abe, S.; Miyaura, N.; Suzuki, A. *Chem Lett.* **1992**, 691. (b) Echavarren, A. M. *Angew. Chem. Int. Ed.* **2005**, *44*, 3962.

9. (a) Netherton, M. R.; Dai, C.; Neuschütz, K.; Fu, G. C. *J. Am. Chem. Soc.* **2001**, *123*, 10099. (b) Kirchhoff, J. H.; Dai, C.; Fu, G. C. *Angew. Chem. Int. Ed.* **2002**, *41*, 1945.

10. Charette, A. B.; Giroux, A.; *J. Org. Chem.* **1996**, *61*, 8718.

11. Charette, A. B.; Freitas-Gil, R. P. D. *Tetrahedron Lett.* **1997**, *38*, 2809.

12. G. A. Molander, T. Ito, *Org. Lett.* **2001**, *3*, 393.

13. Gray, M.; Andrews, I. P.; Hook, D. F.; Kitteringham, J.; Voyle, M.; *Tetrahedron Lett.* **2000**, *41*, 6237.

14. Kataoka, N.; Shelby, Q.; Stambuli, J. P.; Hartwig, J. F. *J. Org. Chem.* **2002**, *67*, 5553.

15. (a) Zou, G.; K. Reddy, Y. K.; Falck, J. R. *Tetrahedron Lett.* **2001**, *42*, 7213. (b) Occhiato, E. G.; Trabocchi, A.; Guarna, A. *J. Org. Chem.* **2001**, *66*, 2459.

16. Andrus, M. B.; Song, C. *Org. Lett.* **2001**, *3*, 3761.

17. (a) Yao, M. L.; Deng, M.-Z. *J. Org. Chem.* **2000**, *65*, 5034. (b) Zhou, S.-M.; Deng, M.-Z.; Xia, L.-J.; Tang, M.-H. *Angew. Chem. Int. Ed. Engl.* **1998**, *37*, 2845.

18. Chen, H.; Deng, M.-Z. *Org Lett.* **2000**, *2*, 1649.

19. Fürstner, A.; Leitner, A. *Synlett* **2001**, 290.

20. Sasaki, M.; Fuwa, H.; Inoue, M.; Tachibana, K. *Tetrahedron Lett.* **1998**, *39*, 9027.

21. Sasaki, M.; Fuwa, H.; Ishikawa, M.; Tachibana, K. *Org Lett.* **1999**, 1, 1075.

22. Balog, A.; Meng, D.; Kamenecka, T.; Bertinato, P.; Su, D-S.; Sorensen, E. J.; S. J. Danishefsky, S. J. *Angew. Chem. Int. Ed.* **1996**, *35*, 2801.

23. (a) H. Takakura, H.; K. Noguchi, K.; M. Sasaki, M.; K. Tachibana, K. *Angew. Chem. Int. Ed.* **2001**, *40*, 1090. (b) Sasaki, M.; Ishikawa, M.; Fuwa, H.; Tachibana, K. *Tetrahedron.* **2002**, *58*, 1889.

24. Narukawa, Y.; Nishi, K.; Onoue, H. *Tetrahedron.* **1997**, *53*, 539.

25. Johns, B. A.; Pan, Y. T.; Elbein, A. D.; Johnson, C. R. *J. Am. Chem. Soc.* **1997**, *119*, 4856.

Norio Miyaura was born in Hokkaido in Japan in 1946. He received his B. Eng. and Dr. Eng. from Hokkaido University. He became a Research Associate and an Associate Professor in the A. Suzuki research group, and then was promoted to the rank of Professor in the same group in 1994. He is now emeritus and a specially appointed Professor after his retirement from Hokkaido University in 2010. In 1981, he joined the J. K. Kochi group at Indiana University as a postdoctoral fellow to study the epoxidation of alkenes catalyzed by metal-salen complexes. His current interests are mainly in the field of metal-catalyzed reactions of organoboron compounds, with emphasis of applications to organic synthesis such as catalyzed hydroboration, palladium-catalyzed cross-coupling reactions of organoboronic acids, rhodium- or palladium-catalyzed conjugate addition reactions of arylboronic acids, and addition and coupling reactions of diborons and pinacolborane for the synthesis of organoboronic esters.

PREPARATION OF ISOPROPYL 2-DIAZOACETYL(PHENYL)CARBAMATE

Submitted by Hubert Muchalski, Amanda B. Doody, Timothy L. Troyer, and Jeffrey N. Johnston.[1]
Checked by John Frederick Briones and Huw M. L. Davies.

1. Procedure

Caution! Diazo compounds are presumed to be toxic and potentially explosive and therefore should be handled with caution in a fume hood. Although 4-acetamidobenzenesulfonyl azide exhibited no impact sensitivity,[2] proper caution should be exercised with all azide compounds. Although in carrying out this reaction numerous times we have never observed an explosion, we recommend that this preparation be conducted behind a safety shield.

Org. Synth. **2011**, *88*, 212-223
Published on the Web 2/8/2011

A. Isopropyl acetyl(phenyl)carbamate (2). A 1000-mL, 3-necked, round-bottomed flask equipped with an overhead mechanical stirrer (teflon paddle, 7 × 2 cm), thermometer (to −100 °C), and 250-mL graduated pressure-equalizing addition funnel fitted with a rubber septum and argon inlet needle (Note 1) is charged with isopropyl chloroformate (252 mL, 252 mmol, 1.09 equiv) (Notes 2 and 3) and cooled to 0–5 °C using an ice–water bath. Aniline (21.2 mL, 232 mmol) (Note 4) is added over a period of 30 min via addition funnel. The resulting suspension is vigorously stirred while triethylamine (35 mL, 252 mmol, 1.09 equiv) (Note 5) is added dropwise via addition funnel over 30 min (Note 6). The suspension is stirred vigorously and allowed to warm to room temperature using a water bath; stirring is continued for 1 h. The suspension is poured into a 1000-mL separatory funnel containing ethyl acetate (300 mL) (Note 4) and 1 N HCl (200 mL), and the aqueous phase is separated and extracted with ethyl acetate (2 x 50 mL). The combined organic layers are washed with 1 N HCl (2 x 100 mL), water (100 mL), and saturated NaCl solution (100 mL), dried over MgSO$_4$, filtered through a fritted-glass funnel of medium porosity, and concentrated by rotary evaporation (50 °C, 40 mmHg), and then dried for 15 h (23 °C, 0.1 mmHg) to afford 40.7 g of white solid (98%) (Notes 7 and 8). The solid is used without further purification.

The product is transferred to a flame-dried, 1000-mL, 3-necked, round-bottomed flask equipped with an overhead mechanical stirrer (teflon paddle, 7 × 2 cm), thermometer, and 250-mL graduated pressure-equalizing addition funnel fitted with a rubber septum and argon inlet needle (Note 1). Anhydrous tetrahydrofuran (273 mL) (Note 9) is added, and the resulting solution is cooled to −70 °C using a dry ice–acetone bath. *n*-Butyllithium (96.3 mL, 241 mmol, 1.06 equiv) (Note 10) is added via addition funnel over a period of 20 min (Note 11). The resulting light brown solution is vigorously stirred for an additional 15 min, and acetic anhydride (24.8 mL, 261 mmol, 1.15 equiv) (Note 4) is added via syringe over a period of 5 min. The resulting suspension is vigorously stirred and warmed to room temperature using a water bath; stirring is continued for 3 h (Note 12). The yellow suspension is poured into a 1000-mL separatory funnel containing diethyl ether (250 mL) and water (250 mL), and the aqueous phase is separated and extracted with diethyl ether (2 x 100 mL). The combined organic layers are washed with saturated sodium bicarbonate solution (2 x 100 mL) and saturated NaCl solution (150 mL), dried over MgSO$_4$, filtered through a fritted-glass funnel of medium porosity, and concentrated by rotary evaporation (50 °C, 40 mmHg), and then dried for 12 h (23 °C, 0.1 mmHg) to afford 50.8 g of yellow solid (99%) (Note 13), which is used without further purification.

   B. *Isopropyl phenyl(4,4,4-trifluoro-3,3-dihydroxybutanoyl)carbamate (3).* A 1000-mL, 3-necked, round-bottomed flask equipped with an overhead mechanical stirrer (teflon paddle, 7 × 2 cm), thermometer, and rubber septum and argon inlet needle (Note 1) is charged with hexamethyldisilazane (24.7 mL, 119 mmol, 1.05 equiv) (Note 2) in tetrahydrofuran (119 mL, 1.0 M) (Note 9) and cooled to −70 °C in a dry ice–acetone bath. *n*-Butyllithium (45.2 mL, 113 mmol, 1.00 equiv) (Note 10) is added via syringe, and the solution is stirred for 15 min. The rubber septum is then replaced with a 250-mL pressure-equalizing addition funnel fitted with a rubber septum with argon inlet needle, and isopropyl acetyl(phenyl)carbamate (25.0 g, 113 mmol, 1.00 equiv) in tetrahydrofuran (377 mL, 0.3 M) (Note 9) is added dropwise over 15 min (Note 14). The mixture is stirred at −70 °C for an additional 30 min, after which time the addition funnel is replaced with a rubber septum, and 2,2,2-trifluoroethyl trifluoroacetate (18.2 mL, 136 mmol, 1.20 equiv) (Note 2) is added in one portion via syringe (Note 15). The reaction mixture is stirred at −70 °C for 10 min, and then quenched at −70 °C with 1 N HCl (50 mL) over 1 min, and while still cold, poured into a 1000-mL separatory funnel containing 1 N HCl (250 mL) and diethyl ether (250 mL, Note 2). The phases are separated, and the organic solution is washed with 1 N HCl (200 mL) then saturated NaCl solution (200 mL). The organic solution is dried over MgSO₄, filtered through a fritted-glass funnel of medium porosity, and concentrated via rotary evaporation (25 °C, 40 mmHg) to afford a white sticky solid (Notes 16, 17, and 18).

   C. *Isopropyl 2-diazoacetyl(phenyl)carbamate (4).* The hydrate is transferred to a 1000-mL, 3-necked, round-bottomed flask equipped with an overhead mechanical stirrer (teflon paddle, 7 × 2 cm), thermometer, and rubber septum and argon inlet needle (Note 1). The solid is dissolved in acetonitrile (377 mL, 0.3 M) (Note 9), then 4-acetamidobenzenesulfonyl azide (27.1 g, 113 mmol, 1.00 equiv) (Note 19) is added in one portion. The stirred reaction mixture is cooled to 0–5 °C in an ice–water bath. Once cooled, triethylamine (23.5 mL, 169 mmol, 1.50 equiv) (Note 5) is added via syringe. The reaction is stirred for 30 min, then the ice–water bath is replaced with a room-temperature bath, and the reaction mixture is stirred for an additional 3 h (Note 20). The orange suspension is then transferred to

214

a 1000-mL separatory funnel containing diethyl ether (350 mL) and 1 M NaOH (300 mL). The layers are separated, and the organic layer is washed with 1 M NaOH (300 mL) and then saturated NaCl solution (300 mL). The organic solution is dried over $MgSO_4$, filtered through a fritted-glass funnel of medium porosity, and concentrated via rotary evaporation (23 °C, 40 mmHg) to give a viscous red oil. The oil is dissolved in dichloromethane (10 mL) (Note 21), and the resulting solution is purified via flash chromatography (Notes 22, 23, and 24) to yield 12.0–13.2 g of yellow solid, which is then recrystallized from warm hexanes (75 mL, 60 °C) (Note 4) to give 8.6–9.2 g (34.7–37.2 mmol, 31–33% for two steps) of pure diazoimide as a yellow crystalline solid (Note 25).

2. Notes

1. The apparatus was oven-dried for 12 h and then maintained under an atmosphere of argon during the course of the reaction.

2. Isopropyl chloroformate (1.0 M solution in toluene), 2,2,2-trifluoroethyl trifluoroacetate (99%), diethyl ether (anhydrous, 99%), and 1,1,1,3,3,3-hexamethyldisilazane (>99%) were purchased from Aldrich Chemical Co. All reagents were used as received.

3. Isopropyl chloroformate is added first to the graduated pressure-equalizing addition funnel via cannula and then added to the round-bottomed flask. The addition funnel is rinsed with anhydrous toluene (2 x 10 mL) (Note 9).

4. Aniline, ethyl acetate (99.9%), hexanes (99.9%) and acetic anhydride (99%) were purchased from Fisher Scientific Company and used as received.

5. Triethylamine (99.5%) was purchased from EMD Chemicals Inc. and distilled from calcium hydride prior to use.

6. The reaction mixture is maintained at 0–5 °C during addition. After ca. 10 min, a precipitate forms. The addition funnel is rinsed with anhydrous toluene (2 x 10 mL) (Note 9).

7. The obtained solid was 99% pure by gas chromatography on an Agilent 7890A gas chromatograph equipped with flame ionization detectors. The column used was HP-5 (30 m × 0.32 mm). The following settings were used: flow = 1.0 mL/min, oven = 60 °C to 220 °C at 20 °C/min; detector = 300 °C, injector = 250 °C; t_r = 6.95 min.

8. Isopropyl phenylcarbamate displayed the following physicochemical properties: mp 82–83 °C; R_f = 0.68 (20% EtOAc/hexanes); IR (neat) 3308, 2979, 1717, 1534, 1232 cm^{-1}; ^{1}H NMR (400 MHz, CDCl$_3$) δ: 1.31 (d, J = 6.4 Hz, 6 H), 5.05 (sept, J = 6.0 Hz, 1 H), 6.89 (br s, 1 H), 7.06 (t, J = 7.2 Hz, 1 H), 7.28–7.32 (m, 2 H), 7.41–7.43 (d, J = 8.0 Hz, 2 H); ^{13}C NMR (100 MHz, CDCl$_3$) δ: 22.3, 68.8, 118.8, 123.3, 129.1, 138.3, 153.6; HRMS (ESI): Exact mass calcd. for C$_{10}$H$_{13}$NO$_2$ 179.0946, found 180.1016 [M+H]$^+$. Anal. calcd. for C$_{10}$H$_{13}$NO$_2$: C, 67.02; H, 7.31; N, 7.82. Found: C, 67.21; H, 7.33; N, 8.01.

9. Tetrahydrofuran (THF), toluene, and acetonitrile were dried by passage through a column of activated alumina as described by Grubbs.[3] The checkers used THF that was freshly distilled from sodium benzophenone ketyl. Toluene and acetonitrile were obtained from a Glass contour solvent purifier system by SG Water USA.

10. n-Butyllithium (2.5 M in hexanes) was purchased from Aldrich Chemical Co. and was titrated prior to use using 2-propanol and 1,10-phenanthroline according to an established procedure.[4]

11. The reaction mixture is maintained below −60 °C during addition.

12. Reaction progress was monitored by gas chromatography on an Agilent 7890A gas chromatograph equipped with flame ionization detectors. The column used is HP-5 (30 m × 0.32 mm). The following settings were used: flow = 1.0 mL/min, oven = 60 °C to 220 °C at 20 °C/min; detector = 300 °C, injector = 250 °C; t_r = 4.90 min.

13. The obtained solid was 98% pure by gas chromatography (Note 12). An analytical sample was obtained by recrystallization from diethyl ether. Mp 85–86 °C; R_f = 0.38 (20% EtOAc/hexanes); IR (neat) 2983, 1734, 1708, 1255 cm^{-1}; ^{1}H NMR (400 MHz, CDCl$_3$) δ: 1.16 (d, J = 6.0 Hz, 6 H), 2.62 (s, 3 H), 4.98 (sept, J = 6.0 Hz, 1 H), 7.09 (d, J = 7.0 Hz, 8 H), 7.43-7.34 (m, 3 H); ^{13}C NMR (100 MHz, CDCl$_3$) δ: 21.5, 26.5, 71.1, 128.0, 128.2, 128.9, 138.3, 153.6, 172.9; HRMS (ESI): Exact mass calcd. for C$_{12}$H$_{15}$NO$_3$ 221.1052, found 222.1246 [M+H]$^+$. Anal. calcd. for C$_{12}$H$_{15}$NO$_3$: C, 65.13; H, 6.84; N, 6.33. Found: C, 65.08; H, 6.98; N, 6.33.

14. The carbamate is first dissolved in THF and then transferred via cannula to the addition funnel. The temperature of the reaction mixture is maintained below –60 °C during addition, and the addition time is adjusted accordingly.

15. The internal temperature is maintained below –60 °C during the addition of 2,2,2-trifluoroethyl trifluoroacetate.

16. ^{1}H NMR of the crude product is used to confirm conversion of the isopropyl acetyl(phenyl)carbamate. Integration of the resonance of the methyl group (2.60 ppm) of the isopropyl acetyl(phenyl)carbamate is calibrated to 3 hydrogens and is compared with resonances at 5.06–4.88 ppm, which correspond to the methine proton of the isopropyl group of all mixture components. In that set of resonances, only one hydrogen corresponds to isopropyl acetyl(phenyl)carbamate. In the event that acceptable conversion is not obtained (below 95%, integration of methine resonances <20), the oil can be triturated with hexanes to remove residual isopropyl acetyl(phenyl)carbamate.

17. To the obtained oil in a 1000-mL round-bottomed flask is added 200 mL of hexanes (Note 4), and the resulting solution is concentrated via rotary evaporation (23 °C, 40 mmHg). To the obtained semi-solid residue is added 200 mL of hexanes and the resulting suspension is concentrated via rotary evaporation (23 °C, 40 mmHg). The solid is filtered through a fritted-glass funnel of medium porosity to afford a white, crystalline solid that displayed the following physicochemical properties: mp 83–84 °C, IR (film) 3373, 2986, 2941, 1742, 1680, 1596, 1491, 1376, 1264 cm^{-1}; ^{1}H NMR (400 MHz, CDCl$_3$) δ: 1.17 (d, $J = 6.3$ Hz, 6 H), 3.51 (s, 2 H), 4.91 (s, 2 H), 5.00 (sept, $J = 6.2$ Hz, 1 H), 7.11 (dd, $J = 1.7$, 1.3 Hz, 1 H), 7.13 (dd, $J = 2.0$, 1.7 Hz, 1 H), 7.47–7.39 (m, 3 H); ^{13}C NMR (100 MHz, CDCl$_3$) δ: 21.4, 38.9, 72.6, 93.3 (q, $J = 33.0$ Hz), 122.3 (q, $J = 284$ Hz), 128.1, 128.7, 129.4, 137.1, 153.2, 174.1; ^{17}F NMR (376 MHz, CDCl$_3$) δ: −87.0 (s, 3F); HRMS (ESI): Exact mass calcd. for C$_{14}$H$_{16}$F$_3$NNaO$_5$ [M+Na]$^+$ 358.0878, found 358.0887. Anal. calcd. for C$_{14}$H$_{16}$F$_3$NO$_5$: C, 50.15; H, 4.81; N, 4.18. Found: C, 50.36; H, 4.70; N, 4.16.

18. The hydrate is not stable in solution at room temperature. If the diazo transfer cannot be performed immediately, the hydrate can be stored dry and under argon at −78 °C for up to 3 months.

19. 4-Acetamidobenzenesulfonyl azide was synthesized using the procedure of Davies.[5] *Caution! The original procedure using methylene chloride as solvent should be avoided because it could produce the highly explosive material, diazidomethane, as a side product.*[6]

20. The yellow color of the clear reaction mixture intensified while warming to room temperature. After approximately 1 h, a white solid precipitated.

21. The oily residue is too viscous to be easily transferred onto silica gel and dilution with dichloromethane is needed.

22. Flash column chromatography was performed on a silica gel column (20 cm length × 5.5 cm width, 250 g of silica gel) (Notes 23 and 24). The product was eluted with hexanes/ethyl acetate/triethylamine, 93:5:2 (ca. 700 mL of eluent are required before collecting fractions). Fractions 5-30 (50 mL each) were collected and analyzed by thin layer chromatography (TLC), eluting with hexanes/ethyl acetate, 4:1 (R_f = 0.38 for isopropyl 2-diazoacetyl (phenyl)carbamate, R_f = 0.27 for decomposition product). Visualization was accomplished with UV.

23. Silica gel was purchased from Sorbent Technologies Company with the following specifications: porosity 60 Å, particle size 40–64 mm, surface area 450–550 m^2/g.

24. Isopropyl 2-diazoacetyl(phenyl)carbamate decomposes during contact with silica gel. Silica gel deactivated with 2% triethylamine is used to avoid decomposition.

25. Isopropyl 2-diazoacetyl(phenyl)carbamate displayed the following physicochemical properties: R_f = 0.38, mp 61–62 °C; IR (neat) 3145, 2983, 2109, 1723 cm^{-1}; ^1H NMR (400 MHz, CDCl$_3$) δ: 1.14 (d, J = 6.0 Hz, 6 H), 4.94 (sept, J = 6.0 Hz, 1 H), 6.62 (s, 1 H), 7.14–7.12 (m, 2 H), 7.42–7.33 (m, 3 H); ^{13}C NMR (100 MHz, CDCl$_3$) δ: 21.6, 51.9, 71.3, 128.1, 128.7, 129.0, 137.7, 153.5, 167.0. HRMS (ESI): Exact mass calcd. for $C_{12}H_{13}N_3O_3$ 247.0957, found 248.1030 [M+H]$^+$. Anal. calcd. for $C_{12}H_{13}N_3O_3$: C, 58.29; H, 5.30; N, 16.99. Found: C, 58.56; H, 5.36; N, 16.73.

Waste Disposal Information

All hazardous materials should be handled and disposed of in accordance with "Prudent Practices in the Laboratory"; National Academy Press; Washington, 1955.

3. Discussion

Diazo compounds and their chemistry have a long history, and the unique diazo functionality translates into their versatile reactivity. Diazoalkanes have been successfully used as carbene precursors in various useful applications such as cyclopropanation and X–H (X=C, Si, O, S, N) insertion reactions.[7,8] They can also be involved in rearrangement reactions,[9] give rise to ylides,[10] or serve as 1,3-dipoles in [3+2] cycloaddition

218

reactions.[11] Additionally, they can be used as nucleophiles in addition reactions with carbonyl and azomethine electrophiles.[12]

Scheme 1.

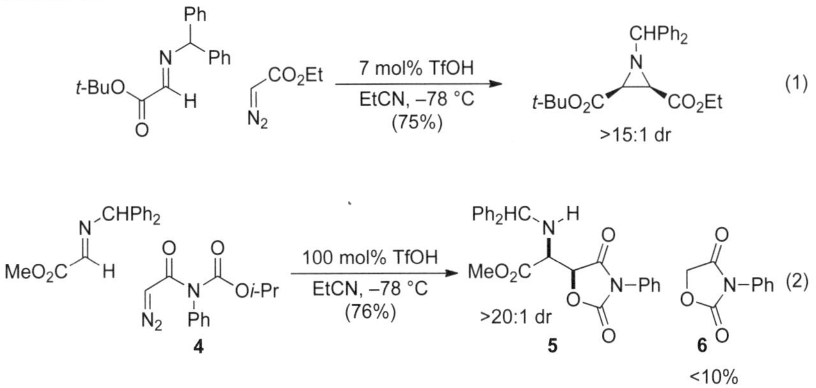

The recent finding that diazo compounds can be effective substrates in carbon-carbon and carbon-heteroatom bond forming reactions when strong Brønsted acids are used as the activating agent (Scheme 1, eq 1)[13] has been followed by the development of a range of new reactions.[14] Moreover, a growing number of examples have revealed that chiral Brønsted acids can promote enantioselective carbon-carbon bond formation using diazoalkane substrates.[15,16] These successes have led to an interest in diazoalkane donors that exhibit reactivity similar to α-diazo esters, but provide new opportunities in product diversity and reaction stereocontrol. We hypothesized that the title diazoalkane (**2**) might provide a new avenue in the reaction map of diazoalkane chemistry. Indeed, we have developed a highly diastereoselective, Brønsted acid promoted *syn*-glycolate Mannich reaction using diazoimide **2** as a donor (Scheme 1, eq 2).[17]

The procedure described here is an optimized synthesis of this reagent on preparative scale. Diazoimide **2** is an acyclic variation of Doyle's oxazolidinone; the conformational mobility of the carbamate allows it to function as an oxygen donor in the *syn*-glycolate Mannich reaction.

Among the plethora of methods reported for obtaining diazo compounds, diazo transfer is one of the most reliable and efficient ways for the synthesis of diazocarbonyl compounds from the corresponding carbonyl compound. This report is an adaptation of the successful procedure reported by Doyle[18] with modifications based on the synthesis of (*E*)-1-diazo-4-phenyl-3-buten-2-one reported by Danheiser.[19]

219

One challenge that was overcome in the development of this preparation is the avoidance of acid-promoted decomposition of the diazoimide to oxazolidine dione **6**. When isopropyl 2-diazoacetyl(phenyl)carbamate (**2**) alone is subjected to triflic acid at room temperature, oxazolidine dione **4** is exclusively formed. A second development challenge was the preparation of fluorinated intermediate **3**. Near complete conversion of precursor **2** must be achieved for the subsequent step to be successful. If unreacted **2** is present in the diazo transfer step, it is impossible to separate it from diazoimide **4**. Straightforward trituration with a nonpolar organic solvent, such as hexanes, effects the removal of **2** should incomplete conversion occur in this step. Careful temperature control in this step normally provides for full conversion to **3**. Additionally, we obtained spectroscopic data for intermediate **3**, in agreement with the hydrate of the trifluoromethyl ketone[20] and not the enol form observed by Doyle.[18]

Although the yield of the diazo transfer (Step C) is lower than reported examples,[18,19] the desired diazoimide can be prepared as a pure crystalline solid, as determined by combustion analysis.

Appendix
Chemical Abstracts Nomenclature; (Registry Number)

Isopropyl chloroformate; (108-23-6)

Aniline: benzeneamine; (62-53-3)

Triethylamine: (121-44-8)

n-Butyllithium: lithium, butyl-; (109-72-8)

Acetic anhydride: acetic acid, anhydride; (108-24-7)

1,1,1,3,3,3-Hexamethyldisilazane: silanamine, 1,1,1-trimethyl-*N*-
(trimethylsilyl)-; (999-97-3)

Isopropyl acetyl(phenyl)carbamate; Carbamic acid, *N*-acetyl-*N*-phenyl-, 1-
methylethyl ester; (5833-25-0)

2,2,2-Trifluoroethyl trifluoroacetate (407-38-5)

4-Acetamidobenzenesulfonyl azide; (2158-14-7)

Carbamic acid, *N*-(2-diazoacetyl)-*N*-phenyl-, 1-methylethyl ester; (1198356-
59-0)

1. Department of Chemistry & Vanderbilt Institute of Chemical Biology, Vanderbilt University, Nashville, TN 37235-1822. E-mail: jeffrey.n.johnston@vanderbilt.edu. This work was supported by the NSF (CHE-0848856). H. M. is grateful for support by a Warren Fellowship (2009).

2. Baum, J. S.; Shook, D. A.; Davies, H. M. L.; Smith, H. D. *Synth. Commun.* **1987**, *17,* 1709.

3. Pangborn, A. B.; Giardello, M. A.; Grubbs, R. H.; Rosen, R. K.; Timmers, F. J. *Organometallics* **1996**, *15,* 1518.

4. Watson, S. C.; Eastham, J. F. *J. Organomet. Chem.* **1967**, *9,* 165.

5. Davies, H. M. L.; Cantrell, W. R.; Romines, K. R.; Baum, J. S. *Org. Synth.* **1992**, *70,* 93.

6. Dean, D. W.; Conrow, R. E. *Org. Proc. Res. Dev.* **2008**, *12*, 1285.

7. (a) Ruppel, J. V.; Gauthier, T. J.; Snyder, N. L.; Perman, J. A.; Zhang, X. P. *Org. Lett.* **2009**, *11,* 2273. (b) Trost, B. M.; Malhotra, S.; Fried, B. A. *J. Am. Chem. Soc.* **2009**, *131*, 1674.

8. For representative leading references, see these enantioselective examples: alcohol: (a) Maier, T. C.; Fu, G. C. *J. Am. Chem. Soc.* **2006**, *128,* 4594. (b) Chen, C.; Zhu, S.-F.; Liu, B.; Wang, L.-X.; Zhou, Q.-L. *J. Am. Chem. Soc.* **2007**, *129,* 12616.; amine: (c) Lee, E. C.; Fu, G. C. *J. Am. Chem. Soc.* **2007**, *129,* 12066. (d) Liu, B.; Zhu, S.-F.; Zhang, W.; Chen, C.; Zhou, Q.-L. *J. Am. Chem. Soc.* **2007**, *129,* 5834.; thiol: (e) Zhang, Y.-Z.; Zhu, S.-F.; Cai, Y.; Mao, H.-X.; Zhou, Q.-L. *Chem. Commun.* **2009**, 5362.

9. (a) Sarpong, R.; Su, J. T.; Stoltz, B. M. *J. Am. Chem. Soc.* **2003**, *125,* 13624. (b) Jiang, N.; Ma, Z.; Qu, Z.; Xing, X.; Xie, L.; Wang, J. *J. Org. Chem.* **2003**, *68,* 893.

10. Sulfonium ylides: (a) Crich, D.; Zou, Y.; Brebion, F. *J. Org. Chem.* **2006**, *71,* 9172. (b) Aggarwal, V. K.; Winn, C. L. *Acc. Chem. Res.* **2004**, *37,* 611.; oxonium ylides: (d) Murphy, G. K.; West, F. G. *Org. Lett.* **2006**, *8,* 4359. (e) Clark, J. S.; Fessard, T. C.; Whitlock, G. A. *Tetrahedron* **2006**, *62,* 73.; ammounium ylides: (f) Roberts, E.; Sancon, J. P.; Sweeney, J. B. *Org. Lett.* **2005**, *7,* 2075. (g) Vanecko, J. A.; West, F. G. *Org. Lett.* **2005**, *7,* 2949.; carbonyl and azomethine ylides: (h) Mejia-Oneto, J. M.; Padwa, A. *Org. Lett.* **2006**, *8,* 3275. (i) Galliford, C. V.; Martenson, J. S.; Stern, C.; Scheidt, K. A. *Chem. Commun.* **2007**, 631.

11. (a) Garcia Ruano, J. L.; Alonso de Diego, S. A.; Martin, M. R.; Torrente, E.; Martin Castro, A. M. *Org. Lett.* **2004**, *6*, 4945. (b) Kano, T.; Hashimoto, T.; Maruoka, K. *J. Am. Chem. Soc.* **2006**, *128*, 2174.
12. Zhang, Z.; Wang, J. *Tetrahedron* **2008**, *64*, 6577.
13. Williams, A. L.; Johnston, J. N. *J. Am. Chem. Soc.* **2004**, *126*, 1612.
14. Johnston, J. N.; Muchalski, H.; Troyer, T. L. *Angew. Chem. Int. Ed.* **2010**, *49*, 2290.
15. Uraguchi, D.; Sorimachi, K.; Terada, M. *J. Am. Chem. Soc.* **2005**, *127*, 9360.
16. Akiyama, T.; Suzuki, T.; Mori, K. *Org. Lett.* **2009**, *11*, 2445.
17. Troyer, T. L., Muchalski, H.; Johnston, J. N. *Chem. Commun.* **2009**, 6195.
18. Doyle, M. P.; Dorow, R. L.; Terpstra, J. W.; Rodenhouse, R. A. *J. Org. Chem.* **1985**, *50*, 1663.
19. Danheiser, R. L.; Miller, R. F.; Brisbois, R. G. *Org. Synth.* **1996**, *73*, 134.
20. Berbasov, D. O.; Soloshonok, V. A. *Synthesis* **2003**, 2005.

Jeffrey N. Johnston completed his B.S. Chemistry degree at Xavier University in 1992 and a Ph.D. in organic chemistry at The Ohio State University in 1997 with Prof. Leo A. Paquette. He then worked as an NIH postdoctoral fellow with Prof. David A. Evans at Harvard University. He began his independent career at Indiana University, where he was promoted to Professor of Chemistry in 2005. In 2006, his research program moved to Vanderbilt University. His group has developed a range of new reactions and reagents that are used to streamline the total synthesis of complex alkaloid natural products. The development of new Brønsted acid-catalyzed reactions, as well as chiral proton catalysts, are ongoing investigational themes.

Hubert Muchalski, a native of Poland, received his B.S./M.S. degree in chemistry from Wrocław University of Technology in 2006. He is currently a graduate student with Prof. Johnston at Vanderbilt University, where he has broadened the scope of the Brønsted acid catalyzed aza-Darzens and syn-glycolate Mannich reactions while exploring new reactions based on diazoimides.

222

Amanda Doody was raised in Martin, Georgia, and completed her B.S. Chemistry degree (summa cum laude) from Wofford College in South Carolina in 2008. During the summer of 2007, she was an NSF-REU student at Columbia University working with Gerard Parkin. She matriculated at Vanderbilt in 2008. Under the mentorship of Prof. Johnston, Amanda's graduate research is focused on the development of Brønsted acid catalyzed additions of diazo-ylides to electrophiles and their applications in alkaloid total synthesis.

Timothy L. Troyer received his B.S. degree in chemistry from Goshen College in 1996. Following a stint in medicinal chemistry at Bristol Myers Squibb (CT), he began graduate studies with Prof. Johnston where he investigated the mechanism of the aza-Darzens reaction and developed a new Brønsted acid catalyzed syn-glycolate Mannich reaction. This work led to a Ph.D. in Chemistry in 2008 from Vanderbilt University. He is currently an Assistant Professor of Chemistry at West Virginia Wesleyan College with interests in new organocarbenes and the isolation of bioactive natural products.

John Frederick Briones was born in 1982 in Laguna, Philippines. He earned his B.S. degree in Chemistry from the University of the Philippines, Los Banos in 2003 and later on pursued his Master's degree at the University of the Philippines, Diliman. He joined the research lab of Prof. Huw Davies in 2007, and currently his research project focuses on Rh(II) catalyzed enantioselective transformations of alkynes.

GRAM SCALE CATALYTIC ASYMMETRIC AZIRIDINATION: PREPARATION OF (2R,3R)-ETHYL 1-BENZHYDRYL-3-(4-BROMOPHENYL)AZIRIDINE 2-CARBOXYLATE

A.

B.

Submitted by Aman A. Desai, Roberto Morán-Ramallal, and William D. Wulff*.[1]

Checked by Shabnam Kouchekzadeh Yazdi, Gavin Haberlin, and Mark Lautens.

1. Procedure

A. *N-(4-Bromobenzylidene)-1,1-diphenylmethanamine (1).*[2] A 250-mL round-bottomed flask with two 24/40 necks is fitted with a magnetic stir bar (20 mm x 8 mm x 8 mm), a rubber septum, and a 24/40 vacuum adapter. The flask is connected to a double manifold vacuum line with a nitrogen ballast. This assembly is flame-dried under high vacuum (1.0 mmHg) and cooled under a slight positive pressure of nitrogen. To this flask is then sequentially added 4-bromobenzaldehyde (11.84 g, 63.95 mmol, 1.1 equiv) and MgSO$_4$ (14.0 g, 116 mmol, 2.1 equiv) (Note 1). This is followed by the addition of 60 mL dry dichloromethane (Note 1), via a plastic syringe fitted with a metallic needle, along the neck and the sides of the flask such that all solids could be rinsed to the bottom of the flask. The mixture is stirred at room temperature (22 °C) under nitrogen to provide a white slurry.

224

Aminodiphenylmethane (10.0 mL, 56.3 mmol at 97% purity, 1.00 equiv) is then added to the solution through the rubber septum, via a plastic syringe fitted with a metallic needle (Note 1). The resulting white slurry is stirred at room temperature (22 °C) for 20 h, at which time the reaction is complete (Note 2). A different 250-mL round-bottomed flask (24/40 single neck) is fitted with a filter adapter and a coarse porosity glass fritted funnel (60 mL, 4.5 × 5.5 cm) packed with Celite (1 cm height). The reaction mixture is filtered through the fritted funnel; the reaction flask is rinsed with dichloromethane (3 x 15 mL) three times, and the rinse is added to the funnel each time (Note 3). The fritted funnel and the Celite bed are then washed twice with dichloromethane (2 x 15 mL). The fritted funnel is removed, the filter adapter is rinsed with 10 mL dichloromethane, and the solution of the crude product is then subjected to rotary evaporation until dryness (40 mmHg, 45 °C) followed by high vacuum (1.0 mmHg) for 3 h. This afforded the crude imine product as an off-white solid (21.36 g). For crystallization, the 250-mL round-bottomed flask containing the crude product is equipped with a magnetic stir bar and fitted with a water condenser, a rubber septum and nitrogen line. A mixture of ethyl acetate:hexanes (1:5) (35 mL) is prepared and a portion is poured into the flask to cover the solids (Note 4). The flask is placed in an oil bath at 80 °C (bath temperature) and brought to reflux. The rest of the ethyl acetate:hexanes (1:5) mixture is then added slowly with continued boiling (total volume 35 mL). At this time, a clear pale yellow solution of the crude product is obtained, and the flask is placed on a wooden cork and left untouched for 19 h. The resulting crystals are then broken up into small pieces by a spatula, and these are collected by filtration on a Büchner funnel (6 cm d × 3.5 cm h). The crystallization flask is rinsed twice with cold (0 °C) hexanes (2 x 10 mL) and the rinse added each time to the Büchner funnel. The resulting crystals are transferred to a 50-mL, single-necked, round-bottomed flask and subjected to high vacuum (1.0 mmHg) for 4 h. The imine product **1** is isolated as white crystals (mp 95–97 °C) in 87% yield (17.21 g, 49.13 mmol) (Notes 5 and 6).

B. *(2R,3R) Ethyl 1-benzhydryl-3-(4-bromophenyl)aziridine-2-carboxylate (2).*[2] A 100-mL glass Schlenk flask fitted with a magnetic stir bar (12 mm x 3 mm x 3 mm) is connected via a rubber tube, attached to the flask's side-arm, to a double manifold vacuum line with an argon ballast (Notes 7, 8 and 9). The flask is flame-dried under high vacuum (0.10 mmHg) and cooled under a slight positive pressure of argon. Through the top of the flask is added sequentially (S)-VANOL (44 mg, 0.10 mmol, 0.005 equiv) and triphenyl borate (116 mg, 0.400 mmol, 0.02 equiv) under a slight positive pressure of argon (Note 10). Thereafter, dry toluene (4 mL) (Note 1) is added along the sides of the Schlenk flask via a plastic syringe fitted with a metallic needle, which had been pre-flushed with nitrogen. This is followed by the addition of water (1.8 µL, 0.10 mmol, 0.005 equiv) via a glass syringe (Note 11). The flask is then sealed, and the mixture is stirred at 80 °C in an oil bath (bath temperature) for 1 h. Thereafter, the valve on the double manifold connected to the Schlenk flask is turned to high vacuum (0.10 mmHg). While stirring is maintained, the threaded valve on the Schlenk flask is carefully and gradually opened to the high vacuum, and the solvent is removed (Note 12). After all solvent is removed, the Schlenk flask is allowed to remain at 80 °C in the oil bath for an additional 0.5 h exposed to high vacuum (Note 13). The flask is removed from the oil bath and cooled to room temperature under a slight positive pressure of argon (*ca.* 20 min) to afford the pre-catalyst as a colorless/off-white oil that adheres to the sides of the Schlenk flask. To this is added through the top of the flask, under a slight positive pressure of argon, imine 1 (7.00 g, 20 mmol, 1 equiv), followed by the addition of 20 mL of dry toluene along the sides of the Schlenk flask via a plastic syringe fitted with a metallic needle (pre-flushed with nitrogen). The magnetic stir bar at this point is stuck to the Schlenk flask, the flask is swirled by hand until the stir bar became free (Note 14). Additional dry toluene (5 mL) is then added along the sides of the Schlenk flask, and the solution is stirred at room temperature for 5-10 min to obtain a clear pale yellow solution of the catalyst-imine complex. Under a slight positive pressure of argon, ethyl diazoacetate (2.5 mL, 24 mmol, 1.2 equiv) is then added to the Schlenk flask via a plastic syringe fitted with a metallic needle (pre-flushed with nitrogen) (Note 10). The clear solution in the Schlenk flask turns dark yellow/orange with this addition, and vigorous nitrogen evolution is observed. Within 1 h, the product aziridine starts to precipitate, and the entire reaction mixture turns into a pale yellow semi-solid mass. The reaction mixture is stirred at room temperature under a slight positive pressure of argon for a total of 8 h, at which point the reaction is complete (Note 15).

Dichloromethane (30 mL) is added to the Schlenk flask (Note 3), and the resulting mixture is stirred to obtain a clear yellow solution of the crude aziridine product. The yellow solution is then added, via a glass funnel, to a pre-weighed 250-mL round-bottomed flask with a single neck (24/40 joint).

226

The Schlenk flask is rinsed with dichloromethane (2 x 20 mL), the plastic funnel with dichloromethane (5 mL), and the rinse added each time to the round-bottomed flask. The solution is then subjected to rotary evaporation (40 mmHg, 45 °C) until *ca.* 25 mL of the crude product solution is left in the flask. Hexanes (50 mL) are added to the flask, and the solution is again subjected to rotary evaporation (40 mmHg, 45 °C) to dryness and finally to high vacuum (1.0 mm Hg) for 12 h to afford the crude aziridine product **2** as an off-white solid (8.52 g). The solid is dissolved in dichloromethane (30 mL) to obtain a clear yellow solution, which is allowed to stand at room temperature for 15 min. A different pre-weighed 150-mL round-bottomed flask (24/40 single neck) is fitted with a filter adapter and a coarse porosity glass fritted funnel (30 mL, 3.5 × 5.0 cm) packed with Celite (1 cm height). The crude product solution is filtered through the fritted funnel; the flask is rinsed with dichloromethane (2 x 20 mL), and the rinses are added to the funnel. The sides of the fritted funnel and the Celite bed are washed with dichloromethane (20 mL). The fritted funnel is removed, the filter adapter is rinsed with dichloromethane (10 mL), and this crude product solution is then subjected to rotary evaporation to dryness (40 mmHg, 40 °C) followed by high vacuum (1.0 mmHg) for 4 h. This process afforded the crude aziridine product **2** as an off-white solid.

For crystallization, the 150-mL round-bottomed flask containing the crude product is equipped with a magnetic stir bar and fitted with a water condenser, a rubber septum and connected to a nitrogen line. A mixture of dichloromethane:hexanes (1:3) (90 mL) is prepared, and a portion is poured into the flask that is just sufficient to cover the solids (Notes 3 and 4). The condenser is turned on and the flask is placed in an oil bath at 80 °C (bath temperature). Once at reflux, more solvent is slowly added via a plastic syringe fitted with a metallic needle until all solids had dissolved (total volume 90 mL) and a clear pale yellow solution of the crude product is obtained. The crude product is placed on a wooden cork and left untouched for 1 h over which time some of the product crystallizes. The condenser is removed, and the crystallization flask is allowed to stand open to air at room temperature for an additional 27 h. After a total of 28 h, the supernatant is carefully decanted, so as to not disturb the crystals, into another 500-mL round-bottomed flask (24/40 single neck) via a plastic funnel. Cold hexanes (–42 °C, 50 mL) is gently added to the crystallization flask, the flask is gently swirled by hand, and the supernatant solution is again added via decantation to the 500-mL round-bottomed flask. The hexane wash is repeated once more. Then, 80 mL of dichloromethane is added to the crystallization flask to completely dissolve the product crystals to afford a clear slightly pale yellow solution. An aliquot (1 mL) of this solution is taken and placed into an HPLC vial. The dichloromethane is evaporated by air. The product is dissolved in 2-propanol (1.5 mL) and the solution is subjected to chiral HPLC analysis, which reveals 99% ee for the first crop of aziridine (Note 16). The dichloromethane solution of the first crop in the 150-mL round-bottomed flask is then subjected to rotary evaporation until dryness (40 °C, 40 mmHg) and finally at high vacuum for 3–4 h to afford the first crop of aziridine **2** as a white solid (mp 154–155 °C) in 43% yield (3.77 g, 8.64 mmol) (Note 16).

The 500-mL round-bottomed flask with the mother liquor and the washes is then subjected to rotary evaporation to dryness (40 °C, 40 mmHg), and the resulting solids are dissolved in dichloromethane (50 mL) and transferred to a different pre-weighed 150-mL round-bottomed flask (24/40 single neck) via a glass funnel. The 500-mL round-bottomed flask is rinsed with dichloromethane (2 x 20 mL), and the glass funnel is rinsed with dichloromethane (10 mL). The rinses are added each time to the 150-mL round-bottomed flask. This mother liquor solution is then subjected to rotary evaporation until dryness (40 °C, 40 mmHg) and finally to high vacuum for 3–4 h to afford the crude aziridine product **2** as a pale yellow solid (6.32 g). A second crop is then taken from this material in the same manner as the first crop, except that a total volume of 35 mL of a 1:4 mixture of dichloromethane:hexanes is used for the crystallization. Two similar cold hexanes (–42 °C) washes are employed as in the first crop, with 25 mL of hexanes being used each time. The second crop of aziridine **2** is obtained as a white/off-white solid (mp 140–143 °C) in 47% yield (4.11 g, 9.42 mmol) and 83% ee. Thus, the overall yield of the reaction is 7.82 g (90%, 17.90 mmol) (Notes 17 and 18).

2. Notes

1. Aminodiphenylmethane (97%) and 4-bromobenzaldehyde (99%) were obtained from Aldrich, used as received and stored under nitrogen on the bench. The checkers obtained $MgSO_4$ (98+%, anhydrous) from ACP Chemicals Inc., which was used as received. The submitters obtained $MgSO_4$ (98+%, anhydrous) from Jade Scientific, which was used as received. Dichloromethane (99.9%) was obtained from Fisher Scientific and

228

dispensed from an MBRAUN solvent purification system. Toluene (99.9%) was obtained from Fisher Scientific and was distilled from sodium under nitrogen. The submitters obtained dichloromethane (99.5+%) from Mallinckrodt Chemicals, which was distilled from calcium hydride under nitrogen. Toluene (99.5+%) was obtained from Mallinckrodt Chemicals and distilled from sodium under nitrogen.

2.   Determined from ^{1}H NMR analysis of the crude reaction mixture. The stirring was stopped and the solids were allowed to settle from the reaction mixture to the bottom of the reaction flask. A small aliquot (<0.5 mL) was then taken from the solution with a glass pipette, which was subjected directly to high vacuum (0.01 mm Hg) for 15 min, and analyzed by ^{1}H NMR. When the disappearance of the singlet corresponding to the methine proton of aminodiphenylmethane was observed (δ = 5.22 ppm, CDCl$_3$) the reaction was judged complete.

3.   The dichloromethane was not dried. Dichloromethane (99.9%) was obtained from Fisher Scientific.

4.   The checkers purchased ethyl acetate (99.9%) from Fisher Scientific, which was used as received. Hexanes (99.9%, containing various methylpentanes, 4.2%) were purchased from Fisher Scientific and used as received. The submitters purchased ethyl acetate (99.5+%) from Mallinckrodt Chemicals, which was used as received. Hexanes (98.5+%, total hexane isomers and methylcyclopentane) was obtained from EMD Chemicals and used as received.

5.   Imine **1** can be stored for a long period of time sealed under nitrogen in a desiccator. Characterization data for imine **1**:[2] white crystals: mp 95–97 °C. ^{1}H NMR (CD$_2$Cl$_2$, 400 MHz) δ: 5.59 (s, 1 H), 7.22–7.26 (m, 2 H), 7.31–7.35 (m, 4 H), 7.39–7.42 (m, 4 H), 7.58 (d, J = 8.0 Hz, 2 H), 7.74 (d, J = 12.0 Hz, 2 H), 8.41 (s, 1 H); ^{13}C NMR (CD$_2$Cl$_2$, 100 MHz) δ: 78.50, 125.65, 127.60, 128.04, 129.00, 130.35, 132.36, 135.91, 144.53, 160.18; IR (thin film) 3026(w), 2844(w), 1643(s), 1486(m), 699(s) cm^{-1}; HRMS (TOF EI) calcd. for C$_{20}$H$_{16}$^{79}BrN (M) m/z 349.0466, found 349.0464. Anal. calcd. for C$_{20}$H$_{16}$BrN: C, 68.58; H, 4.60; N 4.00. Found: C, 68.53; H, 4.71; N, 4.07.

6.   The submitters found a second crop of crystals could be taken to afford imine **1** in 6% yield (1.27 g, 3.63 mmol), but this was found to contain ~1% of 4-bromobenzaldehyde by ^{1}H NMR analysis.

7.   The checkers used a commercial Schlenk flask purchased from Chemglass (Airfree, AF-0094-02). The submitters used a Schlenk flask that was made in a glass-blowing shop by fusing together a high vacuum

threaded Teflon valve (Chemglass, CG-960-03, Valve, Chem-Vac™, Chem-Cap®, Hi-Vac, 1-Arm, 0-12 mm Bore) and a 100 mL recovery flask (Chemglass, CG-622-04, 100 mL Glassblowers Flask Blank, Recovery). The side-arm of the high vacuum valve was modified with a piece of 3/8th inch glass tubing to fit with the rubber tube attached to the double manifold. The submitters report that the use of a large magnetic stir bar (3.8 × 1 × 1 cm) is optimal for efficient stirring during the aziridination reaction. See Picture in Note 8.

8.   Pictures of the set up used by the submitters:

9.   The double manifold had two-way high-vacuum valves, which could be alternated between high vacuum (0.1 mmHg) and an argon supply.

10.   (S)-VANOL is commercially available from Aldrich as well as Strem Chemicals, Inc. It was sealed under argon and stored in a refrigerator away from light. Triphenyl borate was obtained from Aldrich, used as received and stored under nitrogen in a desiccator. Ethyl diazoacetate was obtained from Aldrich, used as received and stored under nitrogen in a refrigerator. Commercially available ethyl diazoacetate usually contains ≤15% dichloromethane, which was the reason behind using 1.2 equiv in the procedure. The checkers found that ethyl diazoacetate decomposes to some extent over the course of one month even when stored in a refrigerator under nitrogen.

11.   It is not problematic if the water sticks to the flask walls instead of falling directly into the reaction mixture. The water will eventually enter the reaction mixture as the vessel is heated.

12. If the threaded valve on the Schlenk flask is not opened with care under high vacuum, the solvent will bump into the manifold and result in loss of catalyst. Great care must be taken, and maintaining stirring reduces bumping.

13. The vacuum is applied to remove the solvent, phenol, and other volatiles, which greatly reduce the yield of the subsequent reaction; therefore, it is essential to use a strong vacuum pump (≤ 1 mmHg) to remove all solvent and volatiles.

14. If needed, the Schlenk flask may be gently tapped on a hard surface to aid in freeing the stir bar stuck inside.

15. Determined from ^{1}H NMR analysis of the crude reaction mixture. A glass pipette was dipped into the Schlenk flask, and a small amount of the semi-solid reaction mass was collected at the tip of the pipette. This was directly rinsed with CDCl$_3$ and analyzed by ^{1}H NMR. The disappearance of the singlets corresponding to the methine protons of imine **1** was monitored ($\delta = 5.59$, 8.36 ppm, CDCl$_3$), and the conversion was judged to be $\geq 95\%$ as determined from the relative integration of the aforementioned imine methine protons vs. the aziridine ring methine protons.

16. The submitters report yields of 62-64% for the first crop of crystals. Aziridine **2** can be stored for a long period of time sealed under nitrogen on the bench. The optical purity of the first crop of (2R,3R)-**2** was determined to be 99% ee by HPLC analysis (Chiralcel OD-H column (column length 25 cm, internal diameter 0.46 cm, particle size 0.5 cm), hexanes/2-propanol 98:2, 222 nm, flow rate 1 mL min^{-1}). The checkers obtained the following retention times: $t_R = 9.1$ min (minor enantiomer) and $t_R = 20.5$ min (major enantiomer). However, the submitters obtained the following retention times: $t_R = 5.5$ min (minor enantiomer) and $t_R = 13.3$ min (major enantiomer). The checkers obtained the reported retention times with three different OD-H columns. Spectral data for (2R,3R)-**2**:[2] ^{1}H NMR (CD$_2$Cl$_2$, 400 MHz) δ: 1.03 (t, J = 8.0 Hz, 3 H), 2.72 (d, J = 6.8 Hz, 1 H), 3.17 (d, J = 6.8 Hz, 1 H), 3.94 (q, J = 8.0 Hz, 2 H), 3.97 (s, 1 H), 7.21 (t, J = 7.3 Hz, 1 H), 7.26–7.41 (m, 9 H), 7.48 (d, J = 8.0 Hz, 2 H), 7.59 (d, J = 8.0 Hz, 2 H); ^{13}C NMR (CD$_2$Cl$_2$, 100 MHz) δ: 14.40, 47.14, 47.81, 61.26, 77.98, 121.76, 127.64, 127.85, 127.95, 128.06, 129.08, 129.10, 130.18, 131.40, 134.97, 143.01, 143.18, 167.81; IR (thin film) 1734(s), 1201(s), 1066(m) cm^{-1}; HRMS (ESI) m/z calcd. for C$_{24}$H$_{23}$BrNO$_2$: 436.0906, found 436.0926; Anal. calcd. for C$_{24}$H$_{22}$BrNO$_2$: C 66.06; H 5.08; N 3.21. Found: C 65.82; H

5.40; N, 3.48; $[\alpha]^{25}_D = +12.0$ ($c = 0.2$, CH_2Cl_2) on 99% ee material; white solid: mp 154–155 °C on 99% ee material.

17. The checkers found that the reaction proceeds at the specified catalyst loading on full scale. At half scale the optimal results were obtained with 1% catalyst loading. The submitters used lower catalyst loading (0.5 mol%) on reduced scale, but used a portion of a stock solution of catalyst.

18. The submitters found during the optimization of this procedure, that one particular run afforded the 1st crop of aziridine 2 in 59% yield (5.13 g, 11.77 mmol) and 99% ee, and the 2nd crop in 24% yield (2.07 g, 4.75 mmol) and 78% ee. At this time, a 3rd crop was taken by simply washing the crude material remaining after the 2nd crop with cold hexanes (–20 °C, 2 × 10 mL) and swirling by hand followed by decantation. Thus, the 3rd crop of aziridine 2 was obtained in 5% yield (0.40 g, 0.92 mmol) and 96% ee. Subjecting the crude material remaining after collection of the 3rd crop to purification by column chromatography on regular silica gel (1:9, EtOAc:hexanes) afforded negligible quantities of pure aziridine 2 (0.07 g, 0.16 mmol, <1% yield).

Hazards and Waste Disposal Information

Ethyl diazoacetate is a diazo compound. The catalyst preparation involves heating a closed glass system at a high temperature and subsequent treatment under high vacuum at the same temperature. Although no mishap has happened in our laboratories during the last 12 years of working on this methodology, due care should be taken with both of the above mentioned aspects of this reaction. All hazardous materials were disposed of in accordance with "Prudent Practices in the Laboratory"; National Academy Press; Washington, DC, 1995.

3. Discussion

Aziridines are important 3-membered heterocycles, found in numerous natural products with promising biological activities.[3] Aziridines are also invaluable building blocks in organic synthesis; by virtue of their inherent ring strain, they participate readily in a multitude of stereoselective ring opening and ring expansion reactions.[4,5] The field of catalytic asymmetric aziridination has seen impressive growth in the last decade, and this growth has been extensively reviewed.[6]

232

In the last decade, our group has developed a protocol for the catalytic asymmetric synthesis of aziridines from benzhydryl imines and ethyl diazoacetate.[7,8] Extensive studies on the original system in subsequent years led to a full report in 2008.[2] The example subjected to a practical scale-up in the present work has been taken from this full report,[2] and in addition, aziridine 2 has been employed in the synthesis of the LFA-1 antagonist BIRT-377.[9] The data in Scheme 1 illustrates the generality of this protocol over a broad spectrum of benzhydryl imines prepared from aromatic as well as 1°, 2° and 3° aliphatic aldehydes.[2] It is remarkable that the catalysts prepared from both the (S)-VANOL and the (S)-VAPOL ligands give essentially the same asymmetric inductions over all 12 of the substrates with an average difference of only 1.2% ee. In addition, all 12 of the benzhydryl aziridines depicted in Scheme 1 are solids, and as indicated by the data in Scheme 1, all can be very readily crystallized up to almost optical purity by a single crystallization with good to excellent recovery in all cases.[2]

Scheme 1. Substrate scope for the catalytic asymmetric aziridination protocol[2]

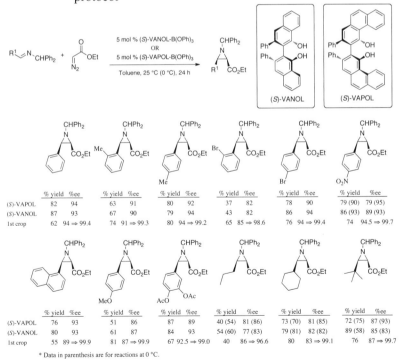

First set of substrates:

	% yield	%ee	% yield	%ee	% yield	%ee	% yield	%ee	% yield	%ee	% yield	%ee
(S)-VAPOL	82	94	63	91	80	92	37	82	78	90	79 (90)	79 (95)
(S)-VANOL	87	93	67	90	79	94	43	82	86	94	86 (93)	89 (93)
1st crop	62	94 ⇒ 99.4	74	91 ⇒ 99.3	80	94 ⇒ 99.2	65	85 ⇒ 98.6	76	94 ⇒ 99.4	74	94.5 ⇒ 99.7

Second set of substrates:

	% yield	%ee	% yield	%ee	% yield	%ee	% yield	%ee	% yield	%ee	% yield	%ee
(S)-VAPOL	76	93	51	86	87	89	40 (54)	81 (86)	73 (70)	81 (85)	72 (75)	87 (93)
(S)-VANOL	80	93	61	87	84	93	54 (60)	77 (83)	79 (81)	82 (82)	89 (58)	85 (83)
1st crop	55	89 ⇒ 99.9	81	87 ⇒ 99.9	67	92.5 ⇒ 99.0	40	86 ⇒ 96.6	80	83 ⇒ 99.1	76	87 ⇒ 99.7

* Data in parenthesis are for reactions at 0 °C.

Table 1. Catalyst Loading Study for the Aziridination of Imine **1**. [a]

entry	ligand	[1] (M)	[1] (mmol)	loading (mol %)	conversion (%) [b]	yield (%) [c]	ee (%) [d]
1	(S)-VANOL	0.5	1	5	100	95	93
2	(S)-VANOL	0.5	5	1	100	83	92
3	(S)-VANOL	0.5	10	0.5	99	85	91
4	(S)-VANOL	0.5	20	0.25	80	72	92
5	(S)-VANOL	1.0	20	0.25	95 [e]	85	93
6	(S)-VANOL	1.0	10	0.5	100 [f]	88	93
7	(R)-VAPOL	0.5	1	5	100	85	92
8	(R)-VAPOL	0.5	5	1	87 [g]	79	90

[a] Unless otherwise specified, all reactions were run with 1.2 equiv of ethyl diazo acetate in toluene at room temperature for 24 h. In all cases the cis:trans ratio was >50:1. The reaction with (R)-VAPOL gave the enantiomer of **2**. [b] Determined by ¹H NMR on the crude reaction mixture. [c] For the 1 mmol scale reaction, all of the product was isolated by column chromatography on silica gel. For all other scales, the yield is the combined yield of the 2-3 crops of crystallized material plus the pure aziridine recovered from the mother liquor after chromatography on silica gel. [d] Determined by chiral HPLC on the combined fractions of aziridine from crystallization and chromatography. [e] Repeat of this reaction only gave a 70% conversion. [f] Reaction is complete in 6 h. [g] Repeat of this reaction gave 83% conversion, 70% yield and 90% ee.

Thus, the magnitude of asymmetric inductions would normally be a minor consideration when choosing between VANOL and VAPOL when contemplating the preparation of an aziridine. Certainly the molecular weight would be a consideration, which would favor use of the VANOL ligand. In addition, a catalyst prepared from the VANOL ligand has been found to be approximately twice as fast as the corresponding VAPOL catalyst for an aziridination reaction.[10] This may or may not be related to the fact that in the present study, the VANOL catalyst was found to give approximately four times as many turnovers as the VAPOL catalyst. This is revealed in a series of catalyst loading studies summarized in Table 1. Reduction of the amount of catalyst did not result in less than complete conversion for the VANOL catalyst until the loading was reduced to 0.25 mol%, which gave only an 80% conversion in a 24 h period (entry 4). Raising the concentration of imine to 1.0 M increased the conversion to 95%, but a repeat of this reaction gave only a 70% conversion (entry 5). Thus the lowest reproducible catalyst loading for the VANOL catalyst was

Org. Synth. **2011**, *88*, 224-237

determined to be 0.5 mol% at 1.0 M in imine (entry 6). The reaction with the catalyst formed from the VAPOL ligand did not quite go to completion with 1 mol% catalyst and thus for the *para*-bromophenyl imine **1**, the VAPOL catalyst does not give as many turnovers as the VANOL catalyst (entry 8).

1. Department of Chemistry, Michigan State University, East Lansing, Michigan 48824, USA. Email: wulff@chemistry.msu.edu. The work was supported by the National Science Foundation (CHE-0750319).

2. Zhang, Y.; Desai, A.; Lu, Z.; Hu, G.; Ding, Z.; Wulff, W. D. *Chem. Eur. J.* **2008**, *14*, 3785-3803.

3. Ismail, F. M. D.; Levitsky, D. O.; Dembitsky, V. M. *Eur. J. Med. Chem.* **2009**, *44*, 3373-3387.

4. (a) Yudin, A. K. *Aziridines and epoxides in organic synthesis*, Wiley-VCH Verlag GmbH & Co. KGaA: Weinheim, 2006. (b) Tanner, D. *Angew. Chem. Int. Ed.* **1994**, *33*, 599-619. (c) Sweeney, J. B. *Chem. Soc. Rev.* **2002**, *31*, 247-258. (d) Watson, I. D. G.; Yu, L.; Yudin, A. K. *Acc. Chem. Res.* **2006**, *39*, 194-206.

5. (a) McCoull, W.; Davis, F. A. *Synthesis* **2000**, 1347-1365. (b) Zwanenburg, B.; Holte, P. T. *Top. Curr. Chem.* **2001**, *216*, 93-124. (c) Hu, X. E. *Tetrahedron* **2004**, *60*, 2701-2743. (d) Mauro, P. *Eur. J. Org. Chem.* **2006**, 4979-4988.

6. (a) Muller, P.; Fruit, C. *Chem. Rev.* **2003**, *103*, 2905-2919. (b) Pellissier, H. *Tetrahedron* **2010**, *66*, 1509-1555.

7. (a) Antilla, J. C.; Wulff, W. D. *J. Am. Chem. Soc.* **1999**, *121*, 5099-5100. (b) Antilla, J. C.; Wulff, W. D. *Angew. Chem. Int. Ed.* **2000**, *39*, 4518-4521.

8. For a review of our early work in this field, see: Zhang, Y.; Lu, Z.; Wulff, W. D. *Synlett* **2009**, 2715-2739.

9. Patwardhan, A. P.; Pulgam, V. R.; Zhang, Y.; Wulff, W. D. *Angew. Chem. Int. Ed.* **2005**, *44*, 6169-6172.

10. Lu, Z.; Zhang, Y.; Wulff, W. D. *J. Am. Chem. Soc.* **2007**, *129*, 7185-7194.

Appendix
Chemical Abstracts Nomenclature; (Registry Number)

N-(4-Bromobenzylidene)-1,1-diphenylmethanamine: Benzenemethanamine,
  N-[(4-bromophenyl)methylene]-α-phenyl-, [N(*E*)]-; (330455-47-5)
4-Bromobenzaldehyde; (1122-91-4)
Aminodiphenylmethane: Benzenemethanamine, α-phenyl-; (91-00-9)
Triphenyl borate: Boric acid (H₃BO₃), triphenyl ester; (1095-03-0)
(*S*)-VANOL: [2,2'-Binaphthalene]-1,1'-diol, 3,3'-diphenyl-, (2*S*)-; (147702-
  14-5)
(2*R*,3*R*)-Ethyl 1-benzhydryl-3-(4-bromophenyl)aziridine-2-carboxylate: 2-
  Aziridinecarboxylic acid, 3-(4-bromophenyl)-1-(diphenylmethyl)-,
  ethyl ester, (2*R*,3*R*)-; (233585-43-8)
Ethyl diazoacetate: Acetic acid, 2-diazo-, ethyl ester; (623-73-4)

Born and raised near Eau Claire, Wisconsin, Professor Wulff obtained his BS degree in chemistry at the University of Wisconsin at Eau Claire in 1971 doing research with Professor Larry Schnack. After completing required service in the US Army, his doctoral education was pursued at Iowa State University under the direction of Professor Thomas Barton. After a postdoctoral stint at Princeton University with Professor Martin Semmelhack, Professor Wulff finally became a tax-payer in 1980 upon assuming a position of assistant Professor of Chemistry at the University of Chicago. In 1999, Professor Wulff took up his present position of Professor of Chemistry at Michigan State University.

Born and raised in Surat, India, Aman Desai received his Bachelor's degree in Chemical Technology from the Institute of Chemical Technology (erstwhile UDCT) in Mumbai, India. Aman then received his PhD in 2010 under the tutelage of Professor William Dean Wulff at Michigan State University, USA. The major focus of his doctoral research was on the development and understanding of a universal asymmetric catalytic aziridination system. Presently, Aman is a Process Research Chemist in the Process Chemistry and Development group at the Dow Chemical Company in Midland, Michigan.

Roberto Morán-Ramallal was born in St. Gallen (Switzerland) in 1981. He studied organic chemistry at the University of Oviedo where he graduated in May 2004. After that he joined the research group of Prof. Gotor where he earned his PhD in 2009. His work was focused on the synthesis of chiral aziridines through enzymatic processes and their use as synthons of high-value-added compounds. During his doctoral studies he did two short stays under the supervision of Prof. Marko Mihovilovic and Prof. William D. Wulff. His current interest is in the use of biocatalysis for the production of high-added-value products.

Shabnam Kouchekzadeh Yazdi completed her undergraduate studies at the University of British Columbia in 2009. There she conducted an undergraduate thesis under the supervision of Professor Jennifer Love. During her BSc she worked at FPInnovations-Paprican and also at the UBC Centre for Drug Research and Development. She is currently pursuing a PhD at the University of Toronto under the supervision of Professor Mark Lautens. Her research focuses on the asymmetric ring opening of oxabicycles.

Gavin G. Haberlin received his B.Sc. in Chemistry at University College Dublin, Ireland in 2008, where he performed his undergraduate research with Professor Paul Murphy, studying glycopids. He was awarded the Hugh Ryan Gold Medal for Chemistry in 2008 and is currently a graduate student at University College Dublin, working with Professor Pat Guiry on natural product synthesis. He received an Ireland Canada University Foundation Scholarship in 2010, studying for a short period at the University of Toronto with Professor Mark Lautens.

Cu-CATALYZED AZIDE-ALKYNE CYCLOADDITION: PREPARATION OF TRIS((1-BENZYL-1H-1,2,3-TRIAZOLYL)METHYL)AMINE

Submitted by Jason E. Hein, Larissa B. Krasnova, Masayuki Iwasaki and Valery V. Fokin.[1]

Checked by Jane Panteleev and Mark Lautens.[2]

1. Procedure

Caution! This procedure employs benzyl azide, which is an energetic and potentially explosive material. All transformations should be performed behind a blast shield in a well-ventilated fume hood.

Tris((1-benzyl-1H-1,2,3-triazolyl)methyl)amine. A 100-mL two-necked, round-bottomed flask equipped with a football-shaped magnetic stir bar (25 x 15 mm), condenser and thermometer (Note 1) is charged with 45 mL of MeCN (Note 2). Copper(II) acetate, monohydrate (0.0610 g, 0.305 mmol, 0.02 equiv) (Note 3) is added as a powder and the reaction mixture is stirred vigorously at room temperature until a bright blue solution is obtained. Tripropargylamine (2.00 g, 15.2 mmol, 1.00 equiv) (Note 3) and benzyl azide (3.67 g, 25.9 mmol, 1.70 equiv) are dissolved in MeCN (10 mL) (Notes 2 and 4) and added to the reaction flask, which is then immersed into a water bath. Sodium ascorbate (0.061 g, 0.31 mmol, 0.02 equiv) (Note 3) is dissolved in water (5 mL) (Note 2), and the resulting solution is added in one portion to the reaction mixture (Note 5). The reaction mixture is stirred at room temperature for 30 min (Note 6) and then heated at 45 °C (internal temperature) for 5 h. A second portion of benzyl azide (3.67 g, 25.9 mmol, 1.70 equiv) is added, and the reaction mixture is

238

heated at 45 °C for an additional 19 h (Notes 7 and 8). The reaction mixture is concentrated to dryness on a rotary evaporator (40 °C, 34 mmHg). The crude residue is taken up in CH_2Cl_2 (75 mL) (Note 2) and treated with conc. NH_4OH (35 mL) (Note 3). The heterogeneous suspension is stirred vigorously on a magnetic stir plate until all solids are dissolved. The suspension is transferred to a 250-mL separatory funnel, and the aqueous layer is extracted with dichloromethane (2 × 35 mL) (Note 2). The combined organic extracts are washed with a solution of conc. aq. NH_4OH and brine (1:1 v/v, 2×15 mL) (Notes 3 and 9), dried over $MgSO_4$ (7 g) (Note 3) and filtered into a 250-mL round-bottomed flask. This solution is concentrated on a rotary evaporator (35 °C, 34 mmHg) to give a crude yellow solid. An egg-shaped Teflon-coated magnetic stir bar is added to the flask, followed by CH_2Cl_2 (50 mL) (Note 2). The suspension is vigorously stirred until a translucent, viscous solution is obtained. Diethyl ether (55 – 75 mL) (Note 2) is added gradually with vigorous stirring, causing the formation of a thick precipitate. The resulting slurry is stirred at room temperature for an additional 5 min, and the off-white precipitate is isolated by filtration (Note 10). The solid is washed with diethyl ether (3 × 15 mL) (Note 2), giving a free-flowing off-white powder (5.78 g, 10.9 mmol). The mother liquor is concentrated on a rotary evaporator, taken up in dichloromethane (25 mL) (Note 2) and precipitated with diethyl ether (50 – 75 mL) (Note 2). The precipitate is again collected by filtration and washed with diethyl ether (10 mL) (Note 2), giving a second batch of a white solid (1.0 g, 1.9 mmol). Both crops are combined to give tris((1-benzyl-1H-1,2,3-triazolyl)methyl)amine as an off-white powder (6.78 g, 12.8 mmol, 84%) (Note 11).

2. Notes

1. All glassware, stir bars and other peripheral equipment were thoroughly washed and rinsed with distilled water prior to reaction.

2. Reactions were performed with deionized water (from institutional facilities), acetonitrile (ACS reagent grade) was purchased from Sigma-Aldrich, dichloromethane (stabilized, certified ACS) was purchased from Fischer Scientific, and diethyl ether (anhydrous, stabilized, ACS reagent grade) was obtained from Caledon. The submitters purchased acetonitrile (certified ACS), dichloromethane (certified ACS) and diethyl ether (stabilized, certified ACS) from Fisher Scientific.

3. Sodium chloride (99%), anhydrous magnesium sulfate (99%) and conc. ammonium hydroxide (ACS reagent grade) were purchased from ACP Chemicals, Inc. and were used as received. The submitters purchased sodium chloride, anhydrous magnesium sulfate, conc. ammonium hydroxide (all certified ACS grade) from Fisher Scientific. Copper(II) acetate monohydrate (98%+, ACS reagent grade) and sodium ascorbate (>99%) were purchased from Sigma Aldrich. The submitters obtained copper(II) acetate monohydrate (99%) and sodium ascorbate (99%) from Acros. Tripropargylamine (98%) was purchased from GFS. Benzyl azide (94%) was purchased from Alfa Aesar and was stored away from light in a refrigerator 0–4 °C. The submitters purchased benzyl azide (99%) from Frinton Laboratories. All other materials were used as received with no other purification or special storage requirements.

4. The azide and alkyne components should be weighed out separately and sequentially dissolved into acetonitrile. The azide and alkyne reagents should not be mixed neat.

5. The solution became colorless on addition of the sodium ascorbate. The reaction solution gradually became light yellow with time and developed into a deep brown/red.

6. The submitters note that if no water bath was used, the internal temperature of the reaction gradually rose, with the maximum of 50.5 °C being observed after 30 minutes.

7. TLC analysis during the reaction allowed each of the intermediates to be visualized: (19:1 CH$_2$Cl$_2$: MeOH) tripropargylamine – R$_f$ = 0.67, N-((1-benzyl-1H-1,2,3-triazol-4-yl)methyl)-N,N-dipropargyl amine – R$_f$ = 0.38, N,N-bis((1-benzyl-1H-1,2,3-triazol-4-yl)methyl)-propargyl amine – R$_f$ = 0.30, tris((1-benzyl-1H-1,2,3-triazol-4-yl)methyl)amine R$_f$ = 0.26. Silica gel plates – Merck (purchased from EMD chemicals), 20 x 5 cm silica gel 60 on glass, 0.25 mm coating. Samples were visualized using UV (254 nm) and KMnO$_4$ stain.

8. The submitters report a 10 h total reaction time and 75% yield. The checkers obtained a yield of 75%, but they observed incomplete consumption of reaction intermediates at 10 h reaction time. A yield of 80-84% could be obtained with a reaction time of 24 h. Addition of 0.3 equiv of benzyl azide at 24 h, and allowing the reaction to proceed for an additional 24 h further increased the yield to 92%.

9. The aqueous layer develops a bright blue color after washing with NH$_4$OH/brine. To ensure efficient removal of all copper salts, this color should be almost undetectable during the final wash.

10. The solid was isolated using a sintered-glass fritted funnel of medium porosity. The submitters isolated the solid using a ceramic Büchner funnel with a P4 Fisher Brand quantitative filter paper.

11. The product exhibits the following physical properties: TLC R_f = 0.26 (19:1 CH$_2$Cl$_2$: MeOH), mp = 140 °C–144 °C (the submitters report 146 °C–148 °C), IR (υ[cm^{-1}]) 3136, 3063, 3032, 2932, 2828, 1497, 1454, 1435, 1331, 1219, 1127, 1049; ^{1}H NMR (400 MHz, CDCl$_3$) δ: 3.70 (s, 6 H), 5.50 (s, 6 H), 7.23–7.28 (m, 6 H), 7.31–7.38 (m, 9 H), 7.66 (s, 3 H); ^{13}C NMR (101 MHz, CDCl$_3$) δ: 47.05, 54.05, 123.68, 127.95, 128.62, 129.03, 134.71, 144.26; EI-MS m/z: 530(1, M$^+$), 359(30), 358(100), 187(13), 173(19), 144(10), 91(96); HRMS (EI): [M$^+$] calcd. for C$_{30}$H$_{30}$N$_{10}$: 530.2655. Found: 530.2662; Anal. calcd. for C$_{30}$H$_{30}$N$_{10}$: C, 67.90; H, 5.70; N, 26.40. Found: C, 67.77; H, 5.95; N, 26.52.

Safety and Waste Disposal Information

All hazardous materials should be handled and disposed of in accordance with "Prudent Practices in the Laboratory": National Academy Press: Washington, DC, 1995.

3. Discussion

This procedure represents a refinement to the synthesis of the tris((triazolyl)methyl)amine family of ligands, as exemplified by the preparation of tris((1-benzyl-1H-1,2,3-triazolyl)methyl)amine. The synthesis of tris((triazolyl)methyl)amine ligands using the Cu(I)-catalyzed dipolar cycloaddition was first reported by Chan et al.[3] The general reaction proceeds via π-coordinated Cu-species 2 to give σ-copper acetylide 3 (Scheme 1). Complexation of the organic azide (4) via the proximal nitrogen gives coordinate complex 5, which undergoes cycloaddition generating copper triazolide 6. Proton transfer liberates the triazole product and regenerates free copper catalyst (7 and 8, respectively).[4]

In general, the rate of the cycloaddition is heavily influenced by numerous off-cycle ligated-copper species, including oligomeric copper acetylide complexes 9.[5,6a] When poly-dentate, coordinating substrates, such as tripropargylamine, are employed networks of extended, thermodynamically-stable oligomeric aggregates are easily formed. Thus, during the synthesis of ligands such as tris((1-benzyl-1*H*-1,2,3-triazolyl)methyl)amine the bulk of the copper-catalyst is tied up off-cycle, leading to a relatively slow rate of reaction. The population of monomeric (and reactive) copper acetylide can be increased by adding supportive, mono-dentate coordination ligands (aliphatic amines, 2,6-lutidine), and by employing polar-coordinating solvents (DMF, NMP, or MeCN; note that MeCN has a high affinity for Cu(I) and therefore inhibits the reaction, requiring the addition of an amine ligand and/or water). The original synthesis of the tris((triazolyl)methyl)amine ligand family used a combination of these two approaches;[3] however, we have found that omitting the supportive ligand greatly facilitates the purification and isolation of the product.

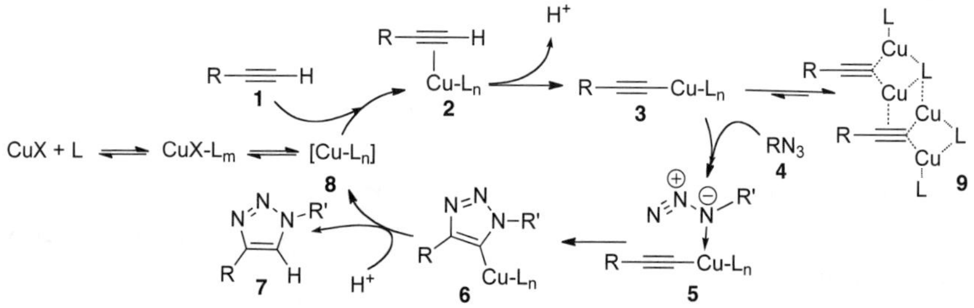

Scheme 1: Mechanism of the Cu-catalyzed azide-alkyne coupling.

Finally, running the reaction at an elevated temperature facilitates ligand exchange at the copper center, preventing the formation of thermodynamically stable, but unreactive metal complexes. However, extended heating also leads to a slow, but competitive decomposition of the azide component. To mitigate this issue benzyl azide is added to the reaction in two separate portions (2 × 1.7 equiv). This treatment obviates a large excess of the azide component and delivers the ligand in very good and reproducible yield.

The reported optimized protocol represents a practical and general method to access numerous members of the tris((triazolyl)methyl)amine ligand family (Figure 1, **10–14**). The efficiency of the cycloaddition permits a vast array of structurally and electronically diverse species to easily be constructed, limited only to the availability of the requisite reagent. Due to this enormous chemical breadth, the physical and chemical properties of the ligand (solubility, polarity, steric environment, etc) can readily be tailored to specific applications.

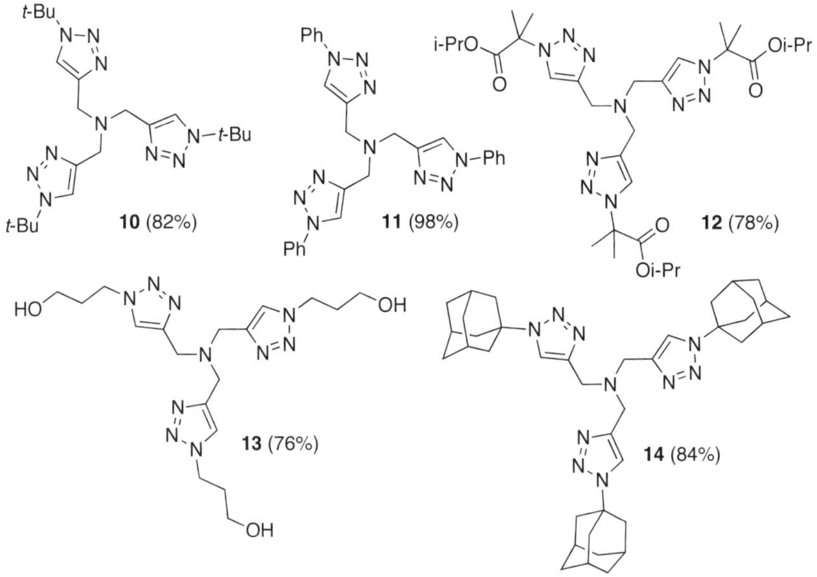

Figure 1: Various tris((triazolyl)methyl)amine ligands.

Our studies identified the tris((triazolyl)methyl))amine core as a particularly effective accelerating ligand for the Cu(I)-catalyzed azide-alkyne cycloaddition (CuAAC).[3,6a] Tris((1-benzyl-1*H*-1,2,3-triazolyl)-methyl)amine (TBTA) was the first and most commonly employed CuAAC ligand; however, other related (triazolyl)methylamines including tris(1-*t*-butyl) (**10**) and tris(1-(3-hydroxypropyl)) (**13**) ligands have since been explored. These latter species have begun to supplant TBTA, as their resultant copper(I)-complexes display broader solvent compatibility while retaining high catalytic activity.

Although the most commonly employed reaction conditions for the CuAAC do not require exogenous ligands,[6] incorporating them may be advantageous when substrates displaying slow reaction profiles are used (*e.g.* low reaction concentrations, low copper loading, chelating substrates etc). Due to this feature, tris((triazolyl)methyl))amine ligands have most often been employed in the field of bioconjugation, where very low substrate and catalyst concentrations are usually encountered.[7]

More recently, tris((triazolyl)methyl)amines have been identified as key ligands in other Cu(I)-catalyzed reactions, such as the cycloaddition between azides and 1-iodoalkynes.[8] Here, both the observed rate and chemoselectivty are strongly dependent on the nature of the ligand, with **10** being the optimal species for this process.

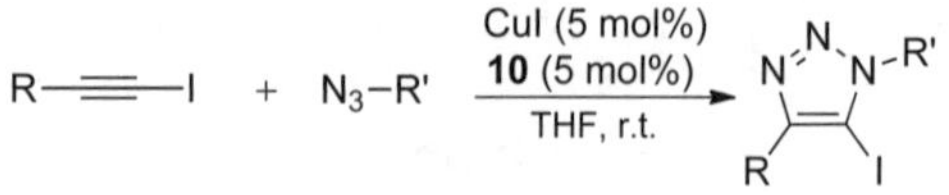

Scheme 2: Cu(I)-catalyzed cycloaddition between azides and 1-ioodalkynes

1. Department of Chemistry, The Scripps Research Institute, La Jolla, CA, 92037, U.S.A. fokin@scripps.edu. This work was supported by the National Institute of General Medical Sciences, National Institutes of Health (GM087620) and National Science Foundation (CHE-0848982).
2. Department of Chemistry, University of Toronto, Toronto, ON, M5S 3H6, Canada. mlautens@chem.utoronto.ca.
3. Chan, T. R; Hilgraf, R.; Sharpless, K. B.; Fokin, V. V. *Org. Lett.* **2004**, *6*,2853-2855.
4. Himo, F.; Lovell, T.; Hilgraf, R.; Rostovtsev, V. V.; Noodleman, L.; Sharpless, K. B.; Fokin, V. V. *J. Am. Chem. Soc.* **2005**, *127*, 210-216.
5. Mykhalichko, B. M.; Temkin, O. N.; Mys'kiv, M. G. *Russ. Chem. Rev.* **2000**, *69*, 957-984.
6. a) Hein, J. E.; Fokin, V. V. *Chem Soc. Rev.* **2010**, *39*, 1302-1315. b) Meldal, M.; Tornøe, C. W. *Chem. Rev.* **2008**, *108*, 2952-3015.
7. a) Hong, V.; Presolski, S. I.; Ma, C.; Finn, M. G. *Angew. Chem. Int. Ed.* **2009**, *48*, 9879-9883. b) Wang, Q.; Chan, T. R.; Hilgraf, R.; Fokin, V. V.; Sharpless, K. B.; Finn, M. G. *J. Am Chem. Soc.* **2003**, *125*, 3192-3193.
8. Hein, J. E.; Tripp, J. C.; Krasnova, L. B.; Sharpless, K. B.; Fokin, V. V. *Angew. Chem. Int. Ed.* **2009**, *48*, 8018-8021.

244

Appendix
Chemical Abstracts Nomenclature; (Registry Number)

Tripropargylamine: 2-Propyn-1-amine, *N,N*-di-2-propynyl: Tri-2-propynylamine; (6921-29-5)

Benzyl azide: (azidomethyl)benzene; (622-79-7)

Copper(II) acetate, monohydrate: Cupric acetate 1-hydrate: Acetic acid, copper(II) salt; (6046-93-1)

Sodium ascorbate: L-Ascorbic acid, sodium salt: Sodium (+)-L-ascorbate; (134-03-2)

Valery Fokin received his undergraduate education at the University of Nizhny Novgorod, Russia, and his Ph.D. degree at the University of Southern California under the tutelage of Prof. Nicos A. Petasis. After a postdoctoral stint with Prof. K. Barry Sharpless at The Scripps Research Institute in La Jolla, California, he joined the Scripps faculty, where he is currently Associate Professor in the Department of Chemistry. His research is centered on the understanding of chemical reactivity of organometallic species and on applying it to the studies of macromolecular and biological phenomena. His research group is working on new reaction development, studies of organic and organometallic mechanisms, medicinal chemistry, synthesis of macromolecular probes for imaging and drug delivery, and smart polymeric materials.

Jason Hein received his B.Sc. in biochemistry in 2000 from the University of Manitoba in Winnipeg, MB, Canada. He then began his Ph.D. studies as an NSERC postgraduate fellow under the guidance of Prof. Philip G. Hultin at the University of Manitoba. In 2005 he completed his graduate work, where he synthesized and studied a new family of soluble-supported chiral auxiliaries. He is currently an NSERC postdoctoral fellow at the Scripps Research Institute. His current interests include the design, development and study of new metal-catalyzed reactions.

After completing her undergraduate degree at the Moscow State University (M.V. Lomonosov), Moscow, Russia, Larissa Krasnova continued her graduate training at the University of Toronto under the supervision of Prof. Andrei K. Yudin. She is currently a postdoctoral associate at The Scripps Research Institute in the group of Prof. Valery V. Fokin where she carries out studies in the field of heterocyclic chemistry with an emphasis on new method development, medicinal chemistry and bioconjugation.

Masayuki Iwasaki was born in Okayama, Japan, in 1982. He obtained his B. Sc. in 2006 from Kyoto University and then began studying to receive his Ph.D. degree under the supervision of Professor Koichiro Oshima. He has been a JSPS research fellow since 2008 and is currently a visiting graduate student at the Scripps Research Institute in the group of Prof. Valery V. Fokin. His research interests include the development of new organic reactions with organometallic reagents.

Jane Panteleev received her Bachelor of Science degree in Biochemistry at Queen's University, in 2007. During this time she had the opportunity to work in the research lab of Prof. Victor Snieckus. She is currently pursuing a Ph.D. degree under the supervision of Prof. Mark Lautens at the University of Toronto. Her current research interests are in the area of asymmetric transition metal catalysis.

LITHIATED PRIMARY ALKYL CARBAMATES FOR THE HOMOLOGATION OF BORONIC ESTERS

Submitted by Matthew P. Webster, Benjamin M. Partridge and Varinder K. Aggarwal.[1]

Checked by Tomohiko Inui and Tohru Fukuyama.

1. Procedure

A. *N,N-Diisopropyl-carbamic acid 3-phenyl-propyl ester* (**1**). A 1 L flask containing a 40 mm magnetic stirring bar is charged with 3-phenylpropan-1-ol (Note 1, 51.0 mL, 51.1 g, 375 mmol, 1.00 equiv), *N,N*-diisopropylcarbamoyl chloride (Note 1, 62.7 g, 383 mmol, 1.02 equiv), triethylamine (Note 1, 53.4 mL, 38.8 g, 383 mmol, 1.02 equiv) and dichloromethane (Note 1, 500 mL). A reflux condenser is fitted and connected to an argon line by way of a gas inlet adaptor. The reaction mixture is stirred and heated at reflux for 24 h. After cooling to room temperature, the solvent is removed *in vacuo* (200 mmHg, 30 °C) using a rotary evaporator and *tert*-butyl methyl ether (Note 2, 500 mL) is added. The mixture is transferred to a separatory funnel and washed with water (2 × 400 mL), and the combined aqueous phases are extracted with *tert*-butyl methyl ether (2 × 300 mL). The combined organic fractions are washed with brine (100 mL) and dried over MgSO$_4$ (15 g) for 2 min. After filtration of the solids, the filtrate is concentrated *in vacuo* (100 mmHg, 30 °C) to give a pale yellow oil. The crude material is transferred to a 250 mL flask equipped with an oval-shaped, Teflon coated, 40 mm magnetic stirring bar. This is fitted with a short path distillation kit incorporating a thermometer. The crude carbamate is distilled (~0.6 mmHg, 122–123 °C) to give the analytically pure *N,N*-diisopropyl-carbamic acid 3-phenyl-propyl ester (83.2–86.3 g, 84–87%) as a colorless oil (Note 3).

B. *Isobutylboronic acid pinacol ester* (**2**). An oven-dried 1 L flask containing a 40 mm magnetic stirrer bar is charged with isobutylboronic acid (Note 4, 25.0 g, 245 mmol 1.0 equiv), pinacol (Note 4, 29.0 g, 245 mmol, 1.0 equiv), magnesium sulfate (Note 4, 44.3 g, 368 mmol, 1.5 equiv) and diethyl ether (Note 4, 300 mL). The suspension is stirred under argon at room temperature for 24 h. After filtration of the solids, the filtrate is concentrated *in vacuo* (200 mmHg, 30 °C). The crude material is dissolved in pentane (Note 5, 700 mL) and washed with H$_2$O (3 × 150 mL), and the organic layer is dried over MgSO$_4$. After filtration of the solids, the filtrate is concentrated *in vacuo* (200 mmHg, 30 °C) to give a pale yellow oil. The crude material is transferred to a 250 mL flask equipped with a 40 mm magnetic stirring bar. This is fitted with a short path distillation kit incorporating a thermometer. The crude boronic ester is distilled (14 mmHg, 71 °C) to give the analytically pure isobutylboronic acid pinacol ester (35.3–37.4 g, 78–83%) as a colorless oil (Note 7).

C. *(R)-5-Methyl-1-phenylhexan-3-ol* (**3**). An oven-dried 1 L three-necked flask equipped with a 40 mm magnetic stirring bar, an oven dried 250 mL pressure equalising dropping funnel, a thermocouple (Note 8) and a gas inlet adaptor is attached to a vacuum manifold. The reaction flask is evacuated and refilled with argon (× 2). Carbamate **1** (13.2 mL, 13.2 g, 50.0 mmol, 1.0 equiv), (–)-sparteine (Note 9, 14.9 mL, 15.2 g, 65.0 mmol, 1.3 equiv, addition by syringe) and anhydrous diethyl ether (Note 10, 250 mL, addition *via* cannula) are charged to the reaction vessel and the resulting solution is cooled to –78 °C (dry ice/acetone bath). A solution of *sec*-butyllithium (Note 11, 61.3 mL, 1.06 M in cyclohexane/hexane (95/5), 65.0 mmol, 1.3 equiv) is added to the reaction mixture *via* dropping funnel over ~30 min whilst maintaining an internal temperature below –70 °C. The reaction mixture is then stirred below –70 °C for 5 h (Note 12). Neat boronic

248

ester **2** (13.9 mL, 12.0 g, 65.0 mmol, 1.3 equiv) is added drop-wise using a syringe over 10 min whilst maintaining an internal temperature below –70 °C. The reaction mixture is then stirred below –70 °C for 30 min before being removed from the cooling bath and allowed to warm to room temperature (Note 13). The center neck of the flask is fitted with an oven-dried reflux condenser attached to an argon line by a gas inlet adaptor. The additional necks of the flask are secured with glass stoppers, and the reaction mixture is heated at reflux for >12 h (Note 14). The reaction mixture is allowed to cool to room temperature and the condenser exchanged with a 250 mL dropping funnel. After being cooled in an ice/water bath, a mixture of H_2O_2 (Note 15, 50 mL, 30 % w/v) and NaOH (100 mL, 2 M) is added *via* dropping funnel over 5 min. The vessel is removed from the cooling bath, and the resulting mixture is stirred at room temperature for 2 h (Note 16). Sodium hydroxide (200 mL, 2 M) is added, and the mixture is partitioned in a separatory funnel. The aqueous phase is extracted with diethyl ether (2 × 300 mL), and the combined organic fractions are washed with $KHSO_4$ (200 mL, 1 M), H_2O (2 × 400 mL), brine (100 mL) and dried over $MgSO_4$ (10 g) for 2 min. After filtration of the solids, the filtrate is concentrated *in vacuo* (150 mmHg, 30 °C) to give a pale yellow oil. The crude material is transferred to a 50 mL flask equipped with a 25 mm magnetic stirring bar. This is fitted with a short path distillation kit incorporating a thermometer. The crude alcohol is distilled (1.2–1.3 mmHg, 101–103 °C) to give the analytically pure alcohol **3** (7.86–8.34 g, 82–87%, *e.r.* = 99:1 (Note 17)) as a colorless oil (Note 18).

2. Notes

1.   3-Phenylpropan-1-ol and *N,N*-diisopropylcarbamoyl chloride were obtained from Sigma-Aldrich, whereas triethylamine (>99.0%) and dichloromethane (>99.5%) were obtained from Kanto Chemical Co., Inc.. The Submitters obtained triethylamine and dichloromethane (Laboratory Reagent Grade) from Fischer Scientific. All reagents were used as purchased.

2.   *tert*-Butyl methyl ether (Reagent Grade) was obtained from Sigma-Aldrich.

3.   In accordance with the literature,[2] the product exhibits the following physiochemical properties: ^{1}H NMR (CDCl$_3$, 400 MHz) δ: 1.22

(d, J = 6.9 Hz, 4 × CH_3, 12 H), 1.98 (tt, J = 7.8, 6.4 Hz, $ArCH_2CH_2$, 2 H), 2.72 (t, J = 7.8 Hz, $ArCH_2$, 2 H), 3.40–4.35 (br. m, NCH, 2 H), 4.12 (t, J = 6.4 Hz, OCH_2, 2 H), 7.16–7.25 (m, ArH, 3 H), 7.25–7.33 (m, ArH, 2 H); ^{13}C NMR (CDCl$_3$, 100 MHz) δ: 21.0 (br), 30.9, 32.5, 45.8 (br), 64.0, 125.9, 128.3, 128.4, 141.5, 155.8; HRMS (ESI) calcd. for $C_{16}H_{25}NNaO_2$: 286.1783. Found: 286.1783; IR v_{max} (neat) 2969, 1692, 1436, 1368, 1309, 1290, 1219, 1157, 1134, 1066 cm^{-1}; Anal. Calcd. for $C_{16}H_{25}NO_2$: C, 72.96; H, 9.57; N, 5.32. Found: C, 72.93; H, 9.56; N, 5.37; Submitters reported: MS (EI): 263 (5, M$^+$), 118 (78), 91 (100).

4.   Isobutylboronic acid was obtained from Frontier Scientific, pinacol was obtained from Alfa-Aesar, $MgSO_4$ (>95.0%) and diethyl ether (>99.0%) were obtained from Wako Pure Chemical Industries, Ltd. The Submitters obtained $MgSO_4$ from Fischer Scientific and diethyl ether (Laboratory Reagent Grade) from VWR. All reagents were used as purchased.

5.   Pentane (>97.0%) was obtained from Wako Pure Chemical Industries, Ltd. and used as purchased. The Submitters obtained pentane (Laboratory Reagent Grade) from Fischer Scientific.

6.   The Submitters note that the boronic ester sublimes at lower temperatures so the material should be heated rapidly to prevent loss of material.

7.   In accordance with the literature,[3] the product exhibits the following physiochemical properties : ^{1}H NMR (CDCl$_3$, 400 MHz), δ: 0.73 (d, J = 7.3 Hz, CH_2, 2 H), 0.93 (d, J = 6.4 Hz, 2 × CHCH_3, 6 H), 1.25 (s, 4 × CCH$_3$, 12 H), 1.86 (t·septet, J = 7.3, 6.4 Hz, CH, 1 H); ^{13}C NMR (CDCl$_3$, 100 MHz) δ: 24.81, 24.81 (4C), 24.85, 25.2, 82.8; HRMS(DART) calcd. for $C_{10}H_{22}BO_2$: 185.1713. Found: 185.1712; IR v_{max} (neat) 2979, 2954, 1371, 1312, 1209, 1146, 972, 848 cm^{-1}; Anal. calcd. for $C_{10}H_{21}BO_2$: C, 65.25; H, 11.50. Found: C, 64.86; H, 11.36; Submitters reported: ^{11}B NMR (CH$_2$Cl$_2$, 128 MHz) δ: 33.9; MS (CI): 185 (17, MH$^+$), 141 (55), 85 (100).

8.   Internal temperature was monitored using a CUSTOM thermocouple (CT-470) fitted with K-type flexible wire probe (passed through a septum). The Submitters used a Hanna thermocouple (HI 93531).

9.   (−)-Sparteine was obtained from the commercially available sulfate salt (>98.0%, Wako Pure Chemical Industries, Ltd.) according to the literature procedure.[4] The submitters obtained (−)-sparteine from Sigma-Aldrich, which was distilled from CaH$_2$ under high vacuum prior to use

(~0.1 mmHg, 98–100 °C, colorless oil). The submitters note that distilled (−)-sparteine can be stored and used for more than 6 months if stored correctly, i.e. in appropriate Schlenk glassware under argon at −20 °C.

10. Anhydrous diethyl ether was obtained using a Grubbs-type system.[5]

11. Solutions of *sec*-butyllithium were purchased from Kanto Chemical Co., Inc. The submitters obtained solutions of *sec*-butyllithium from Acros Organics. The submitters note that the yields of the reactions are critically dependant on the quality of the *sec*-butyllithium. The best results are obtained with fresh solutions (up to 2 months shelf life) that have been stored in a refrigerator and which are colorless to pale yellow in color with minimal precipitated salts. See also Beak, *et al.*[4] The submitters note that 1.3 equiv of *sec*-butyllithium is used with respect to carbamate in the homologation reactions as occasionally lower yields were obtained with less reagent.

12. During the addition of *sec*-butyllithium, a yellow coloration is immediately apparent and increases in intensity over the lithiation period.

13. The yellow color observed during lithiation reduces upon warming to room temperature.

14. After heating at reflux overnight a pale yellow solution is produced.

15. Hydrogen peroxide was obtained from Kanto Chemical Co., Inc. The submitters obtained hydrogen peroxide from Fischer Scientific.

16. Oxidation can be monitored by TLC after partitioning of an aliquot in H_2O/diethyl ether. Eluent: ethyl acetate/hexane, 1:9. R_f boronic ester = 0.7, alcohol = 0.3 (Submitters: Eluent: ethyl acetate/petroleum ether 40–60 °C, 1:9. R_f boronic ester = 0.7, alcohol = 0.2). TLC plate visualized using phosphomolybdic acid [5% $(NH_4)_2Mo_7O_{24} \cdot 4H_2O$ in 95% EtOH (w/v)] followed by heating.

17. *Chiral HPLC conditions for separation of alcohol (R)-3.* Checkers: Column: Daicel Chiralpak AD-H (250 × 4.6 mm) with Chiralpak AD-H Guard Cartridge (10 × 4 mm), eluent: 99.0:1.0 hexane:propan-2-ol (both HPLC grade, hexane was purchased from Wako Pure Chemical Industries, Ltd., propan-2-ol was purchased from Kanto Chemical Co., Inc.), flow rate: 0.7 mL/min, t_R: minor (*S*) 22.4 min, major (*R*) 24.2 min. (Submitters: Column: Daicel Chiralpak IA (250 × 4.6 mm) with Chiralpak IA Guard Cartridge (10 × 4 mm), eluent: 97.5:2.5 hexane:propan-2-ol (both HPLC

grade, purchased from Fisher Scientific), flow rate: 0.7 mL/min, t_R: minor (*S*) 16.6 min, major (*R*) 17.8 min.)

18. In accordance with the literature,[6] the product exhibits the following physiochemical properties. ^{1}H NMR (CDCl$_3$, 400 MHz) δ: 0.90 (d, J = 6.0 Hz, CH$_3$, 3 H), 0.92 (d, J = 6.4 Hz, CH$_3$, 3 H), 1.28 (m, CH(OH)CH*H*CH(CH$_3$)$_2$, 1 H), 1.35 (br. s, OH, 1 H), 1.42 (m, CH(OH)C*H*HCH(CH$_3$)$_2$, 1 H), 1.65–1.87 (m, C*H* + PhCH$_2$C*HH*, 3 H), 2.67 (ddd, J 13.7, 9.6, 6.4 Hz, PhCH*H*, 1 H), 2.80 (ddd, J 13.7, 10.1, 6.0 Hz, PhC*H*H, 1 H), 3.65–3.77 (m, C*H*OH, 1 H), 7.14–7.36 (m, ArH, 5 H); ^{13}C NMR (CDCl$_3$, 100 MHz) δ: 22.1, 23.4, 24.6, 32.0, 39.7, 46.8, 69.5, 125.8, 128.4 (4C), 142.2; HRMS(ESI) calcd. for C$_{13}$H$_{20}$NaO: 215.1412. Found: 215.1416; IR v_{max} (neat) 3349, 2954, 2927, 1496, 1467, 1455, 1367, 1055 cm^{-1}; Anal. Calcd. for C$_{13}$H$_{20}$O: C, 81.20; H, 10.48. Found: C, 80.91; H, 10.58; [α]$_D^{21}$ +4.7 (*c* 1.10, CHCl$_3$), lit. +1.2 (*c* 1.00, CHCl$_3$); Submitters reported: MS (CI): 175 (27, MH$^+$ – OH), 91 (77), 85 (100).

Safety and Waste Disposal Information

All hazardous materials should be handled and disposed of in accordance with "Prudent Practices in the Laboratory"; National Academy Press; Washington, DC, 1995.

3. Discussion

The 1,2-metallate rearrangement of ate-complexes derived from boranes and boronic esters has been used by several groups to produce organoboranes and boronic esters with high enantiomeric purity.[7] The seminal work of Matteson demonstrated the zinc chloride-mediated homologation of boronic esters using dichloromethyllithium. In these examples, asymmetry is introduced by the use of an enantiopure chiral diol (substrate control, Scheme 1) to provide, after 1,2-metallate rearrangement, α-chloroboronic esters with very high diastereomeric purity.[8] Reaction with an alkyl metal provides the corresponding secondary boronic esters with high diastereo- and enantioenrichment.

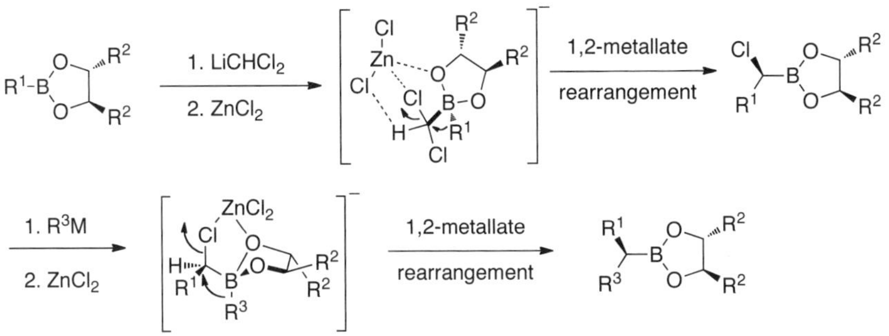

Scheme 1. Matteson-type substrate controlled homologation.

Chiral carbenoids have also been used for the homologation of achiral boranes/boronic esters to give chiral organoboranes and boronic esters (reagent control, Scheme 2). Such reagents include enantioenriched sulfur ylides,[9] Hoffmann-type[10] α-chloroalkylmetal derivatives,[6,11] lithiated epoxides,[12] lithiated aziridines[13] and N-linked carbamates.[14] All have been reported to be highly effective providing the corresponding products in high yields and with excellent levels of stereocontrol.

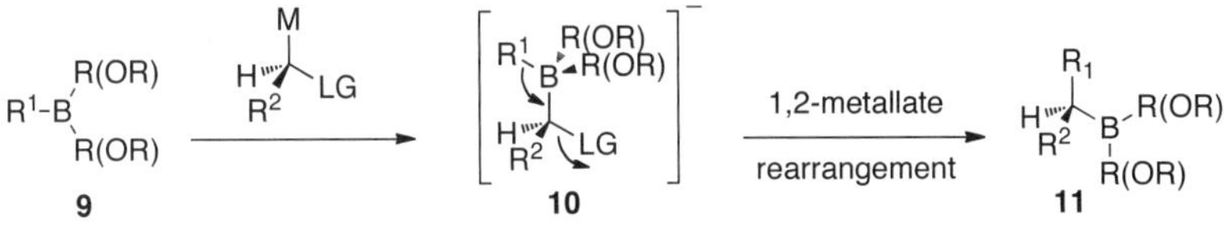

LG = Leaving Group

Scheme 2. Reagent controlled homologation.

The direct reaction of lithiated Hoppe-type carbamates[15] with boranes and boronic esters provides, after 1,2-metallate rearrangement and oxidation, secondary alcohols of high enantiopurity (Table 1).[16] This reaction shows excellent substrate scope and was evaluated for a range of carbamates and boron reagents. Moreover, either enantiomer of the lithiated carbamate is easily prepared by simple choice of (−)-sparteine or O'Brien's (+)-sparteine surrogate.[17]

In this procedure we demonstrate the robust nature of this transformation for the homologation of alkyl boronic esters. As previously reported[18] we note that MgBr₂ is not required to activate the rearrangement for the homologation of alkyl substituted boronic esters. The homologation of phenyl pinacol boronic ester with carbamate **1** has also been achieved on a 50 mmol scale within our laboratory, though not reported herein as a checked procedure. Unlike the case with alkyl boronic esters, in this case it was found that $MgBr_2$ was required to promote the 1,2-metallate rearrangement (Scheme 3).

Table 1. Homologation using Hoppe's lithiated carbamates.

12a R^1 = Ph(CH$_2$)$_2$
12b R^1 = Me$_2$C=CH(CH$_2$)$_2$
12c R^1 = TBSO(CH$_2$)$_2$C(Me)$_2$CH$_2$
12d R^1 = *i*-Pr
12e R^1 = Me

Entry	Carbamate	R^2	$(R^3)_2$	L.A.	Yield (**15**) [%]	e.r.
1	**12a**	Et	Et	-	91	98:2
2		*n*Hex	9-BBN	-	90	98:2
3		*i*Pr	9-BBN	-	81	98:2
4		Ph	9-BBN	-	85	88:12
5		Ph	9-BBN	MgBr$_2$	94	97:3
6		Et	pinacol	MgBr$_2$	90	98:2
7	**12b**	Et	Et	-	90	97:3
8		Ph	9-BBN	MgBr$_2$	71	95:5
9		Et	pinacol	MgBr$_2$	75	97:3
10		Ph	pinacol	MgBr$_2$	73	98:2
11	**12c**	Et	Et	-	67	95:5
12		Ph	9-BBN	MgBr$_2$	65	97:3
13		Ph	pinacol	MgBr$_2$	64	98:2
14	**12d**	Ph	9-BBN	MgBr$_2$	68	96:4
15		Ph	pinacol	MgBr$_2$	70	98:2
16	**12e**	Ph	pinacol	MgBr$_2$	70	97:3

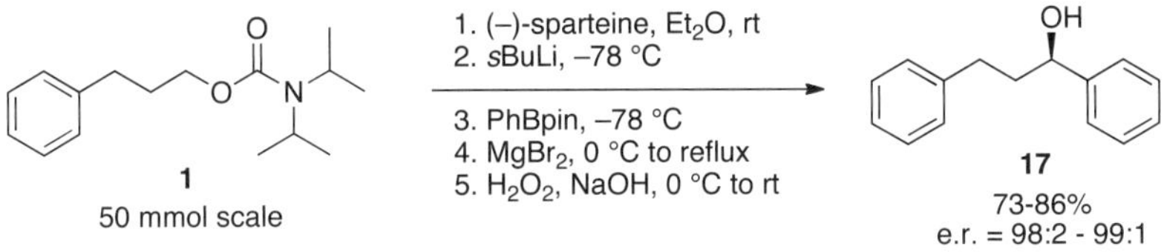

Scheme 3. Homologation of phenyl pinacol boronic ester with carbamate **1**.

It has also been reported that reagent-controlled, multiple homologations of a boronic ester are possible[16a,18] and that these can be carried out in a one-pot fashion without detriment to selectivity.[18] This process has been likened to a molecular assembly line in which successive groups are added to a growing chain with control of relative and absolute stereochemistry.

The stereodivergent use of lithiated chiral secondary benzylic carbamates[19] for the homologation of organoboranes/boronic esters has also been reported.[20] Remarkably, choice of organoborane or boronic ester allows, after oxidation, access to either enantiomer of the tertiary alcohol product from the same enantiomer of carbamate. Excellent yield and chirality transfer was observed in all cases. The benzylic boronic ester products have been isolated, shown to undergo effective protodeboronation[21] and conversion to products bearing all carbon quaternary centers.[22]

The intermediate boron compounds can also be converted to their amine[23a,b] or trifluoroborate counterparts.[23c] Both secondary and tertiary trifluoroborate compounds are easily prepared and have been reported to undergo Rh catalysed[24a] and Lewis-acid catalysed[24b] 1,2-addition to aldehydes with almost complete retention of stereochemistry.

1. School of Chemistry, University of Bristol, Cantock's Close, Bristol, BS8 1TS, UK. Fax: (+44)117-929-8611. E-mail: v.aggarwal@bristol.ac.uk.
    Homepage:
    http://www.bris.ac.uk/chemistry/research/organic/aggarwal-group. The authors thank the EPSRC for support of this work and Frontier Scientific and Syngenta for the generous donation of chemicals.

2. Behrens, K.; Fröhlich, R.; Meyer, O.; Hoppe, D. *Eur. J. Org. Chem.* **1998**, 2397-2403.

3. Shenvi, A. B. US Patent 4537773 (A).

4. Nikolic, N. A.; Beak, P. *Org. Synth.* **1997**, *74*, 23-32.

5. Pangborn, A. B.; Giardello, M. A.; Grubbs, R. H.; Rosen, R. K.; Timmers, F. J. *Organometallics* **1996**, *15*, 1518-1520.

6. Blakemore, P. R.; Burge, M. S. *J. Am. Chem. Soc.* **2007**, *129*, 3068-3069.

7. For a review on the homologation and alkylation of boronic esters and boranes by 1,2-metallate rearrangement of boron ate complexes see: Thomas, S. P.; French, R. M.; Jheengut, V.; Aggarwal, V. K. *Chem. Rec.* **2009**, *9*, 24-39.

8. For reviews on the use of dihalomethyllithium for the homologation of chiral boronic esters and their subsequent alkylation see: (a) Matteson, D. S. *Acc. Chem. Res.* **1988**, *21*, 294-300; (b) Matteson, D. S. *Chem. Rev.* **1989**, *89*, 1535-1551; (c) Matteson, D. S. *Tetrahedron* **1989**, *45*, 1859-1885; (d) Matteson, D. S. *Pure Appl. Chem.* **1991**, *63*, 339-344; (e) Matteson, D. S. *Tetrahedron* **1998**, *54*, 10555-10607; (f) Matteson, D. S. *Chem. Tech.* **1999**, *29*, 6-14; (g) Matteson, D. S., *Boronic Acids*, ed. Hall, D. G., Wiley-VCH Verlag GmbH & Co. KGaA, Weinheim, Germany, **2005**, 305-342.

9. (a) Aggarwal, V. K.; Fang, G. Y.; Schmidt, A. T. *J. Am. Chem. Soc.* **2005**, *127*, 1642-1643; (b) Fang, G. Y.; Wallner, O. A.; Di Blasio, N.; Ginesta, X.; Harvey, J. N.; Aggarwal, V. K. *J. Am. Chem. Soc.* **2007**, *129*, 14632-14639; (c) Fang, G. Y.; Aggarwal, V. K. *Angew. Chem. Int. Ed.* **2007**, *46*, 359-362; (d) Howells, D.; Robiette, R.; Fang, G. Y.; Knowles, L. S.; Woodrow, M. D.; Harvey, J. N.; Aggarwal, V. K. *Org. Biomol. Chem.* **2008**, *6*, 1185-1189.

10. Hoffmann, R. W.; Nell, P. G.; Leo, R.; Harms, K. *Chem. Eur. J.* **2000**, *6*, 3359-3365.

11. Blakemore, P. R.; Marsden, S. P.; Vater, H. D. *Org. Lett.* **2006**, *8*, 773-776.

12. Vedrenne, E.; Wallner, O. A.; Vitale, M.; Schmidt, F.; Aggarwal, V. K. *Org. Lett.* **2009**, *11*, 165-168.

13. Schmidt, F.; Keller, F.; Vedrenne, E.; Aggarwal, V. K. *Angew. Chem. Int. Ed.* **2009**, *48*, 1149-1152.

14. Coldham, I.; Patel, J. J.; Raimbault, S.; Whittaker, D. T. E.; Adams, H.; Fang, G. Y.; Aggarwal, V. K. *Org. Lett.* **2008**, *10*, 141-143.

15. (a) Hoppe, D.; Hintze, F.; Tebben, P. *Angew. Chem. Int. Ed. Engl.* **1990**, *29*, 1422-1424; For reviews see: (b) Hoppe, D.; Hintze, F.; Tebben, P.; Paetow, M.; Ahrens, H.; Schwerdtfeger, J.; Sommerfeld, P.; Haller, J.; Guarnieri, W.; Kolczewski, S.; Hense, T.; Hoppe. I. *Pure Appl. Chem.* **1994**, *66*, 1479-1486; (c) Hoppe, D.; Hense, T. *Angew. Chem. Int. Ed. Engl.* **1997**, *36*, 2282-2316; (d) Basu, A.; Thayumanavan, S. *Angew. Chem. Int. Ed.* **2002**, *41*, 716-738; (e) Hoppe, D.; Marr, F.; Brüggemann, M. *In Organolithiums in Enantioselective Synthesis.*; Hodgson, D. M., Ed.; Springer-Verlag: Berlin Heidelberg, **2003**; Vol. 5, 61-138; (f) Hoppe, D.; Christoph, G. *In Chemistry of Organolithium Compounds* **2004**; Vol. 2, 1055-1164.

16. (a) Stymiest, J. L., Dutheuil, G., Mahmood, A., Aggarwal, V. K. *Angew. Chem. Int. Ed.* **2007**, *46*, 7491-7494; (b) Althaus, M.; Mahmood, A.; Suárez, J. R.; Thomas, S. P.; Aggarwal, V. K. *J. Am. Chem. Soc.* **2010**, *132*, 4025–4028; (c) Binanzer, M.; Fang, G. Y.; Aggarwal, V. K. *Angew. Chem. Int. Ed.* **2010**, *49*, 4264–4268; (d) Robinson, A.; Aggarwal, V. K. *Angew. Chem. Int. Ed.* **2010**, *49*, 6673-6675; (e) Besong, G.; Jarowicki, K.; Kocienski, P. J.; Sliwinski, E.; Boyle, F. T. *Org. Biomol. Chem.* **2006**, *4*, 2193-2207; (f) O'Brien, P.; Bilke, J. L. *Angew. Chem. Int. Ed.* **2008**, *47*, 2734-2736; (g) For related reactions of Grignard reagents with α-carbamoyloxyboronates see: Beckmann, E.; Desai, V.; Hoppe, D. *Synlett* **2004**, 2275-2280.

17. Dearden, M. J.; Firkin, C. R.; Hermet, J.-P. R.; O'Brien, P. *J. Am. Chem. Soc.* **2002**, *124*, 11870-11871.

18. Dutheuil, G.; Webster, M. P.; Worthington, P. A.; Aggarwal, V. K. *Angew. Chem. Int. Ed.* **2009**, *48*, 6317-6319.

19. (a) Hoppe, D.; Carstens, A.; Krämer, T. *Angew. Chem. Int. Ed. Engl.* **1990**, *29*, 1424-1425; (b) Carstens, A.; Hoppe, D. *Tetrahedron* **1994**, *50*, 6097-6108.

20. (a) Stymiest, J. L.; Bagutski, V.; French, R. M.; Aggarwal, V. K. *Nature*, **2008**, *456*, 778-782; (b) Bagutski, V.; French, R. M.; Aggarwal, V. K. *Angew. Chem. Int. Ed.* **2010**, *49*, 5142-5145.

21. Nave, S.; Sonawane, R. P.; Elford, T. G.; Aggarwal, V. K. *J. Am. Chem. Soc.* **2010**, *132*, 17096-17098.

22. Sonawane, R. P.; Jheengut, V.; Rabalakos, C.; Larouche-Gauthier, R.; Scott, H. K.; Aggarwal, V. K. *Angew. Chem. Int. Ed.* **2011**. doi: 10.1002/anie.201008067.

23. (a) Matteson, D. S.; Kim, G. Y. *Org. Lett.* **2002**, *4*, 2153-2155; (b) Bagutski, V.; Elford, T. G.; Aggarwal V. K. *Angew. Chem. Int. Ed.* **2011**, *50*, 1080-1083; (c) Bagutski, V.; Ros, A.; Aggarwal, V. K. *Tetrahedron* **2009**, *65*, 9956-9960.

24. (a) Ros, A.; Aggarwal, V. K. *Angew. Chem. Int. Ed.* **2009**, *48*, 6289-6292; (b) Ros, A.; Bermejo, A.; Aggarwal, V. K. *Chem. Eur. J.* **2010**, *16*, 9741-9745.

Appendix
Chemical Abstracts Nomenclature; (Registry Number)

N,N-Diisopropyl-carbamic acid 3-phenyl-propyl ester; (218601-55-9)
Isobutylboronic acid pinacol ester; (67562-20-3)
(*R*)-5-Methyl-1-phenylhexan-3-ol; (481048-36-6)
3-Phenylpropan-1-ol; (122-97-4)
N,N-Diisopropylcarbamoyl chloride; (19009-39-3)
Triethylamine; (121-44-8)
Isobutylboronic acid: Boronic acid, *B*-(2-methylpropyl)-; (84110-40-7)
Pinacol; (79-09-5)
sec-Butyllithium; (598-30-1)
(−)-Sparteine; (90-39-1)

Varinder K. Aggarwal was born in Kalianpur in North India in 1961 and emigrated to the United Kingdom in 1963. He received his B.A. (1983) and Ph.D. (1986) from Cambridge University, the latter under the guidance of Dr Stuart Warren, and carried out postdoctoral work with Professor Gilbert Stork at Columbia University, NY (1986–1988). He is currently at the University of Bristol where he has held the Chair of Synthetic Chemistry since 2000. His research interests include the development of novel asymmetric synthetic methodology, catalysis and total synthesis.

Matthew P. Webster was born in Bridlington, UK, in 1981. He received a masters degree in chemistry at the University of Hull, UK, in 2005 and subsequently joined the group of Prof. Aggarwal at the University of Bristol. He obtained his Ph.D. in 2010 on the multiple homologations of boronic esters and applications in synthesis. Since January 2010 he has been a postdoctoral scholar in the group of Prof. Alois Fürstner at the Max-Planck-Institut, Mülheim, Germany.

Benjamin M. Partridge was born in Birmingham, UK, in 1985. He graduated in 2007 with a masters degree in chemistry from the University of Bristol. Currently, he is pursuing a Ph.D. degree under the direction of Prof. Aggarwal focussed on the asymmetric total synthesis of medicinally relevant compounds.

Tomohiko Inui was born in Osaka, Japan in 1983. He received his B.S. in 2006 from Kyoto University under the direction of Professor Tamejiro Hiyama. He then moved to the laboratories of Professor Tohru Fukuyama, the University of Tokyo. He received his M.S. in 2008 and he is pursuing a Ph.D. degree. His current research interest is total synthesis of natural products.

PALLADIUM-CATALYZED REDUCTION OF VINYL TRIFLUOROMETHANESULFONATES TO ALKENES: CHOLESTA-3,5-DIENE

Prepared by Sandro Cacchi, Enrico Morera, and Giorgio Ortar.[1]
Original article: Cacchi, S.; Morera, E.; Ortar, G. *Org. Synth.* **1990**, *68*, 138.

Ever since our first reports describing the palladium-catalyzed reduction of vinyl[2] and aryl[3] triflates to alkenes and arenes, this methodology has found a large number of applications ranging from the preparation of fine chemicals to the synthesis of biologically active substances. Some modifications of the original protocol have also been developed. The aim of the present addendum is to provide a brief updated overview of recent developments in this type of chemistry. Reactions have been categorized into three main sections: modifications of the original conditions, reduction of vinyl triflates, and reduction of aryl triflates. The latter two sections are further subclassified by the class of compounds used as substrates.

Modification of the Original Conditions

In general, there have been minor modifications of the original reaction conditions. Though other solvents such as toluene,[4] THF,[5] dioxane,[6] DMSO,[7] MeOH,[8] EtOH/THF,[9] MeOH/THF,[10] EtOH,[11] AcOEt,[12] benzene,[13] MeCN,[14] and EtOH/AcOEt[15] have been used in some cases, DMF has been

260

the solvent of choice. In general, apart from the occasional utilization of K_2CO_3,[14] nitrogen bases have been employed and reductions have been usually carried out in the presence of Et_3N or Bu_3N. The utilization of 2,6-lutidine[5d] and $EtN(i\text{-}Pr)_2$,[7,9,11,16] has also been described. As to the catalyst system, in addition to $Pd(OAc)_2/PPh_3$, $Pd(OAc)_2/dppf$ [1,1'-bis(diphenylphosphino)ferrocene], and preformed $Pd(OAc)_2(PPh_3)_2$ originally used as precatalysts in this chemistry, reduction of organic triflates have been successfully performed in the presence of $Pd(acac)_2/Bu_3P$,[5] $PdCl_2(PPh_3)_2$,[6c,17] $Pd(PPh_3)_4$,[5b-e,6a,c,g,h,14,18] $Pd(PPh_3)_4/dppf$,[13] $Pd_2(dba)_3/P(o\text{-}tol)_3$,[6f] $Pd(OAc)_2/P(OMe)_3$,[19] Pd/C,[8,9,10,11,12a,15,20] $PdCl_2(dppf)$,[15b,21] $Pd(OAc)_2/dppp$ [1,3-bis(diphenylphosphino)propane],[22] $Pd(OAc)_2/dppb$ [1,3-bis(diphenylphosphino)butane],[23] $Pd(OAc)_2/dppp/dppb$,[7b] $Pd(OH)_2/C$,[12] and $PdCl_2(PPh_3)_2/dppp$.[24] The latter combination was found to be particularly useful with electron-rich or highly hindered aryl triflates (Scheme 1).

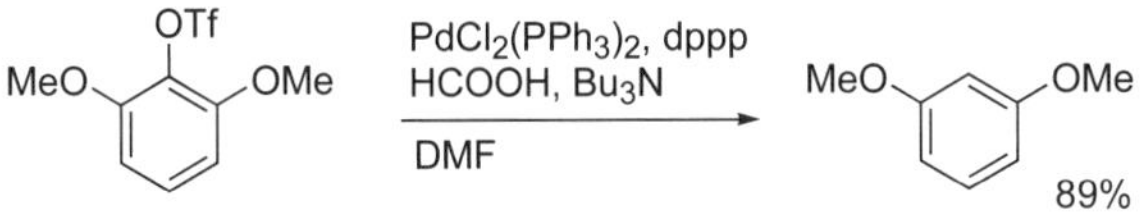

Scheme 1

Formic acid (in the presence of nitrogen bases) has been the reducing agent of choice in the vast majority of applications. However, the utilization of other reducing agents has been described of late. Aryl triflates have been reduced to arenes using HCO_2NH_4,[8,19] Et_3SiH,[5c,13,17g,18,23] H_2,[8a,9,10,11,12,15] (n-Bu)$_3$SnH,[5b,d,6c] $Me_2NH\cdot BH_3$,[14] in the presence of Pd/C, Mg, MeOH, NH_4OAc,[8b,25] and of $Pd(OAc)_2$, BINAP, K_2CO_3 in DMF.[14b] In some cases, pinacolborane was found to act as an hydride source in the reaction with aryl triflates. In one of these examples, the reduction product formed readily in the presence of $PdCl_2(dppf)$ and dioxane at room temperature.[26] With DCO_2D[27] (Scheme 2) or n-BuSnD[5e] the reduction of vinyl and aryl triflates is a powerful tool for substituting a C-D bond for a C-O bond.

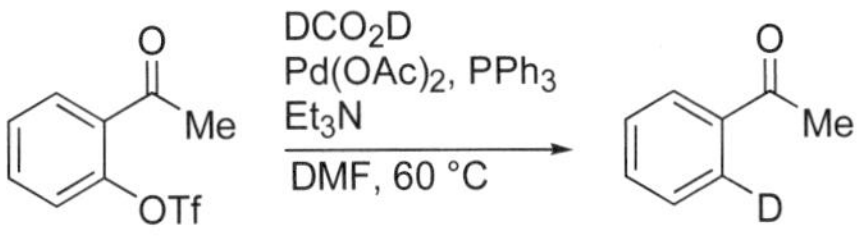

Scheme 2

An interesting variation is the development of polymer-supported triflating agent equivalents which allow for the reduction process to be performed under solid-phase conditions. Indeed, as a result of the general acceptance of combinatorial techniques as a useful tool in drug discovery programs, there is a demand for polymer-supported reagents. Particularly, because of the importance of making molecules lacking any extraneous functionality, traceless linkers represent an exciting aspect of solid-phase synthesis. In this context, a couple of proposals have recently been reported which are based on the utilization of perfluoroalkyl[28] and perfluoroaryl[29] sulfonyl linkers. An example of this chemistry is shown in Scheme 3.[28]

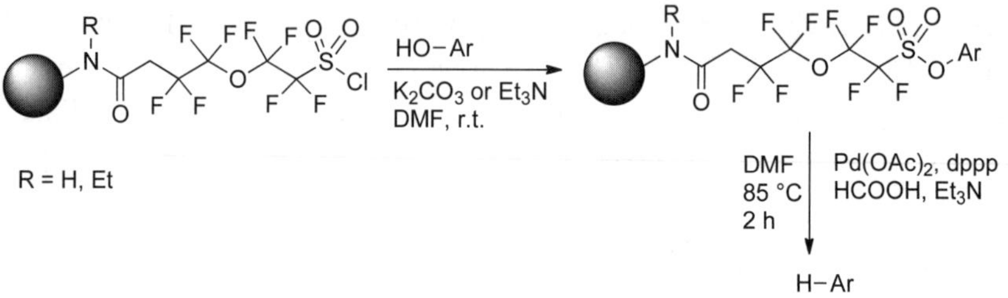

R = H, Et

Scheme 3

Palladium-Catalyzed Reduction of Vinyl Triflates to Alkenes

Alkaloids

The synthesis of alkaloid derivatives has taken particular advantage of the selective palladium-catalyzed reduction of vinyl triflates. The methodology was recently employed in the asymmetric synthesis of the 3-methyl-1-azatricyclo[5.2.1.0]decane substructure,[17f] embedded within the diterpene-derived carbon scaffold of the hetisine alkaloids. Some steps of this synthesis involve conjugate reduction of **1** and triflation of the transient enolate with PhNTf₂ to yield the corresponding vinyl triflate in 70% yield, which was reduced to the alkene **2** in 84% yield (Scheme 4).

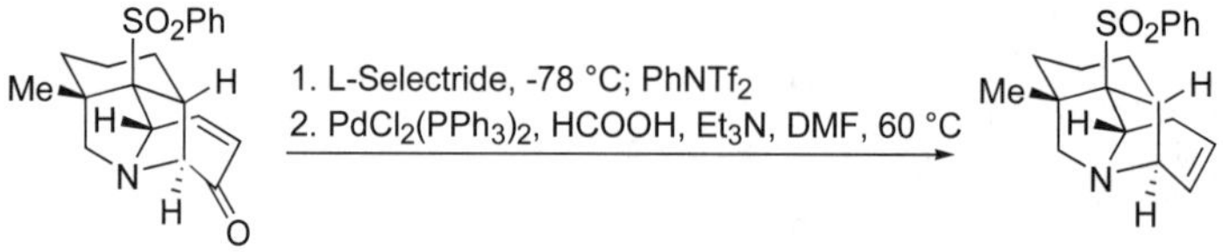

Scheme 4

Other applications in this area involve the preparation of a 'minilibrary' of analogues of manzamine A,[30] an important lead structure for the development of novel antimalarial chemotherapies, the synthesis of 2-azabicyclo[2.2.2]oct-5-enes bearing an endo alkenyl substituent,[31] a study devoted to the investigation of isoquinuclidines as precursors of certain *Gelsemium* alkaloids, a divergent synthesis of tropane (8-azabicyclo[3.2.1]octane)-type alkaloid (+)- and (−)-ferruginine,[32] a concise synthesis of (−)-anatoxin-a,[33] the synthesis of the 2β,3α- and 3β,3β-isomers of 3-(*p*-substituted phenyl)tropane-2-carboxylic acid methyl esters,[34] the enantioselective 20 step total synthesis of the tetracyclic spermidine alkaloid (−)-hispidospermidin,[35] a general synthetic pathway to *Strychnos* indole alkaloids (−)-dehydrotubifoline and (−)-tubifoline,[16a] the reduction of pyridone derivatives,[36] the synthesis of bioavailable 4,4-disubstituted piperidines,[14a] the first total synthesis of the tetracyclic lactone (-)-secu'amamine A, a member of the *Securinega* alkaloid family showing a wide range of biological activities,[37] the synthesis of 1,2-dihydrochromeno[3,4-*f*]quinoline derivatives, selective progesterone receptor modulators (SPRM),[38] the synthesis of Calothrixins B, an antiproliferative agent from the strains of *Calothrix cyanobacterium* with an indolo[3,2-*j*]phenanthridine core structure,[6h] a study on rearrangement of spirocyclic oxindoles, the core skeleton of several naturally occurring alkaloids isolated from *Gelsemium sempervirens*,[39] the asymmetric synthesis of all the known phlegmarine alkaloids, decahydroquinolin-piperidines of significant biological activities,[40] and several syntheses of *Lycopodium* alkaloids,[17c,41] a class of alkaloids containing various structural features which provides challenging targets for total synthesis.

As to the latter research area, an example is given in Scheme 5,[41a] showing the reduction of the triflate **4** to the corresponding alkene. The authors report that the conversion of the enone **3** into the corresponding triflate proved to be more challenging than anticipated. Simple addition of the triflating reagent, *N*-(5-chloro-2-pyridyl)triflimide, to the enolate formed in situ via copper-mediated 1,4-addition to **3** of the Grignard of (chloromethyl)dimethylphenylsilane failed to yield vinyl triflate **4**. Addition of DMPU followed by heating at reflux was necessary to trap the enolate.

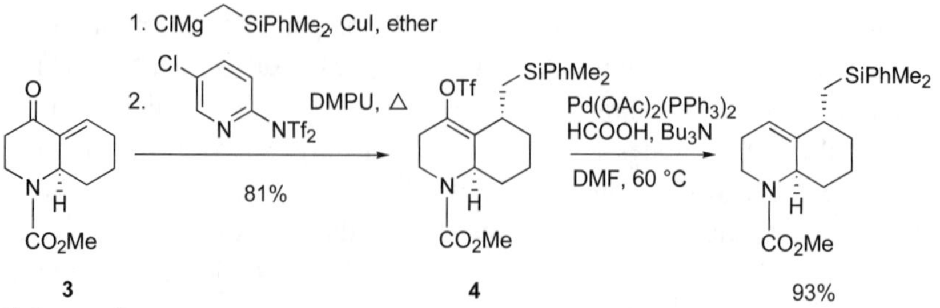

Scheme 5

1. ClMg⌒SiPhMe₂, CuI, ether

2. [chloropyridyl]N(NTf₂) DMPU, Δ

3 → 4 (81%) → (93%)

(Structures: compound **3** with CO₂Me; compound **4** with OTf, SiPhMe₂, CO₂Me; product with SiPhMe₂, CO₂Me)

Conditions for second step: Pd(OAc)₂(PPh₃)₂, HCOOH, Bu₃N, DMF, 60 °C

Sphingosine

Reduction of acyclic vinyl triflates was used in key steps of the synthesis of D-*erythro*-sphingosine - an unsaturated 18-carbon amino alcohol which is a primary component of glycosphingolipids and ceramids - from a chiral β-lactam obtained from D-(−)-tartaric acid[42] and from commercially available and cheap D-*ribo*-phytosphingosine (Scheme 6).[43]

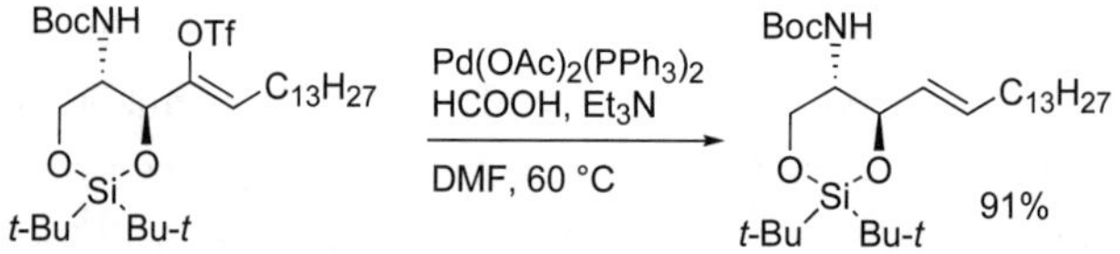

Scheme 6

BocNH, OTf, C₁₃H₂₇, O–Si–O, t-Bu, Bu-t

Pd(OAc)₂(PPh₃)₂, HCOOH, Et₃N, DMF, 60 °C

→ BocNH, C₁₃H₂₇, O–Si–O, t-Bu, Bu-t (91%)

Terpenoids and Steroids

Triflates derived from α-hydroxy-α,β-enones were selectively converted into the corresponding α,β-enones in a study devoted to the preparation of analogues of brusatol, an antitumor quassinoid, which were examined for their cytotoxic activity.[44] The reaction was used in the synthesis of (−)-homogynolide A, a bakkanolide sesquiterpene isolated from *Homogyne alipina* active as insect antifeedant,[45] in the stereoselective synthesis of metasequoic acids B, a new antifungal diterpene recently isolated from *Metasequoia glyptostroboides* which has a labdane skeleton with a cyclopropane ring fused at C(3) and C(4). Compounds with a cyclopropane moiety of this kind are rare among the diterpenoids.[46]

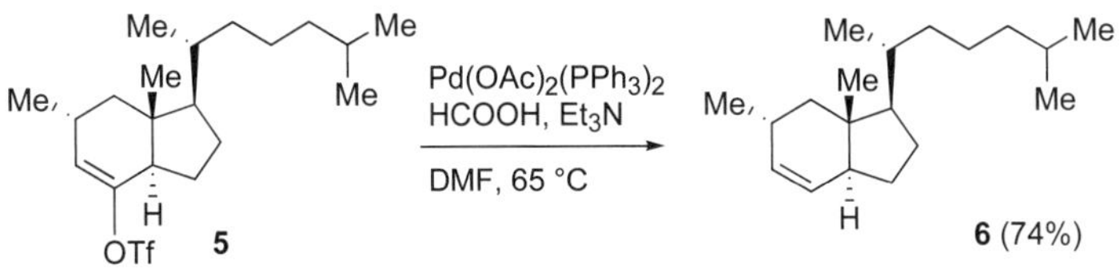

Scheme 7

In a study on the synthesis of C ring modified analogues of 1α,25-dihydroxyvitamin D_3, compound **6** was prepared from **5** in 74% yield (Scheme 6).[47] Attempts to reduce **5** with Bu_3SnH or Et_3SiH in the presence of $Pd(PPh_3)_4$ and LiCl in THF yielded complex reaction mixtures.[48]

Reduction of the vinyl triflate **7** (Scheme 8) was successfully performed under typical conditions in the 31 step asymmetric total synthesis of (−)-merrilactone A (1.1% overall yield),[49] a sesquiterpenoid isolated from *Illicium merrillianum* which was shown to possess neuroprotective and neuritogenic activity in cultures of fetal rat cortical neurons.

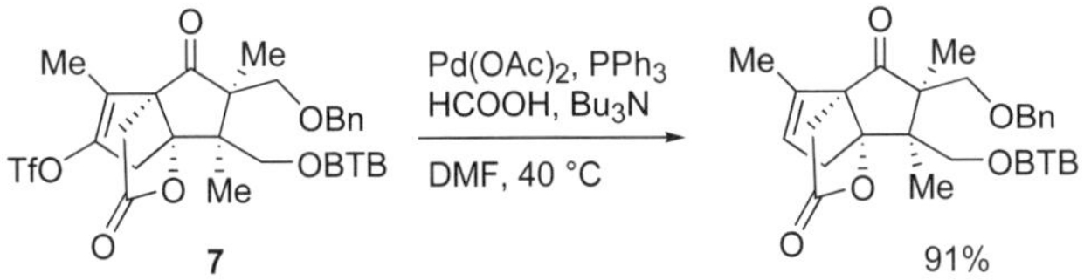

Scheme 8

In the synthesis of a novel sesquiterpene isolated from the pheromone gland of a stink bug, *Tynacantha marginata* Dallas, the tricyclic ketone **8** was converted to the alkene **10** via the vinyl triflate **9** (Scheme 9).[50]

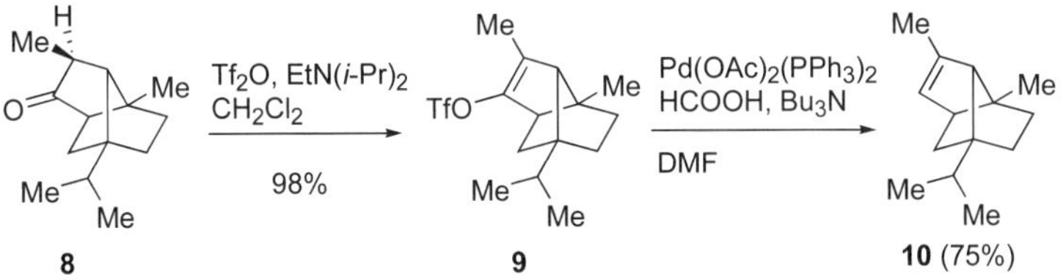

Scheme 9

Enol triflate **11** has been employed as a key intermediate to access to some taiwaniaquinoids, a family of unusual tricyclic diterpenes isolated from East Asian conifers with interesting biological activities. Palladium-catalyzed reduction of **11** gave indene **12** in excellent yield. Because of steric hindrance, the comparatively small ligand trimethyl phosphite was required to effectively carry out his reaction (Scheme 10).[19a]

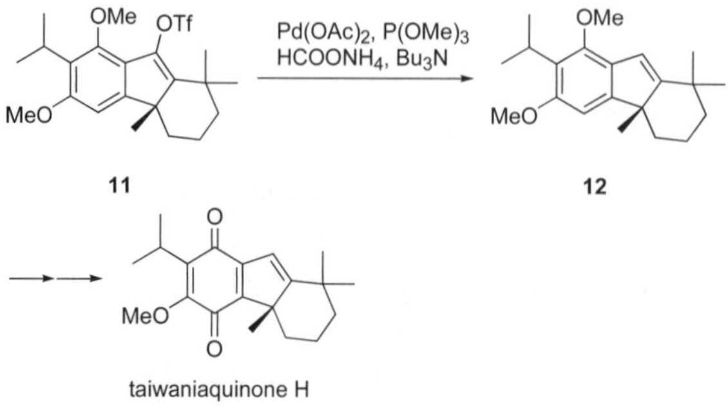

Scheme 10

Some interesting applications to steroid chemistry have also been reported. A series of unsaturated steroids bearing a carboxy substituent at the C-3 position were prepared and assayed in vitro as inhibitors of human and rat prostatic steroid 5α-reductase. Some of the unsaturated 3-carboxysteroids were prepared by triflation of 11-ketone derivatives followed by palladium-catalyzed reduction.[51] The procedure was used to prepare 6,7-didehydroestrogens.[52]

In the synthesis of zymosterol, fecosterol, and related biosynthetic sterol intermediates,[53] multigram quantities of key intermediates containing 8,14-dienic systems were efficiently prepared from the corresponding enones though palladium-catalyzed reduction of vinyl triflates under standard conditions (Scheme 11).[53c]

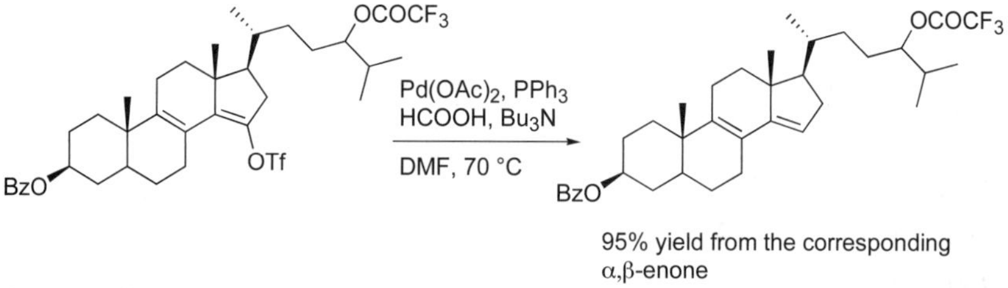

95% yield from the corresponding α,β-enone

Scheme 11

A selective reduction of the 17-vinyl triflate in the presence of the 3-aryl triflate was achieved using Pd(dppf)Cl₂-dichloromethane adduct as the catalyst in the synthesis of four metabolites of the antitumor drug ENMD-

266

1198, a promising analogue of 2-methoxyestradiol for the treatment of various cancers (Scheme 12).[21e]

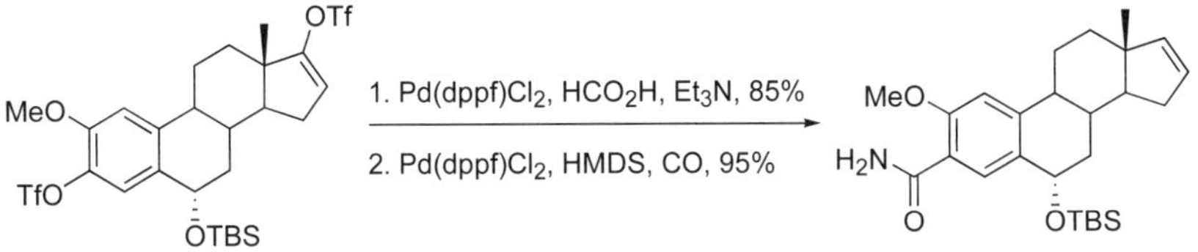

Scheme 12

Further applications in the area of terpenoids and steroids include a stereoselective synthesis of the CD-ring structure of cortistatin A,[54] a straightforward and scaleable synthesis of the sterically congested pheromone of the longtailed mealybug *Pseudococcus longispinus*,[55] a concise diastereoselective and enantiopure route from limonene glycol to the (1*S*,4*R*)- and (1*S*,4*S*)-isomers of 4-isopropyl-1-methyl-2-cyclohexen-1-ol (aggregation pheromones of the ambrosia beetle *Platypus quercivorus*),[16d] and a highly regio- and stereoselective synthesis of C1 hydroxylated 19-norvitamin D analogs from the corresponding 25-hydroxyvitamin D precursors.[56]

Carbocycles

A variety of olefin-containing carbocycles have been prepared from the corresponding ketonic derivatives via palladium-catalyzed reduction of vinyl triflates in the presence of formate anions. This chemistry was applied to bicyclo[2.2.2]octene-2,5-diones,[17a,e] to cyclic 1,2-diketones - used as building blocks for the asymmetric synthesis of cycloalkenones[57] - to the enantioselective total synthesis of (−)-subergorgic acid,[58] to the preparation of intermediates in the synthesis of carba-spirocyclic compounds,[59] to a concise asymmetric synthesis of hamigeran B,[60] and in a key step (Scheme 13) of a stereocontrolled functionalization of the diene system of compactin.[61]

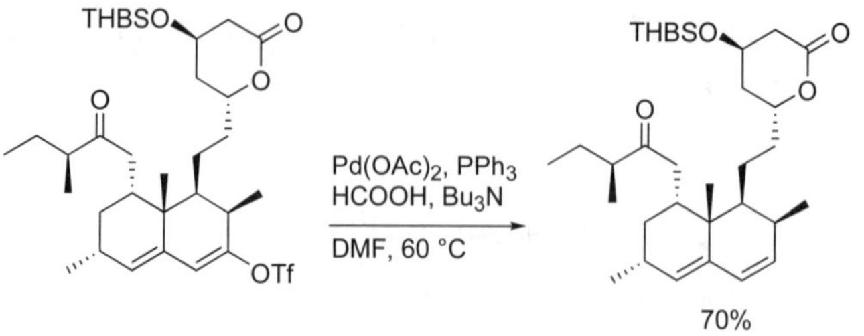

Scheme 13

In a synthetic study devoted to the assembly of the highly functionalized carbocyclic core of CP-263,114 (phomoidride B), vinyl triflate **13** (phomoidride B) was converted into the olefinic derivative **14** as shown in Scheme 14.[5a]

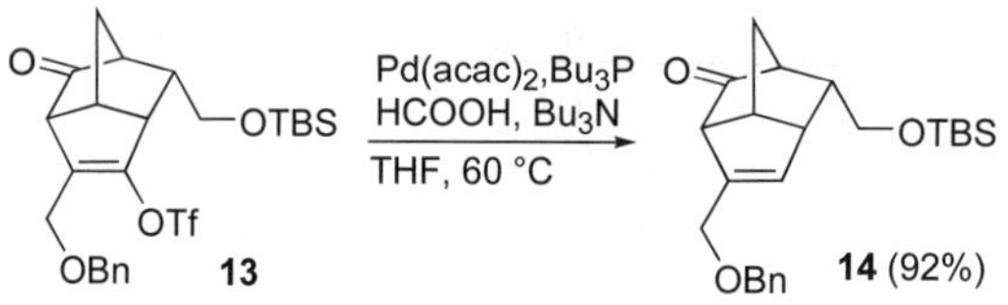

Scheme 14

In a study on the stereocontrolled formal synthesis of (±)-platensimycinin, a potent and broad-spectrum Gram-positive antibacterial agent isolated from *Streptomyces platensis*, the palladium-catalyzed reduction of a vinyl triflate intermediate gave a good yield of the corresponding alkene by using formic acid as hydrogen source, while tributyltin hydride afforded the alkene in only 37% yield.[62]

In a straightforward synthesis of (1-allylcyclohexa-2,5-dienyl)arenes, useful building blocks for the preparation of natural products including amaryllidaceae, strychnos and morphinan alkaloids, the final step consists in palladium-catalyzed formate reduction of the corresponding bis-enol triflate.[63]

The palladium-catalyzed formate reduction has been also applied to an asymmetric synthesis of *endo*-6-aryl-8-oxabicyclo[3.2.1]oct-3-en-2-one, a natural product from *Ligusticum chuanxing* Hort.,[64] to the total synthesis and determination of the absolute configuration of (-)-idesolide, a nitric oxide (NO) production inhibitor isolated from the fruit of *Idesia polycarpa*,[65] and to the catalytic asymmetric synthesis of descurainin, isolated from the seeds of *Descurainia sophia* (L.) Webb ex Prantl and widely used as Chinese traditional medicine.[66]

Palladium-Catalyzed Reduction of Aryl Triflates to Arenes

Alkaloids

The palladium-catalyzed formate reduction of aryl triflates has been extensively applied to the synthesis of aporphines such as 2-fluoro-11-hydroxy-*N*-propylnoraporphine,[67] (±)-aporphine,[68] *R*(-)-*N*-alkyl-11-hydroxynoraporphines,[69] (*R*)-10-hydroxy- and (*R*)-11-hydroxyaporphine,[24b-d,70] 11-substituted (*R*)-aporphines,[71] (*R*)-11-hydroxy-10-methylaporphine[24b,70] esters of *R*(–)-*N*-alkyl-11-hydroxy-2-methoxynoraporphines,[8b] and 2-*O*- or 11-*O*-substituted *N*-alkylnoraporphines[25] in order to evaluate their central serotonergic and dopaminergic effects *in vivo* and *in vitro*. Thus, for example, reduction of the aryl triflate **15** was performed in 65% yield in a simple regioselective synthesis of (*R*)-11-hydroxyaporphine, a D_1 receptor antagonist, from apomorphine (Scheme 15).[24d]

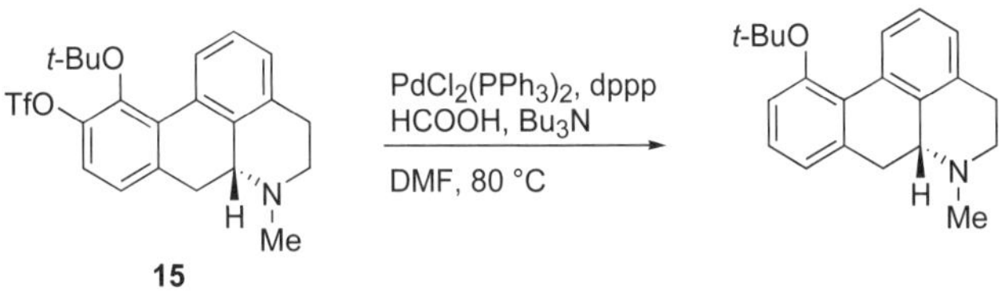

Scheme 15

(–)-Galantamine, an alkaloid isolated from the *Amaryllidaceae* family, is approved in a number of countries for the treatment of Alzheimer's disease as a selective acetylcholinesterase inhibitor. Because of the limited supplies and the high cost of its isolation from natural sources, several syntheses of galantamine exploiting a phenolic oxidative coupling of norbelladine-type derivatives have been reported. In an improved procedure, the extra phenolic hydroxyl group of the intermediate **16** was removed quantitatively by conversion into the triflate **17**, followed by the palladium-catalyzed reduction (Scheme 16).[72]

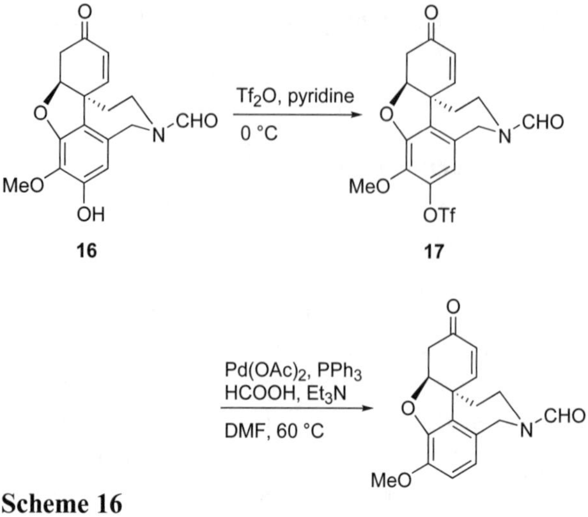

Scheme 16

During the synthesis of analogues of camptothecin, an anticancer alkaloid isolated from Chinese tree *Camptotheca acuminata,* and of camptothecin-related alkaloids, the triflation-reduction sequence has been repeatedly applied to the deoxygenation of phenolic hydroxyl groups of the pentacyclic ring system,[6b,d,e,73] as exemplified in Scheme 17 which outlines the final steps to a close analogue **18** of deoxypumiloside, one of the putative intermediates in the synthesis of camptothecin.[6e]

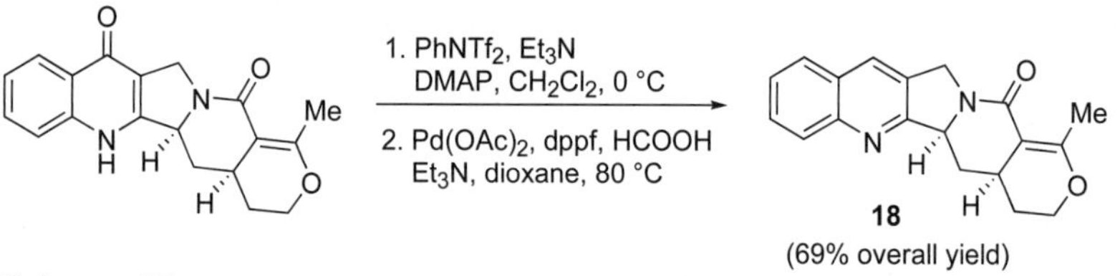

Scheme 17

Subjection of the triflates **19** or **20** to conditions for reduction completed two related syntheses of calothrixin B, a carbazole alkaloid isolated from cell extracts of cyanobacterial *Calothrix* species endowed with antiprotozoal and anticancer properties, and of a calothrixin B isomer (Scheme 18). It is worth noting that these transformations could be effected without the need for reprotection of the indole nitrogen atom.[6g,h]

270

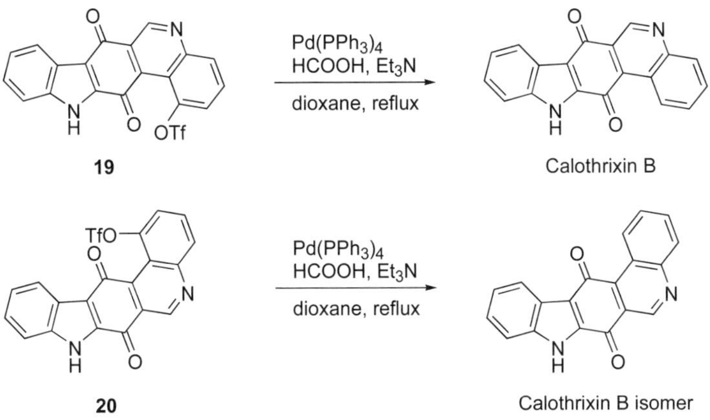

Scheme 18

Palladium-mediated reduction of both vinyl and aryl triflates has been exploited in the synthesis of a series of thrombin receptor antagonists based on himbacine, a tetracyclic piperidine alkaloid isolated from the bark of the Australian pine tree of Galbulimima species (Scheme 19).[74]

Further applications in the area of alkaloids include the synthesis of the ABC ring of safracines,[75] important biogenetic intermediates of saframycins, of analogues of streptonigrin,[76] a 5,8-quinolinedione potent inhibitor of avian myeloblastosis virus reverse transcriptase, of duocarmycin B2,[77] a novel antitumor antibiotic isolated from *Streptomyces* sp, of substituted carbazoles and β-carbolines,[78] of eilatin,[79] a polycyclic marine alkaloid, of koumine and koumidine,[80] contained in the Chinese toxic medicinal plant *Gelsemium elegans*, of calothrixin B,[81] a natural product with impressive in vitro cytotoxicity, obtained from the cell extract of *Calothrix* cyanobacteria, of a calothrixin B isomer,[6h] of onychnine,[17a] the simplest member of the family of 4-azafluorenone alkaloids, of zoanthenol, the sole member of the zoanthamine alkaloids family possessing an aromatic ring,[82] of (±)-8-oxoerymelanthine, an erythrina alkaloid containing a pyridine ring,[22a] of (±)-aurantioclavine, an indole alkaloid characterized by a novel azepino[5.4.3.-*cd*)-indole ring,[8a] and of simplified analogs of naphthylisoquinoline alkaloids[8c] and a SAR study on cassiarin A, a tricyclic alkaloid, isolated from the leaves of *Cassia siamea* which shows powerful antimalarial activity against *Plasmodium falciparum* in vitro.[17g]

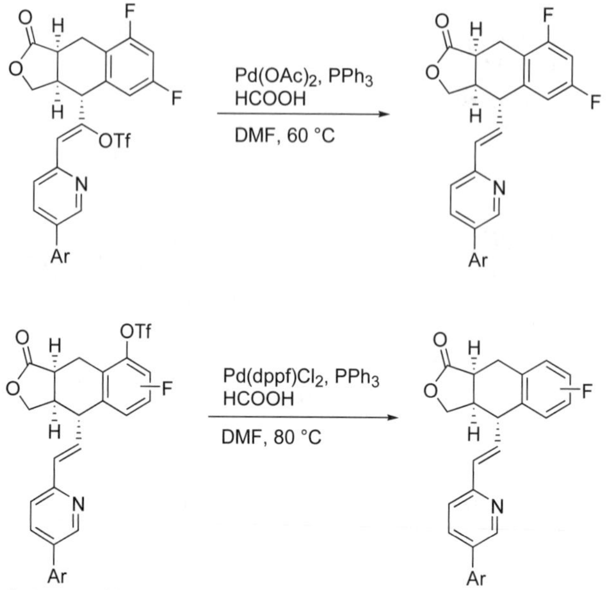

Scheme 19

Enediynes

In a convergent synthetic route to (+)-dynemicin A, a member of the enediyne family of natural products with potent cytotoxicity against a variety of murine and human tumor cell lines, the key intermediate **21** underwent formate reductive cleavage of the triflate group in 97% yield. Tributylstannane proved to be slightly less efficient as alternative reducing agent (Scheme 20).[6a,c]

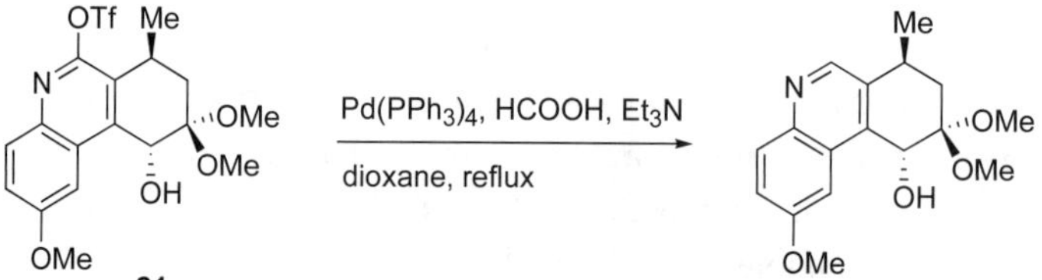

Scheme 20

The triflation-reduction sequence has been also involved in the synthesis of a hybrid analog of the esperamicin and dynemicin cores.[83]

272

Heterocycles

While the 2-and 3-positions of the indole ring system are relatively easy to functionalise, the preparation of indoles bearing complex functionalities on the benzenoid portion remains a challenge. In this view, 5-triflyloxyindoles proved to be excellent participants in palladium-catalyzed reactions, including formate reduction.[84] A series of 3,3-bisaryloxindoles has been synthesized and tested as non steroidal human mineralcorticoid receptor (hMR) antagonists.[24e] The deoxygenation of hindered bis-triflate **22** was achieved in an acceptable yield, using dppp as bidentate ligand (Scheme 21).

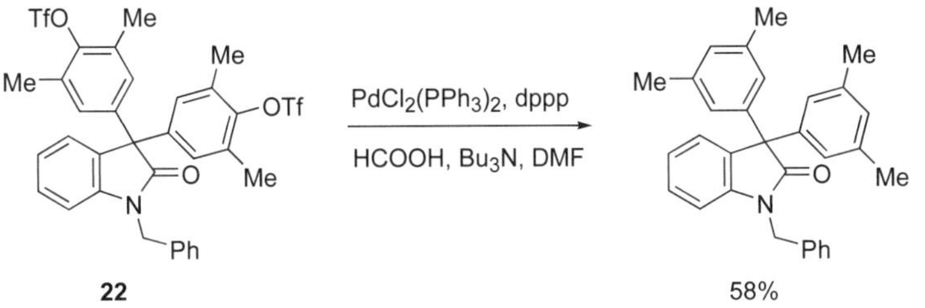

22   58%

Scheme 21

The formate reduction of triflates has been also exploited in a study on the role of the methoxy substituents on the photochromic indolylfulgides[85] and in SAR studies of osthol,[86] a cytotoxic coumarin from *Cnidium monnieri*.

The linearly fused furanocoumarins psoralens are of pharmacological interest owing to their ability to cross-link DNA upon irradiation. Pyridazine analogues of benzopsoralens have been prepared with the aim of improving the stability of the intercalating complex with DNA and convenient substitutions of the pyridazino[4,3-*h*]psoralen skeleton through palladium-catalyzed reactions have been explored.[16c] Although the Pd/C-catalyzed dehalogenation is a well known method for 3-chloropyridazines, all attempts to apply this procedure to compound **23** using either H_2 or ammonium formate were unsuccessful. By contrast, the palladium-catalyzed formate reduction of the triflate **24** afforded the desired compound **25** in 80% yield (Scheme 22).

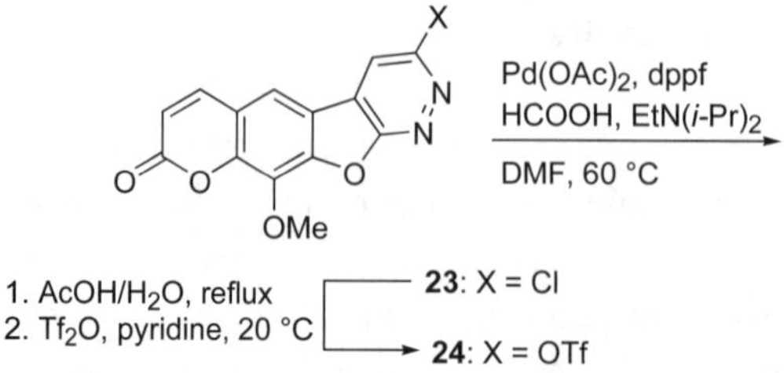

Scheme 22

1. AcOH/H₂O, reflux
2. Tf₂O, pyridine, 20 °C

23: X = Cl
24: X = OTf

25

The key step in the synthesis of angelicin (**28**), the parent compound of the class of furocoumarins with an angular configuration, utilizing the benzannulation reaction of furylcarbene complexes of chromium, was the palladium-catalyzed reduction of the triflate **26** which was found to proceed smoothly to provide the reduced benzofuran **27** in 89% yield (Scheme 23).[87] In a preliminary communication, the same authors compared the efficiencies of triethylammonium formate and sodium borohydride as reducing agents of a series of aryl triflates and found that, in the case of triflates with electron donating substituents, the reduction with formate was clearly superior, while, in the reduction of 1-naphthyl triflate and of triflates bearing electron withdrawing substituents, the two methods were comparable.[88]

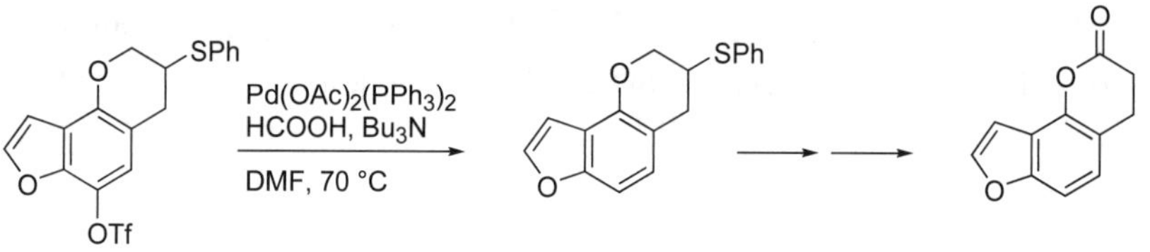

26   27   Angelicin (**28**)

Scheme 23

4-Aryl-furan-3-ols can be readily converted to the corresponding triflates, which in turn can be reduced with formate to furnish tri-substituted furans.[89]

A large class of antimicrobial isoflavonoid phytoalexins contain a substituted pterocarpan ring system. Pterocarpans such as (±)-homopterocarpin (**31**) have been formed directly and efficiently via titanium(IV)-catalyzed cycloaddition of 2*H*-chromenes with 2-alkoxy-1,4-benzoquinones.[90] In the final step of the synthesis of **31**, triflation of **29** followed by a palladium-catalyzed formate reduction of triflate **30** gave (±)-homopterocarpin in 88% overall yield (Scheme 24).[90b]

274

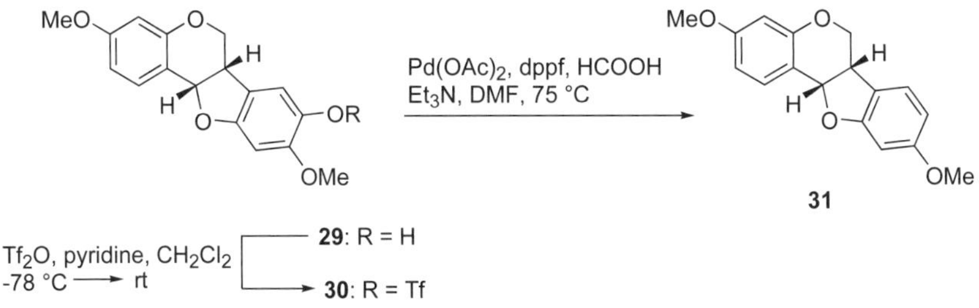

Scheme 24

The herbindole and trikentrin indoles comprise a series of structurally related polyalkylated cyclopent[g]indole natural products with cytotoxic and antibacterial properties, respectively. During a divergent total synthesis of herbindole B and *cis*-trikentrin A, treatment of triflate **32** with ammonium formate under palladium catalysis afforded the deoxygenation intermediate **33** in excellent yield (Scheme 25).[91]

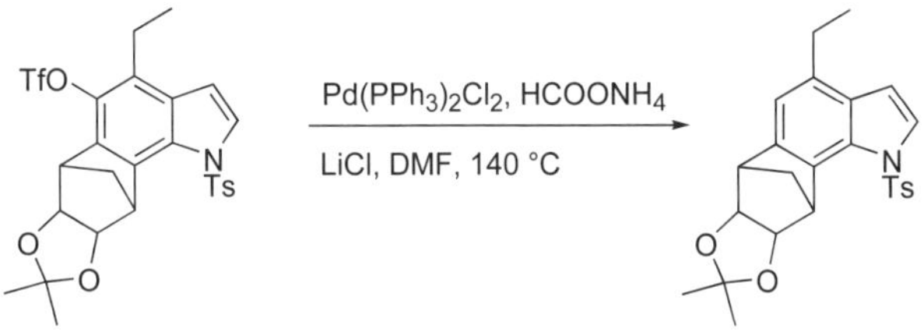

Scheme 25

Removal of phenol groups via triflates has been further employed in the synthesis of tripodal amidopyridine receptors,[17c] tricyclic pyridones,[92] dihydrobenzofurobenzofurans,[93] benzotriazoles,[94] tocopheryl amines and amides,[10] a deoxy derivative of riccardin C, a naturally occurring liver X receptors ligand,[11c] a chiral bisdihydrobenzooxaphosphole ligand for rhodium-catalyzed hydrogenations,[12b] a spirocyclic piperidine related to biologically active molecules such as MK-677,[15] and diazonamide A, a marine natural product with potent antimitotic activity,[20b] in synthetic routes that allow access to either enantiomer of a variety of 2,3-dihydroquinazolinone derivatives,[21f] and to cryptophanes,[21g,24i,j] in a new alternate synthetic route to remove the 4,5-ether bridge of opioids,[23] and in SAR studies on dopamine autoreceptor antagonists substituted (S)-phenylpiperidines,[95] inhibitors of the voltage-gated potassium channel Kv1.3

chalcone derivatives of the natural benzofuran product khellinone,[24h] and 2-(1,1-dioxo-2H-[1,2,4]benzothiadiazin-3-yl)-1-hydroxynaphthalene derivatives as HCV NS5B polymerase inhibitors.[96]

Carbocycles

Alkylphenantrenes (APs) represent a significant class of polycyclic aromatic hydrocarbons useful as synthons for several kind of alkaloids and, in itself, as markers for toxicity. A combined metalation-Suzuki-Miyaura cross-coupling methodology, involving both direct *ortho* (DoM) and remote (DreM) metalation reactions, followed by the palladium-catalyzed hydrogenolysis of the related phenantrols via their triflates, has been adopted to provide short, efficient, and regioselective routes to AP.[97] A similar reductive deoxygenation of phenanthrols was employed in the synthesis of (–)-blestriarene C, a naturally occurring 1,1'-biphenanthrene reported to be active against gram-positive bacteria, *Staphylococcus aureus*, and *Streptococcus mutans*.[98] In a general approach to biologically active angucyclines, natural pruducts featuring the benz(a)anthracene ring system, rubiginone B2 was obtained in 96% yield by a chemoselective triflation of hatomarubigin A, followed by the palladium-catalyzed reductive step (Scheme 26).[99]

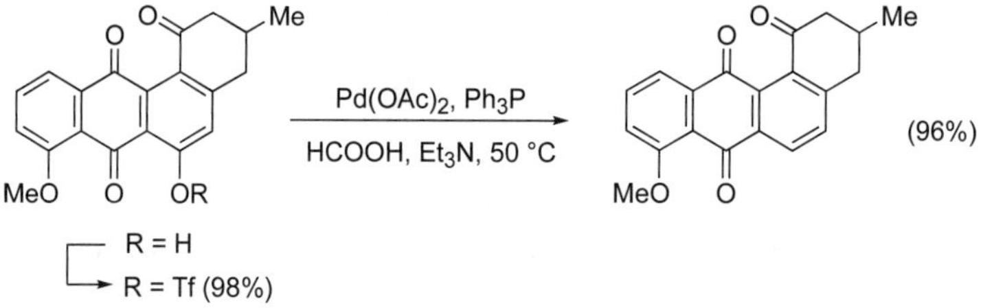

Scheme 26

The triflate reduction procedure has been used in a SAR study on 6-naphthalene-2-carboxylate retinoids for RAR transactivation activity.[100] A complete retention of substrate chirality was observed in the deoxygenation of (–)-1,1'-binaphthalene-2,2'-diol via its bis-triflate, while the same reductive conditions were unsuccessful to convert the monotriflates **34** and **35** to the corresponding monosubstituted derivatives (Scheme 27).[7a]

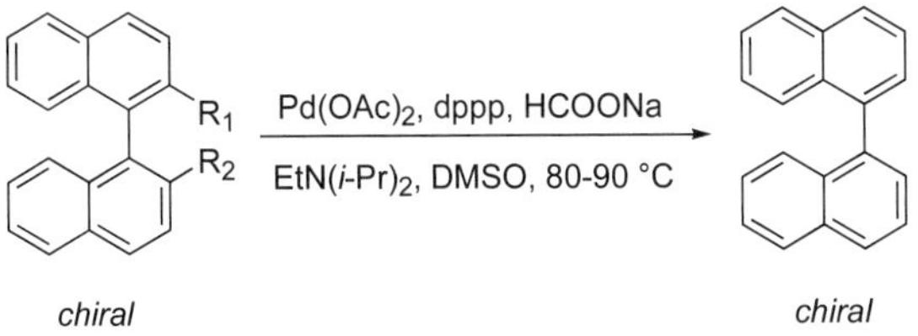

Scheme 27

(-)-R$_1$, R$_2$ = OTf
34: (+)-R$_1$ = OTf, R$_2$ = POPh$_2$ (no reaction)
35: (-)-R$_1$ = OTf, R$_2$ = PO(OEt)$_2$ (no reaction)

Heavily functionalized atropisomeric biphenyl derivatives, useful as liquid crystal dopant or competitive NMDA antagonists, have been synthetized with the use of a Suzuki cross-coupling as the key step, followed by reductive removal of phenolic functional groups.[101] In a versatile route for the synthesis of highly substituted benzenoids, the triflate intermediate was reduced to give trisubstituted benzoate ester in high yield.[102] Alkoxyaromatic compounds, obtained by a regioselective dealkylation of alkyl aryl ethers using niobium(V) pentachloride, were transformed to the corresponding *m*-alkoxytoluenes by palladium-catalyzed reduction of their triflates (Scheme 28).[24g]

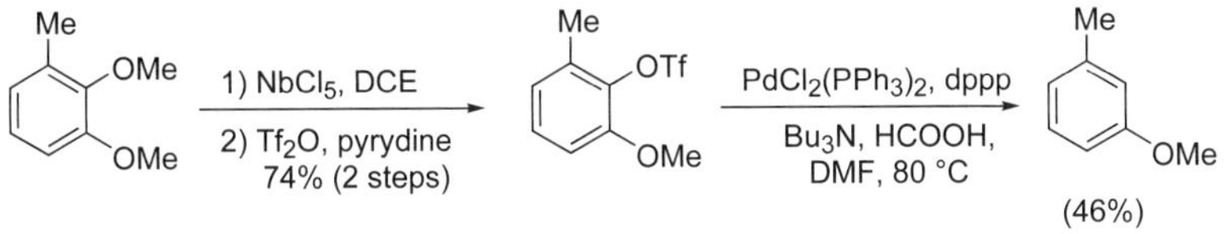

Scheme 28

In several SAR studies on 2-aminotetraline derivatives able to interact with 5-HT$_{1A}$ and/or dopamine D$_2$-D$_3$ receptors, with intrinsic activity profiles ranged from full agonists to antagonists and inverse-agonists, the palladium-catalyzed/formic acid reduction of triflate intermediates has been utilized to produce deoxy derivatives from phenolic precursors.[103] (+)-4-Demethoxydaunomycinone, the aglyconic portion of the potent antitumor agent 4-demethoxydaunorubicin, was obtained in five steps through the regioselective formation of the triflate derivative and its palladium-catalyzed reduction (Scheme 29).[104]

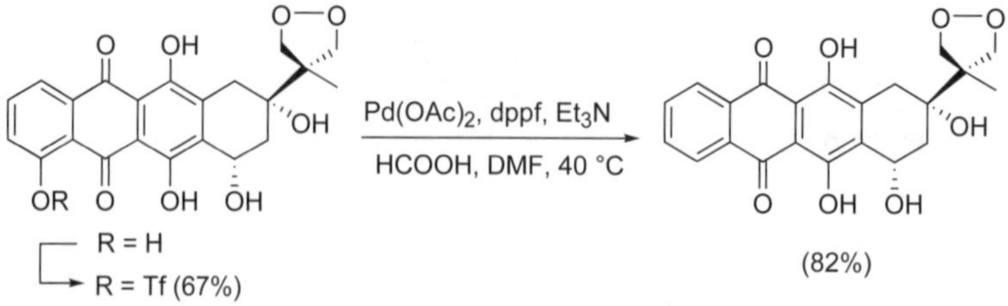

Scheme 29

Benzannulation of a 2,4-dioxygenated aryl Fisher carbene complex, followed by intramolecular Friedel-Crafts cyclization and triflate reduction of the oxygen substituent, provides a synthetic route to a model for the tricyclic core of antitumor antibiotic aureolic acids.[105] In a study of benzobicyclooctanes as novel inhibitors of TNF-a signaling, removal of the aryl hydroxy group was accomplished by the reduction of the corresponding aryl triflate using typical conditions.[106] Use of $Pd_2(dba)_3$ as catalyst, P(o-tolyl)$_3$ as phosphine ligand, and dioxane instead of THF as a solvent were found to be the best reaction conditions to obtain azulene in excellent yield from 2-azulenyl triflate (Scheme 30).[107]

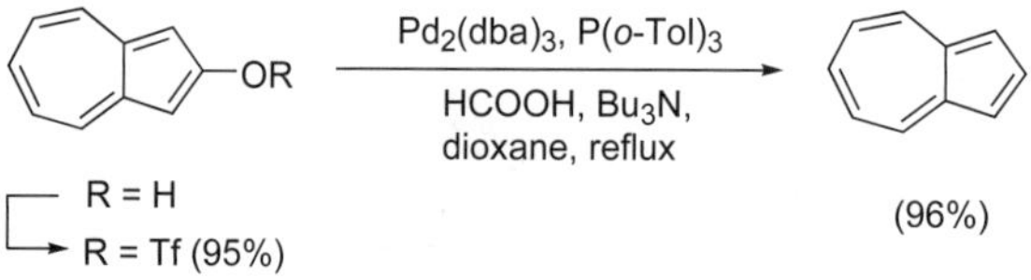

Scheme 30

Stereoselective synthesis of [5.2]metacyclophane was obtained by means of a [2+2] photocycloaddition and conversion of the phenol intermediate to the deoxygenated product.[108] The synthesis of a dimethylene-bridged clip, belonging to the family of molecules termed molecular tweezers due to their concave-convex topology, was reported.[109] In an improved synthesis of (±)-4,12 dihydroxy[2.2]paracyclophane followed by its enantiomeric resolution by enzymatic method, the (–)-hydroxytriflate derivative was reduced to the known (S)-(–)-4-hydroxy[2.2]paracyclophane.[110] A cyclotriveratrylene derivative (CVT), involved in the studies of the optical properties of CVTs and in the synthesis of anti-cryptophane-C, was obtained by the palladium-catalyzed reduction of the bis-triflate precursor, a method claimed to be more appropriate than others to give sizable amounts of deoxygenated CVTs (Scheme 31).[24f]

278

Further applications include a concise ring-expansion route to the compact core of platensimycin, a new natural product which was obtained by screening a large collection of South African soil samples using a novel antibiotic assay approach, [111] the synthesis and resolution of two chiral tetraphenylenes, prepared to address the tetraphenylene inversion barrier problem,[7b] and the synthesis of analogues of tramadol, a centrally acting opioid analgesic structurally related to codeine and morphine,[112] and of dihalogenated pentiptycenes.[13]

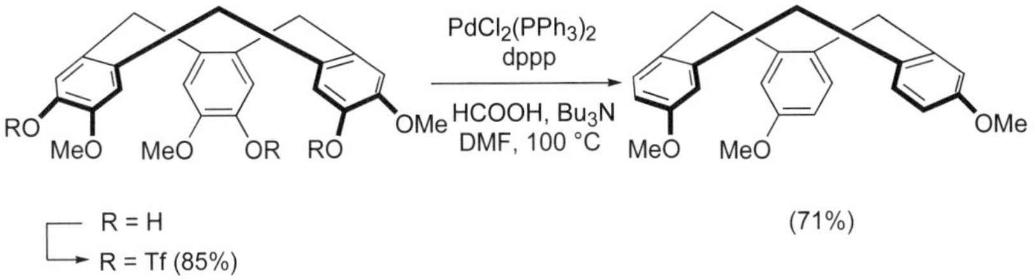

PdCl2(PPh3)2
dppp
HCOOH, Bu3N
DMF, 100 °C
RO
MeO
MeO
OR
RO
OMe
MeO
MeO
OMe
R = H
R = Tf (85%)
(71%)

Scheme 31

Peptides

Palladium-catalyzed reductive deoxygenation of phenols via their triflates was employed in the total synthesis of (*S,S*)-isodityrosine, a naturally occurring key structural subunit of numerous biologically active macromolecules.[113] The palladium-catalyzed formate reduction of aryl triflates was employed in the total synthesis of vancomycin and eremomycin aglycons.[21a-c] In the reduction step on the intermediate **36**, both the aryl triflate on ring 5 and the phenolic allyl protecting group on ring 4 were removed (Scheme 32).[21c]

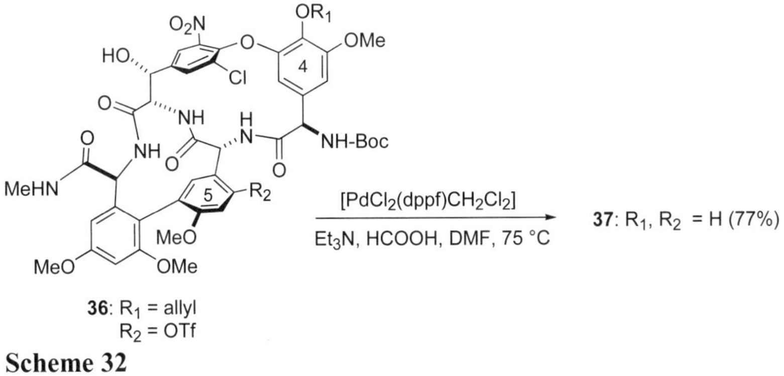

O2N
OR1
OMe
HO
Cl
4
NH
H
N
NH-Boc
MeHN
NH
5
R2
MeO
MeO
OMe
[PdCl2(dppf)CH2Cl2]
Et3N, HCOOH, DMF, 75 °C
37: R1, R2 = H (77%)
36: R1 = allyl
R2 = OTf

Scheme 32

In an investigation on the minimum structure requirements in the aromatic ring moieties for the cytotoxicities of RAs, cyclic hexapeptides originally isolated from the roots of *Rubia akane* and *R. cordifolia* showing significant antileukemic and antitumor activities, tyrosine residues in RA-VII and RA-V were deoxygenated, via the corresponding triflates, in yields ranging from 60 to 91%.[114] In a SAR study on the cyclophilin-binding immunosuppressant sanglifehrin A, no reaction occurred in an attempt to remove the phenolic hydroxy group of **38** by palladium-catalyzed reaction of the corresponding triflate using formic acid as hydrogen donor. The poor reactivity was attributed to the complexation of the palladium catalyst to 1,3-diene systems; indeed, after removal of the double bonds by hydrogenation using palladium on charcoal, the resulting octahydro sanglifehrin A was successfully deoxygenated to **39** by the above method (Scheme 33).[16b]

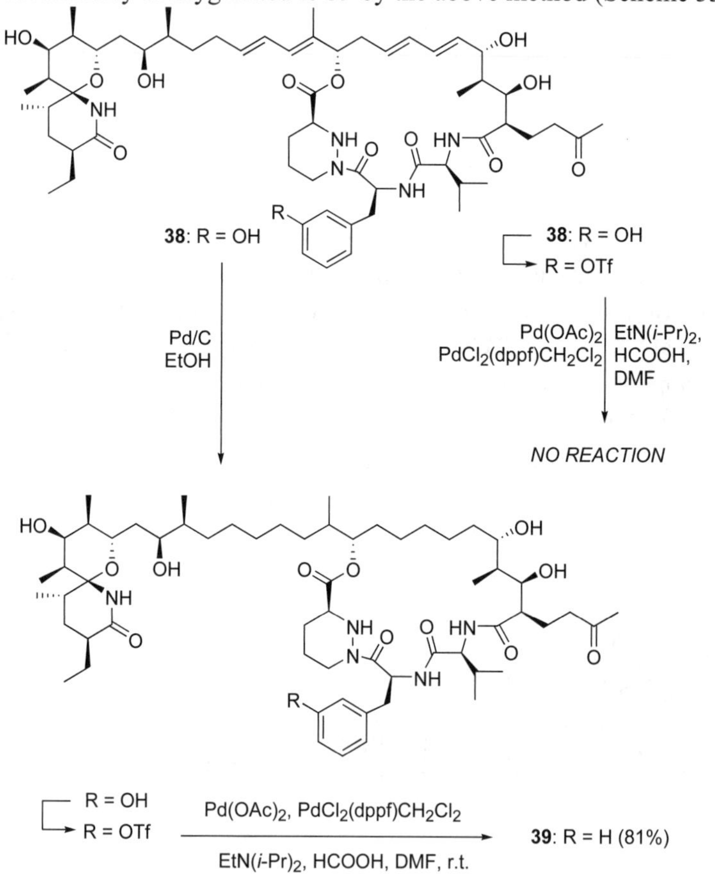

Scheme 33

1. Dipartimento di Chimica e Tecnologie del Farmaco, Sapienza Università di Roma, P.le A. Moro 5, 00185 Roma, Italy

2. Cacchi, S.; Morera, E.; Ortar, G. *Tetrahedron Lett.* **1984**, *25*, 4821.

3. Cacchi, S.; Ciattini, P. G.; Morera, E.; Ortar, G. *Tetrahedron Lett.* **1986**, *27*, 5541.

4. Arcadi, A.; Attanasi, O. A.; Guidi, B.; Rossi, E.; Santeusanio, S. *Eur. J. Org. Chem.* **1999**, 3117.

5. (a) Matsushita, T.; Ashida, H.; Kimachi, T.; Takemoto, Y. *Chem. Commun.* **2002**, 814. (b) Mejia-Oneto, J. M.; Padwa, A. *Helv. Chim. Acta* **2008**, *91*, 285. (c) Manning, J. R.; Davies, H. M. L. *Tetrahedron* **2008**, *64*, 6901. (d) Marchart, S.; Gromov, A.; Mulzer, J. *Angew. Chem. Int. Ed.* **2010**, *49*, 2050. (e) Hours, A. E.; Snyder, J. K. *Organometallics* **2008**, *27*, 410.

6. (a) Myers, A. G.; Fraley, M.; E.; Tom, N. J. *J. Am. Chem. Soc.* **1994**, *116*, 11556. (b) Kitajima, M.; Masumoto, S.; Takayama, H.; Aimi, N. *Tetrahedron Lett.* **1997**, *24*, 4255. (c) Myers, A. G.; Tom, N. J.; Fraley, M. E.; Cohen, S. B.; Madar D. J. *J. Am. Chem. Soc.* **1997**, *119*, 6072. (d) Kitajima, M.; Yoshida, S.; Yamagata, K.; Nakamura, M.; Takayama, H.; Saito, K.; Seki, H.; Aimi, N. *Tetrahedron* **2002**, *58*, 9169. (e) Thomas, O. P.; Zaparucha, A.; Husson H.-P. *Eur. J. Org. Chem.* **2002**, 157. (f) Ito, S.; Yokoyama, R.; Okujima, T.; Terazono,T.; Kubo, T.; Tajiri, A.; Watanabe, M.; Morita, N. *Org. Biomol. Chem.* **2003**, *1*, 1947. (g) Sissouma, D.; Maingot, L.; Collet, S.; Guingant, A. *J. Org. Chem.* **2006**, *71*, 8384. (h) Maingot, l.; Thuaud, F.; Sissouma, D.; Collet, S.; Guingant, A.; Evain, M. *Synlett* **2008**, 263.

7. (a) Kurz, L.; Lee, G.; Morgans, D. Jr.; Waldyke, M. J.; Ward, T. *Tetrahedron Lett.* **1990**, *31*, 6321. (b) Huang, H.; Stewart, T.; Gutmann, M.; Ohhara, T.; Niimura, N.; Li, Y.-X.; Wen, J.-F.; Bau, R.; Wong, H. N. C. *J. Org. Chem.* **2009**, *74*, 359.

8. (a) Yamada, K.; Namerikawa, Y.; Haruyama, T.; Miwa, Y.; Yanada, R.; Ishikura, M. *Eur. J. Org. Chem.* **2009**, 5752. (b) Si, Y.-G.; Choi, Y.-K.; Gardner, M. P.; Tarazi, F. I.; Baldessarini, R. J.; Neumeyer, J. L. *Bioorg. Med. Chem. Lett.* **2009**, *19*, 51. (c) Bringmann, G.; Brun, R.; Kaiser, M.; Neumann, S. *Eur. J. Med. Chem.* **2008**, *43*, 32.

9. Ohta, T.; Fukuda, T.; Ishibashi, F.; Iwao, M. *J. Org. Chem.* **2009**, *74*, 8143.

10. Mahdavian, E.; Sangsura, S.; Landry, G.; Eytina, J.; Salvatore, B. A. *Tetrahedron Lett.* **2009**, 50, 19.

11. (a) Lewis, C. A.; Gustafson, J. L.; Chiu, A.; Balsells, J.; Pollard, D.; Murry, J.; Reamer, R. A.; Hansen, K. B.; Miller, S. J. *J. Am. Chem. Soc.* **2008**, *130*, 16358. (b) Tohyama, S.; Choshi, T.; Azuma, S.; Fujioka, H.; Hibino, S. *Heterocycles* **2009**, *79*, 955. (c) Dodo, K.; Aoyama, A.; Noguchi-Yachide, T.; Makishima, M.; Miyachi, H.; Hashimoto, Y. *Bioorg. Med. Chem.* **2008**, 16, 4272.

12. (a) Clive, D. L. J.; Stoffman, E. J. L. *Org. Biomol. Chem.* **2008**, *6*, 1831. (b) Tang, W.; Qu, B.; Capacci, A. G.; Rodriguez, S.; Wei, X.; Haddad, N.; Narayanan, B.; Ma, S.; Grinberg, N.; Yee, N. K.; Krishnamurthy, D.; Senanayake, C. H. *Org. Lett.* **2010**, *12*, 176.

13. Yang, J.-S.; Yan, J.-L.; Jin, Y.-X.; Sun, W.-T.; Yang, M.-C. *Org. Lett.* **2009**, *11*, 1429.

14. (a) Kazmierski, W. M.; Aquino, C.; Chauder, B. A.; Deanda, F.; Ferris, R.; Jones-Hertzog, D. K.; Kenakin, T.; Koble, C. S.; Watson, C.; Wheelan, P.; Yang, H.; Youngman, M. *J. Med. Chem.* **2008**, *51*, 6538. (b) Pu, J.; Deng, K.; Butera, J.; Chlenov, M.; Gilbert, A.; Kagan, M.; Mattes, J.; Resnick, L. *Tetrahedron* **2010**, *66*, 1963.

15. Arnott, G.; Brice, H.; Clayden, J.; Blaney, E. *Org. Lett.* **2008**, *10*, 3089.

16. (a) Mori, M.; Nakanishi, M.; Kajishima, D.; Sato Y. *Org. Lett.* **2001**, *3*, 1913. (b) Bänteli, R.; Wagner, J.; Zenke, G. *Bioorg. Med. Chem. Lett.* **2001**, *11*, 1609. (c) Gonzáles-Gómez, J. C.; Uriarte, E. *Synlett* **2003**, 2225. (c) Blair, M.; Tuck, K. L. *Tetrahedron: Asymmetry* **2009**, *20*, 2149.

17. Yang, M.-S.; Chang, S.-Y.; Lu, S.-S.; Rao, P. D.; Liao, C.-C. *Synlett* **1999**, 225. (b) Padwa, A.; Heidelbaugh, T. M.; Kuethe, J. T. *J. Org. Chem.* **2000**, *65*, 2368. (c) Yen, C.-F.; Liao, C.-C. *Angew. Chem. Int. Ed.* **2002**, *41*, 4090. (d) Ballester, P.; Capó, M.; Costa, A.; Deya, P.; Gomila, M. R.; Decken, A.; Deslongchamps, G. *J. Org. Chem.* **2002**, *67*, 8832. (e) Lin, Y.-S.; Chang, S.-Y.; Yang, M.-S.; Rao, C. P.; Peddinti, R. K.; Tsai, Y.-F.; Liao, C.-C. *J. Org. Chem.* **2004**, *69*, 447. (f) Peese, K. M.; Gin D. Y. *Org. Lett.* **2005**, *7*, 3323. (g) Morita, H.; Tomizawa, Y.; Deguchi, J.; Ishikawa, T.; Arai, H.; Zaima, K.; Hosoya, T.; Hirasawa, Y.; Matsumoto, T.; Kamata, K.; Ekasari, W.; Widyawaruyanti, A.; Wahyuni, T. S.; Zaini, N. C.; Honda, T. *Bioorg. Med. Chem.* **2009**, *17*, 8234.

18. (a) Nicolaou, K. C.; Ortiz, A.; Zhang, H.; Dagneau, P.; Lanver, A.; Jennings, M. P.; Arseniyadis, S.; Faraoni, R.; Lizos, D. E. *J. Am. Chem. Soc.* **2010**, *132*, 7138. (b) Su, Y.; Xu, Y.; Han, J.; Zheng, J.; Qi, J.; Jiang, T.; Pan, X.; She, X. *J. Org. Chem.* **2009**, *74*, 2743. (c) Pardeshi, S. G.; Ward, D. E. *J. Org. Chem.* **2008**, *73*, 1071.

19. Liang, G.; Xu, Y.; Seiple, I. B.; Trauner, D. *J. Am. Chem. Soc.* **2006**, *128*, 11022.

20. (a) Zhang, A.; van Vliet, S.; Neumeyer, L. *Tetrahedron Lett.* **2003**, *44*, 6459. (b) Mai, C.-K.; Sammons, M. F.; Sammakia, T. *Angew. Chem. Int. Ed.* **2010**, *49*, 2397

21. (a) Evans, D. A.; Dinsmore, C. J.; Evrard, D. A.; DeVries, K. M. *J. Am. Chem. Soc.* **1993**, *115*, 6426. (b) Evans, D. A.; Dinsmore, C. J.; Ratz, A. M.; Evrard, D. A.; Barrow, J. C. *J. Am. Chem. Soc.* **1997**, *119*, 3417. (c) Evans, D. A.; Wood, M. R.; Welsey Trotter, W. B.; Richardson, T. I.; Barrow, J. C.; Katz, J. L. *Angew. Chem. Int. Ed.* **1998**, *37*, 2700. (e) Fang, Z.; Agoston, G. E.; Ladouceur, G.; Treston, A. M.; Wang, L.; Cushman, M. *Tetrahedron* **2009**, *65*, 10535. (f) Chinigo, G. M.; Paige, M.; Grindrod, S.; Hamel, E.; Dakshanamurthy, S.; Chruszcz, M.; Minor, W.; Brown, M. L. *J. Med. Chem.* **2008**, *51*, 4620. (g) Brotin, T.; Cavagnat, D.; Buffeteau, T. *J. Phys. Chem. A* **2008**, *112*, 8464.

22. (a) Yoshida, Y.; Mohri, K.; Isobe, K.; Itoh, T.; Yamamoto, K. *J. Org. Chem.* **2009**, *74*, 6010. (b) Wang, C.; Chen, X.-H.; Zhou, S.-M.; Gong, L.-Z. *Chem. Commun.* **2010**, *46*, 1275.

23. Hupp, C. D.; Neumeyer, J. L. *Tetrahedron Lett.* **2010**, *51*, 2359.

24. (a) Saá, J. M.; Dopico, M.; Martorell, G.; Garcia-Raso, A. *J. Org. Chem.* **1990**, *55*, 991. (b) Hedberg, M. H.; Johansson, A. M.; Nordvall, G.; Yliniemelä, A.; Li, H. B.; Martin, A. R.; Hjorth, S.; Unelius, L.; Sundell, S.; Hacksell, U. *J. Med. Chem.* **1995**, *38*, 647. (c) Kim, J. C.; Bae, S.-D.; Kim, J.-A.; Choi, S.-K. *Org. Prep. Proced. Intern.* **1998**, *30*, 352. (d) Kim, J. C.; Bae, S.-D.; Kim, J.-A. *J. Heterocyclic Chem.* **1998**, *35*, 531. (e) Neel, D. A.; Brown, M. L.; Lander, P. A.; Grese, T. A.; Defauw, J. M.; Doti, R. A.; Fields, T.; Kelley, S. A.; Smith, S.; Zimmerman, K. M.; Steinberg, M. I.; Jadhav, P. K. *Bioorg. Med. Chem. Lett.* **2005**, *15*, 2553. (f) Brotin, T.; Roy, V.; Dutasta, J.-P. *J. Org. Chem.* **2005**, *70*, 6187. (g) Sudo, Y.; Arai, S.; Nishida, A. *Eur. J. Org. Chem.* **2006**, 752. (h) Cianci, J.; Baell, J. B.; Flynn, B. L.; Gable, R. W.; Mould, J. A.; Paul, D.; Harvey, A. J.

Bioorg. Med. Chem. Lett. **2008**, *18*, 2055. (i) Huber, G.; Beguin, L.; Desvaux, H.; Brotin, T.; Fogarty, H. A.; Dutasta, J.-P.; Berthault, P. *J. Phys. Chem. A* **2008**, *112*, 11363. (j) Cavagnat, D.; Buffeteau, T.; Brotin, T. *J. Org. Chem.* **2008**, *73*, 66.

25. Si, Y.-G.; Gardner, M. P.; Tarazi, F. I.; Baldessarini, R. J.; Neumeyer, J. L. *Bioorg. Med. Chem. Lett.* **2008**, *18*, 3971.

26. Bourdreux, Y.; Nowaczyk, S.; Billaud, C.; Mallinger, A.; Willis, C.; Desage-El Murr, M.; Toupet, L.; Lion, C.; Le Gall, T.; Mioskowski, C. *J. Org. Chem.* **2008**, *73*, 22.

27. Park C. P.; Nagle A.; Yoon C. H.; Chen C.; Jung K. W. *J. Org. Chem.* **2009**, *74*, 6231.

28. Pan, Y.; Holmes, C. P. *Org. Lett.* **2001**, *3*, 2769.

29. Cammidge, A. N.; Ngaini, Z. *Chem. Commun.* **2004**, 1914.

30. Winkler, J. D.; Londregan, A. T.; Hamann, M. T. *Org. Lett.* **2006**, *8*, 2591.

31. Choi, Y.; White, J. D. *J. Org. Chem.* **2004**, *69*, 3758.

32. Katoh, T.; Kakiya, K.; Nakai, T.; Nakamura, S.; Nishide, K.; Node, M. *Tetrahedron Asymmetry* **2002**, *13*, 2351.

33. Newcombe, N. J.; Simpkins, N. S. *J. Chem. Soc., Chem. Commun.* **1995**, 831.

34. Keverline, K.; Abraham, P.; Lewin, A. H.; Carroll, F. I. *Tetrahedron Lett.* **1995**, *36*, 3099.

35. Overman, L. E.; Tomasi, A. L. *J. Am. Chem. Soc.* **1998**, *120*, 4039.

36. (a) Kuethe, J. T.; Padwa, A. *Tetrahedron Lett.* **1997**, *38*, 1505. (b) Padwa, A.; Sheehan, S. M.; Straub, C. S. *J. Org. Chem.* **1999**, *64*, 8648.

37. Liu, P.; Hong, S.; Weinreb, S. M. *J. Am. Chem. Soc.* **2008**, *130*, 7562-7563.

38. Pedram, B.; van Oeveren, A.; Mais, D. E.; Marschke, K. B.; Verbost, P. M.; Groen, M. B.; Zhi, L. *J. Med. Chem.* **2008**, *51*, 3696–3699.

39. (a) Rousseau, G.; Robert, F.; Schenk, K.; Landais, Y. *Org. Lett.* **2008**, *10*, 4441-4444. (b) Rousseau, G.; Robert, F.; Schenk, K.; Landais, Y. *Chem. Eur. J.* **2009**, *15*, 11160-11173.

40. Wolfe, B. H.; Libby, A. H.; Al-awar, R. S.; Foti, C. J.; Comins, D. L. *J. Org. Chem.* **2010**, *75*, 8564-8570.

41. (a) Comins, D. L.; Al-awar, R. S. *J. Org. Chem.* **1995**, *60*, 711. (b) Comins, D. L.; Libby, A. H.; Al-awar, R. S.; Foti, C. J. *J. Org. Chem.* **1999**, *64*, 2184.

42. (a) Nakamura, T.; Shiozaki, M. *Tetrahedron Lett.* **1999**, *40*, 9063. (b) Nakamura, T.; Shiozaki, M. *Tetrahedron* **2001**, *57*, 9087.

43. van den Berg, R. J. B. H. N.; Korevaar, C. G. N.; van der Marel, G. A.; Overkleeft, H. S.; van Boom, J. H. *Tetrahedron Lett.* **2002**, *43*, 8409.

44. Hitotsuyanagi, Y.; Kim, I. H.; Hasuda, T.; Yamauchi, Y.; Takeya, K. *Tetrahedron* **2006**, *62*, 4262.

45. Mori, K.; Matsushima, Y. *Synthesis* **1995**, 845.

46. Abad, A.; Agulló, C.; Arnó, M.; Cantín, A.; Cuñat, A. C.; Meseguer, B.; Zaragozá, R. J. *J. Chem. Soc., Perkin Trans. 1* **1997**, 1837.

47. Torneiro, M.; Fall, Y.; Catedo, L.; Mouriño, A. *Tetrahedron* **1997**, *53*, 10851.

48. (a) Scott, W. J.; Crisp, G. T.; Stille, J. K. *J. Am. Chem. Soc.* **1984**, *106*, 4630. (b) Scott, W. J.; Stille, J. K. *J. Am. Chem. Soc.* **1986**, *108*, 3033.

49. (a) Inoue, M.; Sato, T.; Hirama, M. *J. Am. Chem. Soc.* **2003**, *125*, 10772. (b) Inoue, M.; Sato, T.; Hirama, M. *Angew. Chem. Int. Ed.* **2006**, *45*, 4843.

50. Kuwahara, S.; Hamade, S.; Leal, W. S.; Ishikawa, J.; Kodama, O. *Tetrahedron* **2000**, *56*, 8111.

51. Holt, D. A.; Levy, M. A.; Oh, H.-J.; Erb, J. M.; Heaslip, J. I.; Brandt, M.; Lan-Hargest, H.-Y.; Metcalf, B. W. *J. Med. Chem.* **1990**, *33*, 943.

52. Ciattini, P. G.; Morera, E.; Ortar, G. *Synth. Commun.* **1990**, *20*, 1293.

53. (a) Dolle, R. E.; Schmidt, S. J.; Erhard, K. F.; Kruse, L. I. *J. Chem. Soc., Chem. Commun.* 1988, 19. (b) Dolle, R. E.; Schmidt, S. J.; Erhard, K. F.; Kruse, L. I. *Tetrahedron Lett.* **1988**, *29*, 1581. (c) Dolle, R. E.; Schmidt, S. J.; Erhard, K. F.; Kruse, L. I. *J. Am. Chem. Soc.* **1989**, *111*, 278.

54. Kotoku, N.; Sumii, Y.; Hayashi, T.; Kobayashi, M. *Tetrahedron Lett.* **2008**, 49, 7078-7081.

55. Zou, Y. and Millar, J. G. *J. Org. Chem.* **2009**, *74*, 7207.

56. Toyoda, A.; Nagai, H.; Yamada, T.; Moriguchi, Y.; Abe, J.; Tsuchida, T.; Nagasawa, K. *Tetrahedron* **2009**, *65*, 10002.

57. Trost, B. M.; Schroeder, G. M. *J. Am. Chem. Soc.* **2000**, *122*, 3785.

58. Paquette, L. A.; Meister, P. C.: Friedrich, D.; Sauer, D. R. *J. Am. Chem. Soc.* **1993**, *115*, 49.

59. Yamada, S.; Karasawa, S.; Takahashi, Y.; Aso, M.; Suemune, H. *Tetrahedron* **1998**, *54*, 15555.

60.   Trost, B. M.; Pissot-Soldermann, C.; Chen, I. *Chem. Eur. J.* **2005**, *11*, 951.

61.   Senanayake, C. H.; Bill, T. J.; DiMichele, L. M.; Chen, C. Y.; Larsen, R. D.; Verhoeven, T. R.; Reider, P. J. *Tetrahedron Lett.* **1993**, *34*, 6021.

62.   Matsuo, J-i.; Takeuchi, K.; Ishibashi, H. *Org. Lett.* **2008**, *10*, 4049.

63.   Rousseau, G.; Robert, F.; Landais, Y. *Synthesis* **2010**, 1223.

64.   Shimada,N.; Hanari, T.; Kurosaki, Y.; Takeda, K.; Anada, M.; Nambu, H.; Shiro, M.; Hashimoto, S. *J. Org. Chem.* **2010**, *75*, 6039.

65.   Yamakoshi, H.; Shibuya, M.; Tomizawa, M.; Osada, Y.; Kanoh, N.; Iwabuchi, Y. *Org. Lett.* **2010**, *12*, 980.

66.   Shimada, N.; Hanari, T.; Kurosaki, Y.; Anada, M.; Nambu, H.; Hashimoto, S. *Tetrahedron Lett.* **2010**, *51*, 6572.

67.   Zhang, A.; Csutoras, C.; Zong, R.; Neumeyer, J. *Org. Lett.* **2005**, *7*, 3239.

68.   Cuny, G. D. *Tetrahedron Lett.* **2004**, *45*, 5167.

69.   Csutoras, C.; Zhang, A.; Zhang, K.; Kula, N. S.; Baldessarini, R. J.; Neumeyer, J. L. *Bioorg. Med. Chem.* **2004**, *12*, 3553.

70.   Hedberg, M. H.; Johansson, A. M.; Hacksell, U. *J. Chem. Soc. Chem. Commun.* **1992**, 845.

71.   Hedberg, M. H., Linnanen, T.; Jansen, J. M.; Nordvall, G.; Hjorth, S.; Unelius, L.; Johansson, A. M. *J. Med. Chem.* **1996**, *39*, 3503.

72.   Node, M.; Kodama, S.; Hamashima, Y.; Baba, T.; Hamamichi, N.; Nishide, K. *Angew. Chem. Int. Ed.* **2001**, *40*, 3060.

73.   Kingsbury, W. D.; Boehm, J. C.; Jakas, D. R.; Holden, K. G.; Hecht, S. M.; Gallager, G.; Caranfa, M. J.; McCabe, F. L.; Faucette, L. F.; Johnos, R. K.; Hertzberg, R. P. *J. Med. Chem.* **1991**, *34*, 98.

74.   Clasby, M. C.; Chackalamannil, S.; Czarniecki, M.; Doller, D.; Eagen, K.; Greenlee, W. J.; Lin, Y.; Tagat, J. R.; Tsai, H.; Xia, Y.; Ahn, H.-S.; Agans-Fantuzzi, J.; Boykow, G.; Chintala, M.; Hsieh, Y.; McPhail, A. T. *Bioorg. Med. Chem. Lett.* **2007**, *17*, 3647.

75.   Saito, N.; Obara, Y.; Aihara, T.; Harada, S.; Shida, S.; Kubo, A. *Tetrahedron* **1994**, *50*, 3915.

76.   Kiahara, Y.; Nagatsu, M.; Shibano, Y.; Kubo, A. *Chem. Pharm. Bull.* **1997**, *45*, 1697.

77.   Nagamura, S.; Kobayashi, E.; Gomi, K.; Saito, H. *Bioorg. Med. Chem.* **1996**, *4*, 1379.

78. (a) Hagiwara, H.; Choshi, T.; Fujimoto, H.; Sugino, E.; Hibino, S. *Chem. Pharm. Bull.* **1998**, *46*, 1948. (b) Engler, T. A.; Wanner, J. *J. Org. Chem.* **2000**, *65*, 2444. (c) Hagiwara, H.; Choshi, T.; Nobuhiro, J.; Fujimoto, H.; Hibino, S. *Chem. Pharm. Bull.* **2001**, *49*, 881. (d) Duval, E.; Cuny, G. D. *Tetrahedron Lett.* **2004**, *45*, 5411.

79. Nakahara, S.; Tanaka, Y.; Kubo, A. *Heterocycles* **1993**, *5*, 1139.

80. (a) Takayama, H.; Kitajima, M.; Sakai, S. *Heterocycles* **1990**, *30*, 325. (b) Takayama, H.; Sakai, S. *Chem. Pharm. Bull.* **1989**, *37*, 2256.

81. Sissouma, D.; Collet, S. C.; Guingant, A. Y. *Synlett* **2004**, 2612.

82. Behenna, D. C.; Stockdill, J. L.; Stoltz, B. M. *Angew. Chem. Int. Ed.* **2007**, *44*, 4077.

83. Mastarlez, H.; Doyle, T. W.; Kadow, J. F.; Vyas, D. M. *Tetrahedron Lett.* **1996**, *37*, 8687.

84. England, D. B.; Kerr, M. A. *J. Org. Chem.* **2005**, *70*, 6519.

85. Yokoyama, Y.; Sagisaka, T.; Mizuno, Y.; Yokoyama, Y. *Chem. Lett.* **1996**, 587.

86. Hitotsuyanagi, Y.; Kojima, H.; Ikuta, H.; Itokawa, H. *Bioorg. Med. Chem. Lett.* **1996**, *6*, 1791.

87. Wulff, W. D.; McCallum, J. S.; Kunng, F.-A. *J. Am. Chem. Soc.* **1988**, *110*, 7419.

88. Peterson, G. A.; Kunng, F.-A.; McCallum, J. S.; Wulff, W. D. *Tetrahedron Lett.* **1987**, *28*, 1381.

89. Tse, B.; Jones, A. B. *Tetrahedron Lett.* **2001**, *42*, 6429.

90. (a) Engler, T. A.; Combrink, K. D.; Reddy, J. P. *J. Chem. Soc. Chem. Commun.* **1989**, 454. (b) Engler, T. A.; Reddy, J. P.; Combrink, K. D.; Vander Velde, D. *J. Org. Chem.* **1990**, *55*, 1248.

91. Jackson, S. K.; Kerr, M. A. *J. Org. Chem.* **2007**, *72*, 1405.

92. Nadin, A.; Harrison, T. *Tetrahedron Lett.* **1999**, *40*, 4073.

93. Nesvadba, P.; Wendeborn, F.; Schaefer, T.; Schmidhalter, B.; Ricci, A.; Murer, P.; Chebotareva, N. PCT Int. Appl. WO2010046259, 2010.

94. Schaefer, T; Murer, P.; Baudin, G; Kocher, M; Maike, F; Allenbach S; Sift, R; Schmidhalter, B. PCT Int. Appl. WO2008101842, 2008.

95. Sonesson, C.; Lin, C.-H.; Hansson, L.; Waters, N.; Svensson, K.; Carlsson, A.; Smith, M. W.; Wilkström, H. *J. Med. Chem.* **1994**, *37*, 2735.

96. Wang, G.; He, Y.; Sun, J.; Das, D.; Hua, M.; Huang, J.; Ruhrmund. D.; Hooi, L.; Misialek, S.; Rajagopalan, P. T. R.; Stoycheva, A.;

Buckman, B. O.; Kossen, K.; Seiwert, S. D.; Beigelman, L. *Bioorg. Med. Chem. Lett.* **2009**, *19*, 4476.

97. (a) Fu, J.-m.; Sharp, M. J.; Snieckus, V. *Tetrahedron Lett.* **1988**, *29*, 5459. (b) Fu, J.-m.; Snieckus, V. *Can. J. Chem.* **2000**, *78*, 905. (c) Cai, X.; Brown, S.; Hodson, P.; Snieckus, V. *Can. J. Chem.* **2004**, *82*, 195.

98. Hattori, T.; Shimazumi, Y.; Goto, H.; Yamabe, O.; Morohashi, N.; Kawai, W.; Miyano, S. *J. Org. Chem.* **2003**, *68*, 2099.

99. Parker, K. A.; Ding, Q.-j. *Tetrahedron* **2000**, *56*, 10249.

100. Yu, K.-L.; Ostrowski, J.; Chen, S.; Tramposh, K. M.; Reczek, P. R.; Mansuri, M. M.; Starrett Jr., J. E. *Bioorg. Med. Chem. Lett.* **1996**, *6*, 2865.

101. (a) Müller, W.; Bänziger, M.; Kipfer, P. *Helv. Chim. Acta* **1998**, *81*, 729. (b) Cammidge, A. N.; Crpy, K. V. L. *J. Org. Chem.* **2003**, *68*, 6832.

102. Robl, J. A. *Tetrahedron Lett.* **1990**, *24*, 3421.

103. (a) Liu, Y.; Svensson, B. E.; Yu, H.; Cortizo, L.; Ross, S. B.; Lewander, T.; Hacksell, U. *Bioorg. Med. Chem. Lett.* **1991**, *1*, 257. (b) Liu, Y.; Yu, H.; Mohell, N.; Nordvall, G.; Lewander, T.; Hacksell, U. *J. Med. Chem.* **1995**, *38*, 150. (c) Höök, B. B.; Cortizo, L.; Johansson, A. M.; Westlind-Danielsson, A.; Mohell, N.; Hacksell, U. *J. Med. Chem.* **1996**, *39*, 4036. (d) Malmberg, Å.; Höök, B. B.; Johansson, A. M.; Hacksell, U. *J. Med. Chem.* **1996**, *39*, 4421.

104. Cabri, W.; De Bernardinis, S.; Francalanci, F.; Penco, S. *J. Chem. Soc. Perkin Trans. 1* **1990**, 428.

105. Miller, R. A.; Gilbert, A. M.; Xue, S.; Wulff, W. D. *Synthesis* **1999**, 80.

106. Jackson, R. W.; Gelinas, R.; Baughman, T. A.; Cox, T.; Howbert, J. J.; Kucera, K. A.; Latham, J. A.; Ramsdell, F.; Singh, D.; Darwish, I. S. *Bioorg. Med. Chem. Lett.* **2002**, *12*, 1093.

107. Ito, S.; Yokoyama, R.; Okujima, T.; Terazono, T.; Kubo, T.; Tajiri, A.; Watanabe, M.; Morita, N. *Org. Biomol. Chem.* **2003**, *1*, 1947.

108. Okada, Y.; Ishii, F.; Kasai, Y.; Nishimura, J. *Tetrahedron* **1994**, *50*, 12159.

109. Klärner, F.-G.; Panitzky, J.; Bläser, D.; Boese, R. *Tetrahedron* **2001**, *57*, 3673.

110. Braddock, D. C.; MacGilp, I. D.; Perry, B. G. *J. Org. Chem.* **2002**, *67*, 8679.

111. McGrath, N. A.; Bartlett, E. S.; Sittihan, S.; Njardarson, J. T. *Angew. Chem. Int. Ed.* **2009**, *48*, 8543.

112. Shao, L.; Hewitt, M.; Jerussi, T. P.; Wu, F.; Malcolm, S.; Grover, P.; Fang, K.; Koch, P.; Senanayake, C.; Bhongle, N.; Ribe, S.; Bakalec, R.; Curried, M. *Bioorg. Med. Chem. Lett.* **2008**, *18*, 1674.

113. Gupta, A.; Sen, S.; Harmata, M.; Pulley, S. R. *J. Org. Chem.* **2005**, *70*, 7422.

114. Itokawa, H.; Kondo, K.; Hitotsuyanagi, Y.; Nakamura, A.; Morita, H.; Takeya, K. *Chem. Pharm. Bull.* **1993**, *41*, 1266.

Sandro Cacchi graduated in Chemistry at the University of Camerino in 1967. He then moved to the University of Bologna where he was Assistant Professor. In 1972 he joined the University "La Sapienza", Rome, where he became Associate Professor in 1983 and, subsequently, full Professor of Organic Chemistry (1986-2010). Dr. Cacchi made a contribution early in his career with the development of new, selective synthetic procedures, with particular emphasis on the chemistry of molecules of biological interest. In the late seventies, he moved to the exciting new field of palladium chemistry and the utilization of palladium catalysis in organic synthesis became a major goal that he pursued in many ways, with the search for new and selective methodologies being a major thrust even in this area.

Enrico Morera received his degree in Pharmacy in 1974 from University of Rome (Italy). After a fellowship from the Ministry of Education, he was appointed Researcher and then Associate Professor in Medicinal Chemistry at the Faculty of Pharmacy of the University of Rome 'Sapienza'. The research activity was at first devoted to the study of new synthetic methods and their application to the synthesis of steroid, heterocycle and aminoacid derivatives. More recently his research has centered around the field of non-proteinogenic α-aminoacid, and on biological systems such as endocannabinoid system, transient receptor potential channels (TRPs) and on connection between them.

Giorgio Ortar was born in Gorizia in 1943. He graduated in Chemistry at the University of Rome 'Sapienza' in 1967. In 1972 he was appointed researcher of Italian National Council of Researches (CNR) at the Centro di Chimica del Farmaco of Rome and held this position until 1987. From 1987 he is Associate Professor of Medicinal Chemistry at the Faculty of Pharmacy of the University of Rome 'Sapienza' His scientific activity has been concerned initially with studies on oxidation, solvolysis, and rearrangement reactions of steroidal substrates. Then, he devoted himself for many years to studies on palladium-catalyzed reactions with a number of applications to molecules of medicinal interest. In the last seven years his researches have been mainly focused on endocannabinoid system and on transient receptor potential (TRP) superfamily of ion channels.

Discussion Addendum for:
SYNTHESIS OF INDOLES BY PALLADIUM CATALYZED REDUCTIVE *N*-HETEROANNULATION OF 2-NITROSTYRENES: METHYL INDOLE-4-CARBOXYLATE

CO$_2$Me, CH$_3$, NO$_2$ — hv, Br$_2$ cat. (PhCO$_2$)$_2$ CCl$_4$, reflux → CO$_2$Me, Br, NO$_2$ — Ph$_3$P CHCl$_3$ reflux → CO$_2$Me, PPh$_3$Br, NO$_2$

HCHO, Et$_3$N CH$_2$Cl$_2$, rt → CO$_2$Me, NO$_2$ — CO cat. Pd(OAc)$_2$, cat. Ph$_3$P CH$_3$CN, 90 °C → CO$_2$Me, indole N–H

Prepared by Björn C. Söderberg.*[1]
Original article: Söderberg, B.; Shriver, J.; Wallace, J. *Org. Synth.* **2003**, *80*, 75.

 Transition metal catalyzed reductive *N*-heterocyclization of 1-(2-nitroaryl)-1-alkenes, using carbon monoxide as the ultimate reducing agent, is emerging as a powerful methodology for the synthesis of a wide variety of functionalized indoles.[2,3,4,5,6,7,8] Palladium complexes have mainly been used as the catalyst of choice but other transition metals including ruthenium (Ru$_3$(CO)$_{12}$,[8] RuCl$_2$(PPh$_3$)$_2$[6]), rhodium (Rh$_6$(CO)$_{16}$,[8] RhCl(PPh$_3$)$_3$[6]), iron (Fe(CO)$_5$),[8] nickel (NiCl$_2$(PPh$_3$)$_2$),[6] and platinum (PtCl$_2$(PPh$_3$)$_2$)[6] also catalyze this transformation. A molybdenum (MoO$_2$Cl$_2$(dmf)$_2$) catalyzed reaction in the absence of carbon monoxide has also been described.[9] In addition to transition metals, a catalytic amount of elemental selenium in the presence of carbon monoxide can be used.[10] A direct comparison between all of the different catalysts cannot be made, however in general the palladium diacetate – triphenylphosphine catalyst system usually afford superior yield of product at lower temperature and pressure.
 For most substrates, the exclusion of oxygen and water is not required and reagent grade chemicals and solvent can be used with excellent results. The palladium-catalyzed cyclizations are usually free from byproducts derived from the starting material. If observed, byproducts include *N*-

Org. Synth. **2011**, *88*, 291-295
Published on the Web 3/9/2011

hydroxyindoles, indole dimerization products, and reduction of the nitro-group to an amine. The former impurity can be eliminated or minimized by extending the reaction time or increasing the CO pressure. A potential purification problem is triphenylphosphine and the small amounts of triphenylphosphine oxide formed when using Pd(OAc)$_2$-PPh$_3$. This can be particularly problematic on a larger reaction scale. Replacing triphenylphosphine with 1,10-phenanthroline or a related bidentate ligand is a convenient solution to this problem although, these ligands are significantly more expensive.

Davies and Smitrovich *et al.* have more recently found, after extensive optimization using a Parallel Pressure Reactor (PPR$^®$), conditions wherein indoles are formed at a low catalyst loading (1 mol% Pd(OAc)$_2$, 2 mol% 1,10-phenanthroline) under 1 atm of CO at 80 °C in DMF (Scheme 1).[11] To our knowledge, this reaction represents the largest scale used to date for this type of cyclization. An even lower catalyst loading was realized for a specific target substrate employing 0.1 mol% of palladium ditrifluoroacetate, 0.7 mol% of 3,4,7,8-tetramethyl-1,10-phenanthroline under the same CO pressure, solvent, and reaction temperature. Rigorous exclusion of oxygen is necessary for reproducibility using the latter conditions.

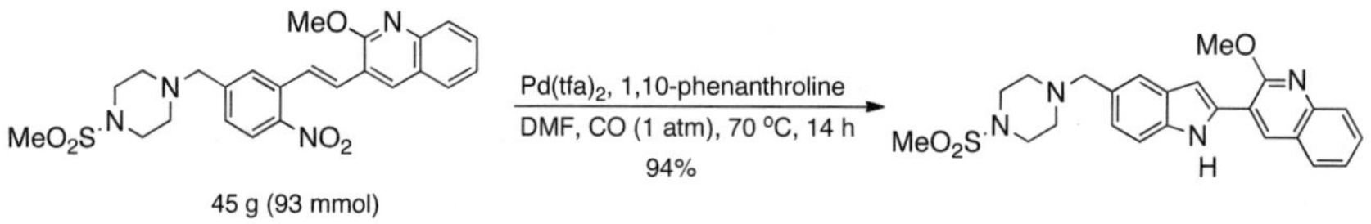

Scheme 1

A wide range of functional groups are compatible with the reaction conditions. Recent applications of the palladium-catalyzed N-heterocyclization include the synthesis of tryptophane derivatives,[12] bicyclic pyrrolo-fused heteroaromatic compounds,[13] a synthesis and revision of the structure of fistulosin,[14] koniamborine,[15] tjipanazoles,[16] 1H-indole-2-yl-1H-quinolin-2-ones,[17] murrayaquinone,[18] bauerine A (Scheme 2),[19] carbazole

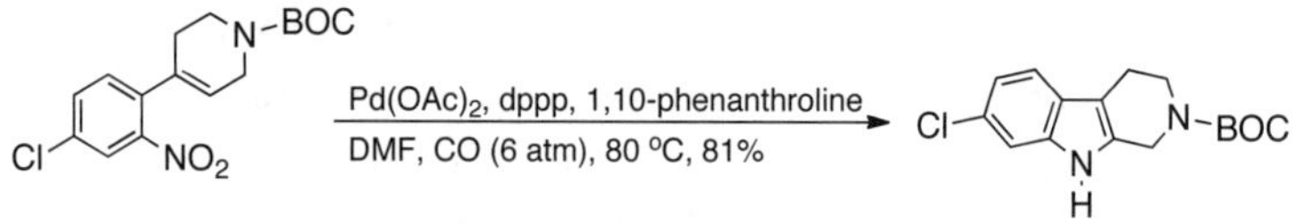

Scheme 2

292

alkaloids (Scheme 3),[20] and mushroom metabolites (Scheme 4).[21] Enhanced reactivity is observed in some cases when two bidentate ligands, bis(diphenylphosphino)propane and 1,10-phenanthroline, are employed. The reason for this is presently unknown.

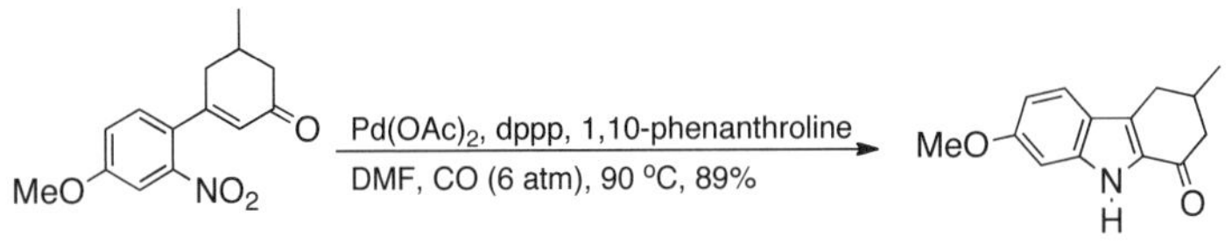

Scheme 3

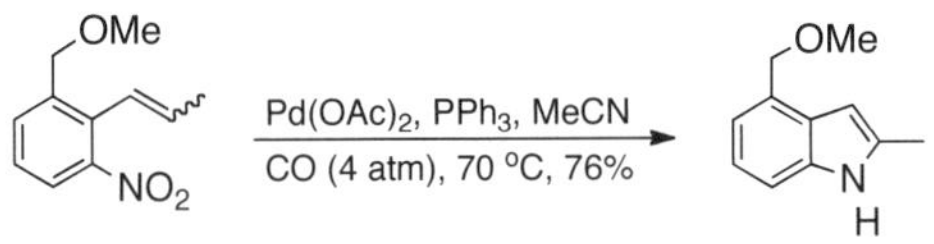

Scheme 4

The reductive *N*-heterocyclization of 1-(2-nitroaryl)-1-alkenes to give indoles is relatively insensitive to the catalyst system used. In contrast, cyclization onto an aromatic ring forming carbazoles and related compounds from 1-aryl-2-nitroaryls is not universal. For example, the Pd(OAc)$_2$-PPh$_3$ catalyst system and reaction conditions used to prepare methyl indole-4-carboxylate do not affect the cyclization of 2-nitrobiphenyl to give carbazole. Ru$_3$(CO)$_{12}$,[22] Fe(CO)$_5$,[23] and MoO$_2$Cl$_2$(dmf)$_2$[9] have been used to prepare carbazoles and related compounds but to date the best results are obtained using Pd(OAc)$_2$-1,10-phenanthroline in DMF at 140 °C and 5 atm of CO (Scheme 5).[24] 1-(2-Nitroaryl)-1-alkenes can also be cyclized to form 3-arylindoles via a cyclization onto an aromatic ring (Scheme 6).[25]

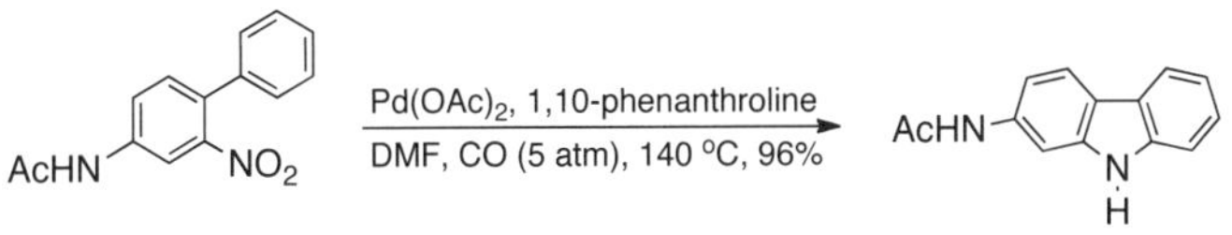

Scheme 5

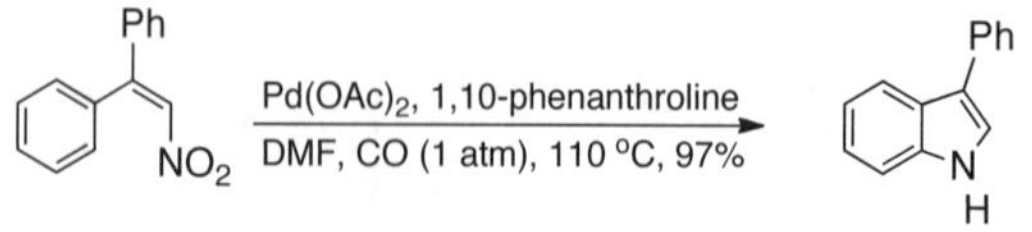

Scheme 6

The transition-metal catalyzed reductive *N*-heterocyclization reaction forming indoles is mechanistically related to reductive cyclizations of nitroaryls using trivalent phosphorous compounds, usually triethylphosphite, at elevated temperatures. However, the palladium-catalyzed reaction offers advantages such as lower reaction temperatures, wide functional group compatibility, and few if any byproducts.

1. C. Eugene Bennett Department of Chemistry, West Virginia University, Morgantown, WV 26506.
2. For an excellent review, see: Ragaini, F.; Cenini, Gallo, E.; Caselli, A.; Fantauzzi, S. *Curr. Org. Chem.* **2006**, *10*, 1479-1510.
3. Clawson Jr., R. W.; Deavers III, R. E.; Akhmedov, N. G.; Söderberg, B. C. G. *Tetrahedron* **2006**, *62*, 10829-10834.
4. Söderberg, B. C. G.; Hubbard, J. W.; Rector, S. R.; O'Neil, S. N. *Tetrahedron* **2005**, *61*, 3637-3649.
5. Söderberg, B. C.; Shriver, J. A. *J. Org. Chem.* **1997**, *62*, 5838-5845.
6. Akazome, M.; Kondo, T.; Watanabe, Y. *J. Org. Chem.* **1994**, *59*, 3375-3380.
7. Tollari, S.; Cenini, S, Crotti, C.; Gianella, E. *J. Mol. Catal.* **1994**, *87*, 203-214.
8. Crotti, C.; Cenini, S.; Rindone, B.; Tollari, S.; Demartin, F. *Chem. Commun.* **1986**, 784-786.
9. Sanz, R.; Escribano, J.; Pedrosa, M. R.; Aguado, R.; Arnaiz, F. J. *Adv. Synth. Catal.* **2007**, *349*, 713-718.
10. Nishiyama, Y.; Maema, R.; Ohno, K.; Hirose, M.; Sonoda, N. *Tetrahedron Lett.* **1999**, *40*, 5717-5720.
11. Davies, I. W.; Smitrovich, J. H.; Sidler, R.; Qu, C.; Gresham, V.; Bazaral, C. *Tetrahedron* **2005**, *61*, 6425-6437.
12. Dacko, C. A.; Akhmedov, N. G.; Söderberg, B. C. G. *Tetrahedron Asymm.* **2008**, *19*, 2775-2783.
13. Gorugantula, S. P.; Carrero-Martinez, G. M.; Dantale, S. W.; Söderberg, B. C. G. *Tetrahedron* **2010**, *66*, 1800-1805.

14. Clawson Jr., R. W.; Dacko, C. A.; Deavers III, R. E.; Akhmedov, N. G.; Söderberg, B. C. G. *Tetrahedron* **2009**, *65*, 8786-8793.

15. Clawson Jr., R. W.; Söderberg, B. C. G. *Tetrahedron Lett.* **2007**, *48*, 6019-6021.

16. Kuethe, J. T.; Wong, A.; Davies, I. W. *Org. Lett.* **2003**, *5*, 3721-3723.

17. Kuethe, J. T.; Wong, A.; Qu, C.; Smitrovich, J.; Davies, I. W.; Hughes, D. L. *J. Org. Chem.* **2005**, *70*, 2555-2567.

18. Scott, T. L.; Söderberg, B. C. G. *Tetrahedron* **2003**, *59*, 6323-6332.

19. Dantale, S. W.; Söderberg, B. C. G. *Tetrahedron* **2003**, *59*, 5507-5514.

20. Scott, T. L.; Yu, X.; Gorunatula, S. P.; Carrero-Martínez, G.; Söderberg, B. C. G. *Tetrahedron* **2006**, *62*, 10835-10842.

21. Söderberg, B. C.; Chisnell, A. C.; O'Neil, S. N.; Shriver, J. A. *J. Org. Chem.* **1999**, *64*, 9731-9734.

22. Crotti, C.; Cenini, S.; Bassoli, A.; Rindone, B.; Demartin, *J. Molec. Catal.* **1991**, *70*, 175-187.

23. Kmiecik, J. E. *J. Org. Chem.* **1965**, *30*, 2014-2020.

24. Smithrowich, J. H.; Davies, I. W. *Org. Lett.* **2004**, *6*, 533-535.

25. Hsieh, T. H. H.; Dong, V. M. *Tetrahedron* **2009**, *65*, 3062-3068.

Björn C. G. Söderberg received his M.S. degree in 1981 and his Ph.D. degree in 1987, both from the Royal Institute of Technology, Stockholm, Sweden. He did postdoctoral research at Colorado State University and started his independent career at the University of South Alabama in 1990. In 1994 he joined the faculty at West Virginia University. Professor Söderberg's research is focused on the discovery, development, and application of transition metal catalyzed or mediated reactions. He has developed palladium-catalyzed reductive cyclization reactions of nitro-aromatic compounds to form indoles, quinoxalines, benzimidazoles, and related systems.

AN ECONOMICAL SYNTHESIS OF 4-TRIMETHYLSILYL-2-BUTYN-1-OL

Submitted by Alexander N. Wein, Rongbiao Tong, and Frank E. McDonald.[1]

Checked by Florian Bächle and Andreas Pfaltz.[2]

1. Procedure

> *Caution! tert-Butyllithium is extremely pyrophoric and must not be allowed to come into contact with the atmosphere. This reagent should only be handled by individuals trained in its proper and safe use. It is recommended that transfers be carried out by using a 20-mL or smaller glass syringe filled to no more than 2/3 capacity, or by cannula. For a discussion of procedures for handling air-sensitive reagents, see Aldrich Technical Bulletin AL-134.*

4-Trimethylsilyl-2-butyn-1-ol (2). A 500-mL three-necked, round-bottomed flask equipped with a two-tap Schlenk adapter connected to a bubbler and an argon/vacuum manifold (Note 1) in the left neck, a scaled 50-mL dropping funnel closed with a rubber septum on top in the middle neck, a rubber septum on the right neck and a magnetic stir bar (3 cm) is flame-dried, vented with argon and then allowed to cool to room temperature under argon atmosphere (Note 2). The rubber septum on the right is replaced by a low temperature thermometer. The flask is charged with anhydrous THF (150 mL) (Note 3), and 2-butyn-1-ol (1, 5.38 mL, 5.00 g, 69.9 mmol) (Note 4) is added. This solution is cooled to −70 °C (internal temperature) with a dry ice-acetone bath. Once cooled, *n*-butyllithium (2.5 M in hexanes, 28.0 mL, 69.9 mmol, 1.0 equiv) (Note 5) is added dropwise over the course of 80 min (Note 6), and afterwards the solution is stirred for an additional 10 min. Next, *tert*-butyllithium (1.7 M in pentane, 45.2 mL, 76.9 mmol, 1.1 equiv) (Note 7) is added dropwise over the course of 60 min (Note 8). The solution develops a yellow color during the addition of the alkyllithium reagents. After the addition of *tert*-butyllithium is completed,

296

the reaction mixture is slowly warmed to 0 °C during two hours by keeping the flask directly over the dry ice-acetone bath. Formation of a precipitate is first observed between –45 °C and –30 °C and the mixture becomes cloudier as the temperature warms towards 0 °C. After two hours, the reaction vessel is maintained at 0 °C for 20 min with an ice-water bath.

The dropping funnel is replaced by a rubber septum, and the reaction vessel is recooled to –70 °C with a dry ice-acetone bath. Trimethylsilyl chloride (17.7 mL, 15.2 g, 140 mmol, 2.0 equiv, Note 9) is added *via* syringe pump at 45 mL / h to the suspension of the dianionic intermediate. During this addition the color changes from dark yellow to light brown and the precipitate dissolves again (Note 10). The reaction mixture is allowed to warm to room temperature and stirred overnight (16-17 h). The solution becomes orange, and a cream colored precipitate is formed during this period.

The reaction is then quenched by adding saturated aqueous ammonium chloride (100 mL) to the reaction mixture at room temperature, and the biphasic mixture is stirred vigorously for 5 min. Water (100 mL) is added to dissolve the precipitated salts, and the mixture is stirred for additional 25 min. Diethyl ether (100 mL) is added to dilute the organic layer, and the layers are separated in a 1-L separatory funnel. The aqueous layer is extracted with diethyl ether (2 x 25 mL), and the combined organic layers are transferred to a round-bottomed flask and concentrated to ~75 mL by rotary evaporation (35 °C, 210 mmHg) (Note 11). To this solution, a magnetic stir bar (3 cm) is added and 8.73 M acetic acid (15 mL, 130 mmol) (Note 12) is added to cleave the trimethylsilyl ether. The reaction mixture is stirred for one hour until full conversion is achieved according to TLC analysis (Note 13). The reaction mixture is then neutralized by portionwise addition of saturated aqueous sodium bicarbonate until the aqueous layer is slightly basic to pH paper (Note 14). The biphasic solution is transferred to a 1-L separatory funnel and the organic layer is collected. The aqueous layer is extracted with diethyl ether (2 x 50 mL), the combined organic layers are washed with brine (1 x 100 mL) and then dried over magnesium sulfate. After vacuum filtration through a medium porosity fritted funnel packed with Celite and washing with diethyl ether (2 x 25 mL), the crude mixture is concentrated under reduced pressure by rotary evaporation (35° C, 150 mmHg) to give a yellow oil (Note 15).

This crude product is purified by vacuum distillation over the course of five to six hours at 75 mmHg. A vacuum distillation setup consisting of a oil bath, a flask charged with compound, a 40 cm silvered and vacuum-jacketed Vigreux column, a distillation head with condenser and an udder connected to four 25-mL round-bottomed flasks immersed in a ice-water bath is used (Note 16). At 45 °C and 75 mmHg the first fraction is collected, which consists of trimethylsilanol. Subsequently the oil bath temperature is slowly increased to 140 °C, and the distillate obtained during this period is combined with the first fraction. The boiling point reaches 115 °C before it then drops. Then the oil bath temperature is increased to 145 °C. At this temperature a mixture of side product and desired product with a boiling point of 121 °C is collected in the second fraction (~ 2 g). After 1.5 hours the oil bath temperature is increased to 150 °C. The desired product is collected in fraction three with a boiling point of 124 °C. A slow increase of the oil bath temperature to 165 °C, followed by a slow decrease of the pressure to 23 mmHg provides additional product in high purity, which is collected in fraction four (Note 17). The combined fractions three and four give the product, 4-trimethylsilyl-2-butyn-1-ol (**2**) in 35-37% yield (3.5-3.7 g) as colorless oil that is stored in the freezer at –20 °C (Notes 18 and 19). The identity of the product is confirmed by ^{1}H NMR, ^{13}C NMR, IR and elemental analysis (Note 20).

2. Notes

1.   A two-tap Schlenk adapter connected to a bubbler and an argon/vacuum manifold is illustrated in Yu, J.; Truc, V.; Riebel, P.; Hierl, E.; Mudryk, B. *Org. Synth.* **2008**, *85*, 64-71.

2.   All steps prior to the aqueous ammonium chloride quench were conducted with careful exclusion of air and moisture, using argon as an inert atmosphere.

3.   Anhydrous tetrahydrofuran was purchased from Sigma-Aldrich and was stored over oven-dried 3Å molecular sieves under argon. Before each reaction, the water content of the THF was determined with a Denver Instruments Model 275KF Colorimetric Karl-Fischer Titrator to ensure less than 10 ppm water.

4.   2-Butyn-1-ol (**1**) was purchased from Sigma-Aldrich (98%) and used without further purification.

5.   *n*-Butyllithium (2.5 M in hexanes) was purchased from Sigma-Aldrich and used as received.

6.   *n*-Butyllithium was added discontinuously in order to keep the internal temperature between −70 °C and −65 °C. Slow addition was necessary since the reaction is highly exothermic. The submitters used a syringe pump with a flow rate of 60 mL / h for the addition of *n*-butyllithium.

7.   *tert*-Butyllithium (1.7 M in pentane) was purchased from Sigma-Aldrich and used as received. *tert*-Butyllithium is highly pyrophoric, and extra care should be taken in handling. *tert*-Butyllithum was transferred *via* cannula transfer to the dropping funnel under argon pressure using a stainless steel cannula.

8.   The internal temperature was kept between −70 °C and −65 °C during addition of *tert*-butyllithium. The submitters used a syringe pump with a flow rate of 60 mL / h for the addition of *tert*-butyllithium. However, *Organic Syntheses* recommends that addition of *tert*-butyllithium by syringe pump be avoided.

9.   Trimethylsilyl chloride was purchased from Sigma-Aldrich (≥98%) and used without further purification. This reagent is corrosive and evolves hydrochloric acid gas in the presence of moisture.

10.   The submitters observed a colorless solution after addition of the first equivalent of trimethylsilyl chloride. After addition of the second equivalent the submitters observed a light green solution.

11.   Pressure was set at 450 mmHg. When most of the diethyl ether was evaporated the pressure was reduced to 210 mmHg. The checkers observed formation of a light grey precipitate during evaporation of the solvent which dissolved again upon addition of aqueous acetic acid.

12.   This corresponds to a 50% aqueous solution of acetic acid. This reagent is corrosive.

13.   Thin layer chromatographic analysis was conducted with silica gel-coated polyester sheets (Macherey-Nagel), using cyclohexane : ethyl acetate (4 : 1) as eluent, and were developed with *p*-anisaldehyde / aqueous sulfuric acid. Under these conditions, the intermediate trimethylsilyl ether exhibited an R_f of 0.9 (and stained red) and the product alcohol (**2**) exhibited an R_f of 0.4 (and stained red). The submitters used hexanes : ethyl acetate (4 : 1) as eluent and observed the same R_f values.

14.   A volume of 150-160 mL of saturated aqueous sodium bicarbonate was added.

15.   The diethyl ether was removed at 450 mmHg. Afterwards the pressure was stepwise reduced to 150 mmHg.

16. The submitters used a 20 cm vacuum-jacketed Vigreux column and a pressure of 150 mmHg for the distillation. However in the checkers' hand a complete separation of the major side product and the product was not possible with this procedure. The removed side product was always obtained as a mixed fraction with the product. Due to the high temperatures during the six-hour distillation, formation of a decomposition product was observed which was difficult to separate. The checkers identified the major side product by ^{1}H NMR, ^{13}C NMR and IR being most likely the isomerized allene (2-(trimethylsilyl)buta-2,3-dien-1-ol).

17. The submitters described an impurity at 119 °C (150 mmHg), followed by the product at 123 °C (150 mmHg).

18. The submitters described the product as a stable, pale yellow oil. However the checkers observed the development of a yellow color and new peaks in ^{1}H NMR upon storage of the product at room temperature.

19. The checkers could isolate another 0.8–1.1 g of product by subjecting the mixed fraction two to flash column chromatography. Flash column chromatography was performed using a 5.5 cm wide and 22 cm high column packed with Fluka silica gel 60 (powder, 0.040–0.063 mm, 250 g) and 20 mL fractions. The eluent used was cyclohexane : ethyl acetate (6 : 1), with KMnO$_4$ as TLC stain (R$_f$allene = 0.24, R$_f$product = 0.18, R$_f$impurity = 0.13). After a 250 mL forerun and five mixed fractions with the allene, pure product appeared in fractions 29-50.

20. For (2): ^{1}H NMR (400 MHz, CDCl$_3$) δ: 0.08 (s, 9 H), 1.47 (t, J = 2.6 Hz, 2 H), 1.93 (br s, 1 H), 4.21 (td, J = 2.6 Hz, 0.5 Hz, 2 H); ^{13}C NMR (101 MHz, CDCl$_3$) δ: –2.0, 7.2, 51.6, 77.3, 84.4; IR (neat film, NaCl): 3548, 3110, 2848, 2216, 1249, 1011, 851 cm^{-1}; Anal. calcd. for C$_7$H$_{14}$OSi: C: 59.09, H: 9.92; found: C: 58.84, H: 9.88.

Safety and Waste Disposal Information

All hazardous materials should be handled and disposed of in accordance with "Prudent Practices in the Laboratory"; National Academy Press: Washington, DC, 1995. Specifically, aqueous and organic phase wastes (residual solvents) were separated before disposal, and solid wastes (MgSO$_4$, Celite) were also separated before disposal. As standard operating procedure, personnel should wear safety glasses, a laboratory jacket, and latex gloves for each step.

3. Discussion

The title compound, 4-trimethylsilyl-2-butyn-1-ol (**2**), was originally prepared by formylation of the metal acetylide arising from 3-trimethylsilyl-1-propyne (**3**, **Figure 1**).[3]

Figure 1. Original synthesis of 4-trimethylsilyl-2-butyn-1-ol (**2**)

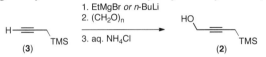

In our hands, we have found that yields are variable in the original procedure, as the formaldehyde (generated by heating paraformaldehyde) condenses and repolymerizes upon addition to the cold (-78 °C) alkynyllithium solution, thus blocking the inlet. Furthermore, 3-trimethylsilyl-1-propyne (**3**) is relatively expensive. The alternative preparation described here begins with commercially available 2-butyn-1-ol (**1**), beginning with sequential *O*- and *C*-deprotonation followed by double silylation of (**5**), and concluding with chemoselective hydrolysis of the silyl ether (**6**, **Figure 2**).[4]

Figure 2. This synthesis of 4-trimethylsilyl-2-butyn-1-ol (**2**), depicting key intermediates

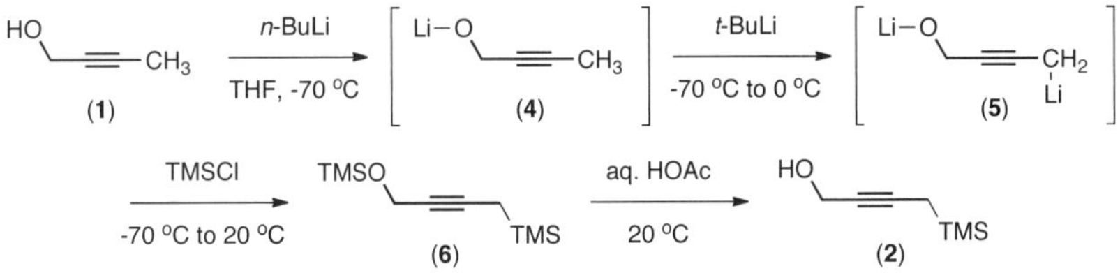

The alcohol of the title compound (**2**) can be converted into several other functional groups, such as the aldehyde (**7**, **Figure 3**),[3b] as well as conversion into electrophilic derivatives for formation of carbon-carbon,[4-6] carbon-nitrogen,[7] carbon-oxygen,[8] and carbon-tin bonds.[9] The internal

Figure 3. Functionalization of the alcohol of (2)

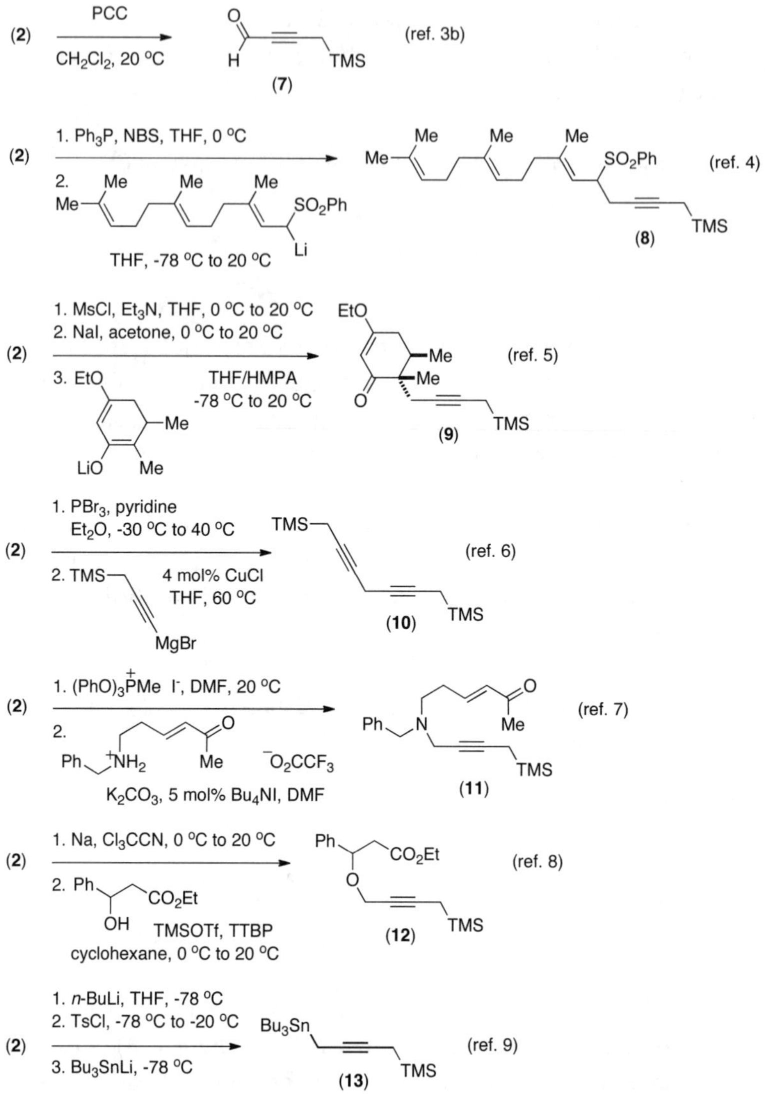

(2) $\xrightarrow[\text{CH}_2\text{Cl}_2,\ 20\ ^\circ\text{C}]{\text{PCC}}$ (7) (ref. 3b)

(2) 1. Ph$_3$P, NBS, THF, 0 °C
2. THF, -78 °C to 20 °C → (8) (ref. 4)

(2) 1. MsCl, Et$_3$N, THF, 0 °C to 20 °C
2. NaI, acetone, 0 °C to 20 °C
3. THF/HMPA -78 °C to 20 °C → (9) (ref. 5)

(2) 1. PBr$_3$, pyridine Et$_2$O, -30 °C to 40 °C
2. 4 mol% CuCl THF, 60 °C → (10) (ref. 6)

(2) 1. (PhO)$_3\overset{+}{\text{P}}$Me I$^-$, DMF, 20 °C
2. K$_2$CO$_3$, 5 mol% Bu$_4$NI, DMF → (11) (ref. 7)

(2) 1. Na, Cl$_3$CCN, 0 °C to 20 °C
2. TMSOTf, TTBP cyclohexane, 0 °C to 20 °C → (12) (ref. 8)

(2) 1. n-BuLi, THF, -78 °C
2. TsCl, -78 °C to -20 °C
3. Bu$_3$SnLi, -78 °C → (13) (ref. 9)

alkyne can also be converted into the *trans*- or *cis*-alkenes (**14, 15, Figure 4**), as functionalized allylic silane reagents.[3c,10] Lewis acid-promoted protonation of (**2**) affords 2,3-butadien-1-ol (**16**).[11] The hydroxyl group directs hydrostannylation to the proximal carbon of the alkyne, to afford the

Org. Synth. **2011**, *88*, 296-308

Z-vinylstannane (**17**) under kinetic conditions.[12]   The propargylic silane serves as a nucleophile upon Lewis acid-promoted reactions with aldehydes to provide functionalized allenes (**18**), (**20**), and (**21, Figure 5**).[3a,8,13]   The corresponding transformations with acetals and *N*-acylaminals have also been reported.[14,15]

Figure 4. Functionalization of the alkyne of (**2**)

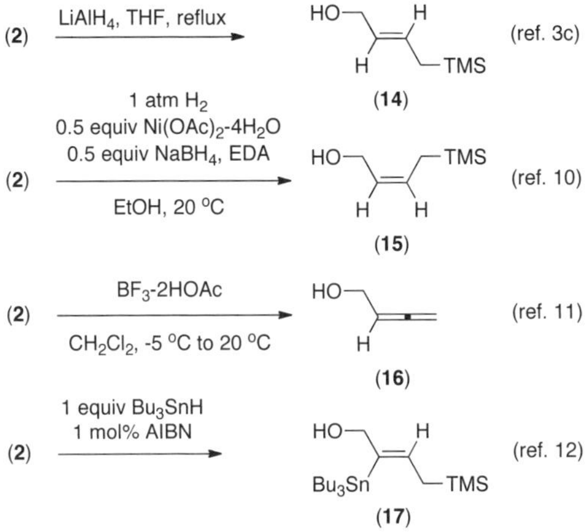

(**2**) — LiAlH$_4$, THF, reflux → (**14**) (ref. 3c)

(**2**) — 1 atm H$_2$, 0.5 equiv Ni(OAc)$_2$·4H$_2$O, 0.5 equiv NaBH$_4$, EDA, EtOH, 20 °C → (**15**) (ref. 10)

(**2**) — BF$_3$·2HOAc, CH$_2$Cl$_2$, -5 °C to 20 °C → (**16**) (ref. 11)

(**2**) — 1 equiv Bu$_3$SnH, 1 mol% AIBN → (**17**) (ref. 12)

Figure 5. Intermolecular additions of propargylic silanes arising from (2)

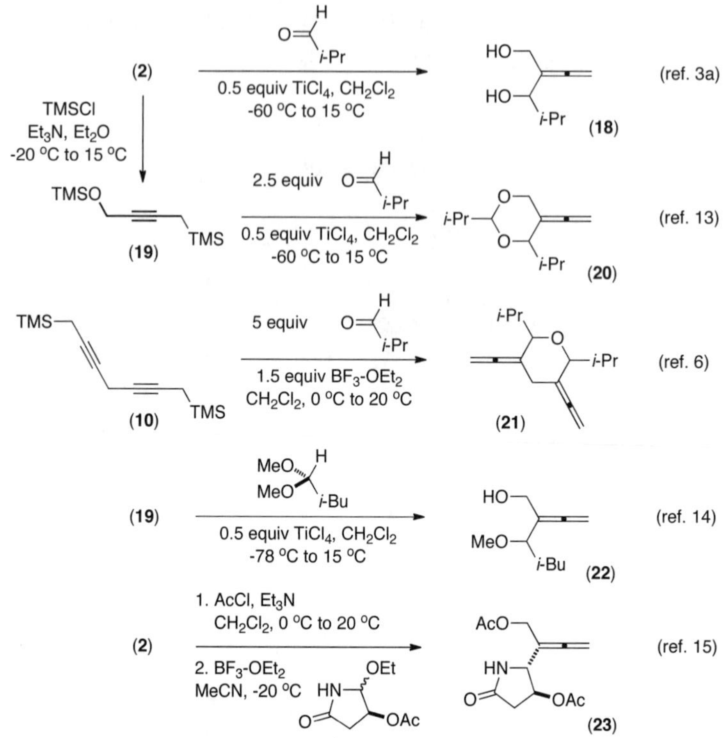

Several Brønsted acid-promoted intramolecular cyclizations have also been accomplished with propargylic silanes arising from (2), tethered to aldehydes (24) or *N*-acylaminals (26), forming the corresponding cyclic structures bearing exocyclic allenes (25, 27, **Figure 6**).[8,16] Moreover, Lewis acid-promoted cyclizations have also been reported in conjugate additions with (11) as well as polyepoxide-alkene cyclizations from (29), to provide the corresponding exocyclic allenes (28, 30).[4,7]

Figure 6. Intramolecular cyclizations of propargylic silanes arising from (**2**)

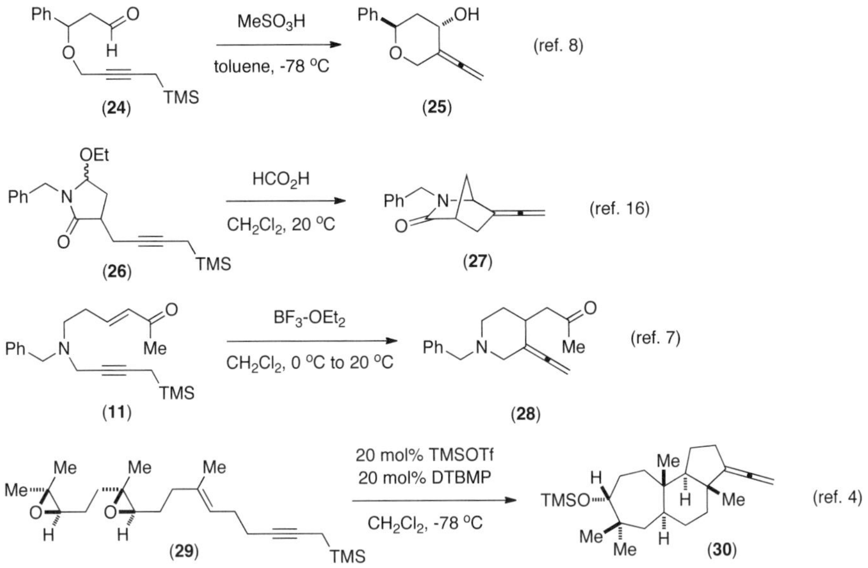

1. Department of Chemistry, Emory University, Atlanta, Georgia 30322 USA Financial support was provided by the National Science Foundation (CHE-0516793).
2. Department of Chemistry, University of Basel, St. Johanns-Ring 19, CH-4056 Basel.
3. (a) Pornet, J.; Randrianoelina, B.; Miginiac, L. *Tetrahedron Lett.* **1984**, *25*, 651-654. (b) Angoh, A. G.; Clive, D. J. L. *J. Chem. Soc., Chem. Commun.* **1984**, 534-536. (c) Mastalerz, H. *J. Org. Chem.* **1984**, *49*, 4092-4094.
4. Tong, R.; Valentine, J. C.; McDonald, F. E.; Cao, R.; Fang, X.; Hardcastle, K. I. *J. Am. Chem. Soc.* **2007**, *129*, 1050-1051.
5. Majetich, G.; Lowery, D.; Khetani, V.; Song, J. S.; Hull, K.; Ringold, C. *J. Org. Chem.* **1991**, *56*, 3988-4001.
6. Aubert, P.; Pornet, J. *J. Organomet. Chem.* **1997**, *538*, 211-221.
7. Solé, D.; García-Rubio, S.; Bosch, J.; Bonjoch, J. *Heterocycles* **1996**, *43*, 2415-2424.
8. Jervis, P. J.; Kariuki, B. M.; Cox, L. R. *Org. Lett.* **2006**, *8*, 4649-4652.

9. Yu, C.-M.; Yoon, S.-K.; Lee, S.-J.; Lee, J.-Y.; Kim, S. S. *Chem. Commun.* **1998**, 2749-2750.

10. Harmata, H.; Ying, W.; Barnes, C. L. *Tetrahedron Lett.* **2009**, *50*, 2326-2328.

11. Pornet, J.; Damour, D.; Miginiac, L. *J. Organomet. Chem.* **1987**, *319*, 333-343.

12. (a) Nativi, C.; Taddei, M. *J. Org. Chem.* **1988**, *53*, 820-826. (b) Nativi, C.; Taddei, M.; Mann, A. *Tetrahedron* **1989**, *45*, 1131-1144.

13. Pornet, J.; Damour, D.; Randrianoelina, B.; Miginiac, L. *Tetrahedron* **1986**, *42*, 2501-2510.

14. Pornet, J.; Miginiac, L.; Jaworski, K.; Randrianoelina, B. *Organometallics* **1985**, *4*, 333-338.

15. Breman, A. C.; Dijkink, J.; van Maarseveen, J. H.; Kinderman, S. S.; Hiemstra, H. *J. Org. Chem.* **2009**, *74*, 6327-6330.

16. (a) Klaver, W. J.; Hiemstra, H.; Speckamp, W. N. *Tetrahedron* **1988**, *44*, 6729-6738. (b) Klaver, W. J.; Moolenaar, M. J.; Hiemstra, H.; Speckamp, W. N. *Tetrahedron* **1988**, *44*, 3805-3818.

Appendix
Chemical Abstracts Nomenclature (Registry Number)

2-Butyn-1-ol; (764-01-2)
n-Butyllithium; (109-72-8)
tert-Butyllithium; (594-19-4)
Chlorotrimethylsilane; (75-77-4)

Frank McDonald received his B.S. degree in chemistry from Texas A&M University in 1984, and completed his Ph.D. degree at Stanford University under the direction of Paul Wender in 1990. After an American Cancer Society postdoctoral fellowship at Yale University with Samuel Danishefsky, he began his independent career as Assistant Professor of Chemistry at Northwestern University, rising to Associate Professor in 1997. In 1998 he moved to Emory University where he is currently Professor of Chemistry. His research interests include the invention of new chemical transformations, explorations in biomimetic synthetic pathways, and applications to the total synthesis of natural products.

Alexander Wein was born in Dalton, Georgia in 1987. He graduated from Emory University *summa cum laude* with B.S. degree in chemistry in 2010, and engaged in research in both inorganic and organic chemistry with Profs. Jack Eichler and Frank McDonald, respectively. He is currently an IRTA fellow at the National Institute of Allergy and Infectious Diseases at the National Institutes of Health in Bethesda, Maryland.

Rongbiao Tong was born in 1976 in Guangdong, China. He obtained his B.S. in chemistry in 2000 at Hunan University, China and then M.S. in organic chemistry in 2003. He subsequently moved to Emory University to continue his research as a doctoral student under the supervision of Professor Frank E. McDonald, working on biomimetic cascade cyclization and total synthesis. After receiving his Ph.D. in 2008, he began his postdoctoral research in the laboratory of Professor Amos B. Smith at the University of Pennsylvania, focusing on anion relay chemistry.

Florian Bächle was born in Bad Säckingen (Germany) in 1983. He studied chemistry at the University of Heidelberg where he obtained his diploma in February 2010 under the supervision of Prof. Günter Helmchen. He joined the group of Prof. Andreas Pfaltz at the University of Basel as a Ph.D. student in April 2010. Currently he is working on the ESI-MS screening of organocatalyzed reactions.

SILVER-CATALYZED REARRANGEMENT OF PROPARGYLIC SULFINATES: SYNTHESIS OF ALLENIC SULFONES

Submitted by Michael Harmata, Zhengxin Cai and Chaofeng Huang.[1]
Checked by Kay M. Brummond and Bo Wen.

1. Procedure

A. But-3-yn-2-yl 4-Methylbenzenesulfinate. A flame-dried 500-mL, three-necked, round-bottomed flask is equipped with a 5.0 cm magnetic stir bar, an internal thermometer, a 125-mL pressure-equalizing addition funnel, and a rubber septum containing nitrogen inlet and outlet needles. The flask is charged with tosyl chloride (9.55 g, 50.1 mmol, 1 equiv) (Note 1), dichloromethane (125 mL) (Note 2), and triethylamine (7.65 mL, 5.56 g, 55.0 mmol, 1.1 equiv) (Note 3), resulting in a colorless solution. The solution is cooled to 19 °C (internal temperature) in a water bath. To this solution is added, dropwise over 70 min by means of the addition funnel, a well mixed yellow solution of 3-butyn-2-ol (4.00 mL of 97% solution, 3.50 g, 50.0 mmol, 1 equiv) (Note 4) and triphenylphosphine (13.1 g, 50.0 mmol, 1 equiv) (Note 5) in 125 mL of dichloromethane, during which time the reaction solution turns a pale yellow color. The rate of the addition is adjusted so as to keep the temperature of the reaction mixture at 19 °C. The reaction is run under an inert nitrogen atmosphere and monitored by TLC. After the disappearance of starting material, the solution is transferred to a 500-mL round-bottomed flask and concentrated on a rotary evaporator (35 °C, 10 mmHg) to a volume of about 50 mL. Ether/hexanes (200 mL/1:4) (Notes 6 and 7) is added to the flask. Upon swirling by hand, a voluminous amount of white solid (triethylamine hydrochloride salt) forms, which is filtered through a short silica gel column (40 g of silica in a 150-mL sintered glass funnel) (Note 8). Diethyl ether (150 mL) is used in three portions to rinse the flask and the solid. The filtrate is concentrated on a rotary evaporator to give a faint yellow oil. The crude oil is purified by silica gel column chromatography. The product is charged on a column (5 x 18 cm) of 150 g of silica gel and eluted with 1:4 ether/hexanes (1000 mL). Collection of 25 mL fractions begins immediately, and the desired product is obtained in fractions 15-38. The product-containing fractions are combined in a 1000-mL round-bottomed flask and concentrated in vacuo by rotary evaporation (35 °C, 10 mmHg) to afford 9.64 g (93%) of a 1:1 mixture of two diastereomers of but-3-yn-2-yl 4-methylbenzenesulfinate as a faint yellow (almost colorless) oil (Note 9).

B. *1-(Buta-1,2-dien-1-ylsulfonyl)-4-methylbenzene.* A 250-mL round-bottomed flask equipped with a 3 cm magnetic stir bar is flame dried under a nitrogen flow. After the flask has cooled to room temperature, it is charged with silver hexafluoroantimonate (266 mg, 0.774 mmol, 0.02 equiv) (Note 10) and protected by a nitrogen atmosphere. The but-3-yn-2-yl 4-methylbenzenesulfinate (8.05 g, 38.7 mmol, 1 equiv) in 77 mL of dichloromethane (0.5 M) is introduced to the flask containing silver hexafluoroantimonate via cannula over 3 min. The reaction is run under a nitrogen atmosphere and monitored by TLC. After the disappearance of starting material (within 10 min), the contents are passed through a short silica gel plug (10 g silica gel in a 60 mL sintered glass filter funnel) and the pad of silica is rinsed with 100 mL of diethyl ether. After concentration of the filtrate on a rotary evaporator (35 °C, 10 mmHg), the residue is put on a high vacuum pump to remove trace amounts of solvent to provide 7.93 g (98.5%) of 1-(buta-1,2-dien-1-ylsulfonyl)-4-methylbenzene as a faint yellow oil. The oil is mixed with 50 mL of 1:4 diethyl ether/hexane (10 mL/40 mL) in a 100-mL round-bottomed flask. The solution is cooled to –20 °C and maintained at that temperature for 22 h, and the resulting crystals are collected by suction filtration on a 60-mL sintered glass funnel quickly so as to not allow the filtrate to warm too much. The flask and the crystals are rinsed using cold hexanes (100 mL). The crystals are transferred to a 100-mL, round-bottomed flask and dried under high vacuum pump to afford 7.73 g (96%) of a white solid (Note 11).

2. Notes

1. *p*-Toluenesulfonyl chloride (99%) was purchased from Acros and used as received.
2. Dichloromethane was purified by passing through alumina using the Sol-Tek ST-002 solvent purification system directly before use.
3. Triethylamine (99%) was purchased from Aldrich and used as received.
4. 3-Butyn-2-ol (97%) was purchased from Aldrich and used as received.
5. Triphenylphosphine (99%) was purchased from Aldrich and used as received.
6. ACS reagent grade ether was purchased from Fischer and used as received.
7. ACS reagent grade hexanes was purchased from Fischer and used as received.
8. Silica gel, standard grade, was purchased from Sorbent Technologies, with 0.040 – 0.063 mm particle size.
9. For characterization purposes, the diastereomers were separated, but the product was used in the next step as a 1:1 mixture. The product displayed the following physical properties. Diastereomer 1: ^{1}H NMR (500 MHz, CDCl$_3$, 298 K) δ: 1.59 (d, J = 6.5 Hz, 3 H), 2.39 (d, J = 2.5 Hz, 1 H), 2.42 (s, 3 H), 4.98 (dq, J = 2.0, 7.0 Hz, 1 H), 7.32 (d, J = 8.0 Hz, 2 H), 7.63 (d, J = 8.5 Hz, 2 H). ^{13}C NMR (125 MHz, CDCl$_3$, 298 K) δ: 21.5, 23.7, 62.0, 74.4, 81.9, 125.4 (2 C), 129.6 (2 C), 141.6, 142.9. Diastereomer 2: ^{1}H NMR (500 MHz, CDCl$_3$, 298 K) δ: 1.52 (d, J = 7.0 Hz, 3 H), 2.42 (s, 3 H), 2.64 (d, J = 2.5 Hz, 1 H), 5.02 (dq, J = 2.0, 7.0 Hz, 1 H), 7.34 (d, J = 8.0 Hz, 2 H), 7.64 (d, J = 8.0 Hz, 2 H). ^{13}C NMR (125 MHz, CDCl$_3$, 298 K) δ: 21.5, 22.9, 63.8, 75.0, 82.2, 125.0 (2 C), 129.7 (2 C), 142.4, 142.9. IR (film) : 3291, 3250, 2989, 2935, 2118, 1596, 1446, 1330, 1138, 1019, 901, 813 cm^{-1}; TLC: R$_f$ = 0.23, 0.29 (for two diastereomers respectively, SiO$_2$, Et$_2$O/hexanes, 1:4); MS m/z (relative intensity) : 209 (15%, M+H), 198 (10%), 215 (11%), 229 (100%), 241 (10%); HRMS m/z : calcd. for C$_{11}$H$_{12}$O$_2$S [M+H] 209.0636, found 209.0636; Anal. calcd. for C$_{11}$H$_{12}$O$_2$S: C, 63.43; H, 5.81. Found C, 63.20; H, 5.84.
10. Silver hexafluoroantimonate (98%) was purchased from Aldrich and used as received.

11. The product displayed the following physical properties: mp 47–48 °C; ^{1}H NMR (500 MHz, CDCl$_3$, 298 K) δ: 1.78 (dd, J = 3.0, 7.5 Hz, 3 H), 2.44 (s, 3 H), 5.80 (dq, J = 6.0, 7.5 Hz, 1 H), 6.14 (dq, J = 6.0, 3.0, 1H), 7.33 (d, J = 8.0 Hz, 2 H), 7.77 (d, J = 8.5 Hz, 2 H); ^{13}C NMR (125 MHz, CDCl$_3$, 298 K) δ: 13.0, 21.6, 95.9, 100.7, 127.6 (2 C), 129.7 (2 C), 138.4, 144.3, 206.1; IR (film): 3020, 2925, 1954, 1596, 1317, 1146, 1085, 815, 767cm^{-1}. TLC: R$_f$ = 0.35 (SiO$_2$, EtOAc/hexanes, 1:1) (Note 12); MS m/z (relative intensity) : 209 (57%, M+H), 201 (100%), 198 (10%), 212 (76%), , 218 (64%), 219 (46%); HRMS m/z : calcd. for C$_{11}$H$_{12}$O$_2$S [M+H] 209.0636, found 209.0642; Anal. calcd. For C$_{11}$H$_{12}$O$_2$S: C, 63.43; H, 5.81. Found C, 63.69; H, 5.75.

12. ACS reagent grade ethyl acetate was purchased from Fischer and used as received.

Safety and Waste Disposal Information

All hazardous materials should be handled and disposed of in accordance with "Prudent Practices in the Laboratory"; National Academy Press; Washington, DC, 1995.

3. Discussion

Allenic sulfones[2] have a rich chemistry that involves such diverse reactions as the carbon-accelerated Claisen rearrangement,[3] metal-catalyzed reactions[4] and DNA cleavage.[5] New, facile approaches to their synthesis are thus in demand.

The preparation of allenic sulfones has been accomplished thermally[6] and under the influence of various metal catalysts including those based on rhodium[7] and palladium.[8] The present method was discovered during a systematic study to use gold catalysts in the rearrangement of propargylic sulfinates to allenic sulfones. Although gold catalysts were effective, we found that the much more economical silver cation was as good or better in effecting the transformation.[9]

The synthesis of propargylic sulfinates used in this study is based on the method of Toru[10] and is related to one published by Sharpless.[11] This approach avoids the preparation of reactive sulfinyl chlorides and broadens the scope of sulfinate ester formation considerably, since many sulfonyl chlorides are either commercially available or very easy to prepare.

312

The procedure succeeds with a variety of propargylic alcohols. Primary, secondary and tertiary propargylic sulfinates rearrange upon exposure to silver cation in near quantitative yields (Figure 1). A chiral, non-racemic alcohol produced a chiral, non-racemic sulfone in the process with no apparent loss of stereochemical integrity. Overall the conversion from alcohol to sulfone is rapid and easy to perform and compares favorably with those methods that use more expensive metal catalysts.

Figure 1

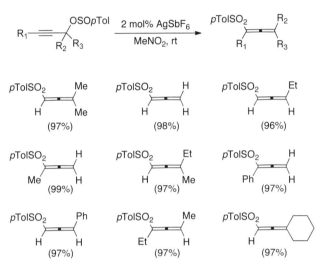

1. Department of Chemistry, University of Missouri-Columbia, Columbia, MO, 65211, email: harmatam@missouri.edu. This work was supported by the National Science Foundation.

2. (a) Back, T. G.; Clary, K. N.; Gao, D. *Chem. Rev.* **2010**, *110*, 4498-4553. (b) Ma, S. *Acc. Chem. Res.* **2009**, *42*, 1679-1688. (c) Back, T. G. *Tetrahedron* **2001**, *57*, 5263-5301.
3. Denmark, S. E.; Harmata, M. A. *J. Am. Chem. Soc.* **1982**, *104*, 4972-4974.
4. (a) Inagaki, F.; Sugikubo, K.; Miyashita, Y.; Mukai, C. *Angew. Chem., Int. Ed.* **2010**, *49*, 2206-2210. (b) Kawamura, T.; Inagaki, F.; Narita, S.; Takahashi, Y.; Hirata, S.; Kitagaki, S.; Mukai, C. *Chem. Eur. J.* **2010**, *16*, 5173-5183. (c) Inagaki, F.; Kawamura, T.; Mukai, C. *Tetrahedron* **2007**, *63*, 5154-5160.

5. (a) Das, S.; Basak, A. *Bioorg. Med. Chem. Lett.* **2009**, *19*, 2815-2818.
    (b) Nicolaou, K. C.; Skokotas, G.; Maligres, P.; Zuccarello, G.;
    Schweiger, E. J.; Toshima, K.; Wendeborn, S. *Angew. Chem., Int. Ed.*
    Engl. **1989**, *28*, 1272-1275.
6. Braverman, S.; Cherkinsky, M. *Top. Curr. Chem.* **2007**, *275*, 67-101.
7. Mukai, C.; Hirose, T.; Teramoto, S.; Kitagaki, S. *Tetrahedron* **2005**, *61*,
    10983-10994.
8. Hiroi, K.; Kato, F. *Tetrahedron* **2001**, *57*, 1543-1550.
9. Harmata, M.; Huang, C. *Adv. Synth. Catal.* **2008**, *350*, 972-974.
10. Watanabe, Y.; Mase, N.; Tateyama, M.-A.; Toru, T. *Tetrahedron:*
    Asymmetry **1999**, *10*, 737-745.
11. (a) Klunder, J. M.; Sharpless, K. B. *J. Org. Chem.* **1987**, *52*, 2598-602.
    (b) Braverman, S.; Pechenick, T.; Zafrani, Y. *ARKIVOC (Gainesville,*
    FL, United States) **2004**, 51-63.

Appendix
Chemical Abstracts Nomenclature; (Registry Number)

But-3-yn-2-yl 4-Methylbenzenesulfinate: Benzenesulfinic acid, 4-methyl-,
    1-methyl-2-propyn-1-yl ester; (32140-54-8)
Tosyl chloride: Benzenesulfonyl chloride, 4-methyl-; (98-59-9)
Triethylamine: (121-44-8)
3-Butyn-2-ol; (2028-63-9)
Triphenylphosphine; (603-35-0)
1-(Buta-1,2-dien-1-ylsulfonyl)-4-methylbenzene: Benzene, 1-(1,2-butadien-
    1-ylsulfonyl)-4-methyl-; (32140-55-9)
Silver hexafluoroantimonate; (26042-64-8)

Michael Harmata was born in Chicago, Illinois in 1959. He received his A.B. degree from the University of Illinois-Chicago (1980) and his Ph.D. from the University of Illinois-Champaign/Urbana (1985). After an NIH postdoctoral fellowship at Stanford University, he began his independent career at the University of Missouri-Columbia (1986). He worked his way through the ranks since that time and was appointed the Norman Rabjohn Distinguished Professor of Chemistry in 2000. He was a Big 12 Faculty Fellow at the University of Texas-Austin (2006), was awarded a black belt in Taekwondo (2009) and was named the first Liebig Professor at the University of Giessen (2010).

Born in Huabei Oilfield, China, in 1982, Zhengxin Cai received his B.S. degree from School of Pharmaceutical Science and Technology, Tianjin University (2005), and has been studying organic synthesis at the University of Missouri under the direction of Michael Harmata since then. While focusing on organic synthesis, his interests extend to anything that is related to drug discovery and development. Besides "cooking" in the hood, he enjoys reading intellectually stimulating books and plays many kinds of sports, including chess, soccer, and table tennis.

Chaofeng Huang was born in Putian, China in 1978. He received his B.S. degree in Chemistry at Tongji University (Shanghai, China, 2001) and M.S. degree in Medicinal Chemistry at Shanghai Institute of Pharmaceutical Industry (Shanghai, China, 2004). In 2009, he earned his Ph.D. degree from the University of Missouri-Columbia. Currently, he is working as a postdoctoral fellow in the Department of Chemistry at the University of Pittsburgh. In October 2010, he will work as a postdoctoral research associate in the Mallinckrodt Institute of Radiology Department at the Washington University School of Medicine in St. Louis.

Bo Wen received his B.S. degree in applied chemistry (2001) from Lanzhou University, China. He then worked in Lanzhou Institute of Chemical Physical, Chinese Academy of Science. In 2005, he moved to West Virginia University for graduate study with Professor Kung Wang, where his research was focused on the synthesis of helical and bowl-shaped polycyclic aromatic compounds. After completing his Ph.D. in 2010, he joined the group of Professor Kay Brummond at the University of Pittsburgh as a postdoctoral research associate and is currently working on the synthesis of biological active guaianolides.

(S_a,S)-N-[2´-(4-METHYLPHENYLSULFONAMIDO)-1,1´-BINAPHTHYL-2-YL]PYRROLIDINE-2-CARBOXAMIDE: AN ORGANOCATALYST FOR THE DIRECT ALDOL REACTION

Submitted by Santiago F. Viózquez,[1] Gabriela Guillena,[1] Carmen Nájera,[1] Ben Bradshaw,[2] Gorka Etxebarria-Jardi,[2] and Josep Bonjoch.[2]
Checked by David Hughes.[3]

1. Procedure

A. (S_a)-N-[2´-Amino-(1,1´-binaphthyl)-2-yl]-4-methylbenzene-sulfonamide (**1**). A 250-mL round-bottomed flask equipped with a 3-cm oval PTFE-coated magnetic stir bar is charged with (S_a)-(−)-1,1´-binaphthyl-2,2´-diamine (3.13 g, 11.0 mmol, 1.0 equiv), dichloromethane (130 mL), and pyridine (10 mL, 124 mmol, 11 equiv). To the stirred solution is added *p*-toluenesulfonyl chloride (2.03 g, 10.7 mmol, 0.97 equiv) in one portion (Note 1). The flask is sealed with a rubber septum through which is inserted an 18-gauge inlet needle, which is connected to a nitrogen line and a gas bubbler, and a thermocouple probe (Note 2). The brown solution is stirred at 22 °C for 10 h (Note 3). The reaction solution is concentrated by rotary evaporation (40 °C bath temperature, 20 mmHg) to an oil that is transferred to a 500-mL separatory funnel with EtOAc (200 mL). The organic layer is washed with 2M HCl (5×30 mL) (Note 4), then vacuum-filtered through a bed of sodium sulfate (40 g) in a 150-mL medium-porosity sintered glass funnel. The filter cake is washed with EtOAc (2 x 40 mL). The filtrate is concentrated in a 500-mL round-bottomed flask by rotary evaporation (40 °C bath, 20 mmHg), then further dried under vacuum (20 mmHg) at room temperature for 14 h to afford **1** as a pink foam (4.7 g, 82% purity, 80% yield) which is used directly in the next step (Notes 5 and 6).

   B.   *(S$_a$,S)-t-Butyl   2-[(2'-(4-methylphenylsulfonamido)-(1,1'-binaphthyl)-2-yl-carbamoyl]pyrrolidine-1-carboxylate*   (**2**).   A   250-mL round-bottomed flask equipped with a 3-cm PTFE-coated magnetic stir bar is charged with (*S*)-*N*-(*t*-butoxycarbonyl)-*L*-proline (3.00 g, 13.9 mmol, 1.6 equiv), anhydrous THF (100 mL), and triethylamine (1.42 g, 14.0 mmol, 1.6 equiv) (Note 7). The flask is sealed with a rubber septum through which is inserted an 18-gauge inlet needle, which is connected to a nitrogen line and a gas bubbler, and a thermocouple probe (Note 2). The mixture is cooled to 3 °C with an ice-water bath and ethyl chloroformate (1.43 g, 13.2 mmol, 1.5 equiv) is added dropwise via a 3-mL syringe over 3 min where upon a fine white precipitate is formed (Note 8). The suspension is stirred 30 min at 0-5 °C and then a solution of **1** (4.7 g, 82 wt%, 3.85 assay g, 88 mmol, 1.0 equiv) in anhydrous THF (25 mL) is added dropwise over 5 min via a 40-mL syringe. After the addition, the rubber septum is replaced with a condenser fitted with a gas adapter connected to a nitrogen line and gas bubbler. The mixture is refluxed using a heating mantle for 12 h (Note 9). At the end of the reaction, the suspension is cooled to room temperature, filtered through a 60-mL medium-porosity sintered glass funnel, and the filter cake is washed with THF (2×25 mL). The combined filtrates are concentrated by rotary evaporation (40 °C bath temperature, 20 mmHg) in a 500-mL round bottomed flask and further dried under vacuum (20 mmHg) for 3 h to provide **2** (7.7 g, estimated 65% purity, 5.0 assay g, 90% yield) as a pink foam which is used directly in the next step (Notes 10-12).

   C.   *(S$_a$,S)-N-[2'-(4-Methylphenylsulfonamido)-1,1'-binaphthyl-2-yl-pyrrolidine-2-carboxamide* (**3**). The 500-mL flask containing crude **2** (7.7 g, 65% purity, 7.9 mmol) from the previous step is equipped with a 3-cm oval

318

PTFE-coated magnetic stir bar and charged with dichloromethane (80 mL) (Note 13). The flask is sealed with a rubber septum through which is inserted an 18-gauge inlet needle, which is connected to a nitrogen line and a gas bubbler, and a thermocouple probe (Note 2). Trifluoroacetic acid (16 mL, 208 mmol, 26 equiv) is added dropwise via a 20-mL syringe over 3 min, and the mixture is stirred at 20-22 °C for 1 h (Notes 14 and 15). At the end of reaction, the solution is cooled to 3 °C using an ice-water bath and the septum is removed and replaced with a 125-mL addition funnel to which is added 2.5 M sodium hydroxide (80 mL). The NaOH solution is added dropwise to the reaction mixture over 10 min (Notes 16 and 17). The mixture is transferred to a 250-mL separatory funnel, and the bottom organic layer is separated. The aq. layer is back extracted with dichloromethane (40 mL). The organic layers are combined and dried with sodium sulfate (100 g) (Note 18), then vacuum filtered through a 150-mL medium porosity sintered glass funnel into a 500-mL round bottomed flask. The filter cake is washed with dichloromethane (2x60 mL). The combined filtrate is concentrated by rotary evaporation (40 °C bath, 20 mmHg) to ~80 mL, then silica gel (30 g) is added, and the mixture is evaporated to a free-flowing powder. The material is purified by column chromatography (Note 19) with a final concentration in a 250-mL round bottom flask. The material is dried under vacuum (room temperature, 20 mmHg, 14 h) to provide 3 as a white solid (3.9–4.1 g). A 3-cm oval PTFE-coated stir bar and dichloromethane (8 mL) are added to the flask and the contents are warmed in a 40 °C water bath to dissolve the product, and then the flask is cooled to room temperature and is equipped with a 60-mL addition funnel. The mixture is stirred as hexanes (40 mL) are added through the addition funnel over 30 min. Crystallization occurs after 10 mL of hexanes is added, and the mixture becomes a thick slurry as the remainder of the hexanes is added. The slurry is stirred 12 h at ambient temperature and then is vacuum-filtered into a 60-mL sintered glass funnel. The filter cake is washed with 5:1 hexanes: dichloromethane (15 mL) and then is air-dried to constant weight to afford (S_a,S)-N-[2'-(4-methylphenylsulfonamido)-1,1'-binaphthyl-2-yl-pyrrolidine-2-carboxamide (3) as a white crystalline solid (3.8–4.0 g, step yield 89–94%, 3-step yield 65–68%) (Notes 20-22).

2. Notes

1. The following reagents and solvents in Step A were used as received: (S_a)-(−)-1,1'-binaphthyl-2,2'-diamine (Sigma-Aldrich, 99%), dichloromethane (Fisher, ACS reagent, 99.5%), EtOAc (Fisher, ACS reagent, 99%), pyridine (Sigma-Aldrich, 99%), and p-TsCl (Acros, 99%).

2. The internal temperature was monitored using a J-Kem Gemini digital thermometer with a Teflon-coated T-Type thermocouple probe (12-inch length, 1/8 inch outer diameter, temperature range −200 to +250 °C). There was no exotherm on addition of p-TsCl.

3. The reaction was monitored by thin layer chromatography on silica gel (EMD, silica gel, grade 60, F_{254}) with 1:1 EtOAc:hexanes as the eluent and visualization with UV. The diamine starting material has $R_f = 0.5$ (blue fluorescence) and the tosyl product has $R_f = 0.6$. The bis-tosyl by-product co-elutes with the mono-tosylate product. The mono- and bis-tosyl products can be separated by TLC by using an eluent of 1:6 EtOAc:hexanes and 3 elutions (bis-Ts, $R_f = 0.45$; mono-Ts, $R_f = 0.40$).

4. The acid wash removes unreacted (S_a)-(−)-1,1'-binaphthyl-2,2'-diamine, which can be recovered as follows. The combined acidic washes are neutralized with 2.5N NaOH until pH 8-10 and then are extracted with dichloromethane (3 × 30 mL). The combined organic layers are washed with brine, dried over sodium sulfate and concentrated by rotary evaporation (40 °C, 20 mmHg) to give 0.30 g (10%) of (S_a)-Binam.

5. The mono-tosylate **1** is approximately 82% pure (80% yield), containing 7 wt% EtOAc and 11 wt% of the bis-tosylate by-product as determined by ^{1}H NMR analysis of the Ts methyl groups of the mono- and bis-tosylated species:^{1}H NMR (400 MHz, CDCl₃); mono-tosylate **1**: δ: 2.32; bis-tosylate: δ: 2.40. This bis-tosylate by-product is difficult to separate by standard column chromatography due to overlap of the peaks using a more polar eluent and tailing of the early eluting bis-tosylate into the mono-tosylate peak using a less polar eluent. The submitters reported preparation of a purified sample by column chromatography on silica gel eluting with hexane/EtOAc, 6:1, following a literature report.[4] The checker purified a 100 mg crude sample by reverse-phase preparative HPLC using the following conditions: column, YMC-pack ODS-AQ, 5um, 150×20 mm I.D; mobile phase, linear gradient elution: 25% MeCN/75% water to 55% MeCN/45% water over 15 min; flow rate, 25mL/min; sample dissolved in MeCN at 10mg/mL; 1 mL per injection. Fractions eluting between 7-9 min

320

were concentrated by rotary evaporation to remove the organic phase (bath temperature 35 °C, 10 mmHg). The remaining aqueous layer was lyophilized, affording 24 mg of mono-tosylate **1**.

6. Mono-tosylate **1** has the following spectroscopic properties: ^{1}H NMR (400 MHz, CDCl$_3$) δ: 2.32 (s, 3 H, CH$_3$), 3.29 (br s, 2 H, NH$_2$), 6.42 (d, J = 8.4 Hz, 1 H), 6.67 (s, 1 H, NH), 6.94–7.09 (m, 5 H), 7.20–7.25 (m, 2 H), 7.37–7.43 (m, 3 H), 7.78 (d, J = 8.0 Hz, 1 H), 7.84 (t, J = 8.7 Hz, 1 H), 7.87 (d, J = 8.0 Hz, 1 H), 7.96 (d, J = 9.0 Hz, 1 H), 8.13 (d, J = 9.0 Hz, 1 H); ^{13}C NMR (100 MHz, CDCl$_3$) δ: 21.7, 109.9, 118.2, 119.7, 121.8, 122.7, 123.6, 125.6, 125.9, 127.35, 127.37, 127.5, 128.37, 128.39, 129.7, 129.9, 130.9, 131.5, 133.0, 133.8, 133.9, 136.5, 142.9, 143.8; LC-MS calcd for [M + H]$^+$ 439.5; found, 439.2.

7. The following reagents and solvents in Step B were purchased from Sigma-Aldrich and used as received: N-(t-butoxycarbonyl)-L-proline (>99%), triethylamine (99.5%), THF (ACS reagent, >99%, inhibited with 250 ppm BHT), and ethyl chloroformate (97%).

8. Addition of ethyl chloroformate results in a slight exotherm from 3 °C to 5 °C.

9. The progress of the reaction can be monitored by TLC (EMD, silica gel, grade 60, F$_{254}$) with 1:1 EtOAc:hexanes and visualization with UV (starting material **1** has R$_f$ = 0.45 and the Boc-proline product **2** has R$_f$ = 0.3); however, the end of reaction cannot be determined by TLC since the unreactive bis-tosylate carried forward from step A co-elutes with the mono-tosylate. NMR of the crude reaction mixture is uninformative due to broad peaks caused by Boc rotamers. Therefore, the end of reaction was assessed by deprotecting the Boc group and determining the amount of mono-tosylate that remained unreacted by ^{1}H NMR. A ~20 mg aliquot of the reaction mixture was evaporated then dissolved in 0.5 mL of CDCl$_3$ followed by addition of 0.2 mL of TFA. The sample was reacted for 15 min at room temperature then analyzed by ^{1}H NMR. The Ts-methyl group was diagnostic for assessing reaction completion: bis-tosylate, δ 2.42; product **3**, δ 2.45; mono-Ts **1**, δ 2.52.

10. Given the broad peaks in the ^{1}H NMR spectrum (Note 12) the purity of the crude material from step B could not be estimated by NMR. The rough purity and yield estimates are based on the 65% recovery of material when subjected to flash chromatography in a separate experiment (Note 11).

11. Compound **2** (2.33 g crude weight) can be purified by column chromatography using 85 g silica gel (Fisher, 230-400 mesh, 60 Å) packed as a slurry with 2:1 hexanes: EtOAc, and eluted with 2:1 hexanes: EtOAc (600 mL), 1:1 hexanes:EtOAc (200 mL), and 1:2 hexanes: EtOAc (200 mL), taking 40 mL fractions. The desired product is obtained in fractions 16-23, (R_f = 0.3, 1:1 hexanes:EtOAc), which are combined and concentrated by rotary evaporation (40 °C, 20 mmHg) to give, after vacuum drying at room temperature to constant weight, 1.68 g of **2** (~90% purity, 65% recovery) as a pink foam.

12. At ambient temperature compound **2** is a mixture of 2 rotamers that cause broad peaks in the ^{1}H NMR and ^{13}C NMR spectra. The following NMR data were collected at 360 K where the rotamers had partially coalesced. ^{1}H NMR (600 MHz, 360 K, DMSO-d_6) δ: 0.76 (br s, 1 H), 1.06 (br s, 1 H), 1.32 (s, 9 H, C(CH$_3$)$_3$], 1.39 (br s, 1 H), 1.67-1.73 (m, 1 H), 2.38 (s, 3 H, CH$_3$), 2.74 (br s, 1 H), 3.08 (app q, J = 8.3 Hz, 1 H), 4.00 (dd, J = 8.9, 3.0, 1 H), 6.70 (d, J = 8.5 Hz, 1 H,), 6.88 (d, J = 8.5 Hz, 1 H), 7.13-7.16 (m, 1 H), 7.20-7.25 (m, 3 H), 7.37 (d, J = 9.0 Hz, 1 H), 7.43-7.46 (m, 2 H), 7.48 (d, J = 8.3 Hz, 2 H), 7.88 (br d, J = 8.6 Hz, 1 H), 7.94 (d, J = 8.2 Hz, 1 H), 7.96-7.99 (m, 2 H), 8.08 (d, J = 8.8 Hz, 1 H), 8.52 (br s, 1 H, NH), 8.78 (br s, 1 H, NH). ^{13}C NMR (150 MHz, 360K, DMSO-d_6) δ: 20.4, 22.1, 27.6, 29.4, 45.8, 60.0, 78.5, 122.4, 123.7, 124.5, 124.7, 124.8, 125.3, 125.93, 125.95, 126.3, 127.4, 127.5, 128.3, 128.7, 128.9, 130.9, 131.0, 131.9, 132.1, 133.3, 134,5, 137.7, 142.5, 153 (br), 171.2.

13. The following reagents and solvents in Step C were used as received: trifluoroacetic acid (Sigma-Aldrich, >99%), dichloromethane (Fisher, ACS reagent, 99.5%), silica gel (Fisher, 230-400 mesh, 60 Å), EtOAc (Fisher, ACS reagent, 99%), and hexanes (Fisher, ACS reagent, >98.5%).

14. During the TFA addition, the temperature decreases from 22 °C to 19 °C.

15. Reaction progress can be monitored by TLC using 2:1 EtOAc:hexanes as eluent and visualized by UV. An aliquot of the reaction mixture is quenched into a mixture of 0.5 mL of 2N NaOH and 0.5 mL of dichloromethane with the bottom organic layer sampled for TLC. R_f product **3**, 0.3; R_f starting material, **2**, 0.8; R_f bis-tosylate, 0.9.

16. Addition of NaOH is exothermic and should be added at a rate to keep the internal temperature below 35 °C to prevent boiling of dichloromethane.

17. At the end of the NaOH addition, the pH is checked by pH paper and should be 8-10. If below 8, additional NaOH is added.

18. The organic layer is hazy due to the retention of a second phase water that is not completely removed upon drying with sodium sulfate.

19. A 6-cm diameter glass column is slurry-packed (2:1 EtOAc:hexanes) with silica gel (200 g). Crude product **3** co-mixed with silica is slurried in 2:1 EtOAc:hexanes and added to the top of the column. The column is topped with 0.5 cm of sand, then eluted with 2:1 EtOAc:hexanes (500 mL), 3:1 EtOAc:hexanes (500 mL), and EtOAc (1 L), taking 100 mL fractions. The chromatography is monitored by TLC (EtOAc, R_f 0.5). The product elutes in fractions 9-15, which are combined and concentrated by rotary evaporation (40 °C water bath, 20 mmHg) in a 1-L flask, then transferred to a 250-mL round-bottomed flask for the final concentration. The white solid is dried under vacuum (20 mmHg) at room temperature for 28 h to constant weight (3.9 – 4.1 g).

20. (S_a,S)-N-[2′-(4-Methylphenylsulfonamido)-1,1′-binaphthyl-2-yl-pyrrolidine-2-carboxamide (**3**) exhibits the following physical and spectroscopic properties: R_f 0.5 (EtOAc); $[\alpha]_D^{25}$ –95 (c 1.0, CHCl$_3$); mp 196-197 °C; Lit[7d] 191-192 °C; ^{1}H NMR (400 MHz, CDCl$_3$) δ: 0.64–0.73 (m, 1 H), 1.15–1.28 (m, 2 H), 1.57–1.64 (m, 1 H), 1.74–1.84 (m, 1 H), 2.20–2.26 (m, 1 H), 2.35 (s, 3 H), 3.32 (dd, J = 4.0, 9.5 Hz, 1 H), 6.35 (br s, 1 H), 6.86 (d, J = 8.4 Hz, 1 H), 6.94 (d, J = 8.5 Hz, 1 H), 7.12 (d, J = 8.1, 2 H), 7.16–7.22 (m, 2 H), 7.37–7.46 (m, 4 H), 7.87 (d, J = 7.9 Hz, 1 H), 7.95 (d, J = 8.2 Hz, 1 H), 8.00 (d, J = 9.1 Hz, 1 H), 8.06 (d, J = 9.0, 1 H), 8.19 (d, J = 9.0 Hz, 1 H), 8.82 (d, J = 9.0 Hz, 1 H), 9.31 (br s, 1 H); ^{13}C NMR (100 MHz, CDCl$_3$) δ: 21.7, 25.4, 30.7, 46.3, 60.7, 117.0, 119.4, 119.6, 120.8, 124.3, 125.3, 125.4, 125.8, 127.6, 127.8, 128.3, 128.8, 129.7, 130.3, 130.7, 130.9, 131.4, 132.3, 132.7, 133.9, 135.9. 136.7, 144.1, 173.5; Anal. calcd. for C$_{32}$H$_{29}$N$_3$O$_3$S: C, 71.75; H, 5.46; N, 7.84; Found: C, 71.41; H, 5.15; N, 7.73.

21. The checkers determined the enantiomeric purity by SFC using a Lux-4 column (150 x 4.6mm, 5um particle size); isocratic elution, 40% MeOH with 25 mM i-butylamine/60% CO$_2$; 3.0 mL/min flow; detection at 210 nm; 200 bar pressure; t$_r$ (S,S)= 4.5 min; t$_r$ (R,R)= 5.5 min; none of the enantiomer was detectable (ee >99%). The submitters determined enantiomeric purity by HPLC analysis at 254 nm using a Chiralpak AD-H column; isocratic elution, 80:20 hexanes: i-PrOH; 1mL/min: t$_r$ (R,R)= 51 min, t$_r$ (S,S)= 105 min. The (R,R)-enantiomer was prepared by the same procedure using (R_a)-(−)-1,1'-binaphthyl-2,2'-diamine and Boc-D-proline

and exhibited the following physical properties: mp 195–197 °C; $[\alpha]_D^{25}$ +93 (c 1.0, CHCl$_3$).

22. The diastereomeric purity (de) was determined to be >99% by ^{1}H NMR analysis in comparison to the (S_a,R)-diastereomer, which was prepared via the same procedure except that Boc-D-proline was used instead of Boc-L-proline. One aromatic proton in the (S_a,R)-diastereomer is upfield (6.54 ppm doublet) relative to the (S_a,S)-diastereomer (6.86 ppm doublet). The 6.54 ppm doublet was undetectible (<0.5%) in the (S_a,S)-diastereomer. The diastereomer (S_a,R)-N-[2′-(4-methylphenylsulfonamido)-1,1′-binaphthyl-2-yl-pyrrolidine-2-carboxamide exhibits the following physical and spectroscopic properties: mp 134–137 °C, Lit^{7d} 152-155 °C; $[\alpha]_D^{25}$ + 6 (c 1.0, CHCl$_3$); ^{1}H NMR (400 MHz, CDCl$_3$) δ: 1.50–1.55 (m, 2 H), 1.86–2.00 (m, 2 H), 2.30–2.34 (m, 1 H), 2.34 (s, 3 H), 2.57–2.62 (m, 1 H), 3.51 (dd, J = 4.5, 9.5 Hz, 1 H), 6.54 (d, J = 8.4 Hz, 1 H), 6.91 (d, J = 8.5 Hz, 1 H), 6.98 (dt, J = 1.1, 7.7 Hz, 1 H), 7.01 (d, J = 8.0 Hz, 2 H), 7.20–7.23 (m, 1 H), 7.34–7.41 (m, 4 H), 7.89 (dd, J = 8.3, 11.0 Hz, 2 H), 7.99 (d, J = 9.0 Hz, 1 H), 8.06 (d, J = 9.0 Hz, 1 H), 8.15 (d, J = 9.0 Hz, 1 H), 8.80 (d, J = 9.0 Hz, 1 H), 9.65 (br s, 1 H, NH); ^{13}C NMR (100 MHz, CDCl$_3$) δ: 21.7, 26.1, 30.9, 47.0, 61.0, 117.6, 118.9, 120.0, 120.3, 124.7, 124.9, 125.6, 127.2, 127.3, 127.6, 128.3, 128.5, 129.8, 130.2, 130.7, 130.9, 131.2, 132.5, 132.9, 134.0, 135.8, 136.4, 144.0, 174.1.

Safety and Waste Disposal Information

All hazardous materials should be handled and disposed of in accordance with "Prudent Practices in the Laboratory"; National Academy Press; Washington, DC, 1995.

3. Discussion

The first generation of BINAM-prolinamides was introduced by several groups in 2006 to use in direct asymmetric aldol reactions[5-7] or other enantioselective processes.[8] (S_a,S)-N-[2-(4-Methylphenylsulfonamido)-1,1′-binaphthyl-2′-yl]-pyrrolidine-2-carboxamide (**3**) is a novel BINAM-prolinamide-type organocatalyst[5] that was developed by Nájera's group[9] and others[7d] almost simultaneously. This (S_a)-binam-L-prolinamide sulfonamide derivative **3**[9] was designed by replacing one proline residue in the first generation catalyst[7] with an acidic sulfonamide group that could activate the carbonyl group of the acceptor through hydrogen-bonding.[10] The efficiency of this catalyst when used with a small amount of benzoic acid as an additive has been proven in several aldol reactions, including the intermolecular aldol reaction between aldehydes and ketones (A, Scheme 1), the cross-aldol reaction between aldehydes (B, Scheme 1), and the intramolecular aldol reaction for the synthesis of the Wieland Miescher ketone (WMK) and related analogues (C, Scheme 1).

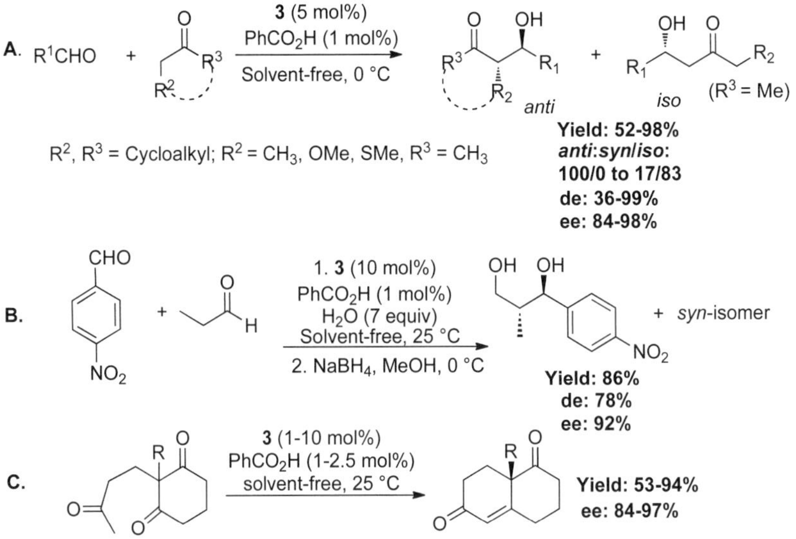

A. R^{1}CHO + [ketone with R^2, R^3]
3 (5 mol%)
PhCO$_2$H (1 mol%)
Solvent-free, 0 °C
→ anti product + iso product (R^3 = Me)

R^2, R^3 = Cycloalkyl; R^2 = CH$_3$, OMe, SMe, R^3 = CH$_3$

Yield: 52-98%
anti:syn/iso:
100/0 to 17/83
de: 36-99%
ee: 84-98%

B. [4-nitrobenzaldehyde] + [propanal]
1. 3 (10 mol%)
PhCO$_2$H (1 mol%)
H$_2$O (7 equiv)
Solvent-free, 25 °C
2. NaBH$_4$, MeOH, 0 °C
→ diol product + syn-isomer

Yield: 86%
de: 78%
ee: 92%

C. [triketone]
3 (1-10 mol%)
PhCO$_2$H (1-2.5 mol%)
solvent-free, 25 °C
→ Wieland-Miescher ketone

Yield: 53-94%
ee: 84-97%

Scheme 1. Aldol processes catalyzed by N-tosyl-(S$_a$)-binam-L-prolinamide **3**

For all the processses, solvent-free reaction conditions could be applied using low catalyst loadings, obtaining the corresponding aldol products with good yields and diastereo- and enantioselectivities comparable to those achieved with other structurally similar catalysts[11] under different reaction conditions. For instance, the large-scale synthesis of the Wieland-Miescher ketone[12] requires only 1 mol% of catalyst **3** (see accompanying article).[13]

The preparation of catalyst **3** described here is a variant of those which already exist,[9,10] and offers the following advantages:

1) Minimal purification steps: only the final product needs to be purified by chromatography.

2) The amide bond formation is efficiently accomplished using ethyl chloroformate, which avoids the use of $SOCl_2$ to form the acid chloride or the need for expensive coupling agents that, moreover, are difficult to remove at the end of the reaction.

This preparation can be also applied to obtain the enantiomer of the desired product, (R_a,R)-N-[2'-(4-methylphenylsulfonamido)-1,1'-binaphthyl-2-yl]pyrrolidine-2-carboxamide, as well as the (R_a,S)- and (S_a,R)-diastereomers.

1. Dpto. Química Orgánica and Instituto de Síntesis Orgánica, Universidad de Alicante, Apdo 99, E-03080 Alicante, Spain. Financial support from the MICINN (projects CTQ2007-62771/BQU, CTQ2010-20387 and Consolider INGENIO CSD2007-0006), FEDER, the Generalitat Valenciana (PROMETEO/2009/038), University of Alicante and European Community (COST Action CM0905: Organocatalysis (ORCA)).

2. Laboratori de Química Orgànica, Facultat de Farmàcia, IBUB, Universitat de Barcelona, Av. Joan XXIII s/n, 08028-Barcelona, Spain. Financial support from the MICINN (projects CTQ2007-61338/BQU, CTQ2010-14846/BQU.

3. The checker thanks Mirlinda Biba for measuring rotations, Zainab Pirzada for developing the chiral SFC assay, Bob Reamer for carrying out the high temperature NMR work on compound 2, and WuXi Pharmatech for the preparative HPLC separation of the mono- and bis-tosylates.

4. Chen, T; Gao, J; Shi, M. *Tetrahedron* **2006**, *62*, 6289-6294.

5. For general reviews of organocatalysis, see: (a) List, B. *Chem. Commun.* **2006**, 819-824. (b) Dondoni, A.; Massi, A. *Angew. Chem. Int. Ed.* **2008**, 47, 4638-4660. (c) Barbas III, C. F. *Angew. Chem., Int. Ed.* **2008**, *47*, 42-47. (d) MacMillan, D. W. C. *Nature* **2008**, *455*, 304-308. (e) Bertelsen, S.; Jørgensen, K. A. *Chem. Soc. Rev.* **2009**, *38*, 2178-2189. (f) Grondal, C.; Jeanty, M.; Enders, D. *Nature Chem.* **2010**, *2*, 167-178.

6. For reviews on the organocatalyzed direct aldol reaction, see: (a) Guillena, G.; Ramón, D. J. *Tetrahedron: Asymmetry* **2006**, *17*, 1465-1492. (b) Zlotin, S. G.; Kucherenko, A. S.; Beletskaya, I. P. *Russ. Chem.*

Rev. **2009**, *78*, 737-784. (c) Trost, B. M.; Brindle, C. S. *Chem. Soc. Rev.* **2010**, *39*, 1600-1632.

7. For pioneering work using BINAM-prolinamides: (a) Gryko, D.; Kowalczyk, B.; Zawadzki, L. *Synlett* **2006**, 1059-1062. (b) Guillena, G.; Hita, M. C.; Nájera, C. *Tetrahedron: Asymmetry* **2006**, *17*, 1493-1497. (c) Guizzetti, S.; Benaglia, M.; Pignataro, L.; Puglisi, A. *Tetrahedron: Asymmetry* **2006**, *17*, 2754-2760. (d) Ma, G.-N.; Zhang, Y.-P.; Shi, M. *Synthesis* **2007**, 197-208.

8. (a) Xiong, Y.; Huang, X.; Gou, S.; Huang, J.; Wen, Y.; Feng, X. *Adv. Synth. Catal.* **2006**, *348*, 538-544. (b) Horillo Martínez, P.; Hultzch, K. C.; Hampel, F. *Chem. Commun.* **2006**, 2221-2223.

9. Guillena, G.; Nájera, C.; Viózquez, S. F. *Synlett* **2008**, 3031-3035.

10. For amide-based bifunctional organocatalysts in asymmetric reactions, see: Liu, X.; Lin, L.; Feng, X. *Chem. Commun.* **2009**, 6145-6158.

11. (a) Guillena, G.; Hita, M. C.; Nájera, C.; Viózquez, S. F. *Tetrahedron: Asymmetry* **2007**, *18*, 2300-2304. (b) Guillena, G.; Hita, M. C.; Nájera, C.; Viózquez, S. F. *J. Org. Chem.* **2008**, *73*, 5933.

12. (a) Bradshaw, B.; Etxeberría-Jardí, G.; Bonjoch, J.; Guillena, G.; Nájera, C.; Viózquez, S. F. *Adv. Synth. Catal.* **2009**, *351*, 2482-2490; (b) Bradshaw, B.; Etxeberría-Jardí, G.; Bonjoch, J. *J. .Am. Chem. Soc.* **2010**, *132*, 5966–5967.

13. Bradshaw, B.; Etxeberría-Jardí, G.; Bonjoch, J.; Viózquez, S. F., Guillena, G.; Nájera, C. *Org. Synth.* **2011**, *88*, 330-341.

Appendix
Chemical Abstracts Nomenclature (Registry Number);

(S_a)-(-)-1,1′-Binaphthyl-2,2′-diamine: (S_a)-(-)-1,1′-Bi(2-naphthylamine); (18531-95-8)

p-Toluenesulfonyl chloride: Benzenesulfonyl chloride, 4-methyl-; (98-59-9)

Pyridine; (110-86-1)

N-(*tert*-Butoxycarbonyl)-*L*-proline; (15761-39-4)

Ethyl chloroformate: Carbonochloridic acid, ethyl ester; (541-41-3)

Triethylamine; (121-44-8)

Trifluoroacetic acid; (76-05-1)

Josep Bonjoch was born in Barcelona (Catalonia, Spain) in 1952. He received his Ph.D. degree (1979) under the supervision of Prof. Joan Bosch at the University of Barcelona, Faculty of Chemistry. He then moved to the Faculty of Pharmacy at the same University, where he was promoted to Associate Professor (1984) and subsequently became Full Professor of Organic Chemistry in 1992. His main research involves the synthesis of complex nitrogen containing natural products, as a motive for developing new synthetic methodology.

Santiago Viózquez was born in Alicante (Spain) in 1981. He received his B.S. degree in chemistry at the Universidad de Alicante in 2006. He is now pursuing his Ph.D. at the Universidad de Alicante under the supervision of G. Guillena and C. Nájera. His research concerns asymmetric organocatalysis with prolinamides derivatives.

Gabriela Guillena received her BSc degree (1993) from University of Alicante. After spending one year as postgraduate student in the group of D. Seebach at the ETH (Zurich), she returned to University of Alicante and received her MSc (1995) and PhD (2000) degrees under the supervision of C. Nájera. After two years as a postdoctoral fellow at research group of G. van Koten (University of Utrecht, Netherlands), she returned to the University of Alicante where she became Assistant Professor in 2003 and Associate Professor in 2008. Her current research interests are focused on new organic methodologies and asymmetric organocatalysis.

Carmen Nájera obtained her B.Sc. (1973) from University of Saragossa and her PhD (1979) at the University of Oviedo under the supervision of J. Barluenga and M. Yus. She performed her postdoctoral work at the ETH (Zurich) with D. Seebach, at the Dyson Perrins Laboratory (Oxford) with J. E. Baldwin, at Harvard University with E. J. Corey, and at Uppsala University with J.-E. Bäckvall. She became Associate Professor in 1985 at the University of Oviedo and Full Professor in 1993 at the University of Alicante. Her scientific contributions are focused on synthetic organic chemistry such as sulfone chemistry, new peptide coupling reagents, oxime-derived palladacycles, asymmetric metal catalysis and organocatalysis.

Ben Bradshaw was born in 1974 in Southport, England. He studied Chemistry at the University of Manchester, where he obtained his PhD in 2001 under the supervision of Professor John Joule. After postdoctoral work with Professor Jim Thomas on the total synthesis of the Bryostatins he joined the group of Professor Josep Bonjoch at the University of Barcelona. In 2008 he was promoted to the position of assistant professor where his research interests include the application of organocatalysis to the total synthesis of complex natural products.

Gorka Etxebarria-Jardí was born in 1981 in Barcelona, Catalonia. He obtained his BSc in Chemistry (2004) and MSc in synthesis of antiretroviral nucleoside drugs (2005) from the University of Barcelona. In 2006, he joined the research group of Prof. Josep Bonjoch and is currently completing his Ph.D in asymmetric catalysis and natural product synthesis.

SYNTHESIS OF (*S*)-8a-METHYL-3,4,8,8a-TETRAHYDRO-1,6-(2*H*,7*H*)-NAPHTHALENEDIONE VIA *N*-TOSYL-(*S*ₐ)-BINAM-L-PROLINAMIDE ORGANOCATALYSIS

Submitted by Ben Bradshaw,[1] Gorka Etxebarria-Jardí,[1] Josep Bonjoch,[1] Santiago F. Viózquez,[2] Gabriela Guillena[2] and Carmen Nájera.[2]
Checked by David Hughes.[3]

1. Procedure

A. 2-Methyl-2-(3-oxobutyl)-1,3-cyclohexanedione (**1**). A 50-mL Erlenmeyer flask (Notes 1 and 2) equipped with a 4 x 0.8 cm PTFE-coated magnetic stir bar is charged with 2-methyl-1,3-cyclohexanedione (10.1 g, 80.0 mmol, 1 equiv) (Note 3), triethylamine (0.12 mL, 0.80 mmol, 0.01 equiv) and methyl vinyl ketone (6.50 g, 92.7 mmol, 1.16 equiv) (Note 4). The flask is capped with a rubber septum through which a thermocouple probe is inserted (Note 5), and the heterogeneous mixture is stirred vigorously for 8 h (Note 6). The initial thick suspension slowly becomes more fluid as the solids dissolve to give a dark brown oil (Note 7). At the end of reaction, the oil is transferred with EtOAc (100 mL) to a 500-mL round-bottomed flask fitted with a 3-cm oval PTFE-coated stir bar. Decolorizing charcoal (3 g) (Note 8) is added, and the mixture is warmed to

40-45 °C using a heating mantle and stirred for 30 min. After cooling to room temperature the mixture is filtered by suction through a 350-mL medium-porosity sintered glass funnel slurry-packed with silica (50 g, 2 cm depth) (Note 9) directly into a 1-L round-bottomed flask. The silica cake is eluted with EtOAc (4 × 75 mL). The filtrate is concentrated by rotary evaporation (40 °C bath, 20 mmHg), then dried under vacuum (22 °C, 20 mmHg) for 12 h to constant weight to afford triketone **1** (15.1–15.6 g, 96–99 %) as a light yellow oil (Note 10).

B. *(S)-8a-Methyl-3,4,8,8a-tetrahydro-1,6-(2H,7H)-naphthalenedione* (**3**). A 50-mL round-bottomed flask fitted with a 2-cm oval PTFE-coated magnetic stir bar is charged with **1** (15.0 g, 76.5 mmol, 1 equiv), catalyst **2** (400 mg, 0.74 mmol, 0.01 equiv) (Note 11) and benzoic acid (230 mg, 1.9 mmol, 0.025 equiv) (Note 12). The flask is capped with a rubber septum and the mixture is stirred at 22 °C for 5 days (Notes 13 and 14). After complete reaction, the solution is transferred with EtOAc (100 mL) to a 250-mL round bottomed flask equipped with a 2-cm PTFE-coated magnetic stir bar. Activated charcoal (3 g) (Note 8) is added, and the mixture is stirred for 15 h at 22 °C. (Note 15). The mixture is diluted with hexanes (100 mL) and filtered with suction through a 350-mL medium-porosity sintered glass funnel slurry-packed with silica (60 g, 2 cm depth) (Note 16) directly into a 2-L round-bottomed flask. The silica cake is eluted with 1:1 EtOAc/hexanes (6 x 100 mL) (Note 17). The filtrate is concentrated by rotary evaporation (40 °C bath temperature, 20 mmHg) and dried under vacuum (22 °C, 20 mmHg) for 6 h to a constant weight, affording 13.5 g (99% yield) of **3** as a brown oil that crystallizes on standing. *t*-Butyl methyl ether (15 mL) is added, and the mixture is warmed to 45 °C with a water bath to dissolve the solids. The solution is cooled to room temperature over 1 h, then placed in a freezer at –15 °C for 12 h, resulting in the formation of large reddish-brown crystals. The supernatant is removed by pipette, and the final traces of solvent are removed by vacuum drying to constant weight (22 °C, 20 mmHg, 16 h) to afford (*S*)-8a-methyl-3,4,8,8a-tetrahydro-1,6-(2*H*,7*H*)-naphthalenedione (**3**) (11.5–11.6 g, 84–85 %) (Notes 18 and 19).

2. Notes

1.   The submitters performed the reaction in a glass vial (3 cm diameter; 10 cm height). The checkers also carried out the reaction in a glass vial (I-Chem 40-mL borosilicate glass vial, 2.6 cm diameter; 8.5 cm height; equipped with a 2.5 cm octagonal PTFE-coated stir bar). Due to poor mixing at the initial stages of the reaction, the total reaction time was 1.5- to 2-fold longer than the reaction carried out in the Erlenmeyer flask, but the yield was similar (96%).

2.   No inert atmosphere or special conditions to exclude water were used.

3.   2-Methyl-1,3-cyclohexanedione was purchased from Sigma-Aldrich (>97%) and Alfa Aesar (>98%). Use of the material as received from each source provided the triketone product **1** in >95% yield in 8-12 h reaction time. The submitters found that on some occasions batches of the impurity-containing starting material had a detrimental effect on the process, not only in the reaction-time to form **1** but also, and most importantly, in the progress of the organocatalyzed formation of **3**. To avoid such problems, the submitters recrystallized 2-methyl-1,3-cyclohexanedione before use. Material recrystallized as follows reacted in 4-6 h to afford triketone product with similar yield and purity.   To a 1-L round bottom flask fitted with a 3-cm PTFE-coated oval stir bar was added 2-methyl-1,3-cyclohexanedione (25.0 g), de-ionized water (620 mL) and absolute ethanol (80 mL, 99.5%, Sigma-Aldrich). The mixture was heated to reflux (93 °C) over 30 min with stirring using a heating mantle to completely dissolve the solids. The mixture was allowed to cool to room temperature over 3 h and then stirred at 22 °C for 4 h. The crystals were collected by filtration, washed with a solution of 7:1 water:EtOH (40 mL), and air dried at room temperature for 12 h to afford 21.5 g (85% recovery) of the dione.

4.   Triethylamine (>99.5%), methyl vinyl ketone (>99%), and EtOAc (ACS reagent, 99%) were purchased from Sigma Aldrich and used as received.

5.   The internal temperature was monitored using a J-Kem Gemini digital thermometer with a Teflon-coated T-Type thermocouple probe (12-inch length, 1/8 inch outer diameter, temperature range −200 to +250 °C). The reaction exothermed to 26 °C during the first hour, then slowly decreased to 23 °C over the course of the reaction.

332

6.   At the initial stages of the reaction, the mixture was stirred at a speed of 700 rpm. Erratic stirring occurred due to the presence of solids. After 15 min, the mixture had become a thick paste and smooth stirring became possible. After 1 h, the mixture had thinned considerably and the stir rate was reduced to 500 rpm, then to 300 rpm after 3 h.

7.   The reaction was monitored by TLC on silica gel with 1:1 hexanes:EtOAc with $KMnO_4$ visualization. The starting material has an R_f = 0.3; product R_f = 0.5. The starting material has a very low solubility in the product oil, so complete reaction is signified by complete dissolution of solids.

8.   Activated carbon (Norit SA3 100 mesh) was purchased from Sigma-Aldrich and was used as received.

9.   Silica gel (50 g, Fisher, 230-400 mesh, 60 Å) was slurried with EtOAc (100 mL) and poured into the sintered glass funnel˘that was connected to the round-bottomed flask using a T-type adaptor. A light vacuum was applied until the EtOAc was absorbed and the silica remained wet and compacted. The filtration and elution of the product were carried out similarly, with a light vacuum applied and the eluting solution taken down to the top of the silica cake for each wash.

10.   Triketone (1) has the following physical and spectroscopic properties: IR (film) cm^{-1}: 1716, 1694; ^1H NMR (500 MHz, $CDCl_3$) δ: 1.23 (s, 3 H, CH_3), 1.88–1.93 (m, 1 H), 1.98–2.05 (m, 1 H), 2.05 (t, 2 H, J = 7.4 Hz), 2.09 (s, 3 H, $COCH_3$), 2.33 (t, 2 H, J = 7.4 Hz), 2.59–2.73 (m, 4 H); ^{13}C NMR (125 MHz, $CDCl_3$) δ: 17.8 (C-5), 20.2 ($COCH_3$), 29.7 (C-1'), 30.1 (C-4'), 37.9 (C-4, C-6), 38.5 (C-2'), 64.5 (CH_3), 207.6 (C-3'), 210.1 (C-1, C-3). To obtain an acceptable elemental analysis, the oil was filtered through a 0.45 micron PTFE syringe filter. Anal. calcd. for $C_{11}H_{16}O_3$: C, 67.32; H, 8.22; Found: C, 67.14; H, 7.84. The material typically contains <1 mol % EtOAc and was >95% pure by NMR analysis.

11.   Catalyst 2 was prepared according to the method described in the preceding procedure.

12.   Benzoic acid (>99.5%), hexanes (ACS reagent, 99%) and methyl t-butyl ether (99%) were purchased from Sigma-Aldrich and used as received.

13.   The initial orange solution darkens to black as the reaction progresses.

14.   The progress of the reaction was monitored by taking an aliquot of the reaction mixture and analyzing the C-8a Me peaks by NMR: ^1H NMR (400 MHz, $CDCl_3$); triketone 1: δ 1.23 (s, CH_3), WMK 3: δ 1.44 (s, CH_3).

Following the reaction by TLC is not feasible since the starting material and product have almost identical R_f values (0.5, EtOAc).

15. The non-aqueous work up is preferred but affords product contaminated with 0.2-0.3 mol % benzoic acid based on NMR analysis. An aqueous work up to remove benzoic acid can be carried out as follows. The reaction mixture is transferred to a 500-mL separatory funnel with EtOAc (100 mL) and then washed with saturated aq. sodium bicarbonate (1 x 10 mL) and brine (10 mL). The aq. layer is back extracted with 25 mL of EtOAc. (The aq. extractions result in the formation of two black layers such that the phase cut cannot be visibly determined. Therefore, the aq. washes are measured and kept to a minimum to facilitate accurate phase cuts.) The combined organic layers are dried over anhydrous sodium sulfate (30 g) and filtered through a 150-mL medium-porosity sintered glass funnel into a 500-mL round-bottomed flask. This solution is taken through the remainder of the workup with activated carbon and silica treatment as described in the body of the procedure.

16. The silica gel was packed with a 1:1 mixture of EtOAc–hexane and the filtration carried out as described in Note 9.

17. The filtrate can be collected in smaller flasks and each fraction checked by NMR. The initial fractions deliver a less-colored product. However all fractions were found to have similar purity by NMR. After the indicated washes are completed further washes can be applied and collected in separate flasks to check that no product remains on the silica pad.

18. (S)-8a-Methyl-3,4,8,8a-tetrahydro-1,6-(2H,7H)-naphthalenedione (3) exhibits the following physical and spectroscopic properties: reddish-brown crystals, mp 49–50 °C, Lit[4] 50–51 °C; R_f 0.5 (EtOAc); enantiomeric purity, 96-97% ee; $[\alpha]_D^{25}$ +93 (c 1.0, toluene), Lit[4b]: $[\alpha]_D^{25}$ +97 (c 1.0, toluene); IR (film) cm^{-1}: 1714 (C(1)=O), 1667 (C(6)=O), 1620 (C=C); ^{1}H NMR (400 MHz, CDCl$_3$) δ: 1.44 (s, 3 H, CH$_3$), 1.70 (app qt, 1 H, J = 13.3, 4.4 Hz, H-3), 2.08–2.18 (m, 3 H, H-3 and 2H-8), 2.42–2.53 (m, 4 H, H-2, H-4 and 2H-7), 2.66–2.77 (m, 2 H, H-2 and H-4), 5.84 (s, 1 H, H-5); ^{13}C NMR (100 MHz, CDCl$_3$) δ: 23.1 (C-3), 23.5 (CH$_3$), 29.9 (C-8), 31.9 (C-4), 33.8 (C-7), 37.8 (C-2), 50.8 (C-8a), 126.0 (C-5), 165.9 (C-4a), 198.4 (C-6), 211.1 (C-1); Anal. calcd. for C$_{11}$H$_{14}$O$_2$: C, 74.13; H, 7.92; O; Found: C, 74.35; H, 7.92.

19. The checkers determined the enantiomeric purity by chiral SFC: Lux-4 column (150 x 4.6mm, 5um particles); gradient elution, MeOH with 25mM i-butylamine/CO$_2$, 4% MeOH hold for 4 min, then ramp at 6%/min to

40% MeOH, hold for 5 min, 15 min total run time; 215 nm detection; 35 °C; 200 bar; (*S*)-isomer t$_r$ = 4.1 min; (*R*)-isomer: t$_r$ = 4.6 min. Enantiomeric purity of **3**: crude mixture, 90-91% ee; crystallized material, 96-97% ee; supernatant from crystallization, 40% ee. The submitters determined the enantiomeric composition by HPLC analysis, Chiralcel OD-H column, 96:4 hexane:*i*-PrOH, 0.8 ml/min, 254 nm: (*S*)-isomer t$_r$ = 23.9 min; (*R*)-isomer: t$_r$ = 26.9 min.

Safety and Waste Disposal information

All hazardous materials should be handled and disposed of in accordance with "Prudent Practices in the Laboratory"; National Academy Press; Washington, DC, 1995.

3. Discussion

The Wieland-Miescher ketone (**3**)[5] is a useful synthetic building block for which a classical asymmetric procedure using (*S*)-proline was published forty years ago.[6] Although this method can be applied on a large scale in the laboratory, it has certain drawbacks, partly due to the modest enantiomeric excess (ee) of the product (70%). To obtain an enantiomerically pure sample from this material, a fractional crystallization procedure[4] allows the undesired enantiomer to be removed by the preferential crystallization of its racemic form, involving the loss of an equal amount of the desired enantiomer. Moreover, the success of the fractional crystallization has been found to depend strongly upon the ee and chemical purity of the partially enantiomerically enriched form of **3**. Thus, although the development of the this procedure enabled a wide application of the enantiopure Wieland-Miescher ketone in natural product synthesis,[7,8] it requires considerable investment of time, material, and experience to achieve satisfactory results and generates considerable waste.

The advent of a plethora of new organocatalysts for aldol reactions has allowed preparation of the Wieland-Miescher ketone (**3**) with high enantiomeric excess without crystallization (86-92% ee),[9] but always conducted on a small scale (0.3–1.0 mmol) due to the use of catalysts that are difficult to prepare (7 to 15 steps) and the requirement for high catalyst loadings (5 to 30 mol %).

Using the method detailed here and the accompanying procedure for the preparation of organocatalyst **2**,[10] 12 g of the Wieland-Miescher ketone (96-97% ee) can be obtained in 83% overall yield from 2-methylcyclohexane-1,3-dione. This streamlined procedure[11,12] has significant practical implications from the synthetic point of view since the sequence is performed with high atom efficiency, low catalyst loading (1%), negligible waste formation, and in both steps a short reaction workup and product purification, which enables a large-scale preparation of this bicyclic enedione.

This new methodology for constructing the Wieland-Miescher ketone has broad application in the synthesis of analogs in which the methyl group is replaced by various other alkyl groups that are not easily obtained using proline as the organocatalyst. Under similar conditions,[11] several triketones undergo cyclization to afford Wieland-Miescher ketone analogs in high yield as shown in Table 1. Among these, a noteworthy compound for terpene synthesis is the allyl derivative,[13] which is prepared in 93% yield and 97% ee.

Furthermore with both enantiomers of the catalyst readily available the respective WMK enantiomers can be easily accessed.

Table 1. Synthesis of a variety of Wieland Meischer Ketone analogs using *N*-Tosyl-(S_a)-binam-L-prolinamide as catalyst.

Entry	Catalyst 2 (mol%)	PhCO$_2$H (mol%)	Time (days)	Product	Yield (%)	ee (%)
1a	2.5	1	5		93	97
1b	1	2.5	6		96	94
2	5	1	10		59	96
3	5	1	10		70	94
4	5	1	4		78	90
5	10	1	4		70	96
6	10	1	8		71	95
7	5	1	4		78	94

1. Laboratori de Química Orgànica, Facultat de Farmàcia, IBUB, Universitat de Barcelona, Av. Joan XXIII s/n, 08028-Barcelona, Spain. Financial support from the MICINN (projects CTQ2007-61338/BQU, CTQ2010-14846/BQU).

2. Dpto. Química Orgánica and Instituto de Síntesis Orgánica, Universidad de Alicante, Apdo-99, E-03080 Alicante, Spain. Financial support from the MICINN (projects CTQ2007-62771/BQU, CTQ2010-20387 and Consolider INGENIO CSD2007-0006), FEDER, the Generalitat Valenciana (PROMETEO/2009/038), University of Alicante and European Community (COST Action CM0905: Organocatalysis (ORCA)).

3. The checker thanks Mirlinda Biba for measuring the rotation of **3** and Zainab Pirzada for developing the chiral SFC assay.

4. (a) Gutzwiller, J.; Bushshacher, P.; Fürst, A. *Synthesis* **1977**, 167-168. (b) Buchschacher, P.; Fürst, A. *Org. Synth.* **1985**, *63*, 37-43. (c) Harada, N.; Sugioka, T.; Uda, H.; Kuriki, T. *Synthesis* **1990**, 53-56.

5. (a) Wieland, O.; Miescher, K. *Helv. Chim. Acta* **1950**, *33*, 2215-2228. (b) Ramachandran, S.; Newman, M. S. *Org. Synth.* **1961**, *41*, 38-41.

6. (a) Eder, U.; Sauer, G.; Wiechert, R. *Angew. Chem. Int. Ed.* **1971**, *10*, 496-497. (b) Hajos, Z. G.; Parrish, D. R. *United States Patent* US 3975442, **1971**.

7. For total syntheses using the WMK up until 2007, see: Guillena, G.; Nájera, C.; Ramón, D. J. *Tetrahedron: Asymmetry* **2007**, *18*, 2249-2293 (references 242-256).

8. For recent applications see: (a) Smith III, A. B.; Kürti. L.; Davulcu, A. H.; Cho, Y. S. *Org. Process. Res. Dev.* **2007**, *11*, 19-24. (b) Hanessian, S.; Boyer, N.; Reddy, G. J.; Deschênes-Simard, B. *Org. Lett.* **2009**, *11*, 4640-4643. (c) Waalboer, D. C. J.; van Kalkeren, H. A.; Schaapman, M. C.; van Delft, F. L.; Rutjes, F. P. J. T. *J. Org. Chem.* **2009**, *74*, 8878-8881. (d) Ma, K.; Zhang, C.; Liu, M.; Chu, Y.; Zhou, L.; Hu, C.; Ye, D. *Tetrahedron Lett.* **2010**, *51*, 1870-1872. (e) Carneiro, V. M. T.; Ferraz, H. M. C.; Vieira, T. O.; Ishikawa, E. E.; Silva, L. F. Jr. *J. Org. Chem.* **2010**, *75*, 2877-2882 (f) Churruca, F.; Fousteris, M.; Ishikawa, Y.; Rekowski, M. W.; Hounsou, C.; Surrey, T.; Giannis, A. *Org. Lett.* **2010**, *12*, 2096-2099.

9. (a) Davies, S. G.; Russell, A. J.; Sheppard, R. L.; Smith, A. D.; Thompson, J. E. *Org. Biomol. Chem.* **2007**, 3190-3200. (b) Kanger, T.; Kriis K.; Laars, M.; Kailas, T.; Müürisepp, A.M.; Pehk, T.; Lopp. M. *J.*

Org. Chem. **2007**, *72*, 5168-5173. (c) D'Elia, V.; Zwicknagl, H.; Reiser, O. *J. Org. Chem.* **2008**, *73*, 3262-3265. (d) Almaşi, D.; Alonso, D. A.; Nájera, C. *Adv. Synth. Catal.* **2008**, *350*, 2467-2472. (e) Guillena, G.; Nájera, C.; Viózquez, S. F. *Synlett* **2008**, 3031-3035. (f) Zhang, X.-M.; Wang. M.; Tu, Y.Q.; Fan, C. A.; Jiang, Y. J.; Zhang, S.Y.; Zhang, F. M. *Synlett* **2008**, 2831-2835.

10. Guillena, G.; Nájera, C.; Viózquez, S. F Bradshaw, B.; Etxeberría-Jardí, G.; Bonjoch, J. *Org. Synth.* **2011**, *88*, 317-329.

11. For the initial procedure see: Bradshaw, B.; Etxebarria-Jardí, G.; Bonjoch, J.; Guillena, G.; Nájera, C.; Viózquez, S. F. *Adv. Synth. Catal.* **2009**, *351*, 2482-2490.

12. The originally published method required the use of 2% of organocatalyst and 0.5% of benzoic acid and provided the product in 94% yield and 94% ee. Both steps (alkylation and cyclization) required flash column chromatography and therefore limited the scale and throughput of the reaction. The modified version has a slightly lower yield (85%) but uses less catalyst (1%) and eliminates the need for column chromatography.

13. (a) Nicolaou, K.C.; Roecker, A. J.; Monenschein, H.; Guntupalli, P.; Fullman, M. *Angew. Chem. Int. Ed.* **2003**, *42*, 3637-3642. (b) Bradshaw, B.; Etxeberría-Jardí, G.; Bonjoch, J. *J .Am. Chem. Soc.* **2010**, *132*, 5966–5967.

Appendix
Chemical Abstracts Nomenclature; (Registry Number)

(*S*)-8a-Methyl-3,4,8,8a-tetrahydro-1,6(2*H*,7*H*)-naphtalenedione:
    1,6(2*H*,7*H*)-Naphthalenedione, 3,4,8,8a-tetrahydro-8a-methyl-, (8aS)- ;
    (33878-99-8)

2-Methyl-2-(3-oxobutyl)-1,3-cyclohexanedione: 1,3-Cyclohexanedione, 2-
    methyl-2-(3-oxobutyl)- ; (85073-65-4)

2-Methyl-1,3-cyclohexanedione: 1,3-Cyclohexanedione, 2-methyl-; (1193-
    55-1)

N-Tosyl-(*S$_a$*)-binam-L-prolinamide: 2-Pyrrolidinecarboxamide, *N*-[(1*S*)-2'-
    [[(4-methylphenyl)sulfonyl]amino][1,1'-binaphthalen]-2-yl]-, (2*S*)-;
    (933782-38-8)

Methyl vinyl ketone: 3-Buten-2-one; (78-94-4)

Josep Bonjoch was born in Barcelona (Catalonia, Spain) in 1952. He received his Ph.D. degree (1979) under the supervision of Prof. Joan Bosch at the University of Barcelona, Faculty of Chemistry. He then moved to the Faculty of Pharmacy at the same University, where he was promoted to Associate Professor (1984) and subsequently became Full Professor of Organic Chemistry in 1992. His main research involves the synthesis of complex nitrogen containing natural products, as a motive for developing new synthetic methodology.

Ben Bradshaw was born in 1974 in Southport, England. He studied Chemistry at the University of Manchester, where he obtained his PhD in 2001 under the supervision of Professor John Joule. After postdoctoral work with Professor Jim Thomas on the total synthesis of the Bryostatins he joined the group of Professor Josep Bonjoch at the University of Barcelona. In 2008 he was promoted to the position of assistant professor where his research interests include the application of organocatalysis to the total synthesis of complex natural products.

Gorka Etxebarria-Jardí was born in 1981 in Barcelona, Catalonia. He obtained his BSc in Chemistry (2004) and MSc in synthesis of antiretroviral nucleoside drugs (2005) from the University of Barcelona. In 2006, he joined the research group of Prof. Josep Bonjoch and is currently completing his Ph.D in asymmetric catalysis and natural product synthesis.

Carmen Nájera obtained her B.Sc.(1973) from University of Saragossa and her PhD (1979) at the University of Oviedo under the supervision of J. Barluenga and M. Yus. She performed her postdoctoral work at the ETH (Zurich) with D. Seebach, at the Dyson Perrins Laboratory (Oxford) with J. E. Baldwin, at Harvard University with E. J. Corey, and at Uppsala University with J.-E. Bäckvall. She became Associate Professor in 1985 at the University of Oviedo and Full Professor in 1993 at the University of Alicante. Her scientific contributions are focused on synthetic organic chemistry such as sulfone chemistry, new peptide coupling reagents, oxime-derived palladacycles, asymmetric metal catalysis and organocatalysis.

Gabriela Guillena received her BSc degree (1993) from University of Alicante. After spending one year as postgraduate student in the group of D. Seebach at the ETH (Zurich), she returned to University of Alicante and received her MSc (1995) and PhD (2000) degrees under the supervision of C. Nájera. After two years as a postdoctoral fellow at research group of G. van Koten (University of Utrecht, Netherlands), she returned to the University of Alicante where she became Assistant Professor in 2003 and Associate Professor in 2008. Her current research interests are focused on new organic methodologies and asymmetric organocatalysis.

Santiago Viózquez was born in Alicante (Spain) in 1981. He received his B.S. degree in chemistry at the Universidad de Alicante in 2006. He is now pursuing his Ph.D. at the Universidad de Alicante under the supervision of G. Guillena and C. Nájera. His research concerns asymmetric organocatalysis with prolinamides derivatives.

STEREOSELECTIVE NICKEL-CATALYZED
1,4-HYDROBORATION OF 1,3-DIENES
[(Z)-Dec-2-en-1-ol]

A.

B.

Submitted by Robert J. Ely and James P. Morken.[1]
Checked by Pamela M. Tadross and Brian M. Stoltz.

Caution! Reactions and subsequent operations involving peracids and peroxy compounds should be run behind a safety shield. Peroxy compounds should be added to the organic material, never the reverse. For relatively fast reactions, the rate of addition of the peroxy compound should be slow enough so that it reacts rapidly and no significant unreacted excess is allowed to build up. The reaction mixture should be stirred efficiently while the peroxy compound is being added, and cooling should generally be provided since many reactions of peroxy compounds are exothermic. New or unfamiliar reactions, particularly those run at elevated temperatures, should be run first on a small scale. Reaction products should never be recovered from the final reaction mixture by distillation until all residual active oxygen compounds (including unreacted peroxy compounds) have been destroyed. Decomposition of active oxygen compounds may be accomplished by the procedure described in Korach, M.; Nielsen, D. R.; Rideout, W. H. Org. Synth. 1962, 42, 50 (Org. Synth. 1973, Coll. Vol. 5, 414).

---

1. Procedure

A. *(E)-deca-1,3-diene* (**1**). In a glove box, a flame-dried one-necked 500-mL round-bottomed flask equipped with a magnetic stir bar (40 mm x 15 mm) is charged with methyltriphenylphosphonium bromide (18.97 g, 53.1 mmol, 1.1 equiv) (Note 1) and potassium *tert*-butoxide (5.96 g, 53.1 mmol, 1.1 equiv) (Note 2). The flask is sealed with a septum, removed from the glove box and transferred to a fume hood. A nitrogen inlet (Note 3) is then attached to the flask and the flask is cooled to 0-5 °C in an ice bath.

Org. Synth. **88**, *2011*, 342-352
Published on the Web 4/5/2011

Tetrahydrofuran (160 mL *via* syringe) (Note 4) is added and the solution turns yellow immediately. After 5 min, the ice bath is removed, and the suspension is allowed to warm to room temperature with stirring over 20–30 min. The suspension is then stirred at room temperature for 30 min. The suspension is then cooled to 0–5 °C in an ice bath, and *trans*-2-nonenal (8.0 mL, 48.3 mmol, 1.0 equiv) (Note 5) is added dropwise *via* syringe over 5 min. After 5 min, the ice bath is removed, and the suspension is allowed to warm to room temperature with stirring over 20–30 min. The suspension is then stirred at room temperature for 2 h. The reaction can be monitored by TLC analysis on SiO_2 (30:1 hexanes:EtOAc as the eluent; visualization with a $KMnO_4$ stain; the starting material aldehyde has a R_f = 0.16, and the product diene has a R_f = 0.72) (Note 6). The light brown heterogeneous solution is concentrated by rotary evaporation (10 mmHg, 25 °C) to approximately 1/3 of the original volume (Note 7), and Et_2O (50 mL) is added to form a white precipitate. The mixture is filtered over a pad of SiO_2 (8.5 cm diameter x 2.5 cm height, 52 g) in a sintered glass funnel (medium porosity). The flask is rinsed with Et_2O (2 x 50 mL), and the SiO_2 is washed with additional Et_2O (300 mL) (Note 8). The filtrate is concentrated to a slurry by rotary evaporation (10 mmHg, 25 °C), and pentane (50 mL) is added to form a white precipitate. The mixture is filtered over a pad of SiO_2 (8.5 cm diameter x 2.5 cm height, 52 g) in a sintered glass funnel (medium porosity). The flask is rinsed with pentane (2 x 50 mL), and the SiO_2 is washed with additional pentane (300 mL). The filtrate is concentrated by rotary evaporation (10 mmHg, 25 °C) to a yellow oil. The oil is transferred to a 50-mL round-bottomed flask equipped with a magnetic stir bar and short-path distillation head (Note 9). The residue is distilled under vacuum (bp 55–58 °C at 12 mmHg, receiver flask cooled in a dry ice/acetone bath throughout the distillation) (Note 10), which provides the desired diene **1** as a clear, colorless oil (5.3–5.8 g, 80–87% yield) (Note 11).

B. *(Z)-dec-2-en-1-ol* (**3**). In a glove box, a flame-dried one-necked 500-mL round-bottomed flask equipped with a magnetic stir bar (40 mm x 15 mm) is charged with $Ni(cod)_2$ (259 mg, 0.94 mmol, 0.025 equiv) (Note 12), tricyclohexylphosphine (264 mg, 0.94 mmol, 0.025 equiv) (Note 13), toluene (120 mL *via* syringe) (Note 14), 4,4,5,5-tetramethyl-1,3,2-dioxaborolane (5.73 mL, 39.5 mmol, 1.05 equiv) (Note 15) *via* syringe, and *(E)-deca-1,3-diene* (**1**) (5.2 g, 37.6 mmol, 1.0 equiv) *via* syringe. The flask containing the diene is rinsed with toluene (30 mL), which is then transferred to the reaction flask. The resulting orange homogeneous solution is capped with a septum, taken out of the glove box, and transferred to a fume hood. A nitrogen inlet is added, and the mixture is stirred at room temperature for 1 h. The reaction can be monitored by TLC analysis on SiO_2 (40:1 hexanes:EtOAc as the eluent; visualization with a $KMnO_4$ stain; the starting material diene has a $R_f = 0.76$, and the allylic boronate ester has a $R_f = 0.16$). The stir bar is removed with a magnetic wand and rinsed with CH_2Cl_2 (2 mL), which turns the solution black. The mixture is concentrated by rotary evaporation (10 mmHg, 25 °C) to a black oil. The intermediate allylic boronate ester (**2**) can be easily purified (Note 16), or oxidized in the same flask. A stir bar is added and the flask is cooled to 0–5 °C in an ice bath followed by the addition of THF (72 mL) (Note 17), 3 M NaOH (8.0 mL *via* syringe) (Note 18), and 30 wt. % chilled hydrogen peroxide in water (8.0 mL dropwise over 3 min *via* syringe; exothermic reaction) (Notes 19 and 20). Following addition, the ice bath is removed and the solution is allowed to warm to room temperature over 10–15 min. The cloudy solution is then stirred vigorously at room temperature for 4.5 h. The oxidation can be monitored by TLC analysis on SiO_2 (7:1 hexanes:EtOAc as the eluent; visualization with a $KMnO_4$ stain; the starting material allylboronate ester has a $R_f = 0.55$, and the product alcohol has a $R_f = 0.23$). The solution is cooled to 0–5 °C in an ice bath and the excess hydrogen peroxide is quenched with a saturated solution of sodium thiosulfate (8.0 mL *via* syringe, dropwise over 3 min) (Notes 21 and 22). The reaction is diluted with deionized water (100 mL), and transferred to a 500-mL separatory funnel, rinsing the flask with EtOAc (2 x 25 mL). The layers are separated, and the aqueous layer is extracted with EtOAc (2 x 80 mL). The combined organic layers are dried over Na_2SO_4 (65 g), filtered, and concentrated by rotary evaporation (10 mmHg, 25 °C) to afford a clear, colorless oil. The crude oil is purified by flash chromatography on SiO_2 (Note 23), eluting with 7:1 hexanes:EtOAc to afford **2** as a clear, colorless oil (5.58–5.61 g, 95–96% yield) (Note 24).

2. Notes

1. Methyltriphenylphosphonium bromide (98%) was purchased from Aldrich and dried under vacuum (1.70 mmHg, 110 °C) 12 h before use.
2. Potassium *tert*-butoxide (95%) was purchased from Aldrich and used as received.

3. All "nitrogen inlet" references are defined as inserting a 16-gauge needle attached to a positive flow of nitrogen through the top of a rubber septum.

4. The submitters used THF (HPLC grade) that was purchased from Fisher Scientific and purified using a Pure Solv MD-4 solvent purification system, from Innovative Technology, Inc., by passing the solvent through two activated alumina columns after being purged with argon. The checkers used THF (HPLC grade) that was purchased from Fisher Scientific and purified by passing the solvent through two activated alumina columns under argon.

5. *trans*-2-Nonenal (97%) was purchased from Aldrich and used as received.

6. The submitters performed thin layer chromatography using 0.25 mm silica gel glass backed plated from Silicycle, with visualization performed with either potassium permanganate (KMnO₄) or phosphomolybdic acid (PMA). The checkers performed thin layer chromatography using E. Merck silica gel 60 F254 precoated glass plates (0.25 mm), with visualization performed with potassium permanganate (KMnO₄).

7. Evaporation of all the solvent will cause the triphenylphosphine oxide to occlude the product, and the yield will be reduced.

8. The submitters reported that occasional stirring of the top layer of the SiO_2 is needed to keep the triphenylphosphine oxide from clogging the filter. The checkers found that this problem could be circumvented by using a filter with a larger diameter.

9. The short-path distillation head and receiver flask were both dried in an oven for 12 h before distillation of compound **1**.

10. The submitters reported distillation of compound **1** at 3.5 mmHg (bp 27–30 °C). The checkers found that somewhat higher yields could be obtained by increasing the pressure to 12 mmHg (bp = 55–58 °C) and cooling the receiver flask in a dry ice/acetone bath for the duration of the distillation.

11. Compound **1** has the following properties: ¹H NMR (500 MHz, CDCl₃) δ: 0.89 (t, J = 7.0 Hz, **CH₃**, 3 H), 1.22–1.45 (m, (**CH₂**)₄, 8 H), 2.08 (dt, J = 6.9, 6.9 Hz, **CH₂**CH=CH, 2 H), 4.95 (ddd, J = 10.1, 1.1, 0.6 Hz, CH=CH*cis***H***trans*, 1 H), 5.09 (ddd, J = 17.0, 1.2, 0.6 Hz, CH=CH*cis***H***trans*, 1 H), 5.71 (dt, J = 15.0, 7.0 Hz, **CH₂**CH=CH, 1 H), 6.05 (dd, J = 15.2, 10.4 Hz, CH₂CH=**CH**, 1 H), 6.32 (dddd, J = 17.0, 10.3, 10.3, 2.0 Hz, CH=**CH***cis***H***trans*,

1 H); ^{13}C NMR (125 MHz, CDCl$_3$) δ: 14.1, 22.6, 28.9, 29.2, 31.7, 32.6, 114.5, 130.8, 135.6, 137.4; IR (neat) 2958 (m), 2926 (s), 2856 (m), 1001 (s), 895 (m) cm^{-1}; HRMS (EI+) m/z calc'd for C$_{10}$H$_{18}$ [M]$^+$: 138.1408, found 138.1423; Anal. calcd. for C$_{10}$H$_{18}$: C, 86.88; H, 13.12. Found: C, 87.02; H, 12.78. The spectral data are in agreement with the reported values.[2]

12. Ni(cod)$_2$ (98%) was purchased from Strem and used as received.

13. Tricyclohexylphosphine (97%) was purchased from Strem and used as received.

14. The submitters used toluene (HPLC grade) that was purchased from Fisher Scientific and purified using a Pure Solv MD-4 solvent purification system, from Innovative Technology, Inc., by passing the solvent through two activated alumina columns after being purged with argon. The checkers used toluene (HPLC grade) that was purchased from Fisher Scientific and was purified by passing the solvent through two activated alumina columns under argon.

15. The submitters reported that 4,4,5,5-tetramethyl-1,3,2-dioxaborolane was generously donated by BASF containing 0.06% dimethyl sulfide, and was used as received. The submitters also noted that 4,4,5,5-tetramethyl-1,3,2-dioxaborolane (97%) purchased from Aldrich can also be used as received. The checkers purchased 4,4,5,5-tetramethyl-1,3,2-dioxaborolane (97%) from Aldrich and used it as received.

16. The checkers performed the hydroboration on a 1.45 mmol (200 mg) scale of diene: after the catalyzed reaction is complete, the stir bar (15 mm x 5 mm) is removed using a magnetic wand, and rinsed with Et$_2$O (2 mL; rinsing with CH$_2$Cl$_2$ causes the solution to turn black and reduces the yield for an unknown reason). The brown residue is then filtered through a short SiO$_2$ plug (8.2 g, 4.5 cm diameter x 1.0 cm height) eluting with 40:1 hexanes:EtOAc (80 mL hexanes, 2 mL EtOAc). The filtrate is then concentrated by rotary evaporation (10 mmHg, 25 °C) and then at 1.70 mmHg to provide the boronate as a clear, colorless oil (359 mg, 93% yield). Compound 2 has the following properties: ^{1}H NMR (500 MHz, CDCl$_3$) δ: 0.88 (t, J = 7.0 Hz, CH$_2$CH$_3$, 3 H), 1.15–1.41 (m, (CH$_2$)$_5$ + (CH$_3$)$_4$, 22 H), 1.67 (d, J = 7.8 Hz, CH$_2$B, 2 H), 2.01 (dt, J = 6.9 Hz, 6.9 Hz, CH$_2$CH=CH, 2 H), 5.39 (dtd, J = 8.5 Hz, 7.0 Hz, 1.0 Hz, CH=CHCH$_2$B, 1 H), 5.49 (dtd, J = 9.5 Hz, 8.0 Hz, 1.5 Hz, CH=CHCH$_2$B, 1 H); ^{13}C NMR (125 MHz, CDCl$_3$) δ: 14.1, 22.7, 24.7, 27.1, 29.3, 29.3, 29.6, 31.9, 83.2, 123.9, 130.0; IR (neat) 2978 (m), 2957 (m), 2926 (s), 2855 (m), 1325 (s), 1145 (s), 968 (w), 848 (w) cm^{-1}; HRMS (EI+) m/z calc'd for C$_{16}$H$_{31}$BO$_2$ [M]$^+$: 266.2417, found

346

266.2419; Anal. calcd. for $C_{16}H_{31}BO_2$: C, 72.18; H, 11.74. Found: C, 72.51; H, 12.01.

17. THF (HPLC grade) for oxidation is purchased from Fisher Scientific and used as received.

18. Sodium hydroxide pellets (A.C.S. grade) are purchased from Fisher Scientific.

19. During the addition of the H_2O_2 solution, an internal thermometer was used to monitor the temperature of the reaction. Over the course of the addition, the checkers observed that the internal temperature of the reaction increased from 5 °C to 30 °C.

20. Hydrogen peroxide (30% wt. in H_2O) is purchased from Aldrich and is stored and used chilled (~4 °C).

21. The submitters reported that quenching of excess peroxide with sodium thiosulfate solution was exothermic. However, the checkers used an internal thermometer to monitor the temperature of the reaction and found no evidence of an exotherm during quenching.

22. Sodium thiosulfate pentahydrate (99.5%) is purchased from Fisher Scientific and is added to deionized water until saturated.

23. Flash chromatography was performed using Silia*Flash* P60 Academic Silica gel (SiO_2, particle size 0.040–0.063 mm) purchased from Silicycle (wet packed in 7:1 hexanes:ethyl acetate, 2.5 cm diameter x 25 cm height, 186 g), eluting with 7:1 hexanes:EtOAc (2.21 L hexanes, 0.29 L EtOAc; 300 mL was passed through, then 200 mL fractions were taken). Fractions 4–9 were collected, concentrated by rotary evaporation (10 mmHg, 25 °C) and then dried under vacuum at 1.70 mmHg to provide a clear, colorless oil.

24. Compound **3** has the following properties: ^{1}H NMR (500 MHz, CDCl$_3$) δ: 0.89 (t, $J = 7.0$ Hz, CH$_2$**CH$_3$**, 3 H), 1.16-1.48 (m, CH$_3$(**CH$_2$**)$_4$, 8 H), 2.07 (dt, $J = 7.3$ Hz, 7.2 Hz, **CH$_2$**CH=CH, 2 H), 4.22 (dd, $J = 6.3$ Hz, 0.7 Hz, **CH$_2$**OH, 2 H), 4.74–5.42 (m, **CH=CH,** 2 H); ^{13}C NMR (125 MHz, CDCl$_3$) δ: 14.1, 22.6, 27.4, 29.1, 29.2, 29.6, 31.8, 58.6, 128.3, 133.3; IR (neat): 3325 (m br), 2956 (m), 2922 (s), 2855 (m), 1465 (m), 1032 (s), 723 (m) cm^{-1}; HRMS (EI+) m/z calc'd for $C_{10}H_{20}O$ [M]$^+$: 156.1514, found 156.1549; Anal. calcd. for $C_{10}H_{20}O$: C, 76.86; H, 12.90. Found: C, 76.86; H, 13.28. The spectral data are in agreement with the reported values.[3]

Safety and Waste Disposal Information

All hazardous materials should be handled and disposed of in accordance with "Prudent Practices in the Laboratory"; National Academy Press, Washington, DC, 1995.

3. Discussion

Allyboronates are synthetically useful intermediates, specifically (Z)-crotylboronate esters. They are commonly used for the allylation of carbonyl compounds to furnish *syn*-homoallylic alcohols, or directly oxidized to provide stereodefined (Z)-allylic alcohols.[4] Current methods to synthesize (Z)-crotylboronates include the trapping of allylic anions using Schlosser's method, but can only be applied to simple crotylboronates.[5] Transmetalation of allylmetal reagents or Matteson homologation of alkenylmetal precursors can provide substituted crotylboronates, but both require the use of a stoichiometric amount of transition metal, and the precursors can be difficult to synthesize.[4] One useful method that has been described in the literature is the 1,4-hydroboration of conjugated dienes. Suzuki and Miyaura found that 1,4-hydroboration can be accomplished with Pd and Rh catalysis.[6] This is an excellent reaction with 2-substituted and 2,3-disubstituted butadienes, but for terminally substituted dienes is limited to cyclic substrates (i.e., cyclohexadiene). Recently, Ritter described an iron catalyst that exhibits remarkable selectivity in the 1,4-hydroboration of 2-substituted dienes, but it is less general for terminally substituted diene substrates that lack substitution at the 2-position.[7] The procedure described here is the only method that is general for terminally substituted dienes.[8]

The Ni-catalyzed 1,4-hydroboration of 1,3-dienes allows for the operationally convenient synthesis of (Z)-crotylboronates and derived (Z)-allylic alcohols. The catalyst is composed of inexpensive and commercially available Ni(cod)$_2$ and PCy$_3$. The stoichiometric reagent, pinacolborane, is commercially available and bench-top stable. The hydroboration is completely regio- and stereoselective for terminal dienes (Table 1), and tolerates several synthetically common functional groups including silyl ethers, benzyl ethers, pthalimides, esters and unprotected alcohols (entries 7-11). Internal dienes are also competent substrates exhibiting high regioselectivity (Table 2).

The Ni-catalyzed 1,4-hydroboration of 1,3-dienes conveniently

348

provides (*Z*)-allylboronates that would be difficult to synthesize by other methods. The allyboronate can be used to synthesize homoallylic alcohols by allylation to carbonyl compounds, or directly oxidize to the stereodefined allylic alcohol.

Table 1.

Reaction scheme: 2.5 mol% Ni(cod)$_2$, 2.5 mol% PCy$_3$, 1.05 equiv pinBH, toluene, rt, 3 h → B(pin); then NaOH, H$_2$O$_2$, THF → OH

entry	substrate	product	yield (%)
1	hexyl	hexyl—OH	85
2	Ph	Ph—OH	91[a]
3	hexyl (Me)	hexyl—OH (Me)	93
4	pentyl (Me)	pentyl—OH (Me)	81
5	Me Me Me	Me Me Me OH	29[b]
6	Me Me Me	Me Me OH Me	63[c]
7	TBDPSO	TBDPSO—OH	56
8	BnO Me Me	Me Me BnO—OH	89
9	EtO$_2$C	EtO$_2$C—OH	81[d]
10	(phthal)N	(phthal)N—OH	61[d]
11	HO	HO—OH	72[e]
12	(cyclohexadiene)	(cyclohexadiene)—OH	60

(a) 1 mol % catalyst Ni(cod)$_2$ and 1 mol % PCy$_3$ employed for this experiment. (b) Reaction 12 h at 60 °C. Product isolated with an equimolar quantity of 4,8-dimethyl-3,7-nonadien-2-ol. (c) Reaction for 12 h at 25 °C. (f) Oxidation with buffered (pH=7) H$_2$O$_2$. (d) 2.1 equiv pinBH employed.

Table 2.

Reaction scheme: R$_1$—CH=CH—CH=CH—R$_2$ (diene), with 2.5 mol% Ni(cod)$_2$, 2.5 mol% PCy$_3$, 1.05 equiv pinBH, toluene, rt, 3 h → R$_1$—CH=CH—CH(B(pin))—R$_2$, then NaOH, H$_2$O$_2$ → R$_1$—CH=CH—CH(OH)—R$_2$

entry	substrate	product	regioselection	yield (%)
1	Ph—diene—pentyl	Ph—CH=CH—CH(OH)—pentyl	10:1	84[a]
2	Ph—diene—OTBS	Ph—CH=CH—CH(OH)—OTBS	>20:1	91[b]
3	Ph—diene—OH	Ph—CH=CH—CH(OH)—OH	>20:1	54[b,c]
4	n-hexyl—diene—Me	n-hexyl—CH=CH—CH(OH)—Me	5:1	61[a]
5	n-hexyl—diene—Me (10:1 E,Z:E,E)	n-hexyl—CH=CH—CH(OH)—Me	>20:1	82[a]
6	Cy—diene—Me (4:1 E,Z:E,E)	Cy—CH=CH—CH(OH)—Me	>20:1	83[a]

(a) Reaction for 12 h. (b) 1 mol % Ni(cod)$_2$ and 1 mol % PCy$_3$ employed for this experiment. (c) 2.1 equiv pinBH employed.

1. Department of Chemistry, Eugene F. Merkert Center, Boston College, Chestnut Hill, MA 02467; E-mail: morken@bc.edu. The submitters acknowledge financial support in the form of NIH GM-64451.
2. Meyers, A. I.; Ford, M. E. *J. Org. Chem.* **1976**, *41*, 1735–1742.
3. Mayer, S. F.; Steinreiber, A.; Orru, R. V. A.; Faber, K. *J. Org. Chem.* **2002**, *67*, 9115–9121.
4. For an excellent review of the allylboration reaction and methods to make allylborons, see: Lachance, H.; Hall, D. G. In *Organic Reactions*; Denmark, S. E., Ed.; Wiley: New York, 2009; Vol. 73.
5. Fujita, K.; Schlosser, M. *Helv. Chim. Acta.* **1982**, *65*, 1258–1263.
6. Satoh, M.; Nomoto, Y.; Miyaura, N.; Suzuki, A. *Tetrahedron Lett.* **1989**,

30, 3789–3792.

7. Wu, J. Y.; Moreau, B.; Ritter, T. *J. Am. Chem. Soc*. **2009**, *131*, 12915–12917.

8. Ely, R. J.; Morken, J. P. *J. Am. Chem. Soc*. **2010**, *132*, 2534–2535.

Appendix
Chemical Abstracts Nomenclature; (Registry Number)

Methyltriphenylphosphonium bromide; (1779-49-3)

Potassium tert-butoxide: 2-Propanol, 2-methyl-, potassium salt (1:1); (865-47-4)

(*E*)-Deca-1,3-diene: 1,3-Decadiene, (3*E*)-; (58396-45-5)

(*Z*)-Dec-2-en-1-ol: 2-Decen-1-ol, (2*Z*)-; (4194-71-2)

Hydrogen peroxide; (7722-84-1)

trans-2-Nonenal: 2-Nonenal, (2*E*)-; (18829-56-6)

Ni(cod)₂Bis(cyclooctadiene)nickel(0); (1295-35-8)

Tricyclohexylphosphine; (2622-14-2)

4,4,5,5-Tetramethyl-1,3,2-dioxaborolane; (25015-63-8)

James P. Morken was born in Concord, CA in 1967. He obtained his B.S. in chemistry in 1989 from UC Santa Barbara working with Prof. Bruce Rickborn, and a Ph.D. from Boston College in 1995 with Prof. Amir Hoveyda. He was an NSF Postdoctoral Fellow with Stuart Schreiber at Harvard University and, in 1997, joined the University of North Carolina at Chapel Hill as an Assistant Professor. He was promoted to Associate Professor in 2002 and in 2006 joined the faculty of Boston College as a Professor of Chemistry. His research focuses on the development of transition-metal-catalyzed asymmetric processes and their use in complex molecule synthesis.

Robert J. Ely was born in 1984 in Grand Junction, CO. He earned a BA in chemistry and biochemistry from the University of Colorado, Boulder in 2007. While there, he interned as a medicinal chemist at Array BioPharma. After graduating, he joined the group of Prof. Morken in 2007. At Boston College he has conducted research involving the Ni-catalyzed hydroboration and diboration of 1,3-dienes.

Pamela M. Tadross was born in Brooklyn, NY in 1983 and received her B.S. degree in chemistry in 2005 from New York University where she conducted research with Professor Marc A. Walters. She subsequently joined the labs of Professor Brian M. Stoltz at the California Institute of Technology in 2005 where she has pursued her Ph.D. as a fellow of the California HIV/AIDS Research Program. Her research interests focus on the exploitation of aryne reactive intermediates in the synthesis of biologically active natural products.

352

TANDEM NUCLEOPHILIC ADDITION / FRAGMENTATION
OF VINYLOGOUS ACYL TRIFLATES:
2-METHYL-2-(1-OXO-5-HEPTYNYL)-1,3-DITHIANE

Submitted by Marilda P. Lisboa, Tung T. Hoang, and Gregory B. Dudley.[1]
Checked by Jimmie Weaver and Jonathan Ellman.

1. Procedure

Caution: *Exercise care in the handling of trifluoromethanesulfonic anhydride (corrosive), pyridine (harmful by inhalation), and butyllithium (flammable, corrosive).*

A. 2-Methyl-3-trifluoromethansulfonyloxy-cyclohex-2-en-1-one. An oven-dried 2-L, three-necked, round-bottomed flask equipped with a 1 inch oval magnetic stir bar, a rubber septum, nitrogen inlet and thermometer is charged with 2-methyl-1,3-cyclohexanedione (15.0 g, 0.115 mol, 1.0 equiv) (Note 1), pyridine (19.0 mL, 0.230 mol, 2 equiv) (Note 2) and methylene chloride (700 mL) (Note 3). The reaction mixture is magnetically stirred, cooled at −78 °C for 20 min (Note 4), and trifluoromethanesulfonic anhydride (23.5 mL, 0.138 mol, 1.2 equiv) (Note 1) is added dropwise via syringe over 10 min (Notes 5 and 6). The light yellow reaction mixture is stirred at −78 °C for 20 min, at 0 °C for 20 min, and at room temperature for 30 min (Note 4), at which time the reaction mixture is a homogeneous red solution. The reaction mixture is acidified using 200 mL of a 1M aqueous solution of hydrochloric acid (Note 7), and the resulting biphasic solution is transferred to a 2-L separatory funnel. The organic phase is removed, and the aqueous layer is extracted with diethyl ether (3 x 200 mL) (Note 8). The organic phases are combined and dried over 100 g of Na_2SO_4 and filtered by vacuum suction through a medium porosity fritted glass funnel, and the solvent is removed under reduced pressure by rotary evaporation (25-45 mmHg, room temperature water bath) providing the crude product as a red oil (Note 9). The crude product is purified by chromatography on silica gel (Note 10) to afford 30.0 g (98%) of the product as a colorless oil (Note 11). 2-Methyl-3-trifluoromethansulfonyloxy-cyclohex-2-en-1-one is indefinitely stable to storage under nitrogen at –13 °C.

B. *2-Methyl-2-(1-oxo-5-heptynyl)-1,3-dithiane.* An oven-dried two-necked, 500-mL, round-bottomed flask equipped with 0.5 inch oval magnetic stir bar, rubber septum, and a nitrogen inlet adapter is charged with 2-methyl-1,3-dithiane (Note 12) (4.64 mL, 38.7 mmol, 1.1 equiv) and tetrahydrofuran (150 mL) (Note 13). The resulting solution is cooled at – 78 °C (Note 4). A solution of *n*-butyllithium (Note 14) (14.5 mL, 35.2 mmol, 1.0 equiv) is added dropwise via syringe over 10 min (Note 15). The light yellow reaction mixture is stirred at –78 °C for 20 min, at 0 °C for 10 min, and at –30 °C for 15 min (Note 4), and then 2-methyl-3-trifluoromethanesulfonyloxy-cyclohex-2-en-1-one (10.0 g, 38.7 mmol, 1.1 equiv) is added dropwise by syringe over 10 min. The light yellow reaction mixture is stirred at –30 °C for 15 min, at 0 °C for 15 min, and at room temperature for 30 min (Note 4), at which time the reaction mixture is a homogeneous red solution (Note 16). A half-saturated solution of aqueous ammonium chloride (100 mL) (Note 17) is then added and the biphasic solution is transferred to a 500-mL separatory funnel and extracted with diethyl ether (3 x 50 mL) (Note 8). The organic phases are combined and dried over 70 g of Na_2SO_4, filtered by vacuum suction through a medium porosity fritted glass funnel, and the solvent is removed under reduced pressure by rotary evaporation (25-45 mmHg, room temperature water bath) providing the crude product as a yellow oil. The crude product is purified by chromatography on silica gel (Note 18) to afford 6.64 g (71%) (Note 19) of the product as a colorless oil (Note 20).

2. Notes

1. 2-Methyl-1,3-cyclohexanedione (97%) and trifluoromethane-sulfonic anhydride (99%) were purchased from Sigma-Aldrich and used without further purification.

2. Pyridine (99%) was purchased from Sigma-Aldrich and stored over KOH pellets.

3. Methylene chloride (HPLC Grade) was purchased from Fischer Scientific and purified using an Innovative Technology solvent purification system.

4. The indicated temperatures at which reaction mixtures are kept refer to the temperature of external cooling baths as follows: (a) –78 °C cooling bath prepared with dry ice and acetone, (b) 0 °C cooling bath prepared with ice and water, and (c) –30 °C cooling bath prepared by adding dry ice to acetone and carefully monitoring the bath temperature. No external bath was employed for reactions conducted at room temperature.

5. The highest internal temperature observed by the checkers was –61 °C.

6. The submitters observed that a small amount of white precipitate forms during the addition of trifluoromethanesulfonic anhydride.

7. Hydrochloric acid (1 M) was prepared by dilution of hydrochloric acid (12.1 M, ACS grade) purchased from EMD Chemicals.

8. Ethyl ether (anhydrous, ACS grade) was purchased from EMD Chemicals and used without further purification.

9. The submitters obtained a yellow oil.

10. The crude product was purified using 300 g of Silica Gel 60 (Geduran® 40-63 µM) wet-loaded as a slurry. The sample was loaded in a minimum volume of eluent, passed through the column using gradient elution (500 mL of 9:1 hexanes:EtOAc, 500 mL of 8:2 hexanes:EtOAc, and 1000 mL of 7:3 hexanes:EtOAc), and collected in 20-mL fractions. Fractions 58–99 were collected, combined, and concentrated under reduced pressure to obtain the final product. The R_f of 2-methyl-3-trifluoromethansulfonyloxy-cyclohex-2-en-1-one = 0.62 and 2-methyl-1,3-cyclohexanedione = 0.22 in 8:2 hexanes:ethyl acetate.

11. The compound exhibited the following physical and spectroscopic data: IR (neat): 2964, 1688, 1416, 1344, 1210, 1135, 1024, 912, 891, 793, 759 cm^{-1}. ^{1}H NMR (500 MHz, CDCl$_3$) δ: 1.77 (d, J = 1.9 Hz, 3 H), 2.05 – 1.97 (apparent pentet, 2 H), 2.40 (t, J = 7.9 Hz, 2 H), 2.66 (dtd, J = 6.1, 4.0, 1.9 Hz, 2 H); ^{13}C NMR (126 MHz, CDCl$_3$) δ: 9.0, 20.5, 28.6, 36.5, 118.2 (q, J_{CF} = 319.8 Hz), 127.9, 162.0, 197.5; HRMS (ESI) calcd. for C$_8$H$_9$HF$_3$O$_4$S[M+H]: 259.0246. Found: 259.0246. Anal. calcd. for C$_8$H$_9$F$_3$O$_4$S: C, 37.21; H, 3.51; F, 22.07; S, 12.42. Found: C, 37.32; H, 3.45; F, 21.84; S, 12.41.

12.   2-Methyl-1,3-dithiane (≥98%) was purchased from Fluka and used without further purification.

13.   Tetrahydrofuran (HPLC Grade) was purchased from EMD Chemicals and purified using an Innovative Technology solvent purification system.

14.   *n*-Butyllithium (2.5 M solution in hexanes) was purchased from Sigma-Aldrich and titrated using (–)-menthol and 1,10-phenanthroline (≥ 99%).

15.   The internal temperature reached –57 °C during addition.

16.   TLC analysis showed formation of product by TLC (R_f (reaction product) = 0.3; R_f (2-methyl-1,3-dithiane) = 0.40, R_f (2-methyl-3-trifluoromethansulfonyloxy-cyclohex-2-en-1-one) = 0.2; elution with 9:1 hexanes:EtOAc; UV and KMnO$_4$/heating visualization). Completion of the reaction is difficult to discern because *n*-butyllithium is the limiting reagent, and therefore both the dithiane and triflate remain even after all of the butyllithium has been consumed.

17.   Half-saturated aqueous ammonium chloride solution was prepared by dilution of a saturated aqueous ammonium chloride solution; ammonium chloride (99.5% ACS) reagent purchased from Sigma-Aldrich.

18.   The crude product was purified using 300 g of Silica Gel 60 (Geduran® 40-63 µM) wet-loaded as a slurry. The sample was loaded in a minimum volume of eluent, passed through the column using isocratic elution (2000 mL of 98:2 hexanes:EtOAc), and collected in 20-mL fractions. Fractions 38–85 were collected, combined, and concentrated under reduced pressure to obtain the final product. The checkers found that excess dithiane complicated isolation of the product. However, most of the excess dithiane could be easily removed by placing the unpurified reaction product under high vacuum overnight. The checkers were then able to isolate the product using 360 g of silica and 3.5 L of 95:5 hexanes:EtOAc. Fractions of 25 mL were collected, and the product was found in fractions 55-93.

19.   The checkers observed that yields were improved substantially using freshly distilled dithiane (55 °C, 1 mmHg), recently chromatographed triflate that was concentrated from anhydrous THF, and a newly opened bottle of butyllithium in hexanes.

20.   The compound exhibited the following physical and spectroscopic data: IR (neat): 2918, 1700, 1446, 1423, 1372, 1359, 1277, 1243, 1217, 1149, 1071, 998, 907, 868, 673 cm^{-1}. ^{1}H NMR (400 MHz, CDCl$_3$) δ: 1.68 (s, 3 H), 1.87 – 1.71 (m, 6 H), 2.14 – 2.04 (m, 1 H), 2.18 (td, J = 6.8, 2.6 Hz, 2

H), 2.62 (dt, J = 7.5, 3.5 Hz, 2 H), 2.79 (t, J = 7.2 Hz, 2 H), 3.10 (t, J = 13.7 Hz, 2 H); ^{13}C NMR (126 MHz, CDCl$_3$) δ: 3.6, 18.2, 24.1, 24.3, 25.0, 28.2, 34.6, 54.8, 76.5, 78.3, 203.9. HRMS (ESI) calcd. for C$_{12}$H$_{18}$HOS$_2$[M+H]: 243.0872. Found: 243.0873. Anal. calcd. for C$_{12}$H$_{18}$OS$_2$: C, 59.46; H, 7.48; S, 26.46. Found: C, 59.48; H, 7.25; S, 26.59.

Safety and Waste Disposal Information

All hazardous materials should be handled and disposed of in accordance with "Prudent Practices in the Laboratory"; National Academy Press; Washington, DC, 1995.

3. Discussion

Alkyne building blocks offer many advantages in chemical synthesis owing to their high enthalpic (chemical potential energy) content and orthogonal reactivity with respect to other common organic functional groups.[2] The advent of late transition metal catalysts that display a preference for coordination to π-systems has expanded the scope of alkyne reactions and further increased their value.[3]

As the scope of alkyne reactions continues to expand, so too does the demand for efficient synthetic methods that produce key alkyne building blocks. However, the energy-rich alkyne π-system is not easy to generate. Functionalized alkynes are typically prepared using methods that make use of simpler alkyne starting materials, like acetylide coupling. Common methods for the generation of alkyne π-bonds from non-acetylenic starting materials include elimination, vinylidene carbene rearrangement, and fragmentation reactions.[4]

Vinylogous acyl triflates (VATs) undergo tandem nucleophilic addition / fragmentation reactions to provide tethered keto-alkynes.[5] The overall sequence achieves the conversion of symmetric, cyclic diones into acyclic alkynyl ketones that comprise orthogonal and non-contiguous functionalities. The stability of the triflate anion (an excellent leaving group) enables the fragmentation process in much the same way as formation of molecular nitrogen is a driving force in the classic Eschenmoser–Tanabe reaction.[6] These mechanistically related reactions are among the few C–C bond cleaving fragmentations that produce alkyne π-systems.[7]

Scheme 1. Alternative strategies for the synthesis of (Z)-6-heneicosen-11-one

(Z)-6-heneicosen-11-one

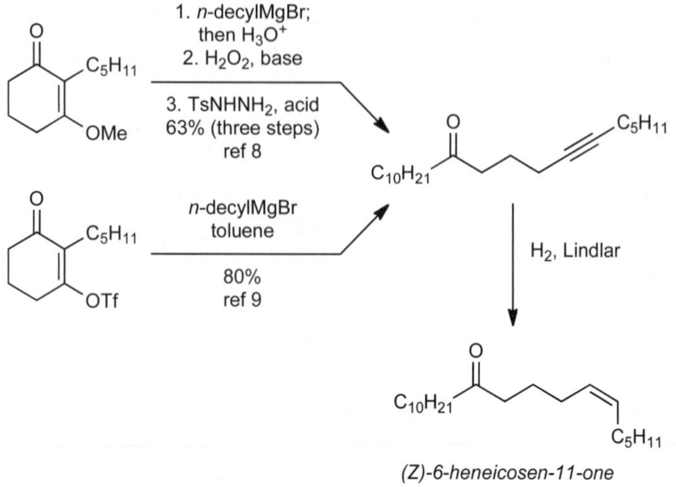

The utility of carbanion-triggered fragmentation of vinylogous acyl triflates as a modern alternative to the Eschenmoser–Tanabe fragmentation is illustrated in the synthesis of (Z)-6-heneicosen-11-one, the sex pheromone of the Douglas fir tussock moth (Figure 1). In an earlier approach,[8] a cyclic vinylogous methyl ester was treated with *n*-decyl Grignard in an addition-hydrolysis sequence to set up the multi-stage Eschenmoser–Tanabe process. Using the current method, nucleophilic addition of the same Grignard reagent to the corresponding vinylogous triflate triggered fragmentation directly.[9] The streamlined entry into the fragmentation pathway provides an increased overall yield in fewer steps, and it avoids acidic conditions and the sometimes-troublesome preparation of cyclic enone oxides.[6b]

The scope of appropriate nucleophiles is illustrated in Table 1. Carbanion reagents were the first nucleophiles to be explored,[5] as these provide alkyl and aryl ketones of the type most closely associated with the Eschenmoser–Tanabe fragmentation (entries 1–4, 10–12). The typical reaction is conducted by treating the nucleophile with the vinylogous acyl triflate in THF at low temperature to achieve the initial 1,2-addition, and then warming the reaction mixture to room temperature or above to promote the (entropically favorable) fragmentation event. It has been observed experimentally that addition / fragmentation reactions using unstabilized

358

Table 1. Scope of the addition/fragmentation reaction

entry	VAT	nucleophile	equiv	solvent	product	yield	Ref
1		PhLi	1	THF		93%	5
2[a]		n-BuMgCl	1	Toluene		63%	9
3		(1,3-dithiane-2-yl)Li	1	THF		74%	5
4		Ph—≡—Li	1	THF		56%	5
5[a]		(EtO)(OLi)C=CH$_2$	2.2	THF		88%	10
6[a]		(EtO)$_2$P(O)CH(Li)Me	1.2	THF		94%	10
7[a]		i-PrNHLi	2.2	THF		87%	5
8		LiBHEt$_3$	2.2	THF		73%	5
9[a]		DIBAL-H; then PhMgBr	1.2; 2.2	THF		84%	11
10[a]		PhLi	1	THF		96%	5
11[a]		PhLi	1	THF		79%	5
12[a]		PhLi	1	THF		61%	5

[a] reaction temp −78 to 60 °C

alkyllithium and alkyl Grignard reagents occur more efficiently in toluene than in THF (entry 2);[9] this observation may be attributable to solvent-dependent aggregation states of the organometallic reagent. In toluene, higher aggregation states of the nucleophile likely attenuate its reactivity and promote selectivity for the desired 1,2-addition over alternative pathways like proton- or electron-transfer.

Addition of stabilized carbanion nucleophiles to vinylogous acyl triflates triggers what we term the VAT-Claisen[10] reaction: a ring-opening fragmentation to give acyclic β-keto ester and related products much like those observed traditionally in the Claisen condensation (entries 5–6). Whereas the Claisen condensation is driven by deprotonation of final product, the VAT-Claisen reaction is rendered irreversible by the fragmentation event. Deprotonation of the final VAT-Claisen product may still occur incidentally, in which case the use of excess nucleophile is required to realize complete conversion (entry 5).

Other suitable nucleophiles for promoting fragmentation include primary lithium amides (entry 7) and lithium triethylborohydride (entry 8).[5] Diisobutylaluminum hydride reduces the VAT carbonyl but does not trigger fragmentation; the resulting alcohol can then be treated with excess Grignard reagents to promote fragmentation and in situ trapping of the resultant aldehyde (entry 9).[11] Other suitable VAT substrates are shown in entries 10–12. The five- and seven-membered ring VATs are less stable than their hexacyclic counterparts, and as such they are best prepared immediately prior to use.

The tandem nucleophilic addition / fragmentation of cyclic VAT substrates provides an alternative entry into the classic Eschenmoser–Tanabe pathway, one of the few Grob-type fragmentations capable of generating alkyne π-bonds. The method outlined above delivers orthogonally functionalized acyclic building blocks — alkynyl ketones and related compounds — from symmetrical cyclic dione starting materials. This methodology continues to evolve; application to the synthesis of natural products,[12] homopropargyl alcohol[13] and homopropargyl amine[14] derivatives from heterocyclic VATs, and other challenging classes of organic compounds[15] are underway at this time.

1. Department of Chemistry and Biochemistry, Florida State University, Tallahassee, FL 32306-4390, USA. E-Mail: gdudley@chem.fsu.edu

This work was supported by a grant from the National Science Foundation (NSF-CHE 0749918). M.P.L. is a recipient of the Capes-Fulbright Graduate Research Fellowship (2008). T.T.H. is a recipient of the Vietnam Education Foundation (VEF) Graduate Fellowship (2009).

2. All-Carbon Functions Polyynes, Arynes, Enynes, and Alkynes. In *Science of Synthesis: Houben–Weyl Methods of Molecular Transformation;* Thomas, E. J., Hopf, H., Eds., Thieme: Stuttgart, 2008; Vol. 43.

3. (a) Tsuji, J. *Transition Metal Reagents and Catalysts: Innovations in Organic Synthesis*; Wiley & Sons: New York, 2000. (b) *Lewis Acids in Organic Synthesis; Yamamoto*, H., Ed.; Wiley-VCH: New York, 2000.

4. (a) Larock, R. C. *Comprehensive Organic Transformations,* 2nd Ed.; Wiley & Sons: New York, 1999; pp 563–583. (b) Brandsma, L. Preparative Acetylenic Chemistry, 2nd Ed.; Elsevier: Amsterdam, 1988.

5. (a) Kamijo, S.; Dudley, G. B. *J. Am. Chem. Soc.* **2005**, *127*, 5028–5029. (b) Kamijo, S.; Dudley, G. B. *J. Am. Chem. Soc.* **2006**, *128*, 6499–6507; Addition/Correction: *J. Am. Chem. Soc.* **2010**, *132*, 8223.

6. (a) Eschenmoser, A.; Felix, D.; Ohloff, G. *Helv. Chim. Acta* **1967**, *50*, 708–713. (b) Tanabe, M.; Crowe, D. F.; Dehn, R. L. *Tetrahedron Lett.* **1967**, 3943–3946. (c) Grob, C. A.; Schiess, P. W. H. *Angew. Chem., Int. Ed. Engl.* **1967**, *6*, 1–15. (d) Weyerstahl, P.; Marschall, H. In *Comprehensive Organic Synthesis*, Trost, B. M., Fleming, I., Eds.; Pergamon Press: Elmsford, NY, 1991; Vol. 6, pp 1041–1070. (e) Prantz, K.; Mulzer, J. *Chem. Rev.* **2010**, *110*, 3741–3766.

7. (a) For an early example driven by formation of carbon dioxide, see: Grob, C. A.; Csapilla, J. Cseh, G. *Helv. Chim. Acta* **1964**, *47*, 1590–1602. (b) For recent examples driven by formation of molecular nitrogen, see: Draghici, C.; Brewer, M. *J. Am. Chem. Soc.* **2008**, *130*, 3766–3767; (c) Draghici, C.; Huang, Q.; Brewer, M. *J. Org. Chem.* **2009**, *74*, 8410–8413; (d) Bayir, A.; Draghici, C.; Brewer, M. *J. Org. Chem.* **2010**, *75*, 296–302.

8. Kocienski, P. J.; Cernigliaro, G. J. *J. Org. Chem.* **1976**, *41*, 2927–2928.

9. Jones, D. M.; Kamijo, S.; Dudley, G. B. *Synlett* **2006**, 936–938.

10. (a) Kamijo, S.; Dudley, G. B. *Org. Lett.* **2006**, *8*, 175–177. (b) Jones, D. M.; Lisboa, M. P.; Kamijo, S.; Dudley, G. B. *J. Org. Chem.* **2010**, *75*, 3260–3267.

11. Kamijo, S.; Dudley, G. B. *Tetrahedron Lett.* **2006**, *47*, 5629–5632.

12. Jones, D. M.; Dudley, G. B. *Synlett* **2010**, 223–226.
13. Tummatorn, J.; Dudley, G. B. *J. Am. Chem. Soc.* **2008**, *130*, 5050–5051.
14. Tummatorn, J.; Dudley, G. B. *Org. Lett.* **2011**, *13*, 158–160.
15. Jones, D. M.; Dudley, G. B. *Tetrahedron* **2010**, *66*, 4860–4866.

Appendix
Chemical Abstracts Nomenclature; (Registry Number)

2-Methyl-2-(1-oxo-5-heptynyl)-1,3-dithiane: 5-Heptyn-1-one, 1-(2-methyl-1,3-dithian-2-yl)-; (892874-94-1)

2-Methyl-1,3-dithiane; (6007-26-7)

n-Butyllithium; (109-72-8)

2-Methyl-3-trifluoromethansulfonyloxy-cyclohex-2-en-1-one: Methanesulfonic acid, 1,1,1-trifluoro-, 2-methyl-3-oxo-1-cyclohexen-1-yl ester; (150765-78-9)

2-Methyl-1,3-cyclohexanedione; (1193-55-1)

Trifluoromethanesulfonic anhydride; (358-23-6)

Gregory B. Dudley received a B.A. from FSU in 1995 and a Ph.D. in 2000 under the direction of Professor Rick Danheiser at MIT. After completing an NIH Postdoctoral Fellowship with Professor Samuel Danishefsky at the Sloan–Kettering Institute for Cancer Research, he returned to FSU as an assistant professor in 2002. Professor Dudley was promoted to associate professor with tenure in 2008. The current mission of his research program is to impact the drug discovery process by contributing fundamental new knowledge in organic synthesis. The addition / fragmentation methodology featured herein has figured prominently in this effort over the past few years.

Marilda P. Lisboa was born in Rio de Janeiro and raised in Minas Gerais, Brazil. She received her B.S. degree in pharmacy in 2005 and M.S. degree in pharmaceutical sciences in 2007 from the Federal University of Minas Gerais, Brazil. In 2008, she was awarded a Capes-Fulbright fellowship to pursue her PhD in Chemistry at Florida State University. Currently, she is a PhD student under the supervision of Professor Gregory Dudley and her research is focused on the synthesis of the natural product palmerolide A.

Tung T. Hoang was born in Ha Long, a small yet beautiful city known for the Ha Long Bay with thousands of islets and its sunshine beach in northeast Vietnam. Inspired by his high school teacher, Tung entered the chemistry program of Vietnam National University where he earned his B.A. degree in 2006. As a recipient of the Vietnam Education Foundation Fellowship cohort 2009, he went to Florida State University to pursue his PhD in organic chemistry. Tung is currently working under the direction of Professor Gregory Dudley with interest in the total syntheses of biologically active natural products.

Jimmie Weaver was born in 1981 in Oklahoma (US). He studied at Southern Nazarene University where he obtained his B.S. degree in chemistry in 2004. After spending the next year working under William Hildebrand (OUHSC) in an immunology lab, he began his graduate studies at the University of Kansas under the supervision of Jon Tunge. The primary focus of his graduate work was the synthesis and decarboxylative coupling of α-sulfonyl acetic esters. He completed his Ph.D. in the spring of 2010 and began a post-doctoral fellowship under the direction of Jon Ellman (Yale) where he is currently working on organocatalysis and drug delivery projects.

SYNTHESIS AND DIASTEREOSELECTIVE ALDOL REACTIONS OF A THIAZOLIDINETHIONE CHIRAL AUXILIARY

A.

B.

C.

Submitted by Michael T. Crimmins, Hamish S. Christie, and Colin O. Hughes.[1]

Checked by Eric M. Phillips and Jonathan A. Ellman.

1. Procedure

Caution! Carbon disulfide has an auto-ignition temperature of 100 °C, so extreme care should be taken to vent excess CS$_2$ away from heat sources.

Org. Synth. **2011**, *88*, 364-376
Published on the Web 4/14/2011

A. *(S)-1-(4-Benzyl-2-thioxothiazolidin-3-yl)propan-1-one* (**3**) To a 1-L, two-necked, round-bottomed flask equipped with a thermometer and magnetic stirring bar (egg shaped, 3 x 1.5 cm) is added water (250 mL) followed by portion wise addition of potassium hydroxide (KOH) (42 g, 750 mmol) (Note 1) over 5 min with stirring (Note 2). The stirring solution is cooled to ambient temperature, (*S*)-phenylalaninol (18.9 g, 125 mmol) (Note 3) and carbon disulfide (CS$_2$) (37.6 mL, 625 mmol) (Note 4) are added to the solution. The mixture rapidly becomes a red/orange color. A Friedrichs condenser is attached to the central neck of the flask and a Claisen adapter is attached to the condenser outlet joint. A hose adapter is attached to one joint of the Claisen adapter, with Tygon® plastic tubing leading to the back of the fume hood (Note 5). A septum with an inert gas inlet needle is placed in the remaining joint. A stream of argon is passed through the needle and the stirring reaction mixture is heated gradually using a heating mantle. After all the excess carbon disulfide has been purged (solution at 102–103 °C, Note 6) the red mixture is heated at reflux (110 °C, internal). After 7-8 h (Note 7), heating is discontinued and the heterogeneous mixture is allowed to cool to room temperature. The mixture is poured into a 1-L separatory funnel and extracted with CH$_2$Cl$_2$ (1 x 125 mL, 2 x 75 mL). The combined organic solution is dried over sodium sulfate (Na$_2$SO$_4$) (20 g), filtered, and evaporated in a 500-mL round-bottomed flask, using a rotary evaporator. The crude solid weighs 24.1 g (Note 8). The flask is equipped with a magnetic stirring bar (egg-shaped, 3 x 1.5 cm) and a rubber septum and flushed with argon. After CH$_2$Cl$_2$ (125 mL) (Note 9) is added, the solution is cooled using an ice/water bath. The solution is stirred vigorously and pyridine (12.3 mL, 153 mmol) (Note 10) is added rapidly via syringe, followed by propionyl chloride (12 mL, 137.5 mmol) (Note 11) via syringe over 5 min (Note 12). After 5 min, the ice/water bath is removed and the solution is allowed to warm to room temperature. After stirring for 1 h at room temperature, methanol (750 μL) is added rapidly. After 20 min the solution is poured into a 1-L separatory funnel and washed with water (100 mL) (aqueous solution back-extracted with CH$_2$Cl$_2$ (25 mL)), then 1 N NaHSO$_4$ (100 mL) (aqueous solution back-extracted with CH$_2$Cl$_2$ (25 mL). The combined organic solution is dried over Na$_2$SO$_4$, and evaporated using a rotary evaporator. The bright-yellow solid is dissolved in hot isopropyl alcohol (175 mL) (Note 13), and the solution is allowed to cool, gradually, to room temperature. After standing overnight, the yellow crystals are collected on a sintered 150-mL coarse glass funnel and washed twice with isopropanol/hexanes (9:1) solution (2 x 63 mL), then dried under vacuum (Note 14), affording 25.2–25.8 g (76–78%) of acylated derivative **3** (Note 15).

B.   *(2S,3R)-3-Hydroxy-1-[(4S)-4-benzyl-2-thioxo-thiazolidin-3-yl]-2,4-dimethyl-pentan-1-one (Evans syn aldol adduct)*. An oven-dried 250-mL single-necked, round-bottomed flask equipped with a magnetic stirring bar (egg-shaped, 3 x 1.5 cm) is charged with thiazolidinethione **3** (5.00 g, 18.8 mmol) and fitted with a rubber septum. After purging the flask with argon, anhydrous CH₂Cl₂ (60 mL) (Note 9) is added via syringe, and the flask is cooled to 0 °C using an ice/water bath. While vigorously stirring the solution, titanium tetrachloride (2.16 mL, 19.8 mmol) (Note 16) is added via syringe over three min (Note 17). After 20 min, (–)-sparteine (5.18 mL, 22.6 mmol) (Note 18) is added via syringe over 3 min (Notes 19 and 20). After another 20 min, N-methyl-2-pyrrolidone (2.16 mL, 22.6 mmol) (Note 21) is added, via syringe over three min, and the black-red mixture is stirred for 10 min before being cooled in a dry ice/acetone bath to –78 °C. To the cooled reaction mixture is added freshly distilled isobutyraldehyde (2.47 mL, 27.1 mmol) (Note 22) as a solution in 5 mL of dry CH₂Cl₂, dropwise using a syringe. After 30 min, the reaction mixture is warmed to 0 °C by immersion in an ice/water bath. After 30 min at 0 °C, saturated NH₄Cl solution (5 mL) is added to quench the reaction. The reaction mixture is quickly poured (Note 23) into a 250-mL separatory funnel, diluted with brine (80 mL) (Note 24), and extracted with CH₂Cl₂ (2 x 50 mL). The combined organic solution is dried over Na₂SO₄, filtered, and concentrated using a rotary evaporator. The resulting yellow oil (6.05 g) is purified using flash chromatography to provide 5.21 g (82%) of pure product **4** (Notes 25-27).

C.   *(2R, 3S)-3-Hydroxy-1-[(4S)-4-benzyl-2-thioxo-thiazolidin-3-yl]-2,4-dimethyl-pentan-1-one (non-Evans syn aldol adduct)*. An oven-dried 500-mL single-necked, round-bottomed flask equipped with a magnetic stirring bar (egg-shaped, 3 x 1.5 cm) is charged with thiazolidinethione **3** (8.00 g, 30.1 mmol) and fitted with a rubber septum. After purging the flask with argon, anhydrous CH₂Cl₂ (120 mL) (Note 9) is added via syringe, and the flask is cooled to 0 °C, using an ice/water bath. While vigorously stirring the solution, titanium tetrachloride (4.97 mL, 45.3 mmol) (Note 16) is added over three min via syringe (Note 17). After 20 min (–)-sparteine (7.63 mL, 33.2 mmol) (Note 18) is added via syringe over three min (Note 19). After 20 min, the flask is cooled to –78 °C in a dry ice/acetone bath. A solution of freshly distilled isobutyraldehyde (3.03 mL, 33.4 mmol) (Note 22) in 10 mL

Org. Synth. **2011**, *88*, 364-376

of anhydrous CH_2Cl_2 is added dropwise via syringe over 3 min. The resulting solution is stirred for 30 min at –78 °C before being warmed to 0 °C by immersion in an ice/water bath. After stirring for 30 min at 0 °C, saturated NH_4Cl solution (10 mL) is added to quench the reaction. The reaction mixture is quickly poured into a 500-mL separatory funnel, diluted with brine (160 mL) (Note 24), and extracted with CH_2Cl_2 (2 x 100 mL). The combined organic solution is dried over Na_2SO_4 (15 g), filtered, and concentrated using a rotary evaporator. The resulting yellow oil (12.5 g) is purified by flash chromatography to afford 8.1 g (80%) of pure product **5**, eluting with 15:85 EtOAc:hexanes (Notes 28 and 29).

2. Notes

1.   Potassium hydroxide was purchased from Fisher Scientific and used as received. A significant loss of color of the solution results during the required reaction time when only 5 equiv of potassium hydroxide is used, as described by le Corre,[2] leading to incomplete conversion.

2.   Dissolution of potassium hydroxide in water evolves heat.

3.   (S)-(–)-2-Amino-3-phenyl-1-propanol [(S)-phenylalaninol)   was purchased from Aldrich Chemical Company and used as received. The submitters note that this reagent can also be prepared and found that Masamune's[3] procedure is the most convenient of the known procedures[4,5] for preparing (S)-phenylalaninol from (S)-phenylalanine.

4.   Carbon disulfide (99.9%, ACS reagent grade) was purchased from Aldrich Chemical Company and used as received. Carbon disulfide has an auto-ignition temperature of 100 °C,[6] *so extreme care should be taken to vent excess CS₂ away from heat sources*.

5.   To ensure that escaping carbon disulfide does not pass over hot surfaces.

6.   The internal temperature reaches 102–103 °C within 1–2 h, at which point any free carbon disulfide, bp = 46 °C, is presumed to be purged from the reaction.

7.   The solution has faded somewhat to an orange color. An aliquot is removed (0.5 mL) and that solution is extracted with CH_2Cl_2 (2 x 0.5 mL). The combined organic solution is dried (Na_2SO_4) and evaporated. This sample is examined by ^1H NMR ($CDCl_3$) to confirm complete conversion (disappearance of dd at 2.8 and 2.5 ppm and the appearance of dddd at 4.45 ppm).

8.   This product is used without further purification, however, the submitters note that recrystallization from ethanol (3 mL per gram of solid) provides colorless blocks, mp 91–92 °C (uncorrected) lit. 84–85 °C.[2] The submitters provided the following characterization data: ^{1}H NMR (400 MHz, CDCl$_3$) the appearance and shift of several resonances varies depending on concentration. δ: 2.92 (dd, J = 13.6, 6.8 Hz, 1 H), 3.01 (dd, J = 13.6, 7.2 Hz, 1 H), 3.5 (dd, J = 11.2, 7.6 Hz, 1 H), 4.43 (dddd, J = 7.2, 7.2, 7.2, 7.2 Hz, 1 H), 7.14–7.18 (band, 2 H), 7.21–7.34 (band, 3 H), 8.17 (variable) (br, s, 1 H); ^{13}C NMR (100 MHz, CDCl$_3$) δ: 37.9, 39.7, 65.0, 127.2, 128.90, 128.92, 135.6, 200.6; [α]$^{24}_D$= –89 (c 4.1, CH$_2$Cl$_2$); MS (ESI) calculated for C$_{10}$H$_{11}$NNaS$_2$ [M+Na]$^+$: m/z 232.0, found m/z 232.0.

9.   Dichloromethane (99%) purchased from Fisher Scientific was passed through an activated alumina column (50 mm x 400 mm) under argon.

10.   Pyridine (anhydrous, 99.8%) was purchased from Aldrich Chemical Company and used as received.

11.   Propionyl chloride (98%) was purchased from Aldrich Chemical Company and used as received.

12.   A yellow color forms immediately, and a precipitate begins to form by the end of the addition.

13.   Isopropanol (99%) was purchased from Fisher Scientific and used as received.

14.   Dried overnight at 0.1 mmHg.

15.   Short, bright yellow needles, mp 103.4–104.8 °C (uncorrected) (mp does not change upon a second recrystallization from *i*-PrOH) ^{1}H NMR (400 MHz, CDCl$_3$) δ: 1.15 (t, J = 7.2 Hz, 3 H), 2.84 (d, J = 11.5 Hz, 1 H), 2.96–3.14 (m, 2 H), 3.18 (dd, J = 13.1, 3.7 Hz, 1 H), 3.33–3.43 (m, 2 H), 5.34 (ddd, J = 10.8, 7.2, 3.8 Hz, 1 H), 7.21–7.32 (m, 5 H); ^{13}C NMR (100 MHz, CDCl$_3$) δ: 8.8, 31.9, 32.3, 36.8, 68.7, 127.2, 128.9, 129.4, 136.6, 174.9, 201.1: [α]$^{20}_D$ = +45 (c 0.01, CH$_2$Cl$_2$); MS (EI) calculated for C$_{13}$H$_{15}$NOS$_2$ [M]$^+$: m/z 265, found: m/z 265. Anal. calcd. for C$_{13}$H$_{15}$NOS$_2$: C, 58.83; H, 5.70; N, 5.28. Found: C, 58.93; H, 5.67; N, 5.12.

16.   Titanium (IV) chloride (99.9%) was purchased from Aldrich Chemical Company and used as received.

17.   This mixture becomes a yellow-orange slurry. Since the viscosity of the solution increases notably, an increased stirring power is often necessary.

18. (-)-Sparteine (99%) was purchased from Aldrich Chemical Company and used as received.

19. Triethylamine, diisopropylethylamine, and diisopropylamine were also used as amine bases, however the diastereomeric ratios were much less than 98:2.

20. The viscosity of the solution decreases and the color becomes dark red/brown.

21. 1-Methyl-2-pyrrolidinone ((99.5%, anhydrous) was purchased from Aldrich Chemical Company and used as received.

22. Isobutyraldehyde (98%) was purchased from Aldrich Chemical Company and distilled over CaH_2 immediately prior to use.

23. If left stirring in saturated NH_4Cl solution for too long, the product will decompose.

24. Although not observed by the checkers, the submitters note that an emulsion may form in the separatory funnel, necessitating the addition of approximately 100 mL of a mixture of hexanes and ethyl acetate (1:1).

25. A 5 cm glass column was packed with 250 g of silica gel (60 Å, 40–60 μm, Sorbent Technologies). The crude product was loaded onto the column and eluted with 2.5 L of 1:5 (v:v) EtOAc:hexanes. The product was collected in 25 mL fractions between fraction numbers 44 and 79. $R_f = 0.15$ (15:85 ethyl acetate: hexanes). The product stains blue with p-anisaldehyde stain and heating. The product should be placed in a freezer for prolonged storage.

26. The subproduct is essentially pure at this stage, but the submitters state that it can be crystallized from 50 mL of toluene/hexanes solution (1:19, seeding required). Seeds can be obtained by allowing the material obtained after chromatography to stand for several days (the checkers were unable to obtain seed crystals).

27. ^{1}H NMR (500 MHz, CDCl$_3$) δ: 0.88 (d, J = 6.8 Hz, 3 H), 1.02 (d, J = 6.6 Hz, 3 H), 1.24 (d, J = 6.8 Hz, 3 H), 1.68 (m, 1 H), 2.65 (d, J = 3.9 Hz, 1 H), 2.90 (d, J = 11.5 Hz, 1 H), 3.05 (dd, J = 13.1, 10.6 Hz, 1 H), 3.23 (dd, J = 13.1, 3.8 Hz, 1 H), 3.40 (dd, J = 11.5, 7.1 Hz, 1 H), 3.54 (ddd, J = 7.4, 3.5, 3.5 Hz, 1 H), 4.70 (dddd, J = 6.9, 6.9, 6.9, 3.2 Hz, 1 H), 5.34 (ddd, J = 10.7, 6.9, 3.8 Hz, 1 H), 7.37–7.27 (m, 5 H); ^{13}C NMR (125 MHz, CDCl$_3$) δ: 10.2, 18.9, 19.0, 31.1, 32.0, 36.7, 41.0, 68.8, 77.4, 127.2, 128.9, 129.4, 136.3, 178.5, 201.1; IR (cm^{-1}) 3494 (br), 2960, 1687 (w); [α]$^{20}_D$= +152 (c 0.017, CH$_2$Cl$_2$); MS (ESI) calculated for C$_{17}$H$_{23}$NNaO$_2$S$_2$ [M+Na]$^+$: m/z

360, found m/z 360. Anal. calcd. for $C_{17}H_{23}NO_2S_2$: C, 60.50; H, 6.87; N, 4.15. Found: C, 60.82; H, 6.81; N, 4.10.

28. The submitters used a 7.5 cm glass column packed with 340 g of silica gel (60 Å, 40 – 60 μm, Sorbent Technologies). The crude product was loaded onto the column and eluted with 4 L of 15:85 (v:v) EtOAc:hexanes. The product was collected in 25 mL fractions between fraction numbers 41 and 80. $R_f = 0.33$ (15:85 ethyl acetate: hexanes).

29. ^1H NMR (400 MHz, CDCl$_3$) δ: 0.88 (d, J = 6.8, 3 H), 1.06 (d, J = 6.5 Hz, 3 H), 1.18 (d, J = 6.8 Hz, 3 H), 1.66–1.75 (m, 1 H), 2.89 (d, J = 11.6 Hz, 1 H), 2.95 (d, J = 2.8 Hz, 1 H) (may appear as a broad singlet), 3.05 (dd, J = 10.5, 13.2 Hz, 1 H), 3.25 (dd, J = 3.9, 13.2 Hz, 1 H), 3.37 (dd, J = 7.1, 11.5 Hz, 1 H), 3.64 (ddd, J = 2.3, 2.3, 8.9 Hz, 1 H), 4.93 (qd, J = 2.1, 7.1 Hz, 1 H), 5.35 (ddd, J = 4.1, 6.9, 10.7 Hz, 1 H) 7.27–7.37 (m, 5 H); ^{13}C (100 MHz, CDCl$_3$) δ: 10.3, 18.8, 19.7, 30.7, 31.7, 36.9, 40.3, 68.9, 76.0, 127.3, 128.9, 129.4, 136.4, 179.2, 201.4; IR v 3550 (br), 2960 (s), 1675 (s) $[\alpha]^{20}_D$= +146 (c 0.026, CH$_2$Cl$_2$); MS (ESI) calculated for $C_{17}H_{23}NNaO_2S_2$ [M+Na]$^+$: m/z 360, found m/z 360. . Anal. calcd. for $C_{17}H_{23}NO_2S_2$: C, 60.50; H, 6.87; N, 4.15. Found: C, 60.78; H, 6.78; N, 4.03.

Safety and Waste Disposal Information

All hazardous materials should be handled and disposed of in accordance with "Prudent Practices in the Laboratory"; National Academy Press; Washington, DC, 1995.

3. Discussion

The preparation and use of thiazolidinethione reagent **3** for the synthesis of aldol products, described in this procedure, is simple and offers many advantages over other available methods. The aldol reaction has found many applications in the total synthesis of natural products.[7] Chiral auxiliary based methods continue to be the most versatile and reliable of the stereoselective aldol reactions. The high selectivities observed with a range of aldehyde reactants, coupled with relatively easy separations, due to the diastereomeric nature of any mixed products, makes these user friendly processes. The Evans syn-aldol procedure employing amino acid-derived oxazolidinone chiral auxiliaries.[4,8] is particularly useful and is widely applied. A useful modification to the Evans procedure has been to replace

370

the oxazolidinone chiral auxiliaries with the thiazolidinethione counter parts.[9] The aldol reaction using these compounds is more user friendly for several reasons: 1) Inexpensive, easily handled, and readily available TiCl$_4$ can be used as the Lewis acid source, rather than Bu$_2$BOTf, required for the Evans procedure. 2) The use of a chlorotitanium enolate also permits a standard acidic work-up to be used, as opposed to the oxidative process needed when alkyl boranes are employed as the Lewis acid. Importantly, as demonstrated in this procedure, the choice of base, and base stoichiometry, allows the preparation of either the typical Evans syn- or the non-Evans syn-aldol product.[9] A variety of functional and protecting groups are tolerated in the reactions (see Tables 1 and 2).

The change in facial selectivity in the aldol additions is postulated to be the result of a switch between chelated and nonchelated transition states as illustrated in Scheme 1. In each case, the aldehyde approaches the enolate from the face opposite the benzyl substituent on the auxiliary and the

Scheme 1

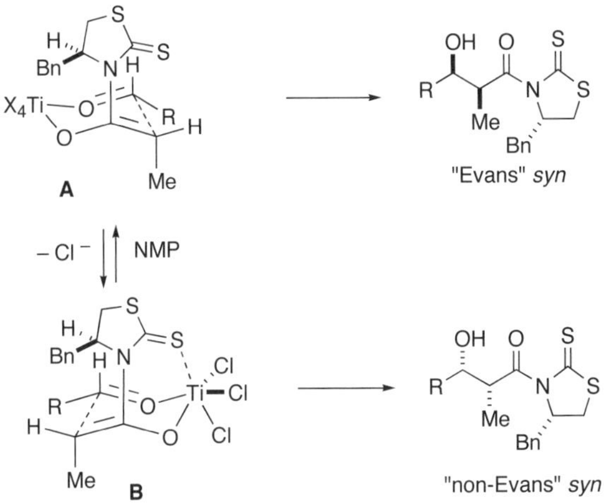

aldehyde carbonyl oxygen is coordinated to the metal center allowing for a six-membered transition state. When the enolate formation and aldol addition are carried out in the absence of excess diamine or another additional ligand for the metal, it is proposed that the nucleophilic thiocarbonyl displaces chloride from the metal center and the reaction proceeds through a chelated transition state **B** giving rise to "non-Evans" *syn* aldol adducts. When NMP or additional diamine are added as an extra

Aldehyde	Evans Syn aldol adduct	yield(%)	ratio
(isobutyl)CHO	OH / O / S ... Me, Bn	74	96:4
(allyl/propenyl)CHO	OH / O / S ... Me, Bn	84	98:2
(vinyl)CHO	OH / O / S ... Me, Bn	97	98:2
Ph–CH=CH–CHO	Ph / OH / O / S ... Me, Bn	98	96:4
Ph–CHO	Ph / OH / O / S ... Me, Bn	99	96:4
OTIPS, Me, TIPSO, CO₂Me, CHO	OTIPS, Me, TIPSO, CO₂Me, OH, O, S ... Me, Bn	94	>95:5
OMe OTES, Me Me, CHO	OMe OTES OH, O, S, Me Me, Me, Bn	65	>95:5
OBn, MeO, CHO	OBn OH, MeO, O, S, Me, Bn	90	>95:5
TBSO, CHO	TBSO, OH, O, S, Me, Bn	78	>95:5

Table 2: non-Evans syn aldol reactions of aldehydes and *N*-propionylthiazolidinethione **3**

Aldehyde	non-Evans Syn aldol adduct	yield(%)	ratio
⌇CHO		91	>95:5
⌇CHO		45	99:1
⌇CHO		49	99:1
Ph–CH=CH–CHO		58	97:3
Ph–CHO		52	99:1
⌇CHO		57	98:2
MeO⌇OBn OTES Me CHO		62	>95:5
⌇CHO		73	>95:5

ligand for the metal center, the chelation between the sulfur of the thiocarbonyl of the auxiliary and titanium is disrupted, leading to a dipole-minimized, non-chelated transition state **A** giving rise to the "Evans" *syn* aldol adduct.

The change in diastereoselectivity through changing reaction conditions enables the use of the same enantiomer of chiral auxiliary to form either enantiomer of a syn-aldol product (after removal of the auxiliary). Once the thiazolidinethione aldol products are formed they are more versatile (than the oxazolidinone products) in the range of transformations that they can undergo. One step reductive removal of the auxiliary to form an aldehyde,[10] and addition of carbon nucleophiles, such as phosphonates[11] and ester enolates[12] are examples of reactions that cannot be directly performed with the equivalent oxazolidinone aldol products.

1. Department of Chemistry, CB 3290, University of North Carolina at Chapel Hill, Chapel Hill, NC 27599-3290; Crimmins@email.unc.edu.
2. Delaunay, D., Toupet, L., Le Corre, M. *J. Org. Chem.* **1995**, 60, 6604–6607.
3. Abiko, A., Masamune, S. *Tetrahedron Lett.* **1992**, 33, 5517–5518.
4. Gage, J. R.; Evans, D. A. *Org. Synth.* **1989**, 68, 77–82.
5. For a list and references to several known procedures see Ref. 2.
6. *The Merck Index, 12th Edition,* **1996**, 295.
7. For example, see: (a) Crimmins, M. T.; Christie, H. S.; Chaudhary, K.; Long, A. *J. Am. Chem. Soc.* **2005**, *127*, 13810–13812. (b) Crimmins, M. T.; DeBaillie, A. C. *J. Am. Chem. Soc.* **2006**, *128*, 4936–4937. (c) Crimmins, M. T.; Slade, D. J. *Org. Lett.* **2006**, *8*, 2191–2194. (d) Crimmins, M. T.; Caussanel, F. *J. Am. Chem. Soc.* **2006**, *128*, 3128–3129. (e) O'Neil, G. W.; Phillips, A. J. *J. Am. Chem. Soc.* **2006**, *128*, 5340–5341. (f) Crimmins, M. T.; Stevens, J. M.; Schaaf, G. M. *Org. Lett.* **2009**, *11*, 3990–3993. (g) Crimmins, M. T; Dechert, A. M. *Org. Lett.* **2009**, *11*, 1635–1638.
8. Gage, J. R.; Evans, D. A. *Org. Synth.* **1989**, 68, 83-90.
9. (a) Crimmins, M. T.; King, B. W.; Tabet, E. A., Chaudhary, K. *J. Org. Chem.* **2001**, *66*, 894–902. (b) Crimmins, M. T.; Chaudhary, K. *Org. Lett.* **2000**, *2*, 775–777.
10. Sano, S.; Kobayashi, Y.; Kondo, T.; Takebayashi, M.; Maruyama, S.; Fujita, T.; Nagao, Y. *Tetrahedron Lett.* **1995**, 36, 2097–2100.
11. (a) Delamarche, I.; Mosset, P. *J. Org. Chem.* **1994**, *59*, 5453–5457. (b) Astles, P. C.; Thomas, E. J. *J. Chem. Soc., Perkin Trans. I* **1997**, 845–856.
12. Smith, T. E.; Djang, M.; Velander, A. J.; Downey, C. W.; Carroll, K. A.; van Alphen, S. *Org. Lett.* **2004**, *6*, 2317–2320.

Appendix
Chemical Abstracts Nomenclature; (Registry Number)

(S)-1-(4-Benzyl-2-thioxothiazolidin-3-yl)propan-1-one: 1-Propanone, 1-
[(4S)-4-(phenylmethyl)-2-thioxo-3-thiazolidinyl]-; (263764-23-4)

(2S, 3R)-3-Hydroxy-1-[(4S)-4-benzyl-2-thioxo-thiazolidin-3-yl]-2,4-
dimethyl-pentan-1-one: 1-Pentanone, 3-hydroxy-2,4-dimethyl-1-[(4S)-4-
(phenylmethyl)-2-thioxo-3-thiazolidinyl]-, (2S,3R)-; (263764-33-6)

(2R, 3S)-3-Hydroxy-1-[(4S)-4-benzyl-2-thioxo-thiazolidin-3-yl]-2,4-
dimethyl-pentan-1-one: 1-Pentanone, 3-hydroxy-2,4-dimethyl-1-[(4S)-
4-(phenylmethyl)-2-thioxo-3-thiazolidinyl]-, (2R,3S)-; (331283-30-8)

(S)-Phenylalaninol: Benzenepropanol, β-amino-, (βS)-: (3182-95-4)

Carbon disulfide; (75-15-0)

Propionyl chloride: Propanoyl chloride: (79-03-8)

Pyridine; (110-86-1)

2-Thiazolidinethione, 4-(phenylmethyl)-, (4S)-; (171877-39-7)

N-Methyl-2-pyrrolidone: 2-Pyrrolidinone, 1-methyl-: (872-50-4)

(–)-Sparteine: 7,14-Methano-2H,6H-dipyrido[1,2-a:1',2'-e][1,5]diazocine,
dodecahydro-, (7S,7aR,14S,14aS)-; (90-39-1)

Isobutyraldehyde: Propanal, 2-methyl-; (78-84-2)

Titanium tetrachloride: (7550-45-0)

Michael T. Crimmins is currently Mary Ann Smith Distinguished Professor of Chemistry and Senior Associate Dean for the Natural Sciences in the College of Arts and Sciences. He received his B.A. degree from Hendrix College, his Ph.D. from Duke and was a postdoctoral associate at the California Institute of Technology. His research interests are in the development of new synthetic methods and their application to the total synthesis of biologically active compounds. Professor Crimmins' research has been recognized by a number of awards including the Charles H. Herty Medal and the Ernest Guenther Award in the Chemistry of Natural Products.

Hamish Christie received his B. Sc. and M. Sc. degrees in Chemistry in 1996 and 1998 respectively, from the University of Adelaide, Australia. He then conducted his Ph.D. studies at the University of California, Berkeley working on the total synthesis of marine alkaloids under the supervision of Clayton Heathcock. In 2003 he began post-doctoral research with Michael Crimmins at the University of North Carolina, Chapel Hill, working on polyketide natural product synthesis. In 2006 he joined the faculty of the University of Arizona. His research interests include the development of small molecule catalysts and the total synthesis of complex organic compounds.

Colin Hughes was born in Berkeley California in 1984. He earned his B.S degree in Chemistry from the University of California at Berkeley in 2006. Later that year he joined the lab of Professor Michael Crimmins at the University of North Carolina at Chapel Hill, where his research has focused on the total synthesis of aldingenin B.

Eric Phillips was born in Grand Rapids, MI in 1983. After receiving his B.S. degree in chemistry from Western Michigan University in 2005, he attended graduate school under the guidance of Prof. Karl Scheidt. Upon graduating in 2010, he became a Ruth Kirschstein NIH post-doctoral fellow in Prof. Jon Ellman's lab at Yale University where his research is focused on transition metal-catalyzed C-H insertion reactions.

SYNTHESIS OF 2,3-DISUBSTITUTED INDOLES VIA PALLADIUM-CATALYZED ANNULATION OF INTERNAL ALK YNES: 3-METHYL-2-(TRIMETHYLSILYL)INDOLE

Submitted by Yu Chen, Nataliya A. Markina, Tuanli Yao, and Richard C. Larock.[1,2]

Checked by Kevin M. Allan and Viresh H. Rawal.

1. Procedure

3-Methyl-2-(trimethylsilyl)indole. A flame-dried 1-L, three-necked round-bottomed flask equipped with a 15 mm × 32 mm football-shaped magnetic stirring bar, a thermometer, a reflux condenser fitted with a nitrogen inlet, and a glass stopper is charged with 2-iodoaniline (**1**, 10.95 g, 50.0 mmol, 1.0 equiv) (Note 1), palladium(II) acetate (281 mg, 1.25 mmol, 0.025 equiv) (Note 2), triphenylphosphine (656 mg, 2.50 mmol, 0.050 equiv) (Note 3), sodium carbonate (13.25 g, 125.0 mmol, 2.5 equiv) (Note 4), tetra-*n*-butylammonium chloride (13.91 g, 50.0 mmol, 1.0 equiv) (Notes 5–8), and *N,N*-dimethylformamide (DMF, 500 mL) (Note 9). The mixture is stirred for 10 min at room temperature under a nitrogen atmosphere resulting in an orange suspension. 1-Trimethylsilyl-1-propyne (**2**, 16.98 g, 22.4 mL, 151.3 mmol, 3.0 equiv) (Note 10) is then added to the flask via syringe. The reaction flask is immersed in a preheated oil bath (at 110 °C) equipped with an external thermometer (Note 11). The reaction mixture is stirred at an internal temperature of 100 °C for 20 h until the reaction is complete (Notes 12 and 13). The resulting dark brown suspension is cooled to room temperature. The glass stopper is replaced with a pressure-equalizing dropping funnel. Saturated aqueous NH₄Cl solution (100 mL) is added to the reaction mixture through the pressure-equalizing dropping funnel over 10 min with vigorous stirring (Note 14). The resulting two-phase mixture is transferred to a 2-L round-bottomed flask equipped with a 15 mm × 32 mm football-shaped magnetic stirring bar. The original three-necked round-bottomed flask is rinsed with diethyl ether (2 × 30 mL) followed by saturated aqueous NH₄Cl solution (50 mL), and each rinse is transferred to the 2 L round-bottomed flask. The resulting mixture is sequentially treated with saturated aqueous NH₄Cl solution (220 mL) and diethyl ether (320 mL). The resulting two-phase mixture is stirred for 5 min at room temperature and transferred to a 2-L separatory funnel. The layers are separated and the aqueous phase is extracted three times with diethyl ether (3 × 150 mL) (Note 15). The combined organic phases are washed two times with water (2 × 100 mL), dried over anhydrous magnesium sulfate (MgSO₄) (Note 16), filtered through a medium porosity fritted glass funnel, rinsed with diethyl ether (50 mL), and concentrated by rotary evaporation (23 °C, 20 mmHg) to give a dark brown oil. The residue is purified by flash column chromatography on silica gel (Note 17) to afford 9.28 g (91%) of indole **3** as an orange oil (Notes 18 and 19).

2. Notes

1. The submitters purchased 2-iodoaniline (**1**, 98%) from TCI America and used it as received. The checkers purchased 2-iodoaniline (**1**, 98%) from Sigma-Aldrich and used it as received.

2. The submitters received palladium(II) acetate donated by Kawaken Fine Chemicals Co., Ltd., and Johnson Matthey, Inc. and used it as received. The checkers purchased palladium(II) acetate (98%) from Sigma-Aldrich and used it as received.

3. The submitters and checkers each purchased triphenylphosphine (PPh₃, 99%) from Sigma-Aldrich and used it as received.

4. The submitters and checkers each purchased sodium carbonate (Na₂CO₃, 99.7%, anhydrous) from Fisher Scientific and used it as received.

5. The submitters purchased tetra-*n*-butylammonium chloride (*n*-Bu₄NCl, >97%) from TCI America and used it as received. The checkers purchased tetra-*n*-butylammonium chloride (*n*-Bu₄NCl, >97%) from Sigma-Aldrich and used it as received.

6. Due to its highly hygroscopic character, the submitters weighed tetra-*n*-butylammonium chloride in a glove box and added it quickly to the reaction flask. The checkers weighed tetra-*n*-butylammonium chloride on the bench top and added it quickly to the reaction flask.

7. In general, LiCl was observed to be more effective and reproducible than n-Bu4NCl in the Larock indole synthesis; however, n-Bu4NCl is superior in the current described reaction. A lower yield was obtained when LiCl was employed instead of n-Bu4NCl.

8. The amount of n-Bu4NCl is critical. More than 1 equiv of n-Bu4NCl favors the formation of multiple insertion products and sharply lowers the yield of the annulation reaction.

9. The submitters purchased N,N-dimethylformamide (DMF, 99.8%, anhydrous) from Sigma-Aldrich and used it as received. The checkers purchased N,N-dimethylformamide (DMF, 99.8%) from Acros Organics and passed it over a column of activated alumina under positive nitrogen pressure prior to use. Dry DMF was collected from the column in a flame-dried 500-mL round-bottomed flask and transferred to the reaction flask via cannula under positive nitrogen pressure.

10. The submitters purchased 1-trimethylsilyl-1-propyne (**2**, 99%) from TCI America and used it as received. The checkers purchased 1-trimethylsilyl-1-propyne (**2**, 99%) from Sigma-Aldrich and used it as received. In order to achieve the best chemical yields, at least 3 equiv of 1-trimethylsilyl-1-propyne are required based on the submitters' experience (boiling point of **2** = 99–100 °C).

11. In order to maintain the internal reaction temperature at 100 °C, the temperature of the oil bath is set at 110 °C. Based on the submitters' experience, there is no significant effect on the result when the oil bath temperature fluctuates between 108 and 112 °C during the course of the reaction.

12. The submitters reported a 22 h reaction time.

13. The completeness of the reaction was judged by the disappearance of 2-iodoaniline by thin-layer chromatography performed on glass-backed pre-coated 60 Å silica gel plates (250 μm) with a UV254 indicator. The submitters obtained TLC plates from Sorbent Technologies and used 30:1 hexane/ethyl acetate as the eluent. (R_f of 2-iodoaniline (**1**) = 0.32; R_f of the product (**3**) = 0.51). The checkers obtained TLC plates from Dynamic Adsorbents and used 20:1 hexane/ethyl acetate as the eluent (R_f of 2-iodoaniline (**1**) = 0.21; the R_f of the product (**3**) = 0.36).

14. In order to dissolve all of the inorganic salts in the reaction mixture, a saturated aqueous NH4Cl solution is added. The addition of aqueous NH4Cl solution is a slightly exothermic process. A slow addition and vigorous stirring of the reaction mixture is recommended. Based on the

submitters' experience, the internal temperature of the two-phase mixture was below 30 °C when the aqueous solution was added over a period of ten minutes.

15. During aqueous work-up, the checkers noted fine black particulate matter suspended at the bottom of the organic phase. The organic layer was decanted away from these solids after each extraction. After washing the combined organic layers with water and decanting, the solids were rinsed with diethyl ether (20 mL), and the rinse was combined with the organic phase before drying over $MgSO_4$.

16. Anhydrous magnesium sulfate was purchased from Fisher Scientific and used as received. To ensure proper dryness, 55 g of $MgSO_4$ was added to the organic phase and the resulting mixture was kept at room temperature for 15 min with occasional swirling.

17. Column chromatography is performed on an 8 cm diameter column, wet-packed with 400 g of silica gel (SiliCycle Silia*Flash* P60 40–63 μm 60Å) in hexanes. The length of silica gel is 25 cm. A gradient of 30:1 hexane/ethyl acetate (2 L) followed by 20:1 hexane/ethyl acetate (2 L) is used as the eluent. Three 200 mL fractions are collected and set aside. The next sixty-four 65 mL fractions are collected. Among the 65 mL fractions, fractions 35–56 contained the desired product and were concentrated by rotary evaporation (23 °C bath, 20 mm Hg), and dried under vacuum (1.0 Torr) at 23 °C for 18 h until a constant weight (9.28 g) was obtained.

18. The physical properties of **3** follow: R_f = 0.36 (TLC analysis performed on glass-backed pre-coated 60 Å silica gel plates (250 μm) with a UV254 indicator obtained from Dynamic Adsorbents; 20:1 hexane/ethyl acetate is used as the eluent; the product is visualized with a 254 nm UV lamp and basic $KMnO_4$ stain); ^1H NMR (500 MHz, $CDCl_3$) δ: 0.40 (s, 9 H), 2.44 (s, 3 H), 7.12 (ddd, J = 7.9, 7.0, 0.9 Hz, 1 H), 7.20 (ddd, J = 8.1, 7.0, 1.1 Hz, 1 H), 7.37 (app d, J = 8.1 Hz, 1 H), 7.59 (app d, J = 7.9 Hz, 1 H), 7.88 (br s, 1 H); ^{13}C NMR (125 MHz, $CDCl_3$) δ: –0.7, 10.6, 110.9, 118.8, 119.2, 120.5, 122.4, 129.7, 133.1, 138.2; IR (neat film, NaCl) 3440, 2954, 1250 cm^{-1}; HRMS m/z calcd. for $C_{12}H_{17}NSi$ [M+H]$^+$ 204.1203, found 204.1189. Anal. calcd. for $C_{12}H_{17}NSi$: C, 70.88; H, 8.43; N, 6.89. Found: C, 70.73; H, 8.39; N, 6.94.

19. The product 3-methyl-2-(trimethylsilyl)indole (**3**) is generally stable under the current described work-up and separation conditions. However, significant decomposition (>90%) of **3** was observed after storing as a dichloromethane solution for four weeks. In case long term storage is

needed, it is recommended that this compound be evaporated to complete dryness before storage and that it be stored at a low temperature (–20 °C).

Safety and Waste Disposal Information

All hazardous materials should be handled and disposed of in accordance with "Prudent Practices in the Laboratory"; National Academy Press; Washington, DC, 1995.

3. Discussion

Palladium-catalyzed annulation processes have proven to be very powerful for the synthesis of a wide variety of heterocycles and carbocycles.[3] The preparation of 3-methyl-2-(trimethylsilyl)indole described here illustrates a general protocol for the palladium-catalyzed coupling of *ortho*-iodoaniline and its derivatives with internal alkynes to produce 2,3-disubstituted indoles.[4] Since this process was first communicated in 1991, it has been subsequently employed by others in the synthesis of potential migraine headache drugs and other indole heterocycles.[5] This approach to 2,3-disubstituted indoles is very versatile. *ortho*-Iodoanilines with a variety of substituents on the nitrogen moiety, such as methyl, acetyl, and tosyl groups, undergo the annulation process successfully (Table 1). On the other hand, 2-bromoaniline and its derivatives are unreactive under our general annulation conditions. However, improved procedures for less reactive halides have more recently been reported.[6] A wide variety of internal alkynes bearing alkyl, aryl, silyl, ester and alcohol-containing groups, and hindered or unhindered substituents have been successfully employed in this process.

Table 1. Palladium-catalyzed annulation of internal alkynes with 2-iodoanilines or *N*-substituted derivatives[a]

entry	2-iodoaniline	alkyne	product	% yield
1		Me—≡—^{t}Bu		82
2		Et—≡—(cyclohexyl-OH)		85
3		Ph—≡—SiMe$_3$		68
4		Me$_2$C=CH—≡—C(Me)$_2$OH		70
5		^{n}Pr—≡—^{n}Pr		71
6		Ph—≡—Me		75
7		Me—≡—SiMe$_3$		70
8		Me—≡—^{t}Bu		86

a. All reactions were run at 100 °C in DMF with 5 mol % of Pd(OAc)$_2$, 1 equiv of ^{n}Bu$_4$NCl or LiCl, 5 equiv of base, and where appropriate, 5 mol % of PPh$_3$. For details, see reference 4b.

It is worth noting that this annulation process is often highly regioselective in the case of unsymmetrical internal alkynes, placing the aryl group of the aniline on the less sterically hindered end of the triple bond and the nitrogen moiety on the more sterically hindered end (Figure 1). Complete regioselectivity is observed in the current example. 3-Methyl-2-(trimethylsilyl)indole is the sole product obtained. It is noteworthy that the 2-silyl-substituted indoles generated by this annulation process can undergo a variety of other synthetically useful substitution processes, such as halogenation, protonolysis, and the Heck reaction, providing a convenient entry into various other substituted indoles.[4] The present procedure is a slight modification of that previously reported by the Larock group.[4] This general process has also been used to prepare indoles tethered to a polymer support.[7]

Figure 1. Steric effects on the regiochemistry of alkyne insertion.

This indole synthesis presumably proceeds via the mechanism displayed in Scheme 1: (1) reduction of the Pd(OAc)$_2$ to Pd(0), (2) coordination of the chloride to form a chloride-ligated zerovalent palladium species, (3) oxidative addition of the aryl iodide to Pd(0) to form intermediate **A**,[8] (4) coordination of the alkyne to the palladium atom of intermediate **A** and subsequent regioselective *syn*-insertion into the arylpalladium bond to form the vinylic palladium intermediate **B**, (5) nitrogen displacement of the halide in intermediate **B** to form a six-membered, heteroatom-containing palladacycle **C**, and (6) reductive elimination to form the indole and regenerate Pd(0).

Scheme 1. Proposed mechanism for the palladium-catalyzed annulation of internal alkynes by 2-iodoanilines or *N*-substituted derivatives.

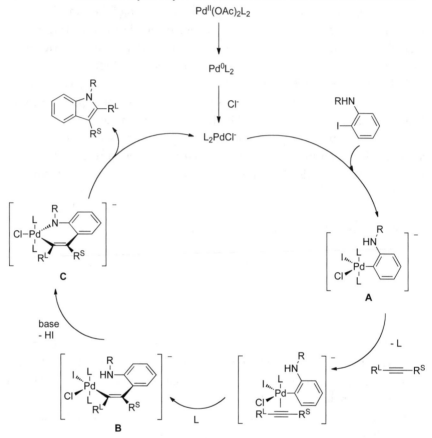

1. Department of Chemistry, Iowa State University, Ames, IA 50011; larock@iastate.edu. We gratefully acknowledge the National Institutes of Health and National Science Foundation for their generous financial support; and Johnson Matthey, Inc., and Kawaken Fine Chemicals Co., Inc., for the palladium reagents.
2. The non-corresponding authors' names are in alphabetical order.
3. For reviews, see: (a) Larock, R. C. *J. Organomet. Chem.* **1999**, *576*, 111-124. (b) Larock R. C.; "Palladium-Catalyzed Annulation" in *Perspectives in Organopalladium Chemistry for the XXI Century*, Ed. J. Tsuji, Elsevier Press, Lausanne, Switzerland, 1999, pp. 111-124. (c) Larock, R. C. *Pure Appl. Chem.* **1999**, *71*, 1435-1442.

384

4. (a) Larock, R. C.; Yum, E. K. *J. Am. Chem. Soc.* **1991**, *113*, 6689-6690.
    (b) Larock, R. C.; Yum, E. K.; Refvik, M. D. *J. Org. Chem.* **1998**, *63*,
    7652-7662.

5. For selected examples, see: (a) Leogane, O.; Lebel, H. *Angew. Chem.,*
    Int. Ed. **2008**, *47*, 350-352. (b) Hong, K. B.; Lee, C. W.; Yum, E. K.
    Tetrahedron Lett. **2004**, *45*, 693-697. (c) Huang, Q.; Larock, R. C. *J.*
    Org. Chem. **2003**, *68*, 7342-7349. (d) Zhang, H.; Larock, R. C. *J. Org.*
    Chem. **2003**, *68*, 5132-5138. (e) Nishikawa, T.; Wada, K.; Isobe, M.
    Biosci. Biotechnol. Biochem. **2002**, *66*, 2273-2278. (f) Mackman, R.
    L.; Katz, B. A.; Breitenbucher, J. G.; Hui, H. C.; Verner, E.; Luong, C.;
    Liu, L.; Sprengeler, P. A. *J. Med. Chem.* **2001**, *44*, 3856-3871. (g)
    Park, S. S.; Choi, J.-K.; Yum, E. K.; Ha, D.-C. *Tetrahedron Lett.* **1998**,
    39, 627-630. (h) Chen, C.; Lieberman, D. R.; Larsen, R. D.; Reamer,
    R. A.; Verhoeven, T. R.; Reider, P. J.; Cottrell, I. F.; Houghton, P. G.
    Tetrahedron Lett. **1994**, *35*, 6981-6984. (i) Wensbo, D.; Eriksson, A.;
    Jeschke, T.; Annby, U.; Gronowitz, S.; Cohen, L. A. *Tetrahedron Lett.*
    1993, *34*, 2823-2826.

6. (a) Shen, M.; Li, G.; Lu, B. Z.; Hossain, A.; Roschangar, F.; Farina, V.;
    Senanayake, C. H. *Org. Lett.* **2004**, *6*, 4129-4132. (b) Liu, J.; Shen, M.;
    Zhang, Y.; Li, G.; Khodabocus, A.; Rodriguez, S.; Qu, B.; Farina, V.;
    Senanayake, C. H.; Lu, B. Z. *Org. Lett.* **2006**, *8*, 3573-3575.

7. (a) Zhang, H.-C.; Brumfield, K. K.; Maryanoff, B. E. *Tetrahedron Lett.*
    1997, *38*, 2439-2442. (b) Smith, A. L.; Stevenson, G. I.; Swain, C. J.;
    Castro, J. L. *Tetrahedron Lett.* **1998**, *39*, 8317-8320.

8. It has been noted by Board members that an alternative role for the
    chloride may be to displace the iodide in intermediate **A**.

Appendix
Chemical Abstracts Nomenclature; (Registry Number)

Palladium acetate; (3375-31-3)
Triphenylphosphine; (603-35-0)
Sodium carbonate; (497-19-8)
Tetra-*n*-butylammonium chloride; (1112-67-0)
N,N-Dimethylformamide; (68-12-2)
1-Trimethylsilyl-1-propyne; (6224-91-5)

Richard C. Larock received his B.S. at the University of California, Davis in 1967. He then joined the group of Prof. Herbert C. Brown at Purdue University, where he received his Ph.D. in 1972. He worked as an NSF Postdoctoral Fellow at Harvard University in Prof. E. J. Corey's group and joined the Iowa State University faculty in 1972. His current research interests include aryne chemistry, electrophilic cyclization, palladium catalysis, and polymer chemistry based on biorenewable resources.

Yu Chen received his B.S. and M.S. degrees at Nankai University in China. He then joined Professor Andrei Yudin's research group at the University of Toronto working in the field of asymmetric catalysis, where he obtained his Ph.D. degree in 2005. After a two-year industrial appointment, he joined Professor Richard Larock's research group at Iowa State University as a postdoctoral fellow in 2007, working on an NIH-funded pilot-scale heterocyclic and carbocyclic library synthesis project. In 2009, he joined Queens College at City University of New York as an assistant professor. His research interests include late transition metal catalysis and asymmetric synthesis.

Nataliya A. Markina received her Specialist (B.S./M.S.) degree at the Higher Chemical College of the Russian Academy of Sciences in Moscow, Russia in 2008. She then joined Professor Richard Larock's research group at Iowa State University, where she is currently pursuing her Ph.D. degree. Her graduate research focuses on NIH-funded heterocyclic library synthesis, multicomponent transition metal-catalyzed processes and chemistry of arynes.

Tuanli Yao earned his B.S. and M.S. degrees in chemistry from Peking University in China, where he carried out research on the synthesis of C(10)-(4-sulfonatophenyl)biliverdin analogs under Professors Jinshi Ma and Sheng Jin. He obtained his Ph.D. in 2005 from Iowa State University working with Professor Richard C. Larock. His graduate research at Iowa State focused on new approaches to heterocycles and carbocycles. After postdoctoral research with Professor Richmond Sarpong at U.C. Berkeley, he joined Deciphera Pharmaceuticals in Lawrence, Kansas.

Kevin M. Allan received his B.S. degree in chemistry in 2004 from U.C. Berkeley where he conducted research with Dr. Ahamindra Jain. He then joined the labs of Professor Brian M. Stoltz at Caltech. His graduate studies focused on the development of new aryne annulation reactions and their application in alkaloid total synthesis. After obtaining his Ph.D. in 2010, he moved to the University of Chicago, where he is currently a postdoctoral researcher in the labs of Professor Viresh H. Rawal.

AN INTRAMOLECULAR AMINATION OF ARYL HALIDES WITH A COMBINATION OF COPPER (I) IODIDE AND CESIUM ACETATE: PREPARATION OF 5,6-DIMETHOXYINDOLE-1,2-DICARBOXYLIC ACID 1-BENZYL ESTER 2-METHYL ESTER

Submitted by Toshiharu Noji[1], Kentaro Okano[1], Tohru Fukuyama[2], and Hidetoshi Tokuyama.[1]

Checked by Mark Lautens and Lei Zhang.

1. Procedure

A.   *(Z)-2-Benzyloxycarbonylamino-3-(2-bromo-4,5-dimethoxyphenyl) acrylic acid methyl ester.* A 300-mL three-necked round-bottomed flask equipped with a Teflon-coated magnetic stir bar (3.5 x 1.0 cm), a rubber septum, a glass stopper, and an argon gas inlet is charged with 6-bromoveratraldehyde (7.35 g, 30.0 mmol) (Note 1) and methyl 2-benzyloxycarbonylamino-2-(dimethoxyphosphinyl)acetate (10.43 g, 31.5 mmol, 1.05 equiv). The flask is evacuated and backfilled with argon and is charged with dry dichloromethane (30 mL) (Note 2). The resulting mixture is cooled to 0 °C with stirring at 300 rpm, and to the suspension is added 1,1,3,3-tetramethylguanidine (4.14 mL, 33.0 mmol, 1.1 equiv) (Note 3) dropwise over 10 min (Note 4). The cooling bath is then removed, and the mixture is stirred at room temperature for 5 h (Note 5). The reaction is quenched with 1 M aq. HCl (30 mL). The reaction mixture is transferred into a 500-mL separatory funnel with the aid of dichloromethane (100 mL) and

388

water (60 mL). After partitioning, the aqueous layer is extracted with dichloromethane (2 x 100 mL). The combined organic extracts are washed with sat. aq. $NaHCO_3$ (1 x 100 mL) and brine (1 x 100 mL), dried over Na_2SO_4 (40 g), and filtered. The filtrate is concentrated on a rotary evaporator under reduced pressure (40 °C, 30 mmHg), and the residue is dried *in vacuo* to afford 15.2 g of the crude product as a pale yellow solid, which is purified by recrystallization to provide 2-benzyloxycarbonylamino-3-(2-bromo-4,5-dimethoxyphenyl)acrylic acid methyl ester as colorless needles (11.2 g, 83%) (Notes 6 and 7).

B. *5,6-Dimethoxyindole-1,2-dicarboxylic acid 1-benzyl ester 2-methyl ester.* A 200-mL three-necked round-bottomed flask equipped with a Teflon-coated magnetic stir bar (3.5 x 1.0 cm), a rubber septum, a glass stopper, and an argon gas inlet is charged with cesium acetate (9.60 g, 50.0 mmol, 2.5 equiv) (Notes 8 and 9), 2-benzyloxycarbonylamino-3-(2-bromo-4,5-dimethoxyphenyl)acrylic acid methyl ester (9.01 g, 20.0 mmol), and copper(I) iodide (761 mg, 4.00 mmol, 0.2 equiv) (Note 10). The flask is evacuated and backfilled with argon and is charged with dry dimethyl sulfoxide (67 mL) (Note 11). After stirring at 300 rpm for 24 h at 30 °C in an oil bath (Note 12), the reaction is quenched by slow addition of 7% NaCl in 10% aq. NH_3 (100 mL) with cooling of the reaction mixture in an ice bath (Note 13). The resulting mixture is transferred into a 500-mL separatory funnel with the aid of EtOAc (100 mL) and water (60 mL). After partitioning, the aqueous layer is extracted with EtOAc (2 x 100 mL). The combined organic extracts are washed with water (1 x 100 mL) and brine (1 x 100 mL), dried over Na_2SO_4 (40 g), and filtered. The filtrate is concentrated on a rotary evaporator under reduced pressure (40 °C, 30 mmHg), and the residue is dried *in vacuo* to afford 8.03 g of the crude product as a pale yellow solid, which is purified by recrystallization to provide 5,6-dimethoxy-indole-1,2-dicarboxylic acid 1-benzyl ester 2-methyl ester as colorless needles (5.58 g, 76%) (Notes 14 and 15).

2. Notes

1. 6-Bromoveratraldehyde (98%) was purchased from Aldrich and

used as received without further purification.

2. Dichloromethane (>99.8%, water content: <0.0001%) was purchased from Sigma-Aldrich Co. and used as received without further purification.

3. 1,1,3,3-Tetramethylguanidine (99.0%) was purchased from Sigma-Aldrich Co. and used as received without further purification.

4. The submitters observed that the internal temperature rose to 9 °C after addition of 1,1,3,3-tetramethylguanidine.

5. The reaction typically requires 5 h to consume 6-bromoveratraldehyde and is monitored by TLC analysis on Merck silica gel 60F-254 plate eluting with hexanes-EtOAc (1:1). The R_f values of the starting aldehyde and the product are 0.58 and 0.31, respectively (visualized with 254 nm UV lamp and stained with an ethanol solution of 2,4-dinitrophenylhydrazine (DNP). After dipping the TLC plate in the DNP solution, the chromatogram is developed by heating on a hot plate).

6. To crystallize the product, the residue is dissolved in hot EtOAc-hexanes, 7:5 (240 mL) using an oil bath (bath temperature: 80 °C) with the flask under a nitrogen atmosphere. The solution is then cooled to room temperature over 5 h. The colorless fine needles are collected by filtration (9.94 g). The second crop of crystals (1.18 g) is provided by dissolving the solid, which was obtained by concentration of the mother liquor, in hot EtOAc (20 mL) using an oil bath (bath temperature: 80 °C), followed by cooling to room temperature over 5 h.

7. The compound exhibits the following physicochemical properties: R_f = 0.31 (hexanes-EtOAc = 1:1); Merck silica gel 60F-254 plate (visualized with 254 nm UV lamp and stained with an ethanol solution of 2,4-dinitrophenylhydrazine (DNP). After dipping the TLC plate in the DNP solution, the chromatogram is developed by heating on a hot plate); mp = 134–136 °C; IR (NaCl, CDCl$_3$): 3328, 2952, 1716, 1597, 1387, 1261, 1210, 1169, 1053, 819, 755 cm^{-1}. ^{1}H NMR (300 MHz, CDCl$_3$) δ: 3.55 (s, 3 H), 3.84 (s, 3 H), 3.85 (s, 3 H), 5.07 (s, 2 H), 6.50 (s, 1 H), 6.99 (s, 1 H), 7.08 (s, 1 H), 7.17 – 7.38 (m, 5 H), 7.44 (s, 1 H). ^{13}C NMR (75 MHz, CDCl$_3$) δ: 52.86, 55.76, 56.14, 67.56, 111.48, 115.28, 116.39, 124.48, 126.11, 128.45, 128.47, 128.58, 129.55, 135.84, 148.01, 150.12, 153.69, 165.57. Anal. calcd.

for $C_{20}H_{20}BrNO_6$: C, 53.35; H, 4.48; N, 3.11. Found: C, 53.49; H, 4.62; N, 3.07.

8. Cesium acetate (>99.99%) was purchased from Sigma-Aldrich Co. and used as received without further purification.

9. Cesium acetate is extremely hygroscopic. The submitters transferred this reagent in a plastic bag filled with argon to prevent rapid absorption of moisture. The checkers transferred the reagent to a sealed vial in a glovebox under nitrogen atmosphere and stored the vial in a dessicator.

10. Copper (I) iodide (99.5%) was purchased from Sigma-Aldrich Co. and used as received without further purification.

11. Anhydrous dimethyl sulfoxide (≥99.9%, water content: <0.005%) was purchased from Sigma-Aldrich Co. and used as received without further purification.

12. The reaction typically requires 24 h to consume almost all of the 2-benzyloxycarbonylamino-3-(2-bromo-4,5-dimethoxyphenyl)acrylic acid methyl ester. The ^{1}H NMR spectrum of the crude material shows that the relative ratio of the signal for the starting enamide (δ 5.08, 2 H) to that for the indole (δ 5.40, 2 H) is less than 0.08. The reaction is also monitored by TLC analysis on Merck silica gel 60F-254 plate eluting with hexanes-EtOAc-toluene (2:1:1). The R_f values of the intermediate and the product are 0.37 and 0.27, respectively (visualized with 254 nm UV lamp and an ethanol solution of 2,4-dinitrophenylhydrazine (DNP). After dipping the TLC plate in the DNP solution, the chromatogram is developed by heating on a hot plate).

13. The solution is prepared by dissolving 35 g of NaCl in 465 g of 10% aq. NH_3. The addition of this solution is exothermic.

14. The submitters observed that the crude material became yellow upon standing at room temperature. Thus, it was immediately purified by recrystallization. To crystallize the product, the residue is dissolved in hot hexanes-EtOAc, 5:2 (140 mL) using an oil bath (bath temperature: 60 °C). The solution is then cooled at 0 °C in an ice bath for 5 h. The colorless needles are collected by filtration (5.26 g). The second crop of crystals (0.92 g) is provided by dissolving the solid, which was obtained by concentration of the mother liquor, in hot hexanes-EtOAc, 3:1 (40 mL) using an oil bath

(bath temperature: 60 °C), followed by cooling at 0 °C in an ice bath for 5 h. The checkers observed that the concentrated mother liquor from the first crop was an oil that was contaminated with ~20% starting material. During the half scale run, the concentrated mother liquor was treated with 20 mL of diethyl ether to form a suspension. The suspension was filtered, and the solids were washed with 20 mL of diethyl ether. The filtrate was concentrated, and the resulting residue was recrystallized by dissolving in hexanes-diethyl ether-EtOAc, 2:1:1 (20 mL) in an oil bath at 35 °C, followed by cooling at 0 °C in an ice bath for 5 h. An additional 0.295 g of the product was isolated.

15. The compound exhibits the following physicochemical properties: R_f = 0.27 (hexanes-EtOAc-toluene = 2:1:1); Merck silica gel 60F-254 plate (visualized with 254 nm UV lamp and stained with an ethanol solution of 2,4-dinitrophenylhydrazine (DNP). After dipping the TLC plate in the DNP solution, the chromatogram is developed by heating on a hot plate); mp = 98–100 °C. IR (NaCl, CDCl$_3$): 2951, 1721, 1538, 1488, 1470, 1436, 1393, 1331, 1301, 1227, 1160, 1126, 1065, 1016, 754 cm^{-1}. ^{1}H NMR (400 MHz, CDCl$_3$) δ: 3.73 (s, 3 H), 3.85 (s, 3 H), 3.91 (s, 3 H), 5.40 (s, 2 H), 6.98 (s, 1 H), 7.08 (s, 1 H), 7.33 – 7.43 (m, 3 H), 7.47 (dd, J = 7.8, 1.6 Hz, 2 H), 7.57 (s, 1 H). ^{13}C NMR (100 MHz, CDCl$_3$) δ: 52.07, 55.93, 56.02, 69.58, 97.83, 102.75, 116.64, 119.99, 128.49, 128.66, 128.78 (2), 132.82, 134.49, 147.15, 150.15, 150.79, 161.84. Anal. calcd. for C$_{20}$H$_{19}$NO$_6$: C, 65.03; H, 5.18; N, 3.79. Found: C, 65.03; H, 5.38; N, 3.75.

Safety and Waste Disposal Information

All hazardous materials should be handled and disposed of in accordance with "Prudent Practices in the Laboratory"; National Academy Press; Washington, DC, 1995.

3. Discussion

Since the indole skeleton is found in a variety of biologically important natural products and therapeutic agents, the development of methods for indole synthesis has been one of the main topics in organic synthesis. We have recently developed a copper-mediated aryl amination[3] using the combination of CuI and CsOAc,[4] which was applied to the formation of *N*-Cbz-2-alkoxycarbonylindoles. The aryl amination proceeds smoothly even at the congested position at room temperature without the addition of any special ligand (Table 1). The procedure provides a facile synthetic method for preparation of a Cbz-protected indole carboxylic acid ester from

Table 1. Synthesis of indole-2-carboxylic acid ester.

Substrate	Product	CuI (equiv)	CsOAc (equiv)	Yield (%)
		1.5	7	98
		2	18	quant
		1	2.5	quant
		1	5	77

Reaction Conditions: CuI (1–2 equiv), CsOAc (2.5–18 equiv), DMSO, rt.

the corresponding *o*-bromobenzaldehyde derivative and readily available Horner-Wadsworth-Emmons reagent in a two-step sequence. Compared to the representative method for thermolysis of an α-azidecinnamic acid ester,[5] this method is able to circumvent regiochemical problems and potential explosion hazards.

The aryl amination was also applicable to the synthesis of indoline bearing a variety of protecting groups on the nitrogen (Table 2). It is notable that o-Ns amides have been found to be a particularly suitable substrate. By utilizing o-Ns amides, tetrahydroquinoline and tetrahydrobenzoazepine were

Table 2. Scope of the intramolecular aryl amination.

Substrate	Product	CuI (mol%)	CsOAc (equiv)	Temp. (°C)	Yield (%)
	R = Boc	200	10	90	82
	Alloc	200	10	90	75
	Bn	10	5	90	83
	Ns	1	2	90	97
		10	1.4	rt	67
		10	5	90	96
		50	5	60	83
		100	5	90	74

Reaction Conditions: CuI (1–200 mol%), CsOAc (1.4–10 equiv), DMSO, rt – 90 °C.

also constructed by this methodology. The reaction conditions are compatible with a wide range of functional groups, including both acid and base labile protective groups. One of the advantages over palladium catalyzed processes[6] is that the bromo group remained unreacted after the reaction, allowing for further functionalization to take place. The aryl

Org. Synth. **2011**, *88*, 388-397

amination was applied to the total syntheses of duocarmycins,[7] (+)-yatakemycin,[8] and PDE-II.[9]

1. Graduate School of Pharmaceutical Sciences, Tohoku University, Aramaki, Aoba-ku, Sendai 980-8578, Japan. E-mail: tokuyama@mail.pharm.tohoku.ac.jp This work was supported by The Ministry of Education, Culture, Sports, Science and Technology, Japan and the Cabinet Office, Government of Japan through its "Funding Program for Next Generation World-Leading Researchers.

2. Graduate School of Pharmaceutical Sciences, University of Tokyo, 7-3-1 Hongo, Bunkyo-ku, Tokyo 113-0033, Japan. E-mail: fukuyama@mol.f.u-tokyo.ac.jp This work was supported by The Ministry of Education, Culture, Sports, Science and Technology, Japan.

3. Ley, S. V.; Thomas, A. W. *Angew. Chem. Int. Ed.* **2003**, *42*, 5400.

4. (a) Yamada, K.; Kubo, T.; Tokuyama, H.; Fukuyama, T. *Synlett* **2002**, 231. (b) Okano, K.; Tokuyama, H.; Fukuyama, T. *Org. Lett.* **2003**, *5*, 4987. (c) Kubo, T.; Katoh, C.; Yamada, K.; Okano, K.; Tokuyama, H.; Fukuyama, T. *Tetrahedron* **2008**, *64*, 11230.

5. Bolton, R. E.; Moody, C. J.; Rees, C. W.; Tojo, G. *J. Chem. Soc., Perkin Trans. 1* **1987**, 931.

6. (a) Hartwig, J. F. *Angew. Chem. Int. Ed.* **1998**, *37*, 2046. (b) Muci, A. R.; Buchwald, S. L. *Top. Curr. Chem.* **2002**, *219*, 131.

7. Yamada, K.; Kurokawa, T.; Tokuyama, H.; Fukuyama, T. *J. Am. Chem. Soc.* **2003**, *125*, 6630.

8. (a) Okano, K.; Tokuyama, H.; Fukuyama, T. *J. Am. Chem. Soc.* **2006**, *128*, 7136. (b) Okano, K.; Tokuyama, H.; Fukuyama, T. *Chem. Asian J.* **2008**, *3*, 296.

9. Okano, K.; Mitsuhashi, N.; Tokuyama, H. *Chem. Commun.* **2010**, *46*, 2641.

Appendix
Chemical Abstracts Nomenclature (Collective Index Number); (Registry Number)

6-Bromoveratraldehyde: Benzaldehyde, 2-bromo-4,5-dimethoxy-; (5392-10-9)

Acetic acid, 2-(dimethoxyphosphinyl)-2-[[(phenylmethoxy)carbonyl]-amino]-, methyl ester: Benzyloxycarbonylamino(dimethoxy-phosphoryl)acetic acid methyl ester; (88568-95-0)
1,1,3,3-Tetramethylguanidine: Guanidine, *N,N,N',N'*-tetramethyl-; (80-70-6)
Dichloromethane: Methane, dichloro-; (75-09-2)
Cesium acetate: Acetic acid, cesium salt (1:1); (3396-11-0)
Copper(I) iodide: Copper iodide (CuI); (7681-65-4)
Dimethyl sulfoxide: Methane, 1,1'-sulfinylbis-; (67-68-5)

Hidetoshi Tokuyama was born in Yokohama in 1967. He received his Ph.D. in 1994 from Tokyo Institute of Technology under the direction of Professor Ei-ichi Nakamura. He spent one year (1994-1995) at the University of Pennsylvania as a postdoctoral fellow with Professor Amos B. Smith, III. He joined the group of Professor Tohru Fukuyama at the University of Tokyo in 1995 and was appointed Associate Professor in 2003. In 2006, he moved to Tohoku University, where he is currently Professor of Pharmaceutical Sciences. His research interest is on the development of synthetic methodologies and total synthesis of natural products.

Toshiharu Noji was born in Fukushima in 1986. He received his B.S. in 2009 from Faculty of Pharmaceutical Sciences, Tohoku University, where he carried out undergraduate research in the laboratories of Professor Hidetoshi Tokuyama. In the same year, he then began his doctoral studies at Graduate School of Pharmaceutical Sciences, Tohoku University under the supervision of Professor Hidetoshi Tokuyama. His graduate research has focused on benzyne chemistry and its application to total synthesis of natural products containing multisubstituted heteroaromatic rings.

Kentaro Okano was born in Tokyo in 1979. He received his B.S. in 2003 from Kyoto University, where he carried out undergraduate research under the supervision of Professor Tamejiro Hiyama. He then moved to the laboratories of Professor Tohru Fukuyama at the University of Tokyo and started his Ph.D. research on synthetic studies toward the antitumor antibiotic yatakemycin by means of the copper-mediated aryl amination strategy. In 2007, he started his academic career at Tohoku University, where he is currently an assistant professor in Professor Hidetoshi Tokuyama's group. His current research interest is natural product synthesis based on the development of new synthetic methodologies.

Tohru Fukuyama received his Ph.D. in 1977 from Harvard University with Yoshito Kishi. He remained in Kishi's group as a postdoctoral fellow until 1978 when he was appointed as Assistant Professor of Chemistry at Rice University. After seventeen years on the faculty at Rice, he returned to his home country and joined the faculty of the University of Tokyo in 1995, where he is currently Professor of Pharmaceutical Sciences. He has primarily been involved in the total synthesis of complex natural products of biological and medicinal importance. He often chooses target molecules that require development of new concepts in synthetic design and/or new methodology for their total synthesis.

Lei Zhang received his B.Sc. at the University of Ottawa in 2009. His is currently pursuing a Ph.D. at the University of Toronto under the supervision of Professor Mark Lautens. His research interests lie in developing new reactions in rhodium and palladium catalysis.

LIGAND-FREE COPPER(II) OXIDE NANOPARTICLES CATALYZED SYNTHESIS OF SUBSTITUTED BENZOXAZOLES

Submitted by Prasenjit Saha, Md Ashif Ali, and Tharmalingam Punniyamurthy.[1]
Checked by Kelvin Sham and Paul Harrington.

1. Procedure

A. N-2-Bromophenylbenzamide. An oven-dried, three-necked, 500-mL round-bottomed flask equipped with an egg-shaped magnetic stirring bar (¾" x 1½"), a rubber septum, a calcium chloride guard tube, and a 25-mL pressure-equalizing addition funnel (capped with a rubber septum) is sequentially charged at room temperature with 2-bromoaniline (16.2 g, 94.0 mmol) (Note 1) and THF (100 mL) (Note 2). The solution is cooled to 0 °C in an ice-water bath. Benzoyl chloride (12.0 mL, 103 mmol, 1.10 equiv) (Note 3) is added dropwise into the flask via the addition funnel over 25 min. The resulting light purple suspension is allowed to warm to ambient temperature (21 °C) over 1 h, and stirring is continued for an additional 20 h. The solvent is removed by a rotary evaporator under reduced pressure (74 mmHg) at 40 °C, and the light pink solid is treated with EtOAc (300 mL). The light pink suspension is transferred into a 500-mL separatory funnel and washed successively with 5% (w/v) aqueous sodium bicarbonate (2 x 100 mL) (Note 4) and saturated aqueous sodium chloride (100 mL) (Note 5) solutions. The organic solution is dried over sodium sulfate (50 g) (Note 6), filtered through a fine-porosity fritted-glass funnel and the sodium sulfate is washed with ethyl acetate (20 mL). The solvent is removed by a

398

Org. Synth. **2011**, *88*, 398-405
Published on the Web 6/2/2011

rotary evaporator under reduced pressure (74 mmHg) at 40 °C to give a tan solid, which is dissolved in 80% aqueous ethanol (150 mL) at 72 °C (Note 7). The solution is allowed to cool to ambient temperature and aged for 17 h. The resulting solid is collected by suction filtration on a Büchner funnel, washed with 10% aqueous ethanol (50 mL), transferred to a filter paper, air dried and then dried under high vacuum (0.11 mmHg) at 80 °C for 16 h to give the title compound as white needles (20.4 g, 79%) (Note 8).

B. *2-Phenylbenzoxazole.* An oven-dried, 100-mL, single-necked, round-bottomed flask equipped with an egg-shaped (1″ x ½″) magnetic stirring bar, a rubber septum, and an argon inlet adapter is charged with *N*-2-bromophenylbenzamide (10.5 g, 38.0 mmol), cesium carbonate (18.6 g, 57.0 mmol, 1.50 equiv) (Note 9), copper oxide nanoparticles (302 mg, 10 mol%) (Note 10) and DMSO (38 mL) (Note 11). The resulting black mixture is stirred in an oil bath at 110 °C until the *N*-2-bromophenylbenzamide is completely consumed as judged by LC-MS analysis (40 h) (Note 12). The mixture is removed from the oil bath, allowed to cool to ambient temperature, and transferred to a 500-mL separatory funnel with the aid of ethyl acetate (150 mL) and water (200 mL). The layers are separated, and the aqueous layer is extracted with ethyl acetate (4 x 100 mL). The combined organic extracts are dried over Na_2SO_4 (50 g) and filtered through a fine-porosity fritted-glass funnel. The Na_2SO_4 is suspended in dichloromethane (100 mL) and stirred at room temperature for 18 h and filtered through a fine-porosity fritted-glass funnel. The filtrates are combined and the solvent is removed by a rotary evaporator under reduced pressure (35 mmHg) at 40 °C to give a black solid. The black solid is dissolved in a minimum amount of dichloromethane and adsorbed onto 15 g of silica gel (Note 13). The solvent is removed under reduced pressure and the resulting solid is charged onto a silica gel chromatography column (80 g RediSep® R$_f$) (Note 13) and eluted using ethyl acetate in hexane to afford the title compound as a white solid (5.90 g, 80%) (Note 14).

2. Notes

1. 2-Bromoaniline was purchased from Aldrich Chemical Company, Inc. and used without further purification.

2. THF (anhydrous, $\geq$ 99.9%) was purchased from the Aldrich Chemical Company, Inc. and used as received.

3. Benzoyl chloride was purchased from the Aldrich Chemical Company, Inc. and used as received.

4. Sodium bicarbonate was purchased from the Aldrich Chemical Company, Inc. and used as received.

5. Saturated sodium chloride solution was purchased from Teknova.

6. Sodium sulfate was purchased from the Aldrich Chemical Company, Inc.

7. Ethanol was purchased from Aaper Alcohol and Chemical Co.

8. The product (white solid) exhibits the following properties: mp 111–113 °C; ^{1}H NMR (400 MHz, CDCl$_3$) δ: 6.98–706 (m, 1 H), 7.38 (t, J = 7.6 Hz, 1 H), 7.47–7.63 (m, 4 H), 7.94 (d, J = 7.2 Hz, 2 H), 8.46 (br s, 1 H), 8.56 (d, J = 8.2 Hz, 1 H); ^{13}C NMR (100 MHz, CDCl$_3$) δ: 113.7, 121.7, 125.2, 127.1, 128.5, 128.9, 132.1, 132.2, 134.6, 135.8, 165.2; IR (ATR: Attenuated Total Reflectance): 3217, 1651, 1575, 1515, 1500, 1428, 1296, 1251, 1044, 1026 cm^{-1}; MS (ESI) m/z: 277.0 (M$^+$+1). Anal. calcd. for C$_{13}$H$_{10}$BrNO: C, 56.55; H, 3.65; N, 5.07. Found: C, 56.38; H, 3.67; N, 4.87.

9. Cesium carbonate (99%) was purchased from the Aldrich Chemical Company, Inc. and used without further purification.

10. Copper oxide nanoparticles (particle size = 28 nm, surface area 33 m^2/g) were purchased from the Aldrich Chemical Company, Inc. The reaction rate appeared to be highly dependent on the size of the copper oxide nanoparticles. For example, under the reaction conditions described above, an 84:13 ratio of N-2-bromophenylbenzamide:2-phenylbenzoxazole was observed by LCMS after 15 h using 40 nm copper oxide nanoparticles, while the same reaction using 28 nm copper oxide nanoparticles yielded a ratio of 45:49.

11. DMSO (anhydrous, ≥99.9%) was purchased from the Aldrich Chemical Company, Inc. and used as received.

12. A small amount of the reaction mixture was taken out and diluted with ethyl acetate and water. The organic extract was analyzed by an Agilent 1100 Series LCMS with UV detection at 254/220 nm and a low resonance electrospray mode (ESI). Using a 5-minute method (Agilent Zorbax SB-C18 column, 3.0 mm x 50 mm, 3.5 mm; A: 0.1% TFA in water and B: 0.1% TFA in acetonitrile; 1.5 mL/min; 0.0-0.2 min: 10% B; 0.2-3.0 min: 10-100% B; 3.0-4.5 min: 100% B; 4.5-5.0 min: 100-10% B; 1.5 min post time; 2.0 μL injection), the retention times of N-2-bromophenylbenzamide and 2-phenylbenzoxazole are determined to be 2.16 min and 2.37 min, respectively.

400

13. Silica gel (200-400 mesh) was purchased from the Aldrich Chemical Company, Inc.

14. The crude product is purified using a CombiFlash® R$_f$ flash chromatography system equipped with an 80 g RediSep® R$_f$ silica column. The product is eluted with 0% to 15% ethyl acetate/hexane and the product separation is monitored by an internal UV detector at 254 nm. The product (white solid) exhibits the following properties: mp 100–102 °C; ^{1}H NMR (400 MHz, CDCl$_3$) δ: 7.30–7.41 (m, 2 H), 7.47–7.63 (m, 4 H), 7.77–7.80 (m, 1 H), 8.19–8.32 (m, 2 H); ^{13}C NMR (100 MHz, CDCl$_3$) δ: 110.6, 120.0, 124.6, 125.1, 127.1, 127.6, 128.9, 131.5, 142.1, 150.8, 163.1; IR (ATR): 1617, 1552, 1472, 1446, 1344, 1241, 1197, 1052, 1021 cm^{-1}; MS (ESI) m/z: 195.9 (M$^+$+1). Anal. calcd. for C$_{13}$H$_9$NO: C, 79.98; H, 4.65; N, 7.17. Found: C, 79.95; H, 4.60; N, 7.11. The second run was run on half the scale and the yield obtained was 79%.

Safety and Waste Disposal Information

All hazardous materials should be handled and disposed of in accordance with "Prudent Practices in the Laboratory"; National Academies Press; Washington, DC, 2011.

3. Discussion

Benzoxazoles are important structural motifs present in numerous natural products and biologically active compounds.[2] The classical methods used for their preparation involve the condensation of 2-aminophenol with either carboxylic acids in the presence of acids or aldehydes under oxidative reaction conditions.[3] These methods, however, have limitations due to non-availability of the suitably substituted starting materials and, in some cases, the requirement for harsh reaction conditions, such as elevated temperature and strong acid. The recent development of cross-coupling reaction using copper-catalysis provides a straightforward route to substituted benzoxazoles by intramolecular C-O cross-coupling under relatively milder conditions.[4]

The present protocol describes the cyclization of N-2-bromophenylbenzamide to afford 2-phenylbenzoxazole via C-O cross-coupling in the presence of 10 mol % of CuO nanoparticles and 1.5 equiv of Cs$_2$CO$_3$ in DMSO at moderate temperature.[5] The catalyst is cheap, commercially available, air stable, recyclable and catalyzes under ligand-free conditions.

Table 1 represents the scope of the procedure. Both the aromatic and aliphatic *N*-2-bromophenylamides undergo cyclization to give the corresponding 2-aryl and 2-alkyl benzoxazoles in good to high yield.

Table 1. CuO Nanoparticles Catalyzed Synthesis of 2-Substituted
          Benzoxazoles

Entry	Substrate	Time (h)	Product	Yield (%)[a]
1		15		86
2		16		83
3		34		55[b]
4		16		68
5		16		73

[a] With the exception of entry 1 (38 mmol scale), all reactions are performed on a 20 mmol scale and the products are purified by column chromatography using hexane and ethyl acetate as eluent. [b] Reaction performed at 120 °C with 2.0 equiv of Cs_2CO_3.

1. Department of Chemistry, Indian Institute of Technology Guwahati, Guwahati-781039, India. Email: tpunni@iitg.ernet.in

2. (a) Easmon, J.; Pürstinger, G.; Thies, K.-S.; Heinisch, G.; Hofmann, J. *J. Med. Chem.* **2006**, *49*, 6343–6350. (b) Sun, L.-Q.; Chen, J.; Bruce, M.; Deskus, J. A.; Epperson, J. R.; Takaki, K.; Johnson, G.; Iben, L.; Malhe, C. D.; Ryan, E.; Xu, C. *Bioorg. Med. Chem. Lett.* **2004**, *14*, 3799–3802. (c) Kumar, D.; Jacob, M. R.; Reynolds, M. B.; Kerwin, S. M. *Bioorg. Med. Chem.* **2002**, *10*, 3997–4004. (d) McKee, M. L.; Kerwin, S. M. *Bioorg. Med. Chem.* **2008**, *16*, 1775–1783. (e) Huang, S.-T.; Hsei, I-J.; Chen, C. *Bioorg. Med. Chem.* **2006**, *14*, 6106–6109.

3. (a) Varma, R. S.; Kumar, D. *J. Heterocycl. Chem.* **1998**, *35*, 1539–1540. (b) Chang, J.; Zhao, K.; Pan, S. *Tetrahedron Lett.* **2002**, *43*, 951–954. (c) Terashima, M; Ishii, M.; Kanaoka, Y. *Synthesis* **1982**, 484–485. (d) Bougrin, K.; Loupy, A.; Soufiaoui, M. *Tetrahedron* **1998**, *54*, 8055–8064. (e) Pottorf, R. S.; Chadha, N. K.; Katkevics, M.; Ozola, V.; Suna, E.; Ghane, H.; Regberg, T.; Player, M. R. *Tetrahedron Lett.* **2003**, *44*, 175–178.

4. (a) Altenhoff, G.; Glorius, F. *Adv. Synth. Catal.* **2004**, *346*, 1661–1664. (b) Barbero, N.; Carril, M.; SanMartin, R.; Domínguez, E. *Tetrahedron* **2007**, *63*, 10425–10432. (c) Evindar, G.; Batey, R. A. *J. Org. Chem.* **2006**, *71*, 1802–1808. (d) Viirre, R. D.; Evindar, G.; Batey, R. A. *J. Org. Chem.* **2008**, *73*, 3452–459. (e) Ueda, S.; Nagasawa, H. *J. Org. Chem.* **2009**, *74*, 4272–4277.

5. Saha, P.; Tamminana, R.; Purkait, N.; Ali, M. A.; Paul, R.; Punniyamurthy, T. *J. Org. Chem.* **2009**, *74*, 8719–8725.

Appendix
Chemical Abstracts Nomenclature;(Registry Number)

2-Phenylbenzoxazole; (833-50-1)

N-2-Bromophenylbenzamide; (70787-27-8)

2-Bromoaniline: Benzenamine, 2-bromo-; (615-36-1)

Benzoyl chloride; (98-88-4)

Cesium carbonate: Carbonic acid, cesium salt (1:2); (534-17-8)

Copper(II) oxide nanoparticles: Copper oxide (CuO); (1317-38-0)

Tharmalingam Punniyamurthy was born in 1964 in Tiruchirapalli, India. He completed his graduate studies at Bharathidasan University and his Ph.D. at IIT Kanpur with Prof. Javed Iqbal. He spent one year with Prof. Mukund Sibi (Fargo), two years with Prof. Tsutomu Katsuki (Fukuoka), one year with Prof. Andre Vioux (Montpellier) and six months with Prof. Joel J. E. Moreau (Montpellier) as a postdoctoral researcher. Since 2001, he has been a member of the faculty at IIT Guwahati and also spent eight months with Dr John M. Brown (Oxford) as a visiting professor. His research interests include new synthetic methods, asymmetric catalysis and natural product synthesis.

Prasenjit Saha was born in Barpeta, India. He received his M.Sc. degree in chemistry from Gauhati University in 2006. Currently, he is pursuing his Ph.D. with Prof. T. Punniyamurthy at the Indian Institute of Technology Guwahati. His research interests are cross-coupling reactions, asymmetric catalysis and molecular recognition.

Md Ashif Ali was born in Burdwan, India. He received his M.Sc. degree in chemistry from the Indian Institute of Technology Guwahati in 2007. He is currently pursuing his Ph.D. at the Indian Institute of Technology Guwahati under supervision of Prof. T. Punniyamurthy. His research interests include cross-coupling reactions, medicinal chemistry and natural product synthesis.

Kelvin Sham was born in Sabah, Malaysia in 1967. He obtained his B.A. in Chemistry from Wartburg College in 1989 and M.S. in Organic Chemistry from Iowa State University in 1992, where he investigated the utility and scope of palladium-catalyzed heteroannulation reactions between o-iodophenol and internal alkynes under the guidance of Professor Richard C. Larock. After a seven-year stint as a medicinal chemist at SmithKline Beecham (now GSK) in King of Prussia, Pennsylvania, Kelvin joined Amgen in 1999 where he is currently a Scientist in the Chemistry Research and Discovery group at Amgen.

Paul Harrington was born in 1974 in Wolfville, Canada. He earned his B.Sc. in Chemistry in 1996 from Acadia University and worked in Professor Michael Kerr's laboratories while there. He attended the University of Hawaii and received his Ph.D. in 2002 under the supervision of Professor Marcus Tius. He did his postdoctoral research with Professor Barry Trost at Stanford University. He is currently a medicinal chemist in the Chemistry Research and Discovery group at Amgen.

(R)-3,3'-BIS(9-PHENANTHRYL)-1,1'-BINAPHTHALENE-2,2'-DIYL HYDROGEN PHOSPHATE

A.

B.

C.

D.

Submitted by Wenhao Hu,*[1a] Jing Zhou,[1a] Xinfang Xu,[1a] Weijun Liu,[1b] and Liuzhu Gong.*[1b]
Checked by Zhanjie Li and Huw M. L. Davies.

Org. Synth. 2011, 88, 406-417
Published on the Web 6/16/2011

1. Procedure

A. (R)-2,2'-Bis(methoxymethoxy)-3,3'-diiodo-1,1'- binaphthalene (**2**). An oven-dried, 500-mL, three-necked, round-bottomed flask is equipped with a thermometer (–50 to 50 °C) with adapter, a rubber septum, a 60-mL pressure-equalizing addition funnel fitted with a rubber septum, and a Teflon-coated magnetic stirring bar (37 mm x 16 mm). The flask is flushed with argon and charged with 3.0 g (8.0 mmol, 1 equiv) of (R)-(+)-2,2'-bis(methoxymethoxy)-1,1'-binaphthyl (Note 1) and 70 mL of dry tetrahydrofuran (THF) (Note 2). The resulting solution is cooled to 2–3 °C in an ice bath. Butyllithium solution (10 mL, 2.5 M, 25 mmol, 3 equiv) (Note 3) is added to the flask with a syringe pump over 30 min. After addition, the ice bath is removed, the resulting brown suspension is stirred at room temperature for 3 h. The suspension is cooled in the ice bath. A solution of 9.0 g (35 mmol, 4.4 equiv) of iodine (Note 4) in 60 mL of dry THF is charged to the addition funnel, and the solution is added to the flask over 2 min. Another 20 mL of dry THF is used to wash the addition funnel, which is also added to the flask over 1 min. After addition, the ice bath is removed. The solution is stirred at room temperature for 20 h. Progress of the reaction is monitored by TLC (Note 5). After the reaction is completed, 100 mL of saturated aqueous ammonium chloride is added (Note 6), followed with 50 g of sodium thiosulfate (Note 7). The resulting mixture is stirred vigorously until the dark color of the mixture changes into light brown, at which time the solution is transferred to a 1-L separatory funnel and extracted with ethyl acetate (2 x 300 mL) (Note 8). The combined organic solution is washed with saturated brine (200 mL) (Note 9) and dried over magnesium sulfate (40 g) (Note 10). After filtration, the solution is concentrated under reduced pressure (40 mmHg) on a rotary evaporator at 40 °C, and the residue is purified by flash chromatography on silica gel (Note 11) to give 2.0 g of **2** as a white solid (40% yield) (Note 12). The purity of **2** is determined to be 99% by HPLC (Note 13).

B. (R)-2,2'-Bis(methoxymethoxy)-3,3'-bis(9-phenanthryl)-1,1'-binaphthalene (**3**). An oven-dried, 100-mL, two-necked, round-bottomed flask is equipped with a thermometer (10 – 260 °C) with adapter, a condenser fitted with a rubber septum, and a Teflon-coated magnetic stirring bar (20 mm x 10 mm). The flask is flushed with argon and charged with (R)-2,2'-bis(methoxymethoxy)-3,3'-diiodo-1,1'-binaphthalene (**2**), (2.0 g, 3.2 mmol, 1.0 equiv), 9-phenanthreneboronic acid (1.42 g, 6.4 mmol, 2.0 equiv) (Note 14), barium hydroxide octahydrate (2.52 g, 8.0 mmol,

2.5 equiv) (Note 15), and *tetrakis*(triphenylphosphine)palladium(0) (0.37 g, 0.32 mmol, 10 mol%) (Note 16). Then the flask is evacuated and backfilled with argon (this sequence is repeated three times). 1,4-Dioxane (24 mL) and water (6.7 mL) are added by syringe (Notes 17 and 18). The reaction mixture is heated with an oil bath to 75-80 °C for 24 h, and the reaction progress is monitored by TLC (Note 19). After the mixture is cooled to room temperature, it is concentrated under reduced pressure (40 mmHg) on a rotary evaporator at 40 °C. The residue is extracted with dichloromethane (2 x 50 mL), and the combined organic layers are washed with 1N aqueous HCl (2 x 20 mL) and saturated brine (40 mL) and then dried over magnesium sulfate (20 g). After filtration, the residue is purified by flash chromatography on silica gel (Note 20) to give 2.1 g (91 %) of **3** as a light yellow solid (Note 21). The purity of **3** is determined to be 99% by HPLC (Note 22).

   C.   *(R)-3,3'-Bis(9-phenanthryl)-1,1'-binaphthalene-2,2'-diol*   (**4**).   A 100-mL, two-necked, round-bottomed flask is equipped with a thermometer (10 – 260 °C) with adapter, a water-cooled condenser, and a Teflon-coated magnetic stirring bar (20 mm x 10 mm). The flask is charged with *(R)-2,2'-bis(methoxy methoxy)-3,3'-bis(9-phenanthryl)-1,1'-binaphthalene* (**3**) (2.0 g, 2.7 mmol, 1 equiv), 1,4-dioxane (27 mL), and 6 N aqueous hydrochloric acid (3.7 mL). The reaction mixture is heated with an oil bath to 98–100 °C for 12 h, and the reaction progress is monitored by TLC (Note 23). After all of **3** is consumed, the solution is cooled to room temperature and carefully neutralized with saturated sodium bicarbonate solution (45 mL). The mixture is transferred to a 250-mL separatory funnel and extracted with dichloromethane (3 x 30 mL). The combined organic solution is concentrated under reduced pressure (40 mmHg) on a rotary evaporator at 40 °C. The residue is dissolved in dichloromethane (150 mL), and the resulting solution is washed with saturated brine (50 mL) and dried over magnesium sulfate (15 g). After filtration, the solution is concentrated under reduced pressure (40 mmHg) on a rotary evaporator at 40 °C. The residue is purified by flash chromatography on silica gel (Note 24) to give 1.74 g (99 %), of **4** as a light yellow solid (Notes 25 and 26).

   D.   *(R)-3,3'-Bis(9-phenanthryl)-1,1'-binaphthalene-2,2'-diyl   hydrogen phosphate* (**5**). A 250-mL round-bottomed flask is equipped with a rubber septum and a Teflon-coated magnetic stirring bar (20 mm x 10 mm) and filled with argon. To the flask is charged with **4** (1.65 g, 2.58 mmol, 1 equiv). The reaction flask is placed in a room temperature water bath, and pyridine (15 mL) (Note 27) is added into the flask *via* syringe followed by

408

slow addition of POCl$_3$ (0.54 mL, 5.80 mmol, 2.25 equiv) (Note 28) over 15 min. After completion of the addition, the mixture is stirred for 12 h at 60 °C, and the reaction progress is monitored by TLC (Note 29). After the mixture is cooled to room temperature, H$_2$O (20 mL) is added, and the resulting mixture is stirred for 5 h. After that, 6 N HCl (80 mL) is added to the mixture in a water-bath, and the mixture is stirred for 30 min. The mixture is extracted with CH$_2$Cl$_2$ (2 x 500 mL). The combined organic layers are washed with 6 N HCl (100 mL), dried over sodium sulfate and solvent is evaporated to give crude product. Purification of the crude product by flash chromatography on silica gel (Note 30) affords the BINOL phosphoric acid as a white solid. The phosphoric acid can be further recrystallized in hot solvents of CH$_2$Cl$_2$ (6 mL) and petrol ether (80 mL) to give white crystals (1.65 g, 91%) (Notes 31 and 32). The product purity is >98% by ^{1}H NMR.

2. Notes

1.   (R)-(+)-2,2'-Bis(methoxymethoxy)-1,1'-binaphthyl was purchased from Strem Chemicals, Inc. with 98% purity.

2.   Tetrahydrofuran (THF) was purchased from Aldrich Chemical Company, Inc. and was distilled under argon (atmospheric pressure) from sodium benzophenone ketyl.

3.   Butyllithium solution (2.5M in hexanes) was purchased from Aldrich Chemical Company, Inc. and was titrated with 1.0 M of *sec*-butanol before use.

4.   Iodine was purchased from Aldrich Chemical Company, Inc. and used as received.

5.   Thin layer chromatography was performed on SCRC silica gel GF 254 plates eluting with 15% ethyl acetate/petroleum ether, and the observed R_f is 0.78 for **2** and 0.62 for the mono-substituted byproduct.

6.   Ammonium chloride was purchased from Aldrich Chemical Company, Inc.

7.   Sodium thiosulfate was purchased from Aldrich Chemical Company, Inc.

8.   Ethyl acetate was purchased from Fisher Scientific and used without further purification.

9.   Sodium chloride was purchased from Aldrich Chemical Company, Inc.

10. Anhydrous magnesium sulfate was purchased from EMD Chemical Company, Inc.

11. Silica gel 60 (230 – 400 mesh) was purchased from Sorbent Technologies. Flash chromatography was performed using 170 g of silica gel 60 (6.5 x 11 cm) and 50-mL fractions were collected. The column was eluted with petroleum ether (150 mL) and then petroleum ether/ethyl acetate: 100:1 (400 mL) and 50:1 (4000 mL). Compound **2** was obtained in fractions 58-86, which were combined and concentrated under reduced pressure (40 mmHg) on a rotary evaporator at 40 °C. The resulting solid was dried at 20 mmHg for 12 h.

12. The physical properties of **2** are as follows: $[\alpha]^{20}_D$ –39.4 (c 1.22, CHCl$_3$); IR (neat): 1559, 1490, 1462, 1446, 1416, 1382, 1345, 1232, 1199, 1157, 1084, 993, 953, 903, 747, 729 cm^{-1}; ^{1}H NMR (400 MHz, CDCl$_3$) δ: 2.61 (s, 6 H), 4.72 (d, J = 5.6 Hz, 2 H), 4.83 (d, J = 6.0 Hz, 2 H), 7.20 (d, J = 8.4 Hz, 2 H), 7.31 (dt, J = 1.2, 7.2 Hz, 2 H), 7.44 (dt, J = 1.2, 7.2 Hz, 2 H), 7.79 (d, J = 8.4 Hz, 2 H), 8.57 (s, 2 H); ^{13}C NMR (100 MHz, CDCl$_3$) δ: 56.6, 92.7, 99.6, 126.0, 126.4, 126.7, 126.9, 127.3, 132.4, 134.0, 140.2, 152.3; HR-MS (+ESI): calcd. for C$_{24}$H$_{20}$I$_2$O$_4$Na ([M+Na$^+$]): 648.9354, found: 648.9341; Anal. calcd. for C$_{24}$H$_{20}$I$_2$O$_4$: C, 46.03; H, 3.22. Found: C, 46.26; H, 3.10.

13. HPLC conditions: Dynamax-60A column, 0.3% isopropanol in hexanes, 1.0 mL/min, UV (226 nm) detector. The t$_R$ of the product is 5.22 min.

14. 9-Phenanthreneboronic acid was purchased from Aldrich Chemical Company, Inc.

15. Barium hydroxide octahydrate was purchased from Fisher Scientific.

16. *Tetrakis*(triphenylphosphine)palladium(0) purchased from Aldrich Chemical Company, Inc.

17. 1,4-Dioxane was purchased from Mallinckrodt Baker, Inc. and used without further purification.

18. 1,4-Dioxane and water were all degassed with the following procedure before use: A 100-mL round-bottomed flask fitted with a adapter was charged with 50 mL of solvent. The flask was placed into a sonicator and the adapter was connected to a filtration pump. It was sonicated for 30 min under reduced pressure (approx 100 mmHg). After that, the solvent was bubbled with argon for 10 min.

19. Thin layer chromatography is eluted with 9% ethyl acetate/petrol ether, and the observed R_f is 0.38 for **3**.

20. Flash chromatography was performed using 175 g of silica gel 60 (6.5 x 11 cm) and 50-mL fractions were collected. The column was eluted

410

with petroleum ether (100 mL) and then petroleum ether/dichloromethane/ethyl acetate: 50:2.5:1 (250 mL), 20:1:1 (800 mL), and 15:1:1 (1800 mL). Pure **3** (1.85 g) was obtained in fractions 25-45, which were combined and concentrated under reduced pressure (40 mmHg) on a rotary evaporator at 40 °C. Fractions 46-58 were also combined and concentrated under the same conditions, and the residue was subjected to a second flash chromatography to give pure **3** (0.26 g). The combined solid was dried at 20 mmHg at 50 °C for 12 h.

21. The physical properties of **3** are as follows: $[\alpha]^{20}_D$ +60.1 (*c* 1.18, CHCl$_3$); IR (neat): 1492, 1448, 1424, 1392, 1351, 1246, 1198, 1155, 1068, 976, 907, 749, 725 cm^{-1}; ^{1}H NMR (600 MHz, CDCl$_3$) δ: 2.10–2.15 (m, 6H), 4.27–4.60 (m, 4 H), 7.32–7.70 (m, 14 H), 7.82–8.09 (m, 10 H), 8.72–8.78 (m, 4 H); ^{13}C NMR (100 MHz, CDCl$_3$) δ: 55.74, 55.82, 55.97, 56.11, 98.52, 98.61, 98.64, 98.81, 122.56, 122.75, 122.81, 122.93, 122.95, 125.41, 125.49, 125.53, 126.41, 126.44, 126.47, 126.51, 126.57, 126.64, 126.67, 126.74, 126.85, 126.93, 127.01, 127.40, 127.78, 128.07, 128.19, 128.40, 128.52, 128.88, 128.96, 130.27, 130.31, 130.34, 130.44, 130.52, 130.60, 130.80, 130.82, 131.14, 131.20, 131.25, 131.61, 131.64, 131.66, 131.73, 131.82, 131.86, 131.91, 131.95, 132.13, 134.10, 134.23, 134.24, 134.39, 134.46, 134.82, 134.89, 135.56, 135.58, 136.36, 136.68, 152.26, 152.29, 152.41; HR-MS (+ESI): calcd for C$_{52}$H$_{38}$O$_4$Na ([M+Na$^+$]): 749.26623, found: 749.26691; calcd. for C$_{52}$H$_{42}$O$_4$N [M+Na]): 749.2662, found: 749.2669; Anal. calcd. for C$_{52}$H$_{38}$O$_4$: C, 85.93; H, 5.27. Found: C, 85.63; H, 5.18.

22. HPLC conditions: Dynamax-60A column, 0.3% isopropanol in hexanes, 1.0 mL/min, UV (226 nm) detector. The t$_R$ of **3** is 6.10 min.

23. Thin layer chromatography is eluted with 15% ethyl acetate/petroleum ether, and the observed R_f is 0.30 for **4**.

24. Column chromatography was performed using 100 g of silica gel 60 (230 – 400 mesh) (6.5 x 7.5 cm) and eluted with petroleum ether (150 mL) and then petroleum ether/dichloromethane/ethyl acetate: 50:2.5:1 (200 mL), 20:1:1 (500 mL), 20:1:2 (1000 mL), and 20:1:2.5 (300 mL). Fractions (50 mL) were collected and the desired product was obtained in fractions 23-35, which were combined and concentrated under reduced pressure (40 mmHg) on a rotary evaporator at 40 °C. The resulting solid was dried under reduced pressure (0.2 mmHg) at 50 °C for 24 h, which contains 1-2% 1,4-dioxane that could not be removed.

25. Compound **4** (50 mg) was further dried under reduced pressure (0.2 mmHg) at 100 °C for 48 h to remove the 1,4-dioxane for analysis purpose. The ^{1}H-NMR didn't show any decomposition comparing with the

sample before the heating. The physical properties of **4** are as follows: mp: 148–150 °C (decomposition); $[\alpha]_D{}^{20}$ +45.5 (*c* 1.0, CHCl$_3$); IR (neat): 3523, 3058, 1621, 1493, 1448, 1425, 1379, 1360, 1258, 1230, 1198, 1143, 905, 768, 747, 726 cm^{-1}; ^{1}H NMR (600 MHz, CDCl$_3$) δ: 5.22, 5.23, 5.28, 5.33, (4 singlets, combined integration: 2H), 7.42-7.80 (m, 15H), 7.88-7.92 (m, 2H), 7.95-8.00 (m, 5H), 8.10-8.12 (m, 2H), 8.75-8.83 (m, 4H); ^{13}C NMR (100 MHz, CDCl$_3$) δ: 113.07, 113.17, 113.48, 122.82, 122.85, 123.10, 123.14, 123.21, 123.24, 124.50, 124.53, 124.71, 124.74, 124.93, 125.00, 126.94, 127.01, 127.05, 127.13, 127.16, 127.22, 127.28, 127.53, 127.59, 128.60, 128.71, 129.01, 129.07, 129.19, 129.41, 129.52, 129.54, 129.66, 129.67, 129.77, 130.63, 130.66, 130.69, 130.76, 130.78, 130.81, 131.29, 131.34, 131.35, 131.39, 131.66, 131.68, 131.70, 131.71, 132.26, 132.32, 133.79, 133.82, 133.92, 133.94, 133.99, 134.20, 150.80, 150.85, 150.87; HR-MS (-ESI): calcd. for C$_{48}$H$_{29}$O$_2$ ([M-H]$^-$): 637.21730, found: 637.21725.

26. HPLC conditions: Dynamax-60A column, 0.5% isopropanol in hexanes, 1.0 mL/min, UV (226 nm) detector. The t$_R$ of **4** is 7.35 min.

27. Pyridine was distilled from KOH before use.

28. POCl$_3$ was purchased from Sinopharm Chemical Reagent Co., Ltd.

29. Thin layer chromatography is eluted with 5% methanol/ CH$_2$Cl$_2$, and the observed R$_f$ are 0.50 for the phosphoric acid product **5** and 1.0 for **4**.

30. Column chromatography was performed using 100 g of 200–300 mesh silica gel 60 (5.5 x 11 cm) and 100-mL fractions were collected (300 mL of CH$_2$Cl$_2$, then 200 mL each of CH$_2$Cl$_2$/methanol, 100:1 and 50:1, and finally 550 mL of CH$_2$Cl$_2$/methanol, 40:1). The desired product was obtained in fractions 8-12, which were combined and concentrated by rotary evaporation under reduced pressure (40 mmHg) at 40 °C to obtain a white solid. The resulting solid was dried at 4 mmHg for 5 h.

31. The phosphoric acid product **5** exhibits the following physicochemical properties: white solid; mp: 359 – 361 °C (decomposition); $[\alpha]^{20}{}_D$ = –11.9 (c=1, CH$_2$Cl$_2$); IR (neat): 3630, 3061, 1622, 1493, 1449, 1416, 1250, 1201, 1102, 1093, 969, 951, 907, 852, 749, 726 cm^{-1}.cm^{-1}; ^{1}H NMR (400 MHz, DMSO-d_6) δ: 7.39–7.73 (m, 16 H), 7.97–8.19 (m, 8 H), 8.88–8.93 (m, 4 H); ^{13}C NMR (100 MHz, DMSO-d_6) δ: 122.43, 122.78, 123.12, 125.09, 126.23, 126.41, 126.51, 126.58, 126.79, 126.87, 128.46, 129.10, 129.40, 129.67, 130.24, 131.27, 131.43, 132.32, 133.30, 134.22, 148.10, 148.19; ^{31}P NMR (162 MHz, DMSO-d_6) δ: 2.61; HRMS (ESI): calcd. for C$_{48}$H$_{30}$O$_4$P (M+H)$^+$ 701.1876, found: 701.1889.

32. HPLC condition: Dynamax-60A column, 7% isopropanol in hexanes, 1.0 mL/min, UV (226 nm) detector. The t$_R$ of **5** is 21.6 min.

Org. Synth. **2011**, *88*, 406-417

Safety and Waste Disposal Information

All hazardous materials should be handled and disposed of in accordance with "Prudent Practices in the Laboratory"; National Academy Press; Washington, DC, 1995.

Discussion

Enantiomerically pure 3,3'-bis(9-phenanthryl)-1,1'-binaphthalene-2,2'-diyl hydrogen phosphate is used as a chiral Bronsted acid for multicomponent reactions. In 2004, the research groups of Akiyama and Terada independently reported activation of electrophiles by way of protonation with moderately strong phosphoric acids derived from chiral BINOLs.[2,3]

Scheme 1. Enantioselective Mannich-type Reaction by Akiyama, et al[2]

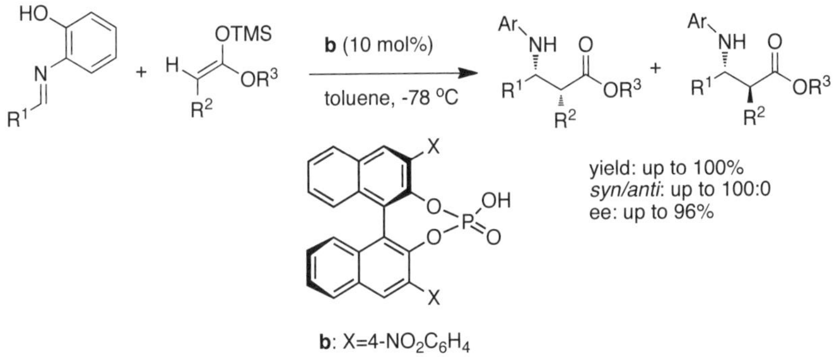

yield: up to 100%
syn/anti: up to 100:0
ee: up to 96%

b: X=4-NO$_2$C$_6$H$_4$

Scheme 2. Direct Mannich Reaction by Terada, et al[3]

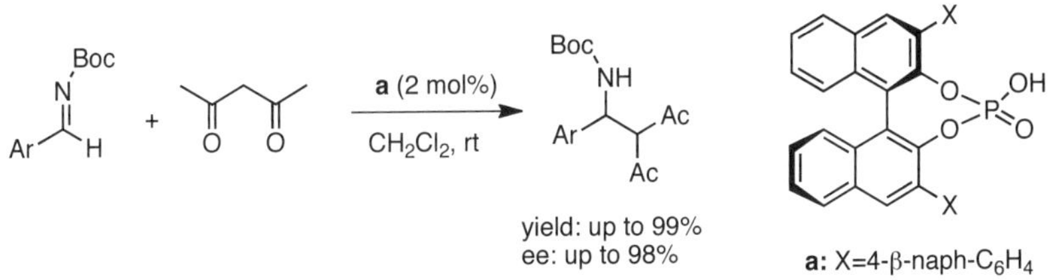

yield: up to 99%
ee: up to 98%

a: X=4-β-naph-C$_6$H$_4$

Due to the suitable acidity and cyclic structure, BINOL phosphates exhibit excellent catalytic activity as a chiral Bronsted acid for other reactions. Nucleophilic addition to aldimines, aza Diels-Alder reactions, and transfer hydrogenations can be catalyzed by BINOL phosphates to give high enantioselectivity with satisfactory yields.[4]

Scheme 3. Aza Diel-Alder Reaction by Akiyama, et al[5]

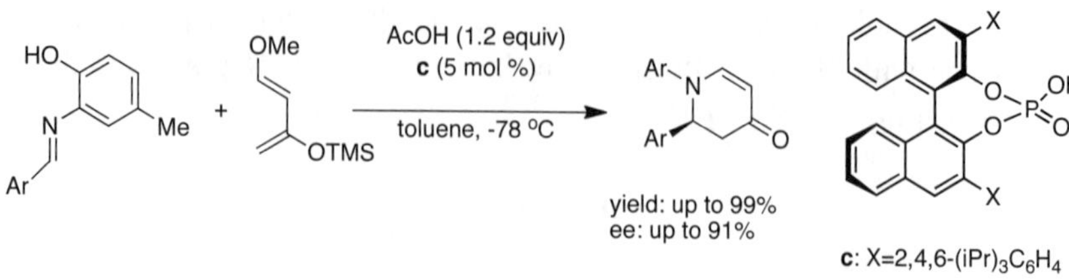

yield: up to 99%
ee: up to 91%

c: X=2,4,6-(iPr)₃C₆H₄

Scheme 4. Transfer Hydrogenation of Quinolines by Rueping, et al[6]

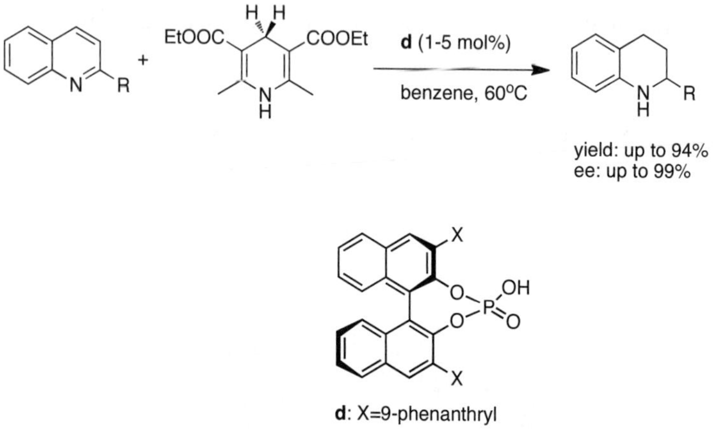

yield: up to 94%
ee: up to 99%

d: X=9-phenanthryl

Suzuki and Grignard cross-coupling is commonly used for the coupling of BINOL and the aryl group. In 1981 Cram, et al. synthesized two optical pure 3,3'-diaryl-substituted BINOLs by a Grignard cross-coupling reaction of 3,3'-dibromo-BINOL dimethyl ether and arylmagnesium bromides employing dichlorobis(triphenylphosphine)nickel(II) as the catalyst.[7] In 1998 Jørgensen, et al. prepared 3,3'-diary BINOLs by a Suzuki cross-coupling reaction of the 3,3'-diboronic acid of BINOL with commercially available aromatic bromides.[8]

The method in this procedure can also be used to synthesize 3,3'-diphenyl or di(2-naphthyl) BINOLs.[9]

1. (a) Department of Chemistry, East China Normal University, Shanghai 200062, China; whu@chem.ecnu.edu.cn; (b) Hefei National Laboratory

Org. Synth. **2011**, *88*, 406-417

for Physical Sciences at the Microscale and Department of Chemistry, University of Science and Technology of China, Hefei 230026, China; gonglz@ustc.edu.cn. Hu thanks the financial support from the National Science Foundation of China (NSFC) (20932003), and from the MOST of China (2011CB808600). Gong thanks NSFC (20732006) for the financial support.

2. Akiyama, T.; Itoh, J.; Yokota, K.; Fuchibe, K. *Angew. Chem. Int. Ed.* **2004**, *43*, 1566–1568.

3. Uraguchi, D.; Terada, M. *J. Am. Chem. Soc.* **2004**, *126*, 5356–5357.

4. Akiyama, T. *Chem. Rev.* **2007**, *107*, 5744–5758.

5. Akiyama, T.; Tamura, Y.; Itoh, J.; Morita, H.; Fuchibe, K. *Synlett.* **2006**, 141–143.

6. Rueping, M.; Antonchick, A. P.; Theissmann, T. *Angew. Chem. Int. Ed.* **2006**, *45*, 3683–3686.

7. Lingenfelter, S. D.; Helgeson, C. R.; Cram, J. D. *J. Org. Chem.* **1981**, *46*, 393–406.

8. Simonsen, B. K.; Gothelf, V. K.; Jorgensen, A. K. *J. Org. Chem.* **1998**, *63*, 7536–7538.

9. Wu, T. R.; Shen, L.-X.; Chong, J. M. *Org. Lett.* **2004**, *6*, 2701–2704.

Appendix
Chemical Abstracts Nomenclature (Collective Index Number); (Registry Number)

(*R*)-(+)-2,2'-Bis(methoxymethoxy)-1,1'-binaphthyl; (1738310-50-0)

n-Butyllithium; (109-72-8)

Iodine; (12190-71-5)

9-Phenanthreneboronic acid; (68572-87-2)

Tetrakis(triphenylphosphine) palladium(0); (14221-01-3)

Phosphorus oxychloride; (10025-87-3)

(*R*)-2,2'-Bis(methoxymethoxy)-3,3'-diiodo-1,1'- binaphthalene ; (189518-78-3)

(*R*)-2,2'-Bis(methoxymethoxy)-3,3'-bis(9-phenanthryl)-1,1'-binaphthalene; (1261302-60-6)

(*R*)-3,3'-Bis(9-phenanthryl)-1,1'-binaphthalene-2,2'-diol; (1058734-56-7)

(*R*)-3,3'-Bis(9-phenanthryl)-1,1'-binaphthalene-2,2'-diyl hydrogen phosphate; (864943-22-6)

Wenhao Hu was born in 1967 in Sichuan Province, China. He received his M.S. degree in Chengdu Institute of Organic Chemistry. He obtained a Ph.D. degree from the Hong Kong Polytechnic University in 1998 under the direction of Professor Albert S. C. Chan, and was a postdoctoral fellow at the University of Arizona with Professor Michael P. Doyle. He then joined GeneSoft Pharm. Inc. at San Francisco as a Staff Scientist (2002-2003). He moved to New Jersey to join Bristol-Myers Squibb Company as a Research Investigator (2003-2006). He returned to China as a Professor in the Department of Chemistry at East China Normal University in 2006. His research interests include development of highly efficient synthetic methods and their application to biologically active compounds.

Jing Zhou was born in 1984 in Shandong Province, China. She received her bachelor's degree in Chemistry in 2007 from East China Normal University, Shanghai. She then began her graduate study in Organic Chemistry at the same university under the mentorship of Professor Wenhao Hu. She performed research on rhodium catalyzed multi-component reactions. Her current research focuses on the synthesis of immunologically active peptidyl disaccharides.

Xinfang Xu was born in 1981 in Zhejiang Province, China. He received his bachelor's degree in Chemistry from East China Normal University in 2005. He then began his graduate studies in Organic Chemistry at the same university, under the supervision of Professor Liping Yang (2005-2006) and Wenhao Hu (2006-present). His current research interest is development of novel asymmetric multi-component reactions.

416

Liu-Zhu Gong was born in October 1970 in Henan, China. He graduated from Henan Normal University (1989) and received his Ph.D. (2000) from Institute of Chemistry, Chinese Academy of Sciences. He was a visiting scholar at the University of Virginia and an Alexander von Humboldt Research Fellow at the University of Munich. He became an associate professor of Chengdu Institute of Organic Chemistry, Chinese Academy of Sciences in 2000. Since 2006 he has been a full professor at the University of Science and Technology of China. His current research interest is focused on organo- and transition metal-catalyzed asymmetric synthesis, the total synthesis of natural products, and chiral conjugated polymers.

Wei-jun Liu was born in 1979 in Hunan, China. He graduated from Xiangtan University (2002) and received his M.S. degree (2005) from Chengdu Institute of Organic Chemistry (Prof. Wen-Hao Hu). He joined WuXiAppTec (Shanghai) Co., Ltd. as a research chemist. He then moved to the research group of Professor Liu-Zhu Gong as a Ph.D. student at the University of Science and Technology of China. His current research interest is development of organo- and Lewis acid-catalyzed asymmetric synthesis.

Zhanjie Li was born in 1975 in Henan, China. He received his Bachelor's degree in 1996 and his Master's degree in 1999. Both were from Lanzhou University in China. After graduation in 1999, he worked at the Guangzhou Institute of Chemistry, Chinese Academy of Sciences. In 2004, he worked with Dr. Scott M. Goodman at State University of New York, College at Buffalo. In 2005 he began his PhD studies in the lab of Prof. Huw Davies at SUNY Buffalo and later on joined him at Emory University in 2008. His PhD research focuses on chiral catalyst design and Rh(II)-catalyzed asymmetric transformations of allylic and propargylic alcohols and donor/acceptor diazo compounds.

ENANTIOSELECTIVE THREE-COMPONENT REACTION FOR THE PREPARATION OF β-AMINO-α-HYDROXY ESTERS

A.

B.

Submitted by Jing Zhou, Xinfang Xu and Wenhao Hu.*[1]

Checked by John Frederick Briones and Huw M. L. Davies.

1. Procedure

A. Methyl 2-diazo-2-(4-methoxyphenyl) acetate. (1c) A 500-mL three-necked flask is equipped with a 100-mL dropping funnel, a rubber septum fitted with argon inlet needle and an egg-shaped 1 ¼ x 5/8 in magnetic stir bar. DBU (Note 1) (14.2 mL, 95.0 mmol, 1.50 equiv) in CH_3CN (60 mL) (Note 2) is added to the dropping funnel. The flask is charged with methyl 2-(4-methoxyphenyl) acetate (Note 3) (11.4 g, 63.3 mmol, 1.00 equiv), *p*-acetamidobenzenesulfonyl azide (*p*-ABSA) (Note 4) (18.2 g, 76.0 mmol, 1.20 equiv) and CH_3CN (120 mL). The DBU solution is added dropwise into reaction mixture. The resulting mixture is stirred over 12 h, and reaction progress is monitored by TLC analysis (Note 5). The reaction mixture is cooled with an ice bath, and saturated aqueous NH_4Cl (100 mL) (Note 6) is then added to quench the reaction. The mixture is extracted with ethyl ether (3 x 100 mL) (Note 7), and the combined organic

418

layers are washed with saturated brine (150 mL) (Note 8), dried over sodium sulfate (10 g) (Note 9) and concentrated by rotary evaporation (23 °C, 40 mmHg) to afford the crude product. Column chromatography purification of the crude product over silica gel (Note 10) affords 7.9 g (60%) of methyl 2-diazo-2-(4-methoxyphenyl) acetate as an orange solid (Notes 11 and 12). The product purity is >99% by HPLC (Note 13).

B. *(2S,3S)-methyl-2-(9-anthryloxymethyl)-2-(4-methoxyphenyl)-3-(4-methoxyphenylamino)-3-phenylpropanoate.* *(7a)* To a 100-mL single-necked, round-bottomed flask equipped with an octagonal-shaped 1 x 5/16 in magnetic stir bar and a rubber septum with argon inlet, is charged in sequence with 9-anthracenemethanol (2.60 g, 12.5 mmol, 1.0 equiv) (Note 14), benzaldehyde-*p*-anisidine imine (2.64 g, 12.5 mmol, 1.0 equiv) (Note 15), Rh$_2$(OAc)$_4$ (110 mg, 0.25 mmol, 0.02 equiv) (Note 16), (*R*)-2,2'-(9-phenanthryl)-BINOL phosphoric acid (87.5 mg, 0.125 mmol, 0.01 equiv) (Note 17), 4 Å molecular sieves (5.0 g) (Note 18) and dry CH$_2$Cl$_2$ (50 mL) (Note 19). The suspension is stirred for 30 min at room temperature (23 °C). The reaction system is cooled to –20 °C in isopropanol bath using Neslab cooling system. Methyl 2-diazo-2-(4-methoxyphenyl) acetate (3.86 g, 18.7 mmol, 1.5 equiv) in 10 mL dry CH$_2$Cl$_2$ is then added over 1 h *via* a syringe pump. After completion of the addition, the reaction mixture is stirred for additional 2 h at –20 °C until completion of the reaction as monitored by TLC analysis (Note 20). The reaction is quenched with 2 mL of saturated aqueous sodium hydrogen carbonate solution (Note 21), and the internal temperature which was measured by a thermometer rises from –20°C to about 0 °C during the time of quenching. The molecular sieves are removed by filtration through a pad of silica gel (10 g) and the filtrate is concentrated by rotary evaporation (40 °C, 40 mmHg) to a volume of approximately 25 mL. To the residue solution is added silica gel (10 g) and the solvent is further removed by rotary evaporation (40 °C, 40 mmHg) to give a dry powder. The crude product in the powder is subjected to column chromatography over silica gel (90 g) to give 6.5 g (87%) of (2*S*, 3*S*)-methyl 2-(9-anthryloxymethyl)-2-(4-methoxyphenyl)-3-(4-methoxy-phenylamino)-3-phenylpropanoate (Note 22) (Note 23). The product purity is 99% by HPLC (Note 24), and the enantiomeric purity is 94% *ee* by HPLC analysis using a chiral column (Notes 25 and 26).

2. Notes

1. DBU (98%) was purchased from Aldrich and used without further purification.

2. Acetonitrile was purchased from EMD Chemicals Inc. and used without further purification.

3. The checkers used methyl 2-(4-methoxyphenyl) acetate (>97%) that was purchased from Alfa Aesar Chemical Company, Inc. and used without further purification.

4. p-Acetamidobenzenesulfonyl azide was prepared from p-acetamidobenzenesulfonyl chloride according to the literature method.[2] p-Acetamidobenzenesulfonyl chloride (98+%) was purchased from Alfa Aesar Chemical Company, Inc. and sodium azide was purchased from Aldrich Chemical Company, Inc.

> Caution! The original procedure using methylene chloride as solvent should be avoided because it can produce the highly explosive material, diazidomethane, as side product.[3]

5. Thin layer chromatography was performed on Whatman precoated 60 Å silica gel plates with fluorescent indicator eluting with 9% ethyl acetate/petrol ether, visualized by a 254-nm UV lamp. The observed R_f values are 0.70 for the diazo product and 0.60 for methyl (4-methoxyphenyl) acetate.

6. NH_4Cl was purchased from Aldrich Chemical Company, Inc.

7. Ethyl ether was purchased from Aldrich Chemical Company, Inc. and used without further purification.

8. Sodium chloride was purchased from Aldrich Chemical Company, Inc.

9. Sodium sulfate was purchased from EMD Chemicals Inc.

10. Silica gel was purchased from Sorbent Technologies Company with the following specifications: porosity 60 Å, particle size 40-64 μm, surface area 450–550 m^2/g.

11. Flash column chromatography was performed on a silica gel column (5.5 cm width x 11cm length, 90 g silica gel). The product was eluted with petroleum ether, then 1500 mL of petroleum ether/ethyl acetate, 50:1. The desired product was obtained in orange fractions 7-13 (50 mL each), which were combined and concentrated by rotary evaporation (23 °C,

420

40 mmHg) to provide an orange solid that was dried for 3 h (25 °C, 0.1 mmHg).

12. The diazo product exhibits the following physicochemical properties: orange solid; IR (neat) 2953, 2837, 2079, 1697, 1511 cm^{-1}; mp 44–45 °C; ^{1}H NMR (400 MHz, CDCl$_3$) δ: 3.81 (s, 3 H), 3.85 (s, 3 H), 6.95 (d, J = 9.2 Hz, 2 H), 7.39 (d, J = 9.2 Hz, 2 H); ^{13}C NMR (100 MHz, CDCl$_3$) δ: 52.2, 55.5, 114.8, 117.1, 126.1, 158.3, 166.3 (C=N, signal missing); HRMS (ESI): calculated for C$_{20}$H$_{21}$N$_2$O$_6$ 385.1321 found 385.1392 [2M-N$_2$+H]$^+$; Anal. calcd. for C$_{10}$H$_{10}$N$_2$O$_3$ C, 58.25; H, 4.89; N, 13.59; Found C, 58.39; H, 4.90; N, 13.45.

13: Dynamax 60A column was used with 10% isopropanol in hexanes as eluent, a flow rate of 0.7 mL/min, and detection by UV (254 nm) detector. The t_R of the product is 5.83 min.

14. The checkers used 9-anthracenemethanol (97%), which was purchased from Aldrich Chemical Company, Inc. and was used without further purification.

15. Benzaldehyde-p-anisidine imine was prepared by condensation of benzaldehyde (>98.5%) with p-anisidine (99%) according to the literature method.[4] Benzaldehyde was purchased from Aldrich Chemical Company, Inc. and used without further purification. p-Anisidine was purchased from Alfa Aesar Chemical Company, Inc. and used without further purification.

16. Rh$_2$(OAc)$_4$ (98+%) was purchased from Johnson Matthey Company.

17. (R)-2,2'-(9-Phenanthryl)-BINOL phosphoric acid was prepared from (R)-BINOL (chiral purity >99%) according to the previous procedure in this volume. (R)-BINOL was purchased from Strem Chemicals Inc.

18. Powder 4 Å molecular sieves were purchased from Acros Organics. They were activated at 200°C in the oven before use.

19. Methylene chloride (HPLC grade) was purchased from Fischer Scientific Company and freshly distilled over calcium hydride before use.

20. Thin layer chromatography was performed on Whatman precoated 60 Å silica gel plates with fluorescent indicator eluting with 15% ethyl acetate/petrol ether and visualized by a 254-nm UV lamp. Observed R$_f$ values are 0.30 for the desired product and 0.32 for the O-H insertion product.

21. Sodium hydrogen carbonate was purchased from Aldrich Chemical Company, Inc.

22. Column chromatography was performed using 90 g of 200-300

mesh silica gel 60 (5.5 x 11 cm), and 50-mL fractions were collected (200 mL of petroleum ether, then 500 mL of petroleum ether/CH₂Cl₂/ethyl acetate, 30:3:1, then 500 mL each of 20:2:1, 15:3:1, 15:5:1, 15:6:1, and finally 500 mL of CH₂Cl₂). The desired product was obtained in fractions 22-54, which were combined and concentrated by rotary evaporation (40 °C, 40 mmHg). The residue was then dissolved in 20 mL of CH₂Cl₂, and 20 mL of petrol ether and concentrated by rotary evaporation (40 °C, 40 mmHg) to obtain white solid. The resulting solid was dried (23 °C, 0.1 mmHg) for 5 h.

23. The product exhibits the following physicochemical properties: white solid; $[\alpha]^{20}_D$= +34.4 (c=1, EtOAc); IR(neat) 3393, 3058, 2950, 2834, 1744, 1608, 1509, 1453, 1406, 1384, 1299, 1242, 1178, 1089, 1035, 819, 735, 702 cm^{-1}; mp 141–142 °C; ^{1}H NMR (400 MHz, CDCl₃) δ: 3.65 (s, 3 H), 3.74 (s, 3 H), 3.93 (s, 3 H), 4.63 (d, J = 9.6 Hz, 1 H), 5.12 (d, J = 9.6 Hz, 1 H), 5.24 (d, J = 10.3 Hz, 1 H), 5.74 (d, J = 10.4 Hz, 1 Hz), 6.41 (d, J = 8.8 Hz, 2 H), 6.62 (d, J = 8.8 Hz, 2 H), 7.02 (d, J = 8.8 Hz, 2 H), 7.15 (m, 5 H), 7.55 (m, 4 H), 7.82 (d, J = 9.2 Hz, 2 H), 8.07 (d, J = 8Hz, 2 H), 8.29 (d, J = 8.8 Hz, 2 H), 8.54 (s, 1 H); ^{13}C NMR (100 MHz, CDCl₃) δ: 51.9, 55.4, 55.7, 60.4, 65.6, 87.7, 113.6, 114.7, 115.5, 125.2, 126.1, 127.6, 128.5, 129.0, 129.2, 129.3, 130.8, 131.2, 131.7, 138.9, 140.5, 152.2, 160.0, 172.0; HRMS(ESI): calcd for C₃₉H₃₆NO₅ 598.2515 found 598.2596 [M+H]$^+$; Anal. calcd. for C₃₉H₃₅NO₅ C, 78.37; H, 5.90; N, 2.34; Found C, 78.45; H, 5.93; N, 2.36.

24. Dynamax 60A column was used with 5% isopropanol in hexanes as eluent, a flow rate of 0.7 mL/min, and detection by UV (254 nm) detector. The t_R of the product is 6.53 min.

25. AD-H column is available from Daicel Chemical Industries, Ltd. A 30-cm column was used with 5% isopropanol in hexanes as a mobile phase, a flow rate of 0.7 mL/min, and detection by UV (254nm). The t_R of the minor isomer (2R,3R) was 27.0 min and the major (2S,3S)-isomer was 18.4 min.

26. The racemic product was prepared using the same procedure as described above. 1,1'-Binaphthyl-2,2'-diyl hydrogen phosphate (95%, purchased from Aldrich Chemical Company, Inc.) was used instead of (R)-2,2'-(9-phenanthryl)-BINOL phosphoric acid.

Safety and Waste Disposal Information

All hazardous materials should be handled and disposed of in accordance with "Prudent Practices in the Laboratory"; National Academy Press; Washington, DC, 1995.

3. Discussion

The catalytic asymmetric version of the three-component reaction of aryldiazoacetates, alcohols, and imines, employs a novel cooperative catalysis strategy by Rh$_2$(OAc)$_4$ and chiral Brønsted acid. The reaction proceeds through oxonium ylide intermediates **IIa** or **IIb**, which are generated in situ from the diazo compounds and the alcohols in the presence of Rh$_2$(OAc)$_4$. This intermediate can be trapped by electrophiles such as imines activated by the chiral Brønsted acid catalyst. As shown in Scheme 1, the oxonium ylide **II** and the activated iminium **III** undergo an enantioselective Mannich-type reaction via proposed transition state **IV** to generate optically active **4**.

The procedure has been employed successfully with other diazo compounds and imines (See Table 1).[5] The Brønsted acid catalyst prepared from (S)-BINOL has also been used to give the *(2R,3R)*-product.

Scheme 1 Proposed Reaction Mechanism of the Title Reaction

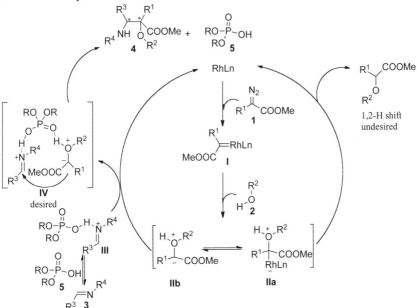

Table 1 Enantioselective Three-Component Reaction with Various Diazo Compounds and Imines

entry[a]	1(Ar₁)	3 or 6 (Ar₃)	7	yield (%)[b]	dr[c]	ee (%)[d]
1	1a (Ph)	6a (*m*-CH₃C₆H₄)	7b	96	>99:1	90
2[e]	1a (Ph)	6b (C₆H₅)	7c	83	>99:1	94
3	1a (Ph)	6c (2,3-Cl₂C₆H₃)	7d	95	>99:1	93
4	1a (Ph)	6d (*o*-CH₃ C₆H₄)	7e	95	>99:1	93
5	1a (Ph)	6e (*p*-CH₃ C₆H₄)	7f	92	>99:1	98
6	1a (Ph)	6f (*o*-ClC₆H₄)	7g	91	>99:1	92
7	1a (Ph)	3b (*p*-BrC₆H₄)	7h	87	>99:1	92
8	1a (Ph)	6g (*p*-ClC₆H₄)	7i	83	>99:1	93
9	1a (Ph)	6h (*p*-ClC₆H₄)	7j	82	>99:1	94
10	1a (Ph)	6i (1-naphthyl)	7k	88	>99:1	95
11	1b (*m*-BrC₆H₄)	3a (C₆H₅)	7l	96	>99:1	84
12	1c (*p*-MeOC₆H₄)	6h (*p*-BrC₆H₄)	7m	97	>99:1	95
13	1d (*p*-BrC₆H₄)	6b (C₆H₅)	7n	84	>99:1	94
14	1d (*p*-BrC₆H₄)	6g (*p*-ClC₆H₄)	7o	95	>99:1	92
15	1d (*p*-BrC₆H₄)	3a (C₆H₅)	7p	84	>99:1	92
16	1e (*o*-BrC₆H₄)	6h (*p*-BrC₆H₄)	7q	91	>99:1	83
17[f]	1c (*p*-MeOC₆H₄)	6b (C₆H₅)	7r	81	>99:1	98

[a] Reactions performed on a 0.25 mmol scale. [b] Isolated yield. [c] Determined by ¹H NMR spectroscopy of the unpurified reaction mixture. [d] Determined by HPLC. [e] Reaction performed on a 2.5 mmol scale with Rh₂(OAc)₄ (0.5 mol%) and chiral phosphoric acid (1 mol%). [f] (*S*)-BINOL phosphoric acid is used.

1. whu@chem.ecnu.edu.cn; Department of Chemistry, East China Normal University, Shanghai 200062, China. We thank the financial support from the National Science Foundation of China (NSFC) (20932003),

and from the MOST of China (2011CB808600).

2. Davies, M. L. H.; Cantrell, R. W.; Jr.; Romines, R. K.; and Baum, S. J.; *Org. Synth.* **1992**, *70*, 93–100; *Coll. Vol. IX* **1998**, 422–426.
3. Conrow, R. E.; Dean, D.W. *Org. Proc. Res. Dev.* **2008**, *12*, 1285-1286.
4. Danheiser, L. R.; Okamoto, I.; Lawlor, D. M.; Lee, W. T.; *Org. Synth.* **2003**, *80*, 160–171. *Org. Synth.* **2009**, *Coll. Vol. 11*, 920–928.
5. Hu, W.-H.; Xu, X.-F.; Zhou, J.; Liu, W.-J.; Huang, H.-X.; Hu, J.; Yang, L.-P.; and Gong, L.-Z. *J. Am. Chem. Soc.* **2008**, *130*, 7782–7783.

Appendix
Chemical Abstracts Nomenclature; (Registry Number)

Benzaldehyde (100-52-7)
p-Anisidine (104-94-9)
Benzaldehyde-*p*-anisidine imine (783-08-4)
9-Anthracenemethanol (1468-95-7)
9-Anthraldehyde (642-31-9)
R-(+)-1,1'-Bi-2-naphthol, (*R*)-BINOL (18531-94-7)
(*R*)-2,2'-(9-Phenanthryl)-BINOL phosphoric acid
Dirhodium tetraacetate dehydrate (15956-28-2)
Methyl (4-methoxyphenyl)diazoacetate
4-Methoxyphenylacetic acid (101-01-8)
(2*S*, 3*S*)-Methyl-2-(9-anthryloxy-methyl)-2-(4-methoxyphenyl)-3-
    (4-methoxy-phenylamino)-3-phenylpropanoate; (1034152-21-0)

Wenhao Hu was born in 1967 in Sichuan Province, China. He received his M.S. degree in Chengdu Institute of Organic Chemistry. He obtained a Ph.D. degree from The Hong Kong Polytechnic University in 1998 under the direction of Professor Albert S. C. Chan, and was a postdoctoral fellow at University of Arizona with Professor Michael P. Doyle. He then joined GeneSoft Pharm. Inc. located in San Francisco as a Staff Scientist (2002-2003). He moved to New Jersey to join Bristol-Myers Squibb Company as a Research Investigator (2003-2006). He returned to China as a Professor in the department of chemistry at East China Normal University in 2006. His research interests include development of highly efficient synthetic methods and their application in the synthesis of biologically active compounds.

Jing Zhou was born in 1984 in Shandong Province, China. She received her bachelor's degree in Chemistry in 2007 from East China Normal University, Shanghai. She then began her graduate study in Organic Chemistry at the same university under the mentorship of Professor Wenhao Hu. She performed research on rhodium catalyzed multi-component reactions. Her current research focuses on the synthesis of immunologically active peptidyl disaccharides.

Xinfang Xu was born in 1981 in Zhejiang Province, China. He received his bachelor's degree in Chemistry from East China Normal University in 2005. He then began his graduate studies in Organic Chemistry at the same university, under the supervision of Professor Liping Yang (2005-2006) and Wenhao Hu (2006-present). His current research interest is the development of novel asymmetric multi-component reactions.

John Frederick Briones was born in 1982 in Laguna, Philippines. He earned his B.S. degree in Chemistry from the University of the Philippines, Los Banos in 2003 and later on pursued his Master's degree at the University of the Philippines, Diliman. He joined the research lab of Prof. Huw Davies in 2007 and currently his research project focuses on Rh(II)-catalyzed enantioselective transformations of alkynes.

426

MILD CONVERSION OF TERTIARY AMIDES TO ALDEHYDES USING Cp₂Zr(H)Cl (SCHWARTZ'S REAGENT)

Submitted by Matthew W. Leighty, Jared T. Spletstoser, and Gunda I. Georg.[1]
Checked by Hirotatsu Umihara and Tohru Fukuyama.

1. Procedure

A. N,N-Diethyl-3,4,5-trimethoxybenzamide. A 300-mL, three-necked, round-bottomed flask equipped with a 4.5-cm rod-shaped, Teflon-coated, magnetic stir bar, an internal thermocouple temperature probe, a calcium chloride drying tube, and a 50-mL pressure-equalizing dropping funnel sealed with a rubber septum is charged with 3,4,5-trimethoxybenzoyl chloride **1** (20.0 g, 86.7 mmol, 1 equiv) under ambient atmosphere (Notes 1 and 2). The acid chloride is dissolved in anhydrous CH₂Cl₂ (120 mL), and the resulting solution is stirred (500 rpm) in an ice-water bath while diethylamine (18.8 mL, 182 mmol, 2.1 equiv) is added dropwise from the dropping funnel over 15 min, such that the internal temperature does not exceed 15 °C (Notes 3 and 4). The reaction mixture is warmed to room temperature and stirred for 15 min at which time TLC analysis shows that no acid chloride **1** remains (Note 5). The reaction mixture is diluted with 2.7 M aqueous HCl (100 mL), and the resulting mixture is transferred to a 500-mL separatory funnel containing CH₂Cl₂ (50 mL) and 2.7 M aqueous HCl (50 mL) (Note 6). The layers are separated, and the aqueous phase is extracted with CH₂Cl₂ (50 mL). The combined organic layers are washed with saturated aqueous NaHCO₃ (100 mL), and the aqueous layer is extracted with CH_2Cl_2 (50 mL). The combined organic layers are dried over Na_2SO_4, filtered into a 1-L one-necked, round-bottomed flask and concentrated on a rotary evaporator (30 °C, 40 mmHg) (Note 7). The residue is purified by silica gel column chromatography (elution with ethyl acetate/n-hexane, 3/1) (Note 8). The combined eluates are concentrated on a rotary evaporator (30 °C, 40 mmHg) and then dried overnight at 20 mmHg at room temperature to afford 22.1 g of amide **2** as a white solid in 95% yield (Note 9).

B. *3,4,5-Trimethoxybenzaldehyde.* A 300-mL, three-necked, round-bottomed flask equipped with a 4.0-cm rod-shaped, Teflon-coated, magnetic stir bar, an internal thermocouple temperature probe, an argon flowing tube, and a rubber septum is charged with amide **2** (10.7 g, 40.0 mmol, 1 equiv). Tetrahydrofuran (THF, 100 mL) is added to the flask, and the resulting mixture is stirred (300 rpm) until homogenous, and then the flask is covered with aluminum foil (Note 10). The rubber septum is briefly removed and Schwartz's reagent (10.8 g, 42 mmol, 1.05 equiv) is added directly to the vigorously stirring solution in one portion followed by recapping the reaction flask with the rubber septum (Notes 11 and 12). An additional 30 mL of THF is added to rinse the reagent on the inside of the flask into the reaction mixture, ensuring complete addition of the zirconium reagent. The suspension is stirred for 30 min. The reaction mixture becomes slowly homogeneous during the course of the reaction at which time TLC analysis shows that no amide **2** remains (Notes 13 and 14). Silica gel (70 g) is added in one portion, and the mixture is stirred (1350 rpm) for one minute under ambient atmosphere (Note 15). The heterogeneous mixture is filtered over a silica plug (10 g) eluting with ethyl acetate (500 mL). The resulting filtrate is then added to a 2-L separatory funnel containing n-hexane (200 mL) and H_2O (400 mL) (Notes 16 and 17). The layers are separated, and the aqueous layer is extracted twice with ethyl acetate (400 mL) and n-hexane (160 mL). The combined organic layers are dried over Na_2SO_4 and concentrated on a rotary evaporator (35 °C, 30 mmHg). The residue is purified by silica gel column chromatography (elution with ethyl acetate/n-hexane, 1/4) (Note 18). The combined eluents are concentrated on a rotary evaporator (35 °C, 30 mmHg) and then dried overnight at 20 mmHg at room temperature to afford 7.20 g of aldehyde **3** as an off-white solid in 92% yield (Note 19).

2. Notes

1. All glassware was flame-dried immediately before use.

2. 3,4,5-Trimethoxybenzoyl chloride (98%) was purchased by the submitters and checkers from Aldrich Chemical Company, Inc. and used as received.

3. The submitters obtained dry CH_2Cl_2 by passing the solvent through an activated alumina column under a nitrogen atmosphere. The checkers followed the same method. Freshly distilled CH_2Cl_2 can also be used.

4. Diethylamine (99.5%) was purchased by the submitters and checkers from Aldrich Chemical Company, Inc. and was used as received. Slow addition allows for minimal heat fluctuation as the amine is added. The reaction mixture becomes slightly heterogeneous during the course of the reaction.

5. TLC analysis was conducted on Merck silica gel 60 F_{254} plates (0.25 mm, glass-backed, visualized with 254 nm UV lamp and stained with p-anisaldehyde) using 50% n-hexane in ethyl acetate as an eluent. Acid chloride **1** had an $R_f = 0.69$ and 0.13 (UV active, dark purple after staining) and amide **2** had an $R_f = 0.17$ (UV active, white after staining).

6. CH_2Cl_2 (ACS grade) used for workup was purchased by the submitters from Fisher Scientific and used as received. The checkers purchased CH_2Cl_2 (ACS grade) from Wako Pure Chemical Industries, Ltd., and used as received.

7. The submitters purchased Na_2SO_4 (anhydrous, $\geq$99%) from Mallinckrodt and used as received. The checkers purchased Na_2SO_4 (anhydrous, $\geq$98.5%) from Nacalai Tesque, Inc. and used as received.

8. Silica gel (acidic) was purchased from Kanto Chemical Co., Inc. (40-100 μm). The crude material was dissolved in the eluent (20 mL), and the solution was then charged onto a column (diameter = 7.5 cm) of 80 g of silica gel. The column was eluted with n-hexane/EtOAc, 1:3, and 100-mL fractions were collected. Fractions 3-16 were collected.

9. Physical characteristics of amide **2**: mp (uncorr.) 59–60 °C, (submitter: 60–62 °C); IR (thin film) 2971, 1632, 1458, 1333, 1236, 1127, 1007 cm^{-1}; ^1H NMR (400 MHz, $CDCl_3$) δ: 1.21 (6 H, br s), 3.31 (2 H, br s), 3.53 (2 H, br s), 3.85 (3 H, s), 3.88 (6 H, s), 6.60 (2 H, s). ^{13}C NMR (100 MHz, $CDCl_3$) δ: 12.06, 13.53, 38.52, 42.60, 55.30, 59.88, 102.74, 131.90, 137.77, 152.41, 169.99. MS (ESI-MS) m/z: (M+H) calcd. for $C_{14}H_{22}NO_4$, 268.33; Found 268.35 (100%); (HR-ESI) m/z: (M+H) calcd. for $C_{14}H_{22}NO_4$,

268.1549; Found 268.1539. The submitters evaluated **2** for purity by LCMS employing a reversed phase HPLC column (ACQUITY UPLC BEH C18 Column (2.1 x 30 mm)), using 95% water/5% MeCN with 0.1% formic acid (solvent A) and 95% MeCN/5% water with 0.1% formic acid (solvent B). A linear gradient of 5% to 95% solvent B in solvent A over 4.5 min was used at a flow rate of 0.25 mL/min; t_R = 3.63 min. The purity was determined to be >98% at 220 nm. The checkers evaluated **2** for purity by elemental analysis. Anal. calcd. for $C_{14}H_{21}NO_4$: C, 62.90; H, 7.92; N, 5.24. Found: C, 62.74; H, 7.95; N, 5.22.

10. The submitters obtained dry THF by passing the solvent through an activated alumina column under a nitrogen atmosphere. The checkers followed the same method. Freshly distilled THF can also be used.

11. $Cp_2Zr(H)Cl$ (95%) was obtained by the submitters and the checkers from Alfa Aesar and used as received. Although this reagent is commercially available from other vendors, the Alfa Aesar reagent appears to be of superior quality. This reagent can also be prepared using a literature procedure.[3]

12. $Cp_2Zr(H)Cl$ is sensitive to prolonged contact to air and light, therefore the reagent was used with minimal exposure to air and light. Before storage, the reagent bottle was flushed briefly with argon, sealed with parafilm, and stored at 0 °C.

13. The reaction was monitored by TLC analysis on Merck silica gel 60 F_{254} plates (0.25 mm, glass-backed, visualized with 254 nm UV lamp and stained with Ce-PMA) using 50% *n*-hexane in ethyl acetate as an eluent. Amide **2** had an R_f = 0.17 (UV active, pale blue after staining) and aldehyde **3** had an R_f = 0.63 (UV active, blue after staining).

14. The submitters reported that an additional portion of $Cp_2Zr(H)Cl$ (2.06 g, 7.99 mmol, 0.2 equiv) was sometimes needed 25 min after the first addition in order for the reaction to reach completion. This is ascribed to variations in the quality of the $Cp_2Zr(H)Cl$.

15. Quenching with silica gel serves to break down the stable zirconacycle intermediate into the aldehyde as well as to destroy any remaining hydride. The use of water to quench the reaction can result in hard to break emulsions.

16. Leaving the aldehyde in contact with the silica gel for prolonged periods can result in a bright yellow or orange-red impurity that is difficult to separate from the product. This impurity is presumed to be a consequence of further reaction of the aldehyde with the zirconium by-products and the

silica gel. Thus, care should be taken to avoid prolonged exposure of the product to the zirconium entities and silica gel after the quench.

16. Ethyl acetate and *n*-hexane used for extraction was purchased by the checkers from Wako Pure Chemical Industries, Ltd., and was used as received.

17. The use of *n*-hexane as a co-solvent for extraction prevented the formation of emulsions.

18. Silica gel (acidic) was purchased from Kanto Chemical Co., Inc. (40–100 μm). The crude material was dissolved in CH_2Cl_2 (15 mL) and then was charged onto a column (diameter = 7.5 cm) of 90 g of silica gel. The column was eluted with *n*-hexane/EtOAc, 4:1, and 100-mL fractions were collected. Fractions 9–24 were collected.

19. Physical characteristics of aldehyde **3**: mp (uncorr.) 72–73 °C (submitter: 75–77 °C); IR (thin film) ν 2971, 1686, 1588, 1459, 1332, 1234, 1128 cm^{-1}; ^{1}H NMR (400 MHz, CDCl$_3$) δ: 3.94 (6 H, s), 3.95 (3 H, s), 7.14 (2 H, s), 9.88 (1 H, s). ^{13}C NMR (100 MHz, CDCl$_3$) δ: 55.86, 60.57, 106.29, 131.40, 143.14, 153.27, 190.71. MS (ESI-MS) *m/z*: (M+H) calcd. for $C_{10}H_{13}O_4$, 197.1; Found 197.0 (100%); MS (HR-ESI) *m/z*: (M+H) calcd. for $C_{10}H_{13}O_4$, 197.0814; Found, 197.0813. The submitters evaluated **3** for purity by LCMS using a reversed phase HPLC column (ACQUITY UPLC BEH C18 Column (2.1 x 30 mm)), using 95%water/ 5% MeCN with 0.1% formic acid (solvent A) and 95% MeCN/ 5% water with 0.1% formic acid (solvent B). A linear gradient of 5% to 95% solvent B in solvent A over 4.5 min was used at a flow rate of 0.25 ml/min; t_R = 3.58 min. The purity was determined to be >98% at 220 nm. The checkers evaluated **3** for purity by elemental analysis. Anal. calcd. for $C_{10}H_{12}O_4$: C, 61.22; H, 6.16. Found: C, 61.14; H, 6.10.

Safety and Waste Disposal Information

All hazardous materials should be handled and disposed of in accordance with "Prudent Practices in the Laboratory"; National Academy Press; Washington, DC, 1995.

3. Discussion

Cp$_2$Zr(H)Cl (Schwartz's reagent) is an efficient reagent for the reduction of tertiary amides to aldehydes.[5] This reaction is conducted at room temperature and is complete in short reaction times. Over-reduction to the alcohol is generally not observed, and the crude reaction mixture is clean enough to require only minimal purification.

One of the remarkable features of this reduction is that it works well on a variety of amide substrates including aromatic and aliphatic amides as well as N,N-dialkyl and N,O-dimethyl (Weinreb) amides. Many of the known methods for the *selective* reduction of amides to aldehydes are substrate specific, generally requiring specialized amide types for successful conversion to the aldehyde.[6] Most other existing methods tend to give over-reduction to the alcohol or amine products.[7] The described method is unique in that nearly all types of amide linkages are compatible substrates and the conditions are remarkably tolerant to a wide range of functional groups. Of considerable note is that the tertiary amide is *selectively* cleaved in the presence of the more easily reduced ester functional group.

Mechanistic evidence[5c] for the reduction suggests that after hydride delivery, the zirconium reagent forms a stable sp^3-hybridized, 18-electron complex with the amide **I** to yield intermediate **II** (Scheme 1). Silica gel addition to the reaction mixture breaks downs the intermediate into the desired aldehyde **III** and zirconium by-products. Therefore, only amides are reduced in the presence of esters and other potentially reducible moieties. Although most amides are reduced using this procedure, important to note is that very sterically-demanding amides generally afford lower yields of aldehyde. The low yields are presumably a consequence of poor hydride delivery/complex formation due to the bulky cyclopentadienyl groups encountering the sterically hindered amide, which results in lower yields. Rawal has developed modified reaction conditions to overcome this limitation.[8]

Scheme 1. Reduction Mechanism

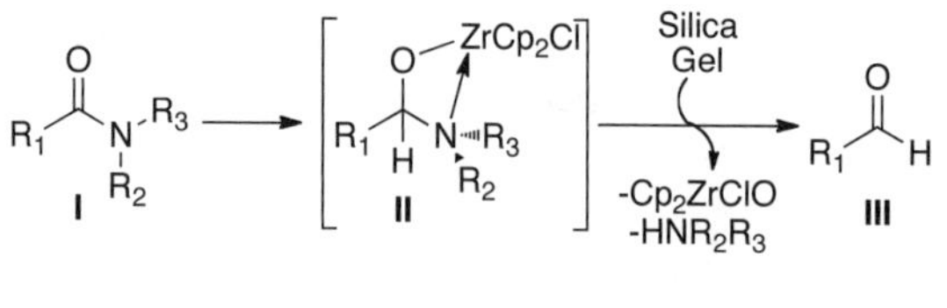

In this protocol, the addition of silica gel is in most cases the best workup procedure. The use of an aqueous quench generally affords problematic emulsions and/or the precipitation of a white solid (assumed to be the zirconium oxide byproducts) that can be difficult to remove. The use of silica gel effectively cleaves the reaction intermediate to the product aldehyde while concomitantly destroying excess reagent and sequestering the zirconium byproducts. The use of an aqueous workup, after filtering off the silica gel, was implemented in order to ensure complete removal of these byproducts as unwanted side reactions were found to occur with prolonged exposure to the silica gel upon scale up. This procedure allows for a very facile and efficient workup procedure that provides a crude product that is easily purified by standard chromatography.

As is evident in Table 1, the reaction is tolerant to a variety of functionality. For example, both neutral and electron-deficient aromatic amides are readily reduced affording the corresponding aldehydes in good yields (entries 1 and 2). As demonstrated in entry 3, *selective* amide reduction occurs in the presence of an ester in good yield. In addition, bulky amides and non-aromatic amides can be reduced to the corresponding aldehydes in excellent yields (entries 4 and 5).

Presented is a mild and operationally simple method for the selective reduction of tertiary amides to aldehydes using $Cp_2Zr(H)Cl$. This method offers an alternative to known methods that are either substrate selective or can result in over-reduction products. In addition, this procedure can be conducted in the presence of esters and results in *selective* reduction of the amide to afford the corresponding aldehyde. Therefore, this procedure is a significant improvement over known amide reduction methods due to its mild and selective nature.

Table 1. Preparative Reduction of Tertiary Amides with Cp$_2$Zr(H)Cl[a]

Entry	Amide	Aldehyde	Scale (mmol)	Yield (%)
1	**4**	**5**	12.9	80
2	**6**	**7**	10.2	82
3	**8**	**9**	10.3	73
4	**10**	**11**	15.8	90
5	**12**	**13**	5.3	72

a) Each amide in the Table was prepared from the corresponding acid chloride or acid, which were purchased from either Aldrich Chemical Company, Inc. or TCI America and used as received. Amides **4**, **6**, **8**, and **10** were synthesized using the procedure described Step A. Amide **12** was prepared from the corresponding acid using standard coupling conditions. All reactions in the Table were conducted for 30 minutes at the indicated scale, using the procedure as described in Step B and purified using MPLC (silica gel) eluting with the appropriate ratio of hexanes and ethyl acetate. The pure aldehydes were characterized by standard methods and are consistent with the proposed structures.

1. Department of Medicinal Chemistry and The Institute of Therapeutics and Drug Development, University of Minnesota, Minneapolis, MN

55414. E-mail: georg@umn.edu. Support is acknowledged from the University of Minnesota through the Vince and McKnight Endowed Chairs.

2. McCabe, E. T.; Barthel, W. F.; Gertler, S. I.; Hall, S. A. *J. Org. Chem.* **1954**, *19*, 493–498.

3. Buchwald, S. L.; LaMaire, S. J.; Nielsen, R. B.; Watson, B. T.; King, S. M. *Org. Synth.* **1998**, *9*, 162–165.

4. Pearl, I. A.; Beyer, D. L. *J. Am. Chem. Soc.* **1952**, *74*, 4262–4263.

5. (a) White, J. M.; Tunoori, A. R.; Georg, G. I. *J. Am. Chem. Soc.* **2000**, *122*, 11995–11996. (b) Spletstoser, J. T.; White, J. M.; Georg, G. I. *Tetrahedron Lett.* **2004**, *45*, 2787–2789. (c) Spletstoser, J. T.; White, J. M.; Tunoori, A. R.; Georg, G. I. *J. Am. Chem. Soc.* **2007**, 3408–3419.

6. (a) Ried, W.; Konigstein, F. J. *Angew. Chem.* **1958**, *70*, 165. (b) Brown, H. C.; Tsukamoto, A. *J. Am. Chem. Soc.* **1961**, *83*, 2016–21017. (c) Brown, H. C.; Tsukamoto, A. *J. Am. Chem. Soc.* **1961**, *83*, 4549–4552. (d) Brown, H. C.; Tsukamoto, A. *J. Am. Chem. Soc.* **1964**, *86*, 1089–1095. (d) Brown, H. C.; Bigley, D. B.; Arora, S. K.; Yoon, N. M. *J. Am. Chem. Soc.* **1970**, *92*, 7161–7167. (e) Nahm, S.; Weinreb, S. M. *Tetrahedron Lett.* **1981**, *22*, 3815–3818. (f) Bower, S.; Kreutzer, K. A.; Buchwald, S. L. *Angew. Chem. Int. Ed.* **1996**, *35*, 1515–1516.

7. Larock, R. C. *Comprehensive Organic Transformations*; 2nd ed.; John Wiley & Sons, Inc.: New York, 1999.

8. (a) McGilvra, J. D.; Unni, A. K.; Modi, K.; Rawal, V. H. *Angew. Chem. Int. Ed.* **2006**, *45*, 6130–6133. (b) Gondi, V. B.; Hagihara, K.; Rawal, V. H. *Chem. Commun.* **2010**, *46*, 904–906.

Appendix
Chemical Abstracts Nomenclature (Registry Number)

Diethylamine: Ethanamine, *N*-ethyl-; (109-89-7)
Zirconocene chloride hydride: Zirconium chlorobis(η^5-2,4-cyclopentadien-1-yl)hydro-; (37342-97-5)
3,4,5-Trimethoxybenzoyl chloride: Benzoyl chloride, 3,4,5-trimethoxy-; (4521-61-3)
3,4,5-Trimethoxybenzaldehyde; (86-81-7)
N,N-Diethyl-3,4,5-trimethoxybenzamide; (5470-42-8)

Gunda I. Georg obtained her doctoral degree from the University of Marburg in Germany. After postdoctoral studies at the University of Ottawa in Canada, and one year on the faculty at the University of Rhode Island, she joined the faculty of the Department of Medicinal Chemistry at the University of Kansas (1984). In 2007 she moved to the University of Minnesota as Head of the Department of Medicinal Chemistry. She holds the Robert Vince Endowed Chair and a McKnight Presidential Chair. She is the Founding Director of the Institute for Therapeutics Discovery and Development at the University of Minnesota.

Matthew W. Leighty (born 1979) graduated with a B.S. in chemistry from Alma College in 2001. He received his Ph.D. in 2009 from the University of Kansas under the guidance of Professor Gunda I. Georg working on the design and synthesis of biologically active compounds. Matthew is currently a postdoctoral research associate in the laboratories of Jeffrey N. Johnston.

Jared Spletstoser was born in Center, ND in 1976 and received his B.S. from the University of North Dakota. He received his Ph.D. in 2004 from the University of Kansas in the labs of Gunda Georg where he worked on Taxol photoprobes and brain delivery, the reduction of amides to aldehydes and the total synthesis of Oximidine II. He then joined the labs of James Leighton where he developed methods for the tandem silylformylation-crotylation of internal alkynes as an American Cancer Society postdoctoral fellow. He is currently a Principal Scientist at GlaxoSmithKline working in Infectious Diseases.

436

Hirotatsu Umihara was born in Kanagawa, Japan in 1988. He received his B.S. in 2011 from the University of Tokyo. In the same year, he began his graduate studies at the Graduate School of Pharmaceutical Sciences, the University of Tokyo, under the guidance of Professor Tohru Fukuyama. His research interests are in the area of the total synthesis of natural products.

CUMULATIVE AUTHOR INDEX FOR VOLUMES 85-88

This index comprises the names of contributors to Volumes **85**, **86**, **87**, and **88**. For authors of previous volumes, see either indices in Collective Volumes I through XI, or the single volume entitled *Organic Syntheses, Collective Volumes I-VIII, Cumulative Indices,* edited by J. P. Freeman.

Burke, M. D., **86**, 344
Butler, C. R., **86**, 274
Butler, J. D., **87**, 339
Buzon, R. A., **87**, 16

Cabart, F., **85**, 147
Cacchi, S., **88**, 260
Caggiano, L., **86**, 121
Cai, Z., **88**, 309
Campbell, L., **85**, 15
Campeau, L. -C., **88**, 22
Carreira, E. M., **87**, 88
César, V., **85**, 34
Chalker, J. M., **87**, 288
Chaloin, O., **85**, 147
Chandrasekaran, P., **86**, 333
Chang, S., **85**, 131
Chang, Y., **87**, 245
Charette, A. B., **87**, 115, 170
Chen, Q.-Y., **87**, 126
Chen, Y., **88**, 377
Chiong, H. A., **86**, 105
Cho, S. H., **85**, 131
Chouai, A., **86**, 141, 151
Christie, H. S., **88**, 364
Clososki, G. C., **86**, 374
Coates, G. W., **86**, 287
Coste, A., **87**, 231
Counceller, C. M., **88**, 33
Couty, F., **87**, 231
Crimmins, M. T., **88**, 364

Dai, P., **86**, 236
Daugulis, O., **86**, 105, **87**, 184
Davies, S. G., **87**, 143
Davis, B. G., **87**, 288
de Alaniz, J. R., **87**, 350
DeBerardinis, A. M., **87**, 68
Delaude, L., **87**, 77
DeLuca, R. J., **88**, 4
Deng, X., **85**, 179
Denmark, S. E., **86**, 274; **88**, 102
Desai, A. A., **88**, 224
Ding, K., **87**, 126

Do, H.-Q., **87**, 184
Do, N., **85**, 138
Donahue, J. P., **86**, 333
Doody, A. B., **88**, 212
Drago, C., **86**, 121
Du, H., **86**, 315, **87**, 263
Duchêne, A., **85**, 231
Dudley, G. B., **88**, 353
Dudley, M. E., **86**, 172

Ebner, D. C., **86**, 161
Eichman, C. C., **88**, 33
Eisenberger, P., **88**, 168
Ekoue-Kovi, K., **87**, 1
Ellman, J. A., **86**, 360
Ely, R. J., **88**, 342
Endo, K., **86**, 325
Erkkilä, A., **87**, 201
Etxebarria-Jardi, G., **88**, 317, 330
Evano, G., **87**, 231

Fagnou, K., **88**, 22
Fei, X.-S., **87**, 126
Fidan, M., **86**, 47
Fleming, M. J., **85**, 1
Fletcher, A. M., **87**, 143
Fokin, V. V., **88**, 238
Franckevičius, V., **85**, 72
Fu, G. C., **87**, 299, 310, 317, 330
Fu, R., **87**, 263
Fujimoto, T., **86**, 325
Fujiwara, H., **86**, 130
Fukuyama, T., **86**, 130; **88**, 152, 388
Fürstner, A., **85**, 34; **86**, 298

Gálvez, E., **86**, 70, 81
Gaspar, B., **87**, 88
Georg, G. I., **88**, 427
Giguère-Bisson, M., **88**, 14
Gillis, E. P., **86**, 344
Glasnov, T. N., **86**, 252
Glorius, F., **85**, 267
Gong, L., **88**, 406
Gooßen, L. J., **85**, 196

Goss, J. M., **86**, 236
Goudreau, S. R., **87**, 115
Graham, T. H., **88**, 42
Greszler, S., **86**, 18
Guichard, G., **85**, 147
Guillena, G., **88**, 317, 330
Gulder, T. A. M., **88**, 70
Gulder, T., **88**, 70

Haddadin, M. J., **87**, 339
Hahn, B. T., **85**, 267
Hans, M., **87**, 77
Harada, S., **85**, 118
Harmata, M., **88**, 309
Hartwig, J. F., **88**, 4
Hein, J. E., **88**, 238
Henry-Riyad, H., **88**, 121
Hierl, E., **85**, 64
Hill, M. D., **85**, 88
Hoang, T. T., **88**, 353
Hodgson, D. M., **85**, 1
Horning, B. D., **88**, 42
Hossain, M. M., **86**, 172
Hu, W., **88**, 406, 418
Huang, C., **88**, 309
Huang, D. S., **88**, 4
Huang, K., **87**, 26, 36
Huard, K., **86**, 59
Hughes, C. O., **88**, 364
Humphreys, P. G., **85**, 1
Hwang, S. J., **85**, 131
Hwang, S., **86**, 225

Ichikawa, J., **88**, 162
Iwasaki, M., **88**, 238

Jackson, R. F. W., **86**, 121
Javed, M. I., **85**, 189
Johnson, J. S., **85**, 278
Johnston, J. N., **88**, 212
Ju, L., **87**, 218

Kakiuchi, F., **87**, 209
Kang, H. R., **86**, 225

Kappe, C. O., **86**, 252
Kerr, M. S., **87**, 350
Kieltsch, I., **88**, 168
Kim, S., **86**, 225
Kirai, N., **87**, 53
Kitazawa, K., **87**, 209
Kitching, M. O., **85**, 72
Knauber, T., **85**, 196
Knochel, P., **86**, 374
Kochi, T., **87**, 209
Kocienski, P. J., **85**, 45
Koenig, S. G., **87**, 275
Koller, R., **88**, 168
Kong, J., **87**, 137
Kozmin, S. A., **87**, 253
Kramer, J. W., **86**, 287
Krasnova, L. B., **88**, 238
Krause, H., **85**, 34, **86**, 298
Krout, M. R., **86**, 181, 194
Kuethe, J. T., **86**, 92
Kurth, M. J., **87**, 339
Kwon, O., **86**, 212; **88**, 138

La Vecchia, L., **85**, 295
Landais, Y., **86**, 1
Langenhan, J. M., **87**, 192
Langle, S., **85**, 231
Larock, R. C., **87**, 95; **88**, 377
Lathrop, S. P., **87**, 350
Lautens, M., **85**, 172, **86**, 36
Lazareva, A., **86**, 105
Lebel, H., **86**, 59, 113
Lebeuf, R., **86**, 1
Lee, H., **87**, 245
Leighty, M. W., **88**, 427
Leogane, O., **86**, 113
Lesser, A. B., **88**, 109
Ley, S. V., **85**, 72
Li, B., **87**, 16
Li, C. -J., **88**, 14
Linder, C., **85**, 196
Lisboa, M. P., **88**, 353
List, B., **86**, 11
Liu, J. H. -C., **88**, 102

442

Liu, W., **88**, 406
Longbottom, D. A., **85**, 72
Lou, S., **86**, 236; **87**, 299, 310, 317, 330
Lu, C. -D., **85**, 158
Lu, K., **86**, 212

MacMillan, D. W. C., **88**, 42
Mani, N. S., **85**, 179
Mans, D. J., **85**, 238, 248
Marcoux, D., **87**, 115
Marin, J., **85**, 147
Markina, N. A., **88**, 377
Marshall, A.-L., **87**, 192
Matsunaga, S., **85**, 118
Maw, G., **85**, 219
McAllister, G. D., **85**, 15
McDermott, R. E., **85**, 138
McDonald, F. E., **88**, 296
McNaughton, B. R., **85**, 27
Meletis, P., **86**, 47
Meng, T., **87**, 137
Meyer, H., **85**, 287, 295
Miller, B. L., **85**, 27
Miyaura, N., **88**, 79, 202, 207
Mohr, J. T., **86**, 181, 194
Montchamp, J.-L., **85**, 96
Moore, D. A., **85**, 10
Morán-Ramallal, R., **88**, 224
Morera, E., **88**, 260
Morken, J. P., **88**, 342
Morra, N. A., **85**, 53
Morshed, M. M., **86**, 172
Mosa, F., **85**, 219
Mousseau, J. J., **87**, 170
Movassaghi, M., **85**, 88
Muchalski, H., **88**, 212
Mudryk, B., **85**, 64
Müller-Hartwieg, J. C. D., **85**, 295
Murakami, K., **87**, 178

Nájera, C., **88**, 317, 330
Nakamura, E., **86**, 325
Nakamura, M., **86**, 325
Ngi, S. I., **85**, 231

444

Sepulveda, D., **88**, 121
Shi, F., **87**, 95
Shi, Y., **86**, 263, 315
Shibasaki, M., **85**, 118
Shubinets, V., **87**, 253
Simanek, E. E., **86**, 141, 151
Singh, S. P., **87**, 275
Sirois, L. E., **88**, 109
Slatford, P. A., **86**, 28
Smith, C. R., **85**, 238, 248
Snaddon, T. N., **85**, 45
Snyder, S. A., **88**, 54
Söderberg, B. C., **88**, 291
Solano, D. M., **87**, 339
Sperotto, E., **85**, 209
Spletstoser, J. T., **88**, 427
Stambuli, J. P., **88**, 33
Stanek, K., **88**, 168
Stepanenko, V., **87**, 26
Stevens, K. L., **86**, 18
Stoltz, B. M., **86**, 161, 181, 194
Storgaard, M., **86**, 360
Struble, J. R., **87**, 362
Sugai, J., **88**, 79
Sun, H., **88**, 87, 181

Takita, R., **85**, 118
Takizawa, M., **88**, 79
Tambar, U. K., **86**, 161
Taylor, R. J. K., **85**, 15
Thibonnet, J., **85**, 231
Thirsk, C., **85**, 219
Thompson, A. L., **87**, 288
Tian, W.-S., **87**, 126
Ting, P., **87**, 137
Togni, A., **88**, 168
Tokuyama, H., **86**, 130; **88**, 152, 388
Tong, R., **88**, 296
Treitler, D. S., **86**, 287, **88**, 54
Troyer, T. L., **88**, 212
Truc, V., **85**, 64
Tseng, N. -W., **85**, 172
Turlington, M., **87**, 59, 68

Urpí, F., **86**, 70, 81

van Klink, G., P.M. **85**, 209
van Koten, G., **85**, 209
Vandenbossche, C. P., **87**, 275
Vaultier, M., **85**, 219
Venditto, V. J., **86**, 141, 151
Vinogradov, A., **87**, 104
Viózquez, S. F., **88**, 317, 330
Vora, H. U., **87**, 350

Wagner, A. J., **86**, 374
Walker, E.-J., **86**, 121
Wang, Y., **87**, 126
Waser, J., **87**, 88
Watson, I. D. G., **87**, 161
Webster, M. P., **88**, 247
Wein, A. N., **88**, 296
Welin, E. R., **88**, 33
Wender, P. A., **88**, 109
Whiting, A., **85**, 219
Whittlesey, M. K., **86**, 28
Williams, J. M. J., **86**, 28
Williams, R. M., **86**, 262; **88**, 197
Wolf, C., **87**, 1
Wong, J., **87**, 137
Woodward, S., **87**, 104
Wray, B. C., **88**, 33
Wulff, W. D., **88**, 224

Xu, X., **88**, 406, 418

Yamamoto, Y., **87**, 53
Yamamoto, Y., **88**, 79
Yang, J. W., **86**, 11
Yao, T., **88**, 377
Yoo, W. -J., **88**, 14
Yorimitsu, H., **87**, 178
Yu, J., **85**, 64
Yudin, A. K., **87**, 161

Zakarian, A., **85**, 158
Zaragoza, F., **87**, 226
Zhang, A., **85**, 248

446

Zhang, H., **85**, 147
Zhang, Z., **87**, 16
Zhao, B., **86**, 263, 315
Zhao, H., **87**, 275
Zhong, Yong-L. **87**, 8
Zhou, J., **88**, 406, 418
Zimmermann, B., **85**, 196

CUMULATIVE SUBJECT INDEX FOR VOLUMES 85-88

This index comprises subject matter for Volumes **85**, **86**, **87**, and **88**. For subjects in previous volumes, see either the indices in Collective Volumes I through XI or the single volume entitled *Organic Syntheses, Collective Volumes I-VIII, Cumulative Indices,* edited by J. P. Freeman. The index lists the names of compounds in two forms. The first is the name used commonly in procedures. The second is the systematic name according to Chemical Abstracts nomenclature, accompanied by its registry number in parentheses. Also included are general terms for classes of compounds, types of reactions, special apparatus, and unfamiliar methods.

Most chemicals used in the procedure will appear in the index as written in the text. Entries are generally included for all starting materials, reagents, important by-products, and products, which are indicated by the use of italics.

4-Acetamidobenzenesulfonyl azide; (2158-14-7) **85**, 131, 278
Acetal **86**, 81, 130, 262; **88**, 121, 152, 181, 212
Acetal Formation, **85**, 287; **88**, 152
4-Acetamidobenzenesulfonyl azide; (2158-14-7) **88**, 212
Acetic acid (64-19-7) **85**, 138
Acetic acid, copper(1+) salt; (598-54-9) **87**, 53
Acetic acid, 2,2-difluoro-2-(fluorosulfonyl)-, methyl ester; (680-15-9) **87**, 126
Acetic acid, 2-(dimethoxyphosphinyl)-2-[[(phenylmethoxy)carbonyl]- amino]-, methyl ester: Benzyloxycarbonylamino(dimethoxy- phosphoryl)acetic acid methyl ester; (88568-95-0) **88**, 388
Acetic anhydride; (108-24-7) **85**, 34; **87**, 275; **88**, 212
Acetic formic anhydride; (2258-42-6) **85**, 34
2-Acetylbenzoic acid; (577-56-0) **85**, 196
Acetyl chloride; (75-36-5) **86**, 262; **88**, 42
2-Acetyl-4'-methylbiphenyl; (16927-79-0) **85**, 196
3-Acetyl pyridine; (350-03-8) **87**, 36
N-Acetylsulfanilyl chloride; (121-60-8) **85**, 278
Acylation **85**, 34, 147, 158, 219, 295; **86**, 59, 70, 81, 141, 161, 194, 236, 315, 374; **87**, 275; **88**, 212, 247, 317, 364, 427
Adamantine-1-carboxylic acid: Tricyclo[3.3.1.13,7]decane-1-carboxylic acid; (828-51-3) **86**, 113
1-Adamantylamine: Tricyclo[3.3.1.13,7]decan-1-amine; (768-94-5) **85**, 34
1-Adamantyl-3-mesityl-4,5-dimethylimidazolium tetrafluoroborate, **85**, 34
Addition **85**, 231; **87**, 26, 53, 68, 88, 143, 161, 317; **88**, 87, 168, 207, 330
L-Alanine methyl ester hydrochloride; (2491-20-5) **88**, 42
Aldol **86**,11, 81, 92; **88**, 121, 330, 364
Alkene Formation, **85**, 1, 15, 248
Alkenylation **87**, 143, 170, 201, 231; **88**, 138, 291, 342, 388

Alkylation **85**, 10, 34, 45, 88, 158, 295; **86**, 1, 28, 47, 161, 194, 262, 298, 325; **87**, 36, 59, 339, 350, 362; **88**, 87, 121, 168, 181, 247

ALLENES **88**, 138, 309

Allyl alcohol: 2-Propen-1-ol; (107-18-6) **86**, 194; **87**, 226

Allyl bromide; (106-95-6) **85**, 248 ; **86**, 262

(S)-(–)-2-Allylcyclohexanone: Cyclohexanone, 2-(2-propen-1-yl)-, (2S)-; (36302-35-9) **86**, 47

(+)-B-Allyldiisopinocampheylborane ((+)-(Ipc)₂B(allyl) or (ᶦIpc)₂B(allyl)); (106356-53-0) **88**, 87

Allyl methyl carbonate: Carbonic acid, methyl 2-propen-1-yl ester; (35466-83-2) **86**, 47

Allyl 1-methyl-2-oxocyclohexanecarboxylate: Cyclohexanecarboxylic acid, 1-methyl-2-oxo-, 2-propenyl ester; (7770-41-4) **86**, 194

Allylmagnesium bromide; (1730-25-2) **88**, 87

(S)-2-Allyl-2-methylcyclohexanone: Cyclohexanone, 2-methyl-2-(2-propen-1-yl)-, (2S)-; (812639-07-9) **86**, 194

(S)-2-(2-Allyl-2-methylcyclohexylidene)hydrazinecarboxamide: Hydrazinecarboxamide, 2-[(2S)-2-methyl-2-(2-propenyl)cyclohexylidene]-, (2E)-; (812639-25-1) **86**, 194

(3R,7aR)-7a-Allyl-3-(trichloromethyl)tetrahydropyrrolo[1,2-c]oxazol-1(3H)-one (220200-87-3) **86**, 262

Amide formation **86**, 92; **87**, 1, 218, 350; **88**, 14, 42, 398, 427

(S)-2-Amido-pyrrolidine-1-carboxylic acid benzyl ester (34079-31-7) **85**, 72

2'-Aminoacetophenone; (551-93-9) **88**, 33

2-Aminobenzyl alcohol: Benzenemethanol, 2-amino-; (5344-90-1) **87**, 339

(Sₐ)-N-[2'-Amino-(1,1'-binaphthyl)-2-yl]-4-methylbenzene-sulfonamide; (933782-32-2) **88**, 317

Aminodiphenylmethane: Benzenemethanamine, α-phenyl-; (91-00-9) **88**, 224

(1R, 2S)-(+)-cis-1-Amino-2-indanol; (136030-00-7) **87**, 350

(2S)-2-Amino-3-methyl-1,1-diphenyl-butanol; (78603-95-9) **87**, 26

(S)-3-Amino-4-methylpentanoic acid; (s)-β-homovaline; (s)-β-leucine (40469-85-0) **87**, 143

(R)-2-Amino-4-phenylbutan-1-ol: Benzenebutanol, β-amino-, (βR)-; (761373-40-4) **87**, 310

(S)-2-Amino-4-phenylbutan-1-ol: Benzenebutanol, β-amino-, (βS)-; (27038-09-1) **87**, 310

(2S)-(-)-2-Amino-3-phenyl-1-propanol (3182-95-4) **85**, 267

Ammonium bicarbonate; (1066-33-7) **85**, 72

Amination **87**, 1, 143, 263, 362

(E)-1-(2-Aminophenyl)ethanone oxime; (4964-49-2) **88**, 33

Androst-4-ene-3,17-dione; (63-05-8) **87**, 126

Androst-4-ene-3,17-dione, 4-bromo-; (19793-14-7) **87**, 126

Androst-4-ene-3,17-dione, 4-(trifluoromethyl)-; (201664-30-4) **87**, 126

Aniline: benzeneamine; (62-53-3) **88**, 212

Annulation **88**, 330

9-Anthraldehyde; (642-31-9) **88**, 418

9-Anthracenemethanol; (1468-95-7) **88**, 418

Antimony pentachloride; (7647-18-9) **88**, 54

Aqueous HBF₄, Borate(1-), tetrafluoro-, hydrogen (1:1): (16872-11-0) **85**, 34

Arylation **87**, 178, 209

7-Aza-bicyclo[4.1.0]heptane: Cyclohexene imine; (286-18-0) **87**, 161

Azides **87**, 161

Aziridines **87**, 161; **88**, 224

Benzaldehyde; (100-52-7) **85**, 118; **86**, 212; **87**, 1, 231; **88**, 14, 418
Benzaldehyde-*p*-anisidine imine; (783-08-4) **88**, 418
Benzenamine, 2-(2,2-dibromoethenyl)-; (167558-54-5) **86**, 36
Benzene, 1-(2,2-dibromoethenyl)-2-nitro-; (253684-24-1) **86**, 36
(β R)-Benzenepropanoic acid, α-acetyl-β-[[[(phenylmethyl)amino]carbonyl]amino]-,
methyl ester; (865086-76-6) **86**, 236
(β R)-Benzenepropanoic acid, α-acetyl-β-[[(2-propen-1-yloxy)carbonyl]amino]-, methyl
ester; (921766-57-6) **86**, 236
Benzenesulfinic acid sodium salt; (873-55-2) **86**, 360
5-Benzo[1,3]dioxol-5-yl-3-(4-chloro-phenyl)-1-methyl-1H-pyrazole **85**, 179
3-Benzofurancarboxylic acid, 5-chloro-, ethyl ester; (899795-65-4) **86**, 172
Benzophenone; (119-61-9) **85**, 248 ; **87**, 245
Benzophenone hydrazone (5350-57-2) **85**, 189
Benzoyl chloride; (98-88-4) **88**, 398
N-Benzoyl pyrrolidine: Methanone, phenyl-1-pyrrolidinyl-; (3389-54-6) **87**, 1
Benzyl alcohol; (100-51-6) **86**, 28
Benzylamine; (100-46-9) **85**, 106
(1R,2R)-trans-2-(N-Benzyl)amino-1-cyclohexanol; (141553-09-5) **85**, 106
(1S,2S)-trans-2-(N-Benzyl)amino-1-cyclohexanol; (322407-34-1) **85**, 106
Benzyl azide: (Azidomethyl)benzene; (622-79-7) **88**, 238
Benzyl bromide: Benzene, (bromomethyl)-; (100-39-0) **87**, 36, 137, 170
N-Benzyl-2-bromo-*N*-phenylbutanamide: Butanamide, 2-bromo-*N*-phenyl-*N*-
(phenylmethyl)-; (851073-30-8) **87**, 330
Benzyl carbamate: Carbamic acid, phenylmethyl ester; (621-84-1) **85**, 287; **88**, 152
Benzyl chloroformate, carbobenzoxy chloride; (501-53-1) **88**, 42
Benzyl chloromethyl ether: Benzene, [(chloromethoxy)methyl]-; (3587-60-8) **85**, 45
1-Benzyl-3-(4-chloro-phenyl)-5-p-tolyl-1H-pyrazole (908329-95-3) **85**, 179
(S)-N-Benzyl-7-cyano-2-ethyl-n-phenylheptanamide: Heptanamide, 7-cyano-2-ethyl-N-
phenyl-N-(phenylmethyl)-, (2S)-; (851073-44-4) **87**, 330
Benzylhydrazine dihydrochloride; (20570-96-1) **85**, 179
Benzyl hydroxymethyl carbamate: Carbamic acid, (hydroxymethyl)-, phenylmethyl ester;
(31037-42-0) **85**, 287
(E)-N-Benzylidene-4-methylbenzenesulfonamide: Benzenesulfonamide, 4-methyl-*N*-
(phenylmethylene)-, [*N(E)*]-: (51608-60-7) **86**, 212; **88**, 138
Benzyl isopropoxymethyl carbamate **85**, 287
1 (2R,4S) [2-Benzyl-3-(4-isopropyl-2-oxo-5,5-diphenyl-3-oxazolidinyl)-3-
oxopropyl]carbamic acid benzyl ester (218800-56-7) **85**, 295
Benzylmagnesium chloride: Magnesium, chloro(phenylmethyl)-; (6921-34-2) **88**, 54
(Z)-2-Benzyloxycarbonylamino-3-(2-bromo-4,5-dimethoxyphenyl) acrylic acid methyl
ester: 2-Propenoic acid, 3-(2-bromo-4,5-dimethoxyphenyl)-2-
[[(phenylmethoxy)carbonyl]amino]-, methyl ester, (2Z)-; (873666-89-8) **88**, 388
(R)-2-(Benzyloxycarbonylaminomethyl)-3-phenylpropanoic acid: Benzenepropanoic
acid, α-[[[(phenylmethoxy)carbonyl]amino]methyl]-, (aR)-; (132696-47-0) **85**,
295
Benzyloxymethoxy-1-hexyne: Benzene, [[(1-hexyn-1-yloxy)methoxy]methyl]-; (162552-
11-6) **85**, 45
Benzyloxymethoxy-2,2,2-trifluoromethyl ether: Benzene, [[(2,2,2-
450

trifluoroethoxy)methoxy]methyl]-: (153959-88-7) **85**, 45

(*R*)-*N*-Benzyl-*N*-(α-methylbenzyl)amine; (*R*)-*N*-benzyl-α-phenylethylamine; (*R*)-*N*-benzyl-1-phenylethylamine (38235-77-7) **87**, 143

Benzyl propagyl ether; (4039-82-1) **86**, 225

(S)-1-(4-Benzyl-2-thioxothiazolidin-3-yl)propan-1-one: 1-Propanone, 1-[(4S)-4-(phenylmethyl)-2-thioxo-3-thiazolidinyl]-; (263764-23-4) **88**, 364

(*R*)-(+)-1,1'-Bi(2-naphthol); (18531-94-7) **85**, 238

(*R*)-BINOL; (18531-94-7) **85**, 238; **88**, 418

(*S*)-BINOL: [1,1'-Binaphthalene]-2,2'-diol, (1*S*)-: (18531-99-2) **85**, 118

(R)-(1,1'-Binaphthalene-2,2'-dioxy)chlorophosphine: (R)-Binol-P-Cl; (155613-52-8) **85**, 238

[1,1'-Binaphthalene]-2,2'-diol, (1*S*)-: (18531-99-2) **85**, 118

(R)-2,2-Binaphthoyl-(S,S)-di(1-phenylethyl)aminoylphosphine (415918-91-1) **85**, 238

(*S*ₐ)-(-)-1,1′-Binaphthyl-2,2′-diamine: (*S*ₐ)-(-)-1,1′-Bi(2-naphthylamine); (18531-95-8) **88**, 317

(2-Biphenyl)dicyclohexylphosphine; (247940-06-3) **86**, 344

Bis(4-tert-butylphenyl)iodonium triflate; (84563-54-2) **86**, 308

Bis[1,2:5,6-η-(1,5-cyclooctadiene)]nickel: [bis(1,5-cyclooctadiene)nickel (0)]; (1295-35-8) **85**, 248

[*N*,*N*′-Bis(3,5-di-*tert*-butylsalicylidene)-1,2-phenylenediamino-chromium-di-tetrahydrofuran]tetracarbonylcobaltate (1); (909553-60-2) **86**, 287

4,5-Bis(Diphenylphosphino)-9,9-dimethylxanthene (Xantphos); (161265-03-8) **86**, 28

Bis(diphenylphosphino)methane; (2071-20-7) **85**, 196

Bis(4-Fluorophenyl)Difluoromethane (339-27-5) **87**, 245

Bis(4-fluorophenyl) ketone: 4,4'-Difluorobenzophenone; (345-92-6) **87**, 245

Bis-(Hydroxymethyl)-cyclopropane; (2345-75-7) **85**, 15

(*R*)-(+)-2,2'-Bis(methoxymethoxy)-1,1'-binaphthyl; (1738310-50-0) **88**, 406

(R)-2,2'-Bis(methoxymethoxy)-3,3'-bis(9-phenanthryl)-1,1'-binaphthalene; (1261302-60-6) **88**, 406

(R)-2,2'-Bis(methoxymethoxy)-3,3'-diiodo-1,1'- binaphthalene; (189518-78-3) **88**, 406

(S)-3,3'-Bis-morpholinomethyl-5,5',6,6',7,7',8,8'-octahydro-1,1'-bi-2-naphthol: [1,1'-Binaphthalene]-2,2'-diol, 5,5',6,6',7,7',8,8'-octahydro-3,3'-bis(4-morpholinylmethyl)-, (1S)-; (758698-16-7) **87**, 59, 68

(R)-3,3'-Bis(9-phenanthryl)-1,1'-binaphthalene-2,2'-diol; (1058734-56-7) **88**, 406

(R)-3,3'-Bis(9-phenanthryl)-1,1'-binaphthalene-2,2'-diyl hydrogen phosphate; (864943-22-6) **88**, 406Bis(triphenylphosphine)palladium(II) chloride; (13965-03-2) **88**, 197

(-)-Bis[(*S*)-1-phenylethyl]amine (56210-72-1) **85**, 238

(-)-Bis[(*S*)-1-phenylethyl]amine hydrochloride (40648-92-8) **85**, 238

Bis(Pyridine)Iodonium(I) tetrafluoroborate: Iodine(1+), bis(pyridine)-, tetrafluoroborate(1-) (1:1); (15656-28-7) **87**, 288

Bis[rhodium(α,α,α′,α′-tetramethyl-1,3-benzenedipropionic acid)]; (819050-89-0) **87**, 115

3,5-Bis(trifluoromethyl)bromobenzene; (328-70-1) **85**, 248

Boc₂O: Dicarbonic acid, *C,C′*-bis(1,1-dimethylethyl) ester; (244424-99-5) **86**, 113, 374

9-Borabicyclo[3.3.1]nonane: 9-BBN; (280-64-8) **87**, 299; **88**, 207

(5-(9-Borabicyclo[3.3.1]nonan-9-yl)pentyloxy)triethylsilane: 9-Borabicyclo[3.3.1]-nonane, 9-[5-[(triethylsilyl)oxy]pentyl]-; (157123-09-6) **87**, 299

Borane-dimethylsulfide complex: Boron, trihydro[thiobis[methane]]-(T-4)-; (13292-87-0) **88**, 102

Borane tetrahydrofuran complex solution, 1.0 M in tetrahydrofuran; (14044-65-6) **87**, 36

Boron **87,** 26, 299; **88,** 79, 87, 181, 202, 207, 247
Boron trifluoride etherate: BF₃·OEt₂ (109-63-7) **86,** 81, 212
Boronic acid, phenyl-; (98-80-6) **87,** 53
Bromination, **85,** 53
Bromine; (7726-95-6) **85,** 231; **87,** 126
2-[3-(3-Bromophenyl)-2H-azirin-2-yl]-5-(trifluoromethyl)pyridine **86,** 18
3'-Bromoacetophenone: 1-acetyl-3-bromobenzene; (2142-63-4) **86,** 18
2-Bromoaniline: Benzenamine, 2-bromo-; (615-36-1) **88,** 398
3-Bromoanisole: Benzene, 1-bromo-3-methoxy-; (2398-37-0) **88,** 1
4-Bromoanisole; (104-92-7) **88,** 197
4-Bromobenzaldehyde; (1122-91-4) **88,** 224
Bromobenzene; (108-86-1) **87,** 26, 178
2-Bromobenzoyl chloride: Benzoyl chloride, 2-bromo-; (7154-66-7) **86,** 181
N-(4-Bromobenzylidene)-1,1-diphenylmethanamine: Benzenemethanamine, N-[(4-
    bromophenyl)methylene]-α-phenyl-, [N(E)]-; (330455-47-5) **88,** 224
1-Bromo-8-chlorooctane: Octane, 1-bromo-8-chloro-; (28598-82-5) **87,** 299
Bromodiethylsulfonium Bromopentachloroantimonate: Sulfonium, bromodiethyl-, (OC-6-
    22)-bromopentachloroantimonate(1-) (1:1); (1198402-81-1) **88,** 54
6-Bromohexanenitrile: Hexanenitrile, 6-bromo-; (6621-59-6) **87,** 330
8-Bromo-1-octanol: 1-Octanol, 8-bromo-; (50816-19-8) **87,** 299
1-Bromo-2-naphthoic acid: 2-Naphthoic acid, 1-bromo-; (20717-79-7) **88,** 70
N-2-Bromophenylbenzamide; (70787-27-8) **88,** 398
4-Bromophenylboronic acid; (5467-74-3) **86,** 344
4-Bromophenylboronic MIDA ester **86,** 344
p-Bromophenyl methyl sulfide: Benzene, 1-bromo-4-(methylthio)-; (104-95-0)
    86, 121
(S)-(−)-p-Bromophenyl methyl sulfoxide **86,** 121
(*S*)-(−)-*p*-Bromophenyl methyl sulfoxide: Benzene, 1-bromo-4-[(*S*)-methylsulfinyl]-;
    (145266-25-0) **86,** 121
1-(3-Bromophenyl)-2-[5-(trifluoromethyl)-2-pyridinyl]ethanone **86,** 18
2-(3-Bromophenyl)-6-(trifluoromethyl)pyrazolo[1,5-a]pyridine **86,** 18
(1Z)-1-(3-Bromophenyl)-2-[5-(trifluoromethyl)-2-pyridinyl]-ethanone oxime **86,** 18
2-Bromopropene; (557-93-7) **85,** 1, 172
2-Bromopyridine: Pyridine, 2-bromo-; (109-04-6) **88,** 79
N-Bromosuccimide: NBS; (128-08-5) **85,** 53, 267; **86,** 225; **87,** 16
4-Bromo-5-(thiophen-2-yl)oxazole: Oxazole, 4-bromo-5-(2-thienyl)-; (959977-82-3) **87,**
    16
(Z)-β-Bromostyrene; (103-64-0) **88,** 202
4-Bromotoluene; (106-38-7) **85,** 196 ; **88,** 22
1-Bromo-1-phenylthioethene; (80485-53-6) **88,** 207
6-Bromoveratraldehyde: Benzaldehyde, 2-bromo-4,5-dimethoxy-; (5392-10-9) **88,** 388
1-(Buta-1,2-dien-1-ylsulfonyl)-4-methylbenzene: Benzene, 1-(1,2-butadien-1-ylsulfonyl)-
    4-methyl-; (32140-55-9) **88,** 309
2-Butanone; (78-93-3) **86,** 333
trans-Butene; (624-64-6) **88,** 181
2-tert-Butoxycarbonylamino-4-(2,2-dimethyl-4,6-dioxo-[1,3]dioxan-5-yl)-4-oxo-butyric
    acid tert-butyl ester; (10950-77-9) **85,** 147
(2S)-2-[(tert-Butoxycarbonyl)amino]-2-phenylethyl methanesulfonate (110143-62-9) **85,**
    219
2-[3,3'-Di-(*tert*-butoxycarbonyl)-aminodipropylamine]-4,6-dichloro-1,3,5-triazine; 12-

Oxa-2,6,10-triazatetradecanoic acid, 6-(4,6-dichloro-1,3,5-triazin-2-yl)-13,13-dimethyl-11-oxo-, 1,1-dimethylethyl ester; (947602-03-1) **86**, 141, 151

N-α-*tert*-Butoxycarbonyl-L-aspartic acid α-*tert*-butyl ester (Boc-L-Asp-O*t*-Bu); (34582-32-6) **85**, 147

1-tert-Butoxycarbonyl-2,3-dihydropyrrole: 1H-Pyrrole-1-carboxylic acid, 2,3-dihydro-, 1,1-dimethylethyl ester; (73286-71-2) **85**, 64

2-(*tert*-Butoxycarbonyloxyimino)-2-phenylacetonitrile; (58632-95-4) **86**, 141

N-(*tert*-Butoxycarbonyl)-piperazine; (57260-71-6) **86**, 141

N-(*tert*-Butoxycarbonyl)-L-proline; (15761-39-4) **88**, 317

N-(*tert*-Butyloxycarbonyl)pyrrolidin-2-one; (85909-08-6) **85**, 64

tert-Butyl acetoacetate; (1694-31-1) **85**, 278

N-*tert*-Butyl adamantan-1-yl-carbamate: Carbamic acid, tricyclo[3.3.1.13,7]dec-1-yl-, 1,1-dimethylethyl ester; (151476-40-3) **86**, 113

tert-Butanol: 2-Methyl-2-propanol; (75-65-0) **87**, 8

tert-Butylamine: 2-Propanamine, 2-methyl-; (75-64-9) **86**, 315

t-Butyl 3-amino-4-methylpentanoate (202072-47-7) **87**, 143

tert-Butylbenzene; (98-06-6) **86**, 308

(E)-tert-Butyl benzylidenecarbamate **86**, 11

(E)-*tert*-Butyl benzylidenecarbamate: Carbamic acid, *N*-(phenylmethylene)-,1,1-dimethyl-ethyl ester, [*N*(*E*)]-; (177898-09-2) **86**, 11

t-Butyl 3-[*N*-benzyl-*N*-(α-methylbenzyl)amine]-4-methylpentanoate (38235-77-7) **87**, 143

(-)-2-tert-Butyl-(4S)-benzyl-(1,3)-oxazoline: 4,5-Dihydrooxazole, (4S)-benzyl, 2-tert-butyl; (75866-75-0) **85**, 267

tert-Butyl bromoacetate; (5292-43-3) **85**, 10

tert-Butyl *tert*-butyldimethylsilylglyoxylate **85**, 278

(S)-tert-Butyl (4-chlorophenyl)(thiophen-2-yl)methylcarbamate **86**, 360

tert-Butyl (1R)-2-cyano-1-phenylethylcarbamate (126568-44-3) **85**, 219

Butyl di-1-adamantylphosphine; (321921-71-5) **86**, 105

tert-Butyl diazoacetate; (35059-50-8) **85**, 278

tert-Butyldimethylsilyl chloride: Silane, chloro(1,1-dimethylethyl)dimethyl-; (18162-48-6) **86**, 130

tert-Butyldimethylsilyl trifluoromethanesulfonate; (69739-34-0) **85**, 278

t-Butyl diethylphosphonoacetate (27784-76-5) **87**, 143

(2R,5S)-2-tert-Butyl-3,5-dimethylimidazolidin-4-one: 4-Imidazolidinone, 2-(1,1-dimethylethyl)-3,5-dimethyl-, hydrochloride (1:1), (2R,5S)-: (1092799-01-3) **88**, 42

(2R,3R)-1-(t-Butyldiphenylsilyloxy)-2-methylhex-5-en-3-ol; (112897-06-0) **88**, 87

(R)-3-(*tert*-Butyldiphenylsilyloxy)-2-methylpropanal: Propanal, 3-[[(1,1-dimethylethyl) diphenyl-silyl]oxy]-2-methyl-, (2*R*)-; (112897-04-8) **88**, 87

tert-Butyl ethyl phthalate: 1,2-Benzenedicarboxylic acid, 1,2-bis[2-[3,5-bis(1,1-dimethylethyl)-4-hydroxyphenyl]ethyl] ester; (259254-67-6) **86**, 374

3-Butyl-2-fluoro-1-tosylindole: 1H-Indole, 3-butyl-2-fluoro-1-[(4-methylphenyl)sulfonyl]-; (195734-36-2) **88**, 162

t-Butyl hydroperoxide: Hydroperoxide, 1,1-dimethylethyl; (75-91-2) **87**, 1, 88; **88**, 14

tert-Butyl (1S)-2-hydroxy-1-phenylethylcarbamate (117049-14-6) **85**, 219

tert-Butyl-hypochlorite: Hypochlorous acid, 1,1-dimethylethyl ester; (507-40-4) **86**, 315; **88**, 168

n-Butyllithium; (109-72-8) **85**, 1, 45, 53, 158, 238, 248, 295; **86**, 47, 70, 262; **87**, 137, 143; **88**, 1, 79, 102, 212, 296, 353, 406 *sec-Butyllithium; (598-30-1)* **88**, 247

*tert-*Butyllithium: Lithium, (1,1-dimethylethyl)-; (5944-19-4) **85**, 1, 209; **88**, 296

*tert-*Butyl *(E)*-4-methylbut-2-enoate (87776-18-9) **87**, 143

(Sₐ,S)-t-Butyl 2-[(2'-(4-methylphenylsulfonamido)-(1,1'-binaphthyl)-2-yl-carbamoyl]pyrrolidine-1-carboxylate; (933782-35-5) **88**, 317

3-Butyl-4-methyl-2-triisopropylsiloxy-cyclobut-2-enecarboxylic acid methyl ester (731853-28-4) **87**, 253

*tert-*Butyl *phenyl(phenylsulfonyl)methylcarbamate* **86**, 11

*tert-*Butyl phenyl(phenylsulfonyl)methylcarbamate: Carbamic acid, *N*-[phenyl(phenyl-sulfonyl)methyl]-1,1-dimethylether ester; (155396-71-7) **86**, 11

*tert-*Butyl *phenylsulfonyl(thiophen-2-yl)methylcarbamate: Carbamic acid, N-[(phenylsulfonyl)-2-thienylmethyl]-, 1,1-dimethylethyl ester;* (479423-34-2) **86**, 360

(S)-tert-ButylPHOX: Oxazole, 4-(1,1-dimethylethyl)-2-[2-(diphenylphosphino)phenyl]-4,5-dihydro-, (4S)-; (148461-16-9) **86**, 181

2-Butyn-1-ol; (764-01-2) **88**, 296

3-Butyn-2-ol; (2028-63-9) **88**, 309

But-3-yn-2-yl 4-Methylbenzenesulfinate: Benzenesulfinic acid, 4-methyl-, 1-methyl-2-propyn-1-yl ester; (32140-54-8) **88**, 309

Carbamates **86**, 11, 59, 81, 113, 141, 151, 236, 333; **88**, 152, 212, 247, 317

Carbamic acid, N-(2-diazoacetyl)-N-phenyl-, 1-methylethyl ester; (1198356-59-0) **88**, 212

Carbamic acid, (hydroxymethyl)-, phenylmethyl ester; (31037-42-0) **85**, 287

Carbamic acid, *N*-[(1*S*,2*S*)-2-methyl-3-oxo-1-phenylpropyl]-1,1-dimethylethyl ester; (926308-17-0) **86**, 11

Carbamic acid, N-[phenyl(phenylsulfonyl)methyl]-, 2-propen-1-yl ester; (921767-12-6) **86**, 236

Carbamic acid, 2-propen-1-yl ester; (2114-11-6) **86**, 236

Carbobenzyloxy-L-proline: 1,2-Pyrrolidinedicarboxylic acid, 1-(phenylmethyl) ester, (2*S*)-; (1148-11-4) **85**, 72

(Carboethoxymethylene)triphenylphosphorane; (1099-45-2) **85**, 15

(2-Carbomethoxy-6-nitrobenzyl)triphenylphosphonium bromide: Phosphonium, [[2-(methoxycarbonyl)-6-nitrophenyl]methyl]triphenyl-, bromide; (195992-09-7) **88**, 291

Carbon disulfide; (75-15-0) **86**, 70; **88**, 364

Carbon monoxide; (630-08-0) **86**, 287; **88**, 291

Carbon tetrabromide: Methane, tetrabromo-; (558-13-4) **87**, 231

Carbonyl(dihydrido)tris(triphenylphosphine)ruthenium (II); (25360-32-1) **86**, 28

1,1'-Carbonyldiimidazole: Methanone, di-1*H*-imidazol-1-yl-; (530-62-1) **86**, 58

Catecholborane: 1,3,2-benzodioxaborole; (274-07-7) **88**, 202

Cesium acetate: Acetic acid, cesium salt (1:1); (3396-11-0) **88**, 388

Cesium carbonate: Carbonic acid, cesium salt (1:2); (534-17-8) **86**, 181; **87**, 231; **88**, 398

Cesium fluoride; (13400-13-0) **86**, 161

Chloroacetonitrile (107-14-2) **86**, 1

4-Chlorobenzaldehyde; (104-88-1) **85**, 179

Chlorobenzene; (108-90-7) **86**, 105; **87**, 362

(3-Chlorobutyl)benzene; (4830-94-8) **87**, 88

Chloro(chloromethyl)dimethylsilane; (1719-57-9) **87**, 178

Chlorodicarbonylrhodium dimer: Di-μ-chloro-bis(dicarbonylrhodium): Tetracarbonyldi-μ-chlorodirhodium; (14523-22-9) **88**, 109

1-Chloro-1,3-dihydro-3,3-dimethyl-1,2-benziodoxole; (69352-04-1) **88**, 168

2-Chloro-6,7-dimethoxy-1,2,3,4-tetrahydroisoquinoline **87**, 8

(Chloromethyl)dimethylphenylsilane; (1833-51-8) **87**, 178

2-Chloro-5-(3-methylphenyl)-thiophene; (1078144-58-7) **87**, 178, 184

4-Chlorophenylboronic acid: Boronic acid, B-(4-chlorophenyl)-; (1679-18-1) **86**, 360

2-Chloropyridine; (109-09-1) **85**, 88; **88**, 121

5-Chlorosalicylaldehyde: Benzaldehyde, 5-chloro-2-hydroxy-; (635-93-8) **86**, 172

2-Chlorothiophene; (96-43-5) **87**, 178, 184

(13-Chlorotridecyloxy)triethylsilane: Silane, [(13-chlorotridecyl)oxy]triethyl-; (374754-99-1) **87**, 299

Chlorotriethylsilane: Silane, chlorotriethyl-; (994-30-9) **87**, 299

2-Chloro-5-(trifluoromethyl)pyridine; (52334-81-3) **86**, 18

m-Chloroperbenzoic acid; Peroxybenzoic acid, *m*-chloro- (8); Benzocarboperoxoic acid, 3-chloro- (9); (937-14-4) **86**, 308

Chlorotrimethylsilane: Silane, Chlorotrimethyl-; (75-77-4) **86**, 252; **88**, 296

Cholesta-3,5-diene; (747-90-0) **88**, 260

Cholesta-3,5-dien-3-yl trifluoromethanesulfonate; (95667-40-6) **88**, 260

Cholest-4-en-3-one; (601-57-0) **88**, 260

Cinnamyl alcohol: 3-Phenyl-2-propen-1-ol; (104-54-1) **85**, 96

Cinnamyl-H-phosphinic acid: [(2E)-3-phenyl-2-propenyl]-Phosphinic acid; (911128-46-6) **85**, 96

(±)-Citronellal: ((±)-3,7-dimethyl-6-octenal); (106-23-0) **87**, 201

Cobalt(II) tetrafluoroborate hexahydrate; (15684-35-2) **87**, 88

Condensation **85**, 27, 34, 179, 248, 267; **86**, 11, 18, 92, 121, 212, 252, 262; **87**, 36, 59, 77, 88, 115, 143, 192, 201, 218, 275, 310, 339, 362; **88**, 33, 42, 138, 212, 224. 247, 338

Copper(II) acetate, monohydrate: Cupric acetate 1-hydrate: Acetic acid, copper(II) salt; (6046-93-1) **88**, 238

Copper(I) bromide; (7787-70-4) **85**, 196

Copper-catalyzed reactions **87**, 53, 126, 184, 231; **88**, 14, 79, 388, 398

Copper chloride: Cuprous chloride; (7758-89-6) **85**, 209 ;

Copper cyanide; (544-92-3) **85**, 131

Copper iodide; (1335-23-5) **86**, 181; **88**, 79

Copper(I) iodide: Cuprous iodide; (7681-65-4) **86**, 225; **87**, 126, 178, 184, 231; **88**, 388

Copper(II) oxide nanoparticles: Copper oxide (CuO); (1317-38-0) **88**, 398

Coupling **85**, 158, 196; **86**, 105, 225, 274; **87**, 184, 299, 317, 330; **88**, 4, 14, 22, 70,79, 102, 162, 197, 202, 207, 377, 388, 398, 406

Cuprous chloride; (7758-89-6) **85**, 209

(S)-2-Cyano-pyrrolidine-1-carboxylic acid benzyl ester: (63808-36-6) **85**, 72

Cyanuric chloride: 2,4,6-Trichloro-1,3,5-triazine; (108-77-0) **85**, 72; **86**, 141

Cyclen: 1,4,7,10-Tetraazacyclododecane; (294-90-6) **85**, 10

Cyclization **86**, 18, 92, 172, 181, 194, 212, 236, 252, 262, 333; **87**, 16, 77, 161, 310, 339, 350, 362; **88**, 33, 42, 54, 162, 291, 377, 388, 398

Cycloaddition, **85**, 72, 131, 138, 179; **87**, 95, 253, 263; **88**, 109, 121, 138, 224, 238

Cycloheptane-1,3-dione (1194-18-9) **85**, 138

Cyclohexanecarboxaldehyde; (2043-61-0) **87**, 68

1,4-Cyclohexanedione mono-ethylene ketal: 1,4-Dioxaspiro[4.5]decan-8-one; (4746-97-8) **88**, 121

Cyclohexanemethanol; (100-49-2) **86**, 58
Cyclohexanone; (108-94-1) **86**, 47
Cyclohexene imine: 7-Aza-bicyclo[4.1.0]heptane; (286-18-0) **87**, 161
Cyclohexene oxide; (286-20-4) **85**, 106; **87**, 161
9-[2-(3-Cyclohexenyl)ethyl]-9-BBN: 9-Borabicyclo[3.3.1]nonane, 9-[2-(3-cyclohexen-1-yl)ethyl]-; (69503-86-2) **88**, 207
4-(3-Cyclohexenyl)-2-phenylthio-1-butene; (155818-88-5) **88**, 207
(R)-(+)-α-Cyclohexyl-3-methoxy-benzenemethanol; (1036645-45-0) **87**, 68
Cyclohexylmethyl N-hydroxycarbamate **86**, 58
Cyclohexylmethyl *N*-hydroxycarbamate: Carbamic acid, hydroxy-, cyclohexylmethyl ester; (869111-30-8) **86**, 58
Cyclohexylmethyl N-tosyloxycarbamate **86**, 58
Cyclohexylmethyl *N*-tosyloxycarbamate: Benzenesulfonic acid, 4-methyl-, [(cyclohexylmethoxy)carbonyl]azanyl ester; (869111-41-1) **86**, 58
Cyclopropanation **85**, 172; **87**, 115
Cyclopropanecarboxylic acid, 2-bromo-2-methyl-, ethyl ester; (89892-99-9) **85**, 172
Cyclopropanecarboxylic acid, 2-methylene-, ethyl ester; (18941-94-1) **85**, 172

DABCO: 1,4-Diazabicyclo[2.2.2]octane; (280-57-9) **87**, 104
(E)-Deca-1,3-diene: 1,3-Decadiene, (3E)-; (58396-45-5) **88**, 342
(Z)-Dec-2-en-1-ol: 2-Decen-1-ol, (2Z)-; (4194-71-2) **88**, 342
(*R,R*)-deguPHOS: Pyrrolidine, 3,4-bis(diphenylphosphino)-1-(phenylmethyl)-, (3*R*,4*R*)-; (99135-95-2) **86**, 360
Dehydration, **85**, 34, 72
Dendrimer **86**, 151
Deoxo-Fluor: Bis(2-methoxyethyl)aminosulfur trifluoride: Ethanamine, 2-ethoxy-N-(2-ethoxyethyl)-*N*-(trifluorothio)-; (202289-38-1) **87**, 245
Deprotection **87**, 143; **88**, 317, 406
Diallyl pimelate: Pimelic acid, diallyl ester; (91906-66-0) **86**, 194
Diamination **87**, 275
3,3'-Diaminodipropylamine; (56-18-8) **86**, 141
1,8-Diazabicyclo[5.4.0]undec-7-ene (DBU); (6674-22-2) **87**, 170
Diazo compounds **88**, 212, 418
Di-μ-bromobis(tri-*tert*-butylphosphine)dipalladium; (185812-86-6) **88**, 1
(E)-2,3-Dibromobut-2-enoic acid: (2-Butenoic acid, 2,3-dibromo-, (2E)- (9); (24557-17-3) **85**, 231
2,5-Dibromo-1,1-dimethyl-3,4-diphenylsilole: Silacyclopenta-2,4-diene, 2,5-dibromo-1,1-dimethyl-3,4-diphenyl-; (686290-22-2) **85**, 53
(2,2-Dibromoethenyl)benzene; (7436-90-0) **87**, 231
1-(2,2-Dibromoethenyl)-2-nitrobenzene **86**, 36
Di(μ-bromo)bis(η-allyl)nickel(II): [allylnickel bromide dimer]; (12012-90-7) **85**, 248
2-(2,2-Dibromo-vinyl)-phenylamine **86**, 36
Di-tert-butyldiaziridinone: 3-Diaziridinone, 1,2-bis(1,1-dimethylethyl)-; (19656-74-7) **86**, 315; **87**, 263
Di-*tert*-butyl dicarbonate: Dicarbonic acid, C,C'-bis(1,1-dimethylethyl) ester; (24424-99-5) **85**, 72, 219; **86**, 113, 374
Di(tert-butyl) (2S)-4,6-dioxo-1,2-piperidinedicarboxylate; (653589-10-7) **85**, 147
3,5-Di-*t*-butyl-2-hydroxybenzaldehyde: Benzaldehyde, 3,5-bis(1,1-dimethylethyl)-2-hydroxy-; (37942-07-7) **87**, 88

456

2-(3,5-Di-*tert*-butyl-2-hydroxybenzylideneamino)-2,2-diphenylacetic acid potassium salt, (SALDIPAC); (858344-69-1) **87**, 88

Di(tert-butyl) (2S,4S)-4-hydroxy-6-oxo-1,2-piperidinedicarboxylate; (653589-16-3) **85**, 147

trans-1,3-Di-*tert*-butyl-4-phenyl-5-vinyl-imidazolidin-2-one: 2-Imidazolidinone, 1,3-bis(1,1-dimethylethyl)-4-ethenyl-5-phenyl-, (4*R*,5*R*)-rel-; (927902-91-8) **86**, 315; **87**, 263

2-(Di-*tert*-butylphosphino)biphenyl: Phosphine, [1,1'-biphenyl]-2-ylbis(1,1-dimethylethyl)-; (224311-51-7) **86**, 274

Di-tert-butylurea: Urea, N,N'-bis(1,1-dimethylethyl)-; (5336-24-3) **86**, 315

Dichloroacetyl chloride (79-36-7) **85**, 138

1,3-Dichloro-1,3-bis(dimethylamino)propenium chloride; (34057-61-9) **86**, 298

Dichloro[1,3-bis(diphenylphosphino)propane]palladium: Palladium, dichloro[1,1'-(1,3-propanediyl)bis[1,1-diphenylphosphine-κP]]-, (SP-4-2)-; (59831-02-6) **88**, 79

Dichlorobis(triphenylphosphine)palladium(II): Bis(triphenylphosphine)palladium(II) Dichloride; (13965-03-2) **86**, 225; **88**, 202

(*S*)-(−)-5,5'-Dichloro-6,6'-dimethoxy-2,2'-bis(diphenylphosphino)-1,1'-biphenyl: Phosphine, [(1*S*)-5,5'-dichloro-6,6'-dimethoxy[1,1'-biphenyl]-2,2'-diyl]bis[diphenyl-; (185913-98-8) **86**, 47

Dichlorodimethylsilane; (75-78-5) **85**, 53

Dichloromethylen-dimethyliminium chloride; (33842-02-3) **86**, 298

Dicyclohexylmethylamine: Cyclohexanamine, *N*-cyclohexyl-*N*-methyl-; (7560-83-0) **85**, 118

Dichloro(*N,N,N',N'*-tetramethylethylenediamine)zinc; (28308-00-1) **87**, 178

Dichlorotriphenylphosphorane: Phosphorane, dichlorotriphenyl-; (2526-64-9) **87**, 299, 317

Dicyclohexylamine: Cyclohexanamine, *N*-cyclohexyl-; (101-83-7) **88**, 1

Dicyclohexylcarbodiimide: Carbodiimide, dicyclohexyl-; Cyclohexanamine, *N,N'*-methanetetraylbis-; (538-75-0) **88**, 70

2-Dicyclohexylphosphino-2',4',6'-triisopropylbiphenyl: X-Phos:; (564483-18-7) **87**, 104

Diene formation, **85**, 1; **88**, 260, 342

Diethylamine: Ethanamine, *N*-ethyl-; (109-89-7) **88**, 427

2-[(Diethylamino)methyl]benzene thiolato-copper(I) **85**, 209

Diethyl chlorophosphate: Phosphorochloridic acid, diethyl ester; (814-49-3) **88**, 54

(±)-Diethyl (*E,E,E*)-cyclopropane-1,2-acrylate, **85**, 15

Diethyl trans-1,2-cyclopropanedicarboxylate; (3999-55-1) **85**, 15

N,N-Diethyl-3,4,5-trimethoxybenzamide; (5470-42-8) **88**, 427

Diethyl(2-[(trimethylsilanyl)sulfanyl]benzyl)amine **85**, 209

Diethylzinc; (557-20-0) **87**, 68

o-(1,1-Difluorohex-1-en-2-yl)aniline: Benzenamine, 2-[1-(difluoromethylene)pentyl]-; (134810-59-6) **88**, 162

o'-(1,1-Difluorohex-1-en-2-yl)-p-toluenesulfonanilide: Benzenesulfonamide, N-[2-[1-(difluoromethylene)pentyl]phenyl]-4-methyl-; (195734-33-9) **88**, 162

3,4-Dihydronaphthalene-1(2*H*)-one oxime; (3349-64-2) **87**, 275

N-(3,4-Dihydronaphthalene-1-yl)acetamide: Acetamide, *N*-(3,4-dihydro-1-naphthalenyl)-; (213272-97-0) **87**, 275

(5a*R*,10b*S*)-5a,10b-Dihydro-2-(2,4,6-trimethylphenyl)-4H,6H-indeno[2,1-*b*]-1,2,4-triazolo[4,3-*d*]-1,4-oxazinium chloride; (903571-02-8) **87**, 362

3,5-Diiodosalicylaldehyde: Benzaldehyde, 2-hydroxy-3,5-diiodo-; (2631-77-8) **86**, 121

(S)-(−)-2-(N-3,5-Diiodosalicyliden)amino-3,3-dimethyl-1-butanol [(S)-1] **86**, 121

(S)-(–)-2-(N-3,5-Diiodosalicyliden)amino-3,3-dimethyl-1-butanol; (477339-39-2) **86**, 121
Diisopropylamine: 2-Propanamine, N-(1-methylethyl)-; (108-18-4) **86**, 47, 262; **87**, 137
N,N-Diisopropyl-carbamic acid 3-phenyl-propyl ester; (218601-55-9) **88**, 247
N,N-Diisopropylcarbamoyl chloride; (19009-39-3) **88**, 247
Diisopropylethylamine: 2-Propanamine, N-ethyl-N-(1-methylethyl)-; (7087-68-5) **85**, 64,
    158, 278; **86**, 81; **87**, 339; **88**, 121
(S,S)-Diisopropyl tartrate ((S,S)-DIPT): Diisopropyl D-tartrate (D-DIPT); (62961-64-2)
    88, 181
(S,S)-Diisopropyl tartrate (E)-crotylboronate; (99687-40-8) **88**, 181
Dimedone: 1,3-cyclohexanedione, 5,5-dimethyl-; (126-81-8) **86**, 252
N,N'-Dimesitylethylenediamine dihydrochloride: 1,2-Ethanediamine, N,N'-bis(2,4,6-
    trimethylphenyl)-, dihydrochloride; (258278-23-8) **87**, 77
N,N'-Dimesitylethylenediimine: Benzenamine, N,N'-1,2-ethanediylidenebis[2,4,6-
    trimethyl-; (56222-36-7) **87**, 77
1,3-Dimesitylimidazolinium chloride: 1H-Imidazolium, 4,5-dihydro-1,3-bis(2,4,6-
    trimethylphenyl)-, chloride (1:1); (173035-10-4) **87**, 77
5,6-Dimethoxyindole-1,2-dicarboxylic acid 1-benzyl ester 2-methyl ester: 1H-Indole-1,2-
    dicarboxylic acid, 5,6-dimethoxy-, 2-methyl 1-(phenylmethyl) ester; (873666-97-
    8) **88**, 388
(4-(Dimethoxymethyl)phenoxy)(tert-butyl)dimethylsilane **86**, 130
[(E)-3,3-Dimethoxy-2-methyl-1-propenyl]benzene: Benzene, [(1E)-3,3-dimethoxy-2-
    methyl-1-propen-1-yl]-: (137032-32-7) **86**, 81
6,7-Dimethoxy-1,2,3,4-tetrahydroisoquinoline hydrochloride: Isoquinoline, 1,2,3,4-
    tetrahydro-6,7-dimethoxy-, hydrochloride (1:1); (2328-12-3) **87**, 8
(2R,3S,4E)-N,3-Dimethoxy-N,2,4-trimethyl-5-phenyl-4-pentenamide
(3,5-Dimethoxy-1-phenyl-cyclohexa-2,5-dienyl)-acetonitrile **86**, 1
Dimethyl acetamide; (127-19-5) **86**, 298
Dimethylamine; (124-40-3) **86**, 298
4-(Dimethylamino)benzoic acid: (619-84-1) **87**, 201
N-(3-Dimethylaminopropyl)-N'-ethylcarbodiimide hydrochoride (EDC·HCl); (25952-53-
    8) **85**, 147
4-Dimethylaminopyridine: 4-Pyridinamine, N,N-dimethyl-; (1122-58-3) **85**, 64; **86**, 81
1,3-Dimethyl-6H-benzo[b]naphtho[1,2-d]pyran-6-one: 6H-Benzo[b]naphtho[1,2-
    d]pyran-6-one, 1,3-dimethyl-; (138435-72-0) **88**, 70
9,9-Dimethyl-4,5-bis(diphenylphosphino)xanthene: Xantphos; (161265-03-8) **85**, 96
Dimethyl-bis-phenylethynyl silane: Benzene, 1,1'-[(dimethylsilylene)di-2,1-
    ethynediyl]bis-; (2170-08-3) **85**, 53
(2R,3R)-2,3-Dimethylbutane-1,4-diol: (2R,3R) 2,3-Dimethyl-1,4-butanediol; (127253-15-
    0) **85**, 158
3,3-Dimethylbutyryl chloride: Butanoyl chloride, 3,3-dimethyl-; (7065-046-05) **88**, 138
2,2-Dimethyl-1,3-dioxane-4,6-dione (Meldrum's acid); (2033-24-1) **85**, 147
4,5-Dimethyl-1,3-dithiol-2-one; (49675-88-9) **86**, 333
N,N'-Dimethylethylenediamine: 1,2-Ethanediamine, N1,N2-dimethyl-; (110-70-3) **86**,
    181; **87**, 231
(Z)-1,1-Dimethyl-1-heptenylsilanol: Silanol, (1Z)-1-heptenyldimethyl-; (261717-40-2) **88**,
    102
N,O-Dimethylhydroxylamine hydrochloride: Methanamine, N-methoxy-, hydrochloride;
    (6638-79-5) **86**, 81
Dimethyl malonate; (108-59-8) **87**, 115

(4S)-N-[(2R,3S,4E)-2,4-Dimethyl-3-methoxy-5-phenyl-4-pentenoyl]-4-isopropyl-1,3-
   thiazolidine-2-thione **86**, 81
3,7-Dimethyl-2-methylene-6-octenal; (22418-66-2) **87**, 201
*4,4-Dimethyl-3-oxo-2-benzylpentanenitrile (*875628-78-7) **86**, 28
4,4-Dimethyl-3-oxopentanenitrile; (59997-51-2) **86**, 28
3,5-Dimethylphenol: Phenol, 3,5-dimethyl-; (108-68-9) **88**, 70
3,5-Dimethylphenyl-1-bromo-2-naphthoate: 2-Naphthalenecarboxylic acid, 1-bromo-,
   3,5-dimethylphenyl ester; (138435-66-2) **88**, 70
Dimethyl 2-phenylcyclopropane-1,1-dicarboxylate (3709-20-4) **87**, 115
7,7-Dimethyl-3-phenyl-4-p-tolyl-6,7,8,9-tetrahydro-1H-pyrazolo[3,4-b]-quinolin-5(4H)-
   one: 5H-Pyrazolo[3,4-b]quinolin-5-one, 1,4,6,7,8,9-hexahydro-7,7-dimethyl-4-(4-
   methylphenyl)-3-phenyl-; (904812-68-6) **86**, 252
(S,E)-2-(2,2-Dimethylpropylidenamino)-N-methylpropanamide
N, N-Dimethyl-4-pyridinamine: (1122-58-3) **85**, 64
(2R,3R)-2,3-Dimethylsuccinic acid; (5866-39-7) **85**, 158
Dimethyl sulfoxide: Methyl sulfoxide; Methane, sulfinybis-; (67-68-5) **85**, 189 ; **88**, 388
(–)-(S)-1-(1,3,2-Dioxaborolan-2-yloxy)-3-methyl-1,1-diphenylbutan-2-amine; (879981-
   94-9) **87**, 26, 36
(1,4-Dioxaspiro[4.5]dec-7-en-8-yloxy)trimethylsilane; (144810-01-5) **88**, 121
2-(1,3-Dioxolan-2-yl)ethyl bromide: 1,3-Dioxolane, 2-(2-bromoethyl)-; (18742-02-4) **87**,
   317
2-[2-(1,3-Dioxolan-2-yl)ethyl]zinc bromide: Zinc, bromo[2-(1,3-dioxolan-2-yl)ethyl]-;
   (864501-59-7) **87**, 317
1,3-Diphenylacetone *p*-tosylhydrazone: Benzenesulfonic acid, 4-methyl-, [2-phenyl-1-
   (phenylmethyl)ethylidene]hydrazide; (19816-88-7) **85**, 45
Diphenyldiazomethane (883-40-9) **85**, 189
(4-((4R,5R)-4,5-Diphenyl-1,3-dioxolan-2-yl)phenoxy)(tert-butyl)dimethylsilane **86**, 130
(1R,2R)-1,2-Diphenylethane-1,2-diol **86**, 130
α,α-Diphenylglycine: Benzeneacetic acid, α-amino-α-phenyl-; (3060-50-2) **87**, 88
Diphenylphosphine: Phosphine, diphenyl-; (829-85-6) **86**, 181
(S)-(–)-1,3-Diphenyl-2-propyn-1-ol: Benzenemethanol, α-(2-phenylethynyl)-, (αS)-;
   (132350-96-0) **85**, 118
Dirhodium tetraacetate dihydrate; (15956-28-2) **88**, 418

Elimination **85**, 45, 172; **86**, 11, 18, 212; **87**, 95, 231; **88**, 162, 207
Enamines **88**, 388
1,2-Epoxydodecane: Oxirane, decyl-; (2855-19-8) **85**, 1
1,2-Epoxy-3-phenoxypropane: oxirane, 2-(phenoxymethyl)-; (122-60-1) **86**, 287
Esterification **86**, 194; **88**, 70
Ethyl (*R*)-2-amino-3-mercaptopropanoate hydrochloride; (868-59-7) **88**, 168
Ethyl (R)-2-amino-3-(trifluoromethylthio)propanoate hydrochloride; (931106-56-8) **88**,
   168
(2R,3R)-Ethyl 1-benzhydryl-3-(4-bromophenyl)aziridine-2-carboxylate: 2-
   Aziridinecarboxylic acid, 3-(4-bromophenyl)-1-(diphenylmethyl)-, ethyl ester, (2
   R,3R)-; (233585-43-8) **88**, 224
Ethyl benzoate: Benzoic acid, ethyl ester; (93-89-0) **86**, 374
Ethyl 1-benzyl-4-fluoropiperidine-4-carboxylate **87**, 137
Ethyl 1-benzylpiperidine-4-carboxylate (24228-40-8) **87**, 137
Ethyl 4-bromobenzoate: (5798-75-4) **87**, 104

Imides **86**, 70, 81; **88**, 109, 212
Imines **86**, 11, 121, 212; **87**, 77, 88; **88**, 42, 224
Iminodiacetic acid; (142-73-4) **86**, 344
5H-Indazolo-[3,2-b]benzo[d]-1,3-oxazine: 5H-Indazolo[2,3-a][3,1]benzoxazine; (342-86-9) **87**, 339
1H-Indazole-3-carboxylic acid, ethyl ester (4498-68-4) **87**, 95
Indium bromide: (13465-09-3) **85**, 118
Indium(III) tris(trifluoromethanesulfonate): Methanesulfonic acid, trifluoro-, indium(3+) salt; (128008-30-0) **86**, 325
Insertion **87**, 209; **88**, 418
Iodine (7553-56-2) **85**, 219, 248; **86**, 1, 308; **87**, 288; **88**, 102, 406
o-Iodoaniline: Benzenamine, 2-iodo-; (615-43-0) **88**, 162
3-Iodoanisole: Benzene, 1-iodo-3-methoxy-; (766-85-5) **87**, 68
4-Iodoanisole: Benzene, 1-iodo-4-methoxy-; (696-62-8) **88**, 102
(Z)-1-Iodo-1-Heptene: 1-Heptene, 1-iodo-, (1Z)-; (63318-29-6) **88**, 102
1-Iodo-1-heptyne: 1-Heptyne, 1-iodo-; (54573-13-6) **88**, 102
Iodomethane: Methane, iodo-; (74-88-4) **86**, 194
1-Iodo-3-methylbenzene; (625-95-6) **87**, 178, 184
2-(2-Iodophenyl)propan-2-ol; (69352-05-2) **88**, 168
Iodosobenzene diacetate; (3240-34-4) **87**, 115
(E)-(2-Iodovinyl)benzene; (42599-24-6) **87**, 170
Isobutylboronic acid: Boronic acid, *B*-(2-methylpropyl)-; (84110-40-7) **88**, 247
Isobutylboronic acid pinacol ester; (67562-20-3) **88**, 247
2-Isobutylthiazole; (18640-74-9) **86**, 105
Iso-butyraldehyde; 2-methylpropionaldehyde; 2-methylpropanal (78-84-2) **87**, 143; **88**, 364
Isopropyl acetyl(phenyl)carbamate; Carbamic acid, N-acetyl-N-phenyl-, 1-methylethyl ester; (5833-25-0) **88**, 212
Isopropyl chloroformate; (108-23-6) **88**, 212
iPrMgCl·LiCl: isopropylmagnesium chloride lithium chloride complex; (807329-97-1) **86**, 374
(4*S*)-4-Isopropyl-5,5-diphenyloxazolidin-2-one (DIOZ): 2-Oxazolidinone, 4-(1-methylethyl)-5,5-diphenyl-, (4*S*)-; (184346-45-0) **85**, 295
(4S)-4-Isopropyl-5,5-diphenyl-3-(3-phenyl-propionyl)oxazolidin-2-one: 2-Oxazolidinone, 4-(1-methylethyl)-3-(1-oxo-3-phenylpropyl)-5,5-diphenyl-, (4*S*)-; (213887-81-1) **85**, 295
(4*S*)-Isopropyl-2-oxazolidinone: (4*S*)-4-(1-Methylethyl)-2-oxazolidinone; (17016-83-0) **85**, 158
O-Isopropyl S-3-oxobutan-2-yl dithiocarbonate: carbonodithioic acid, O-(1-methylethyl) S-(1-methyl-2-oxopropyl) ester; (958649-73-5) **86**, 333
Isopropyl phenyl(4,4,4-trifluoro-3,3-dihydroxybutanoyl)carbamate; 88, 212
(S)-4-Isopropyl-N-propanoyl-1,3-thiazolidine-2-thione **86**, 70
(*S*)-4-Isopropyl-*N*-propanoyl-1,3-thiazolidine-2-thione: 2-Thiazolidinethione, 4-(1-methylethyl)-3-(1-oxopropyl)-, (4*S*)-; (102831-92-5) **86**, 70
(4S)-Isopropyl-3-propionyl-2-oxazolidinone: (4S)-4-(1-Methylethyl)-3-(1-oxopropyl)-2-oxazolidinone; (77877-19-1) **85**, 158
(S)-4-Isopropyl-1,3-thiazolidine-2-thione **86**, 70
(*S*)-4-Isopropyl-1,3-thiazolidine-2-thione: 2-Thiazolidinethione, (4*S*)-4-(1-methylethyl)-; (76186-04-4) **86**, 70

(L)-*tert*-Leucine: L-Valine, 3-methyl-; (20859-02-3) **86**, 181
(L)-(+)-*tert*-leucinol: 1-Butanol, 2-amino-3,3-dimethyl-, (2*S*)-; (112245-13-3) **86**, 121
Lithium (7439-93-2) **85**, 53; **86**, 1
Lithium aluminum hydride; (16853-85-3) **85**, 158; **87,** 310
Lithium bis(trimethylsilyl)amide; (4039-32-1) **87,** 16
Lithium *t*-butoxide; (1907-33-1) **87,** 178, 184
Lithium chloride; (7447-41-8) **86**, 47
Lithium hydroxide monohydrate; (1310-66-3) **85**, 295
*Lithium 2-pyridiltriolborate:Borate(1-), [2-[(hydroxy-kO)methyl]- 2-methyl-1,3-
   propanediolato(3-)-kO1,kO3]-2-pyridinyl-, lithium (1:1), (T-4)-; (1014717-10-2)*
   88, 79
Lithium triethylborohydride; (22560-16-3) **85**, 64

(*R*)-(-)-Mandelic acid; (611-71-2) **85**, 106
(*S*)-(+)-Mandelic acid; (17199-29-0) **85**, 106
(R)-Mandelic acid salt of (1S,2S)-trans-2-(N-benzyl)amino-1-cyclohexanol; (882409-00-
   9) **85**, 106
(S)-Mandelic acid salt of (1R,2R)-trans-2-(N-benzyl)amino-1-cyclohexanol; (882409-01-
   0) **85**, 106
Manganese(IV) oxide; (1313-13-9)
Magnesium; (7439-95-4) **87,** 178
Mesitylamine: Benzenamine, 2,4,6-trimethyl-; (88-05-1) **85**, 34; **87,** 77
*3-(Mesitylamino)butan-2-one: 2-Butanone, 3-[(2,4,6-trimethylphenyl)-amino]-; (898552-
   96-0) **85**, 34
Mesitylene (108-67-8) **85**, 196
2-Mesitylhydrazinium chloride; (76195-82-9) **87,** 362
*N-Mesityl-N-(3-oxobutan-2-yl)formamide: Formamide, N-(1-methyl-2-oxopropyl)-N-
   (2,4,6-trimethylphenyl)-; (898553-01-0) **85**, 34
Metal complexation **87,** 26, 104
Metallation, **85,** 1, 45, 209; **86,** 374; **87,** 317, 330; **88,** 79, 181
Methanesulfonyl chloride; (124-63-0) **85**, 219; **86**, 181; **88**, 33
trans-4-Methoxy-3-buten-2-one; (51731-17-0) **87,** 192
(+)-*B*-Methoxydiisopinocampheylborane ((+)-(Ipc)₂BOMe); (99438-28-5) **88,** 87
(4*S*)-3-[(2*R*,3*S*,4*E*)-3-Methoxy-2,4-dimethyl-1-oxo-5-phenyl-4-pentenyl]-4-(1-
   methylethyl)-2-thiazolidinethione; (332902-42-8) **86**, 81, 181
1-(2-Methoxyethoxy)-1-vinylcyclopropane: Cyclopropane, 1-ethenyl-1-(2-
   methoxyethoxy)-; (278603-80-8) **88,** 109
4-Methoxy-4'-nitrobiphenyl,1,1'-Biphenyl, 4-methoxy-4'-nitro-; (2143-90-0) **88**, 197
4-Methoxyphenylacetic acid; (101-01-8) **88**, 418
Methyl acetoacetate: Butanoic acid, 3-oxo-, methyl ester; (105-45-3) **86**, 161
*Methyl 2-(2-acetylphenyl)acetate: Benzeneacetic acid, 2-acetyl-, methyl ester; (16535-
   88-9) **86**, 161
*(R)-Methyl 2-allylpyrrolidine-2-carboxylate hydrochloride (112348-46-6) **86**, 262
Methylamine; (74-89-5) **88**, 42
Methyl anthranilate; (134-20-3) **87,** 226
*(2S,3S)-Methyl-2-(9-anthryloxymethyl)-2-(4-methoxyphenyl)-3-(4-
   Methoxyphenylamino)-3-phenylpropanoate; (1034152-21-0) **88**, 418

464

3-Phenylpropan-1-ol; (122-97-4) **88**, 247

3-Phenylpropanoyl chloride: Benzenepropanoyl chloride; (645-45-4) **85**, 295

5-Phenyl-1H-pyrazol-3-amine: 1*H*-pyrazol-3-amine, 5-phenyl-; (1571-10-7) **86**, 252

Phenylpyruvic acid: α-Oxobenzenepropanoic acid, 2-Oxo-3-phenylpropionic acid; (156-06-9) **87**, 218

Phenylsilane: Benzene, silyl-; (694-53-1) **87**, 88

8-*Phenyl-1-tetralone* (501374-10-3) **87**, 209

N-[1-Phenyl-3-(trimethylsilyl)-2-propyn-1-ylidene]-benzeneamine; (77123-64-9) **85**, 88

Phenyl vinyl sulfide; (1822-73-7) **88**, 207

Phosphine, tris(1,1-dimethylethyl)-, tetrafluoroborate(1-) (1:1); (155026-77-0) **88**, 22

Phosphoric triamide, *N*,*N*,*N'*,*N'*,*N''*,*N''*-hexamethyl- ; (680-31-9) **87**, 126

Phosphorus oxychloride; (10025-87-3) **88**, 406

Phosphorus trichloride (7719-12-2) **85**, 238; **87**, 263; **88**, 152

Phosphorylation **87**, 263; **88**, 54, 152, 406

2-Picoline borane: Boron, trihydro(2-methylpyridine)-, (T-4)-; (3999-38-0) **87**, 339

Pimelic acid: Heptanedioic acid; (111-16-0) **86**, 194

Pinacol; (79-09-5) **88**, 247

Pinacolone: 2-Butanone, 3,3-dimethyl-; (75-97-8) **87**, 209

Piperidinium acetate; (4540-33-4) **86**, 28

Pivaldehyde, trimethylacetaldehyde; (630-19-3) **88**, 42

Potassium *tert*-butoxide: 2-Propanol, 2-methyl-, potassium salt (1:1); (865-47-4) **86**, 315; **88**, 181, 342

Potassium carbonate; (584-08-7) **85**, 287; **86**, 58

Potassium hydroxide; (1310-58-3) **85**, 196; **87**, 115

Potassium *O*-isopropylxanthate: Carbonodithioic acid, O-(1-methylethyl) ester, potassium salt (1:1); (140-92-1) **86**, 333

Potassium trimethylsilanolate; (10519-96-7) **86**, 274

Propanol, 2-amino-, 3-phenyl, (*S*); (3182-95-4) **85**, 267

N-Propargylphthalimide: Phthalimide, *N*-2-propynyl-: 1*H*-Isoindole-1,3(2*H*)-dione, 2-(2-propynyl)-; (7223-50-9) **88**, 109

Propionyl chloride : Propanoyl chloride; (79-03-8) **85**, 158; **86**, 70; 88. 364

(*S*)-Proline; (147-85-3) **86**, 262

2-Propenoic acid, 3,3'-(1,2-cyclopropanediyl)bis-, diethyl ester, [1α(E),2β(E)]-; (58273-88-4) **85**, 15

(*R*)-*i*-Pr-Pybox: Pyridine, 2,6-bis[(4*R*)-4,5-dihydro-4-(1-methylethyl)-2- oxazolyl]-; (131864-67-0) **87**, 330

1*H*-Pyrazole, 5-(1,3-benzodioxol-5-yl)-3-(4-chlorophenyl)-1-methyl-; (908329-89-5) **85**, 179

2,6-Pyridinedicarbonitrile; (2893-33-6) **87**, 310

Pyridine *N*-oxide: Pyridine, 1-oxide; (694-59-7) **88**, 22

(*E*)-1-Pyridin-3-yl-ethanone *O*-benzyl-oxime: Ethanone, 1-(3-pyridinyl)-, *O*-(phenylmethyl)oxime, (1*E*)-; (1010079-98-7) **87**, 36

(*E*)-1-Pyridin-3-yl-ethanone oxime; (106881-77-0) **87**, 36

(*S*)-1-Pyridin-3-yl-ethylamine: 3-Pyridinemethanamine, α-methyl-, (α*S*)-; (27854-9) **87**, 36

(*S*)-1-*Pyridin-3-yl-ethylamine hydrochloride*: 3-Pyridinemethanamine, *α-methyl-, dihydrochloride, (S)-*; (40154-84-5) **87**, 36

(4*S*)-5-*Pyrimidinecarboxylic acid, 1,2,3,4-tetrahydro-6-methyl-2-oxo-4-phenyl-1-(phenylmethyl)-, methyl ester; (865086-56-2)* **86**, 236

Pyrrolidine (123-75-1) **87**, 1, 201

1-Pyrrolidinecarboxylic acid, 2-(aminocarbonyl)-, phenylmethyl ester, (2*S*)-; (34079-31-7) **85**, 72

1,2-Pyrrolidinedicarboxylic acid, 1-(phenylmethyl) ester, (2*S*)-; (1148-11-4) **85**, 72

2-Pyrrolidinone; (616-45-5) **87**, 350

2-Pyrrolidinone, 1-methyl-; (872-50-4) **87**, 126

(S)-5-Pyrrolidin-2-yl-1H-tetrazole: 2H-Tetrazole, 5-[(2S)-2-pyrrolidinyl]-; (33878-70-5) **85**, 72

Quinidine: Cinchonan-9-ol, 6'-methoxy-; (56-54-2) **88**, 121

Quinoline; (91-22-5) **85**, 196

Rearrangements **86**, 18, 113, 172; **88**, 309

Reduction, **85**, 15, 27, 53, 64, 72, 138, 147, 158, 219, 248; **86**, 1, 28, 36, 92, 181, 344; **87**, 36, 77, 143, 275, 310, 362; **88**, 22, 70, 102, 202, 260, 291, 309, 427

Resolution, **85**, 106

Rhodium (II) tetrakis(triphenylacetate);(142214-04-8) **86**, 58

Rhodium **86**, 59; **87**, 115; **88**, 109, 418

Rhodium acetate dimer: Rhodium, tetrakis[μ-(acetato-κO:κO')]di-, (Rh-Rh); (15956-28-2) **85**, 172; **86**, 58

[RhCl(cod)]₂; (chloro(1,5-cyclooctadiene)rhodium(I) dimer) (12092-47-6) **86**, 360

Ring Expansion, **85**, 138

RuH₂(CO)(PPh₃)₃: Ruthenium, carbonyldihydrotris(triphenylphosphine); (25360-32-1) **87**, 209

Saponification, **85**, 27

Semicarbazide hydrochloride: Hydrazinecarboxamide, hydrochloride (1:1); (563-41-7) **86**, 194

Silanes **87**, 178, 253, 299; **88**, 102, 296, 377

Silver **88**, 14, 309

Silver carbonate: Carbonic acid, silver(1+) salt (1:2); (534-16-7) **87**, 253, 288

Silver(I) fluoride; (7775-41-9) **86**, 225

Silver hexafluoroantimonate; (26042-64-8) **88**, 309

Silver iodate: iodic acid (HIO₃), silver(1+) salt (1:1); (7783-97-3) **88**, 14

Silver nitrate; (7761-88-8) **86**, 225

Silylation **87**, 178, 299; **88**, 102, 121, 296

Sodium; (7440-23-5) **86**, 262

Sodium acetate; (127-09-3) **86**, 105, 194

Sodium acetate trihydrate; (6131-90-4) **85**, 10

Sodium amide; (7782-92-5) **86**, 298

Sodium ascorbate: L-Ascorbic acid, sodium salt: Sodium (+)-L-ascorbate; (134-03-2) **88**, 238

Sodium azide; (26628-22-8) **85**, 72, 278; **86**, 113; **87**, 161

Sodium bis(trimethylsilyl)amide; (1070-89-9) **87**, 170

Sodium borohydride: Borate(1-), tetrahydro-, sodium (1:1); (16940-66-2) **85**, 147; **86**, 181; **87**, 77

Sodium carbonate; (497-19-8) **86**, 36; **88**, 377

Sodium cyanide; (143-33-9) **85**, 219

Sodium hydride; (7646-69-7) **85**, 45, 172; **86**, 18, 194; **87**, 36
Sodium hypochlorite; (7681-52-9) **87**, 8
Sodium nitrite; (7632-00-0) **87**, 362
Sodium tetrafluoroborate; (13755-29-8) **85**, 248
Sodium tetrakis[(3,5-trifluoromethyl)phenyl]borate; (79060-88-1) **85**, 248
Sodium thiosulfate (7772-98-7) **86**, 1
(−)-Sparteine: 7,14-Methano-2*H*,6*H*-dipyrido[1,2-a:1',2'-e][1,5]diazocine, dodecahydro-, (7*S*,7a*R*,14*S*,14a*S*)-; (90-39-1) **88**, 247, 364
4-Spirocyclohexyloxazolidinone **86**, 58
4-Spirocyclohexyloxazolidinone: 3-Oxa-1-azaspiro[4.5]decan-2-one: (81467-34-7) **86**,58
*(1'S,5'R)-Spiro[1,3-dioxolane-2,3'-[6]oxabicyclo[3.2.0]heptan]-7'-one; (*360794-30-5)
 88, 121
Styrene; (100-42-5) **87**, 115
Substitution **86**, 18, 141, 151, 181, 333; **87**, 104, 126, 299, 317; **88**, 33, 54, 87, 152, 168, 181, 291
Sulfination **88**, 168, 309
Sulfonation **86**, 11; **87**, 231; **88**, 260, 317, 353
Sulfur; (7704-34-9) **85**, 209; **87**, 231
Super-Hydride®: Lithium triethylborohydride; (22560-16-3) **85**, 64

1,4,7,10-Tetraazacyclododecane-1,4,7-triacetic acid, Tri-tert-butyl Ester Hydrobromide:
 1,4,7,10-Tetraazacyclododecane-1,4,7-tricarboxylic acid, 1,4,7-tris(1,1-
 dimethylethyl) ester; (175854-39-4) **85**, 10
Tetrabromomethane; (558-13-4) **86**, 36
Tetrabutylammonium bromide; (1643-19-2) **85**, 278
Tetra-*n*-butylammonium chloride; (1112-67-0) **88**, 377
Tetrabutylammonium fluoride; (429-41-4) **87**, 95; **88**, 162
Tetrabutylammonium fluoride trihydrate: 1-Butanaminium, *N, N, N*-tributyl, fluoride, trihydrate; (87749-50-6) **88**, 102
Tetrahydro-1,3-dimethyl-2(1H)-pyrimidinone; (7226-23-5) **87**, 178, 184
4,4a,9,9a-Tetrahydro-1-oxa-4-azafluoren-3-one: Indeno[2,1-*b*]-1,4-oxazin-3(2*H*)-one, 4,4a,9,9a-tetrahydro-, (4a*R*,9a*S*)-; (862095-79-2) **87**, 350
Tetrafluoroboric acid: Borate(1-), tetrafluoro-, hydrogen (1:1); (16872-11-0) **87**, 288
Tetrafluoroboric acid diethyl etherate; (67969-82-8) **86**, 172
Tetrakis(dimethylamino)allene; (42928-64-3) **86**, 298
1,1,3,3-Tetrakis(dimethylamino)propenium tetrafluoroborate; (125254-01-5) **86**, 298
Tetrakis(triphenylphosphine)palladium; (14221-01-3) **86**, 315; **88**, 207, 406
α-Tetralone: 1(2*H*)-Naphthalenone, 3,4-dihydro-; (529-34-0) **87**, 209, 275
4,4,5,5-Tetramethyl-1,3,2-dioxaborolane; (25015-63-8) **88**, 342
1,1,3,3-Tetramethylguanidine: Guanidine, *N,N,N',N'*-tetramethyl-; (80-70-6) **88**, 388
2,2,6,6-Tetramethylpiperidine; (768-66-1) **86**, 374; **87**, 263
1,3,5,7-Tetramethyl-1,3,5,7-tetravinylcyclotetrasiloxane; (2554-06-5) **86**, 274
*(S)-2-(1H-Tetrazol-5-yl)-pyrrolidin-1-carboxylic acid benzyl ester: 1-
 Pyrrolidinecarboxylic acid, 2-(2H-tetrazol-5-yl)-, phenylmethyl ester, (2S)-;*
 (33876-20-9) **85**, 72
Tetrolic acid: 2-Butynoic acid (9); (590-93-2) **85**, 231
2-Thiazolidinethione, 4-(phenylmethyl)-, (4*S*)-; (171877-39-7) **88**, 364
2-Thiophene-carboxaldehyde; (98-03-3) **86**, 360; **87**, 16
5-(Thiophen-2-yl)oxazole: Oxazole, 5-(2-thienyl)-; (70380-70-0) **87**, 16

Thiourea; (62-56-6) **87**, 115
Tin (II) chloride; (7772-99-8) **85**, 27
Tin(II) chloride dihydrate; (10025-69-1) **87**, 362
Titanium tetrachloride; (7550-45-0) **85**, 295; **86**, 81; **88**, 34
p-Tolualdehyde: benzaldehyde, 4-Methyl-; (104-87-0) **86**, 252
p-Toluenesulfonamide: Benzenesulfonamide, 4-methyl-; (70-55-3) **86**, 212
p-Toluenesulfonic acid: Benzenesulfonic acid, 4-methyl-; (104-15-4) **85**, 287
p-Toluenesulfonic acid monohydrate: Benzenesulfonic acid, 4-methyl-, hydrate (1:1);
    (6192-52-5) **86**, 194
(*S*)-3-[*N*-(*p*-Toluenesulfonyl)amino]-4-methylpentanoic acid (936012-07-6) **87**, 143
p-Toluenesulfonyl chloride: Benzenesulfonyl chloride, 4-methyl-; (98-59-9) **86**, 58; **87**,
    88, 231; **88**, 309
p-Tolylboronic acid; (5720-05-8) **86**, 344
4-(p-Tolyl)-phenylboronic acid **86**, 344
4-(p-Tolyl)-phenylboronic acid MIDA ester **86**, 344
N-Tosyl-(S_a)-binam-L-prolinamide: 2-Pyrrolidinecarboxamide, *N*-[(1S)-2'-[[(4-
    methylphenyl)sulfonyl]amino][1,1'-binaphthalen]-2-yl]-, (2S)-; (933782-38-8) **88**,
    330
Tosylmethylisocyanide: Benzene, [(isocyanomethyl)sulfonyl]-; (36635-63-9) **87**, 16
1,3,5-*Triacetylbenzene*; (779-90-8) **87**, 192
Tribasic potassium phosphate; (7778-53-2) **86**, 105
Tributyl(4-methoxyphenyl)stannane; (70744-47-7) **88**, 197
Tri-*n*-butylphosphine: Phosphine, tributyl-; (998-40-3) **86**, 212; **88**, 138
Tributyltin chloride; (1461-22-9) **88**, 197
2,2,2-Trichloro-1-ethoxyethanol; (515-83-3) **86**, 262
Trichloromethane (1:1); (52522-40-4) **88**, 162
(3R,7aS)-3-(Trichloromethyl)tetrahydropyrrolo[1,2-c]oxazol-1(3H)-one; (97538-67-5)
    86, 262
Tricyclohexylphosphine: Phosphine, tricyclohexyl-; (2622-14-2) **87**, 299; **88**, 342
Triethylamine: Ethanamine, *N,N*-diethyl-; (121-44-8) **85**, 131, 189, 219, 295; **86**, 58, 81,
    252, 315; **87**, 231; **88**, 42, 121, 138, 212, 247, 309, 317
Triethylamine hydrochloride: Ethanamine, *N,N*-diethyl-, hydrochloride (1:1); (554-68-7)
    85, 72
Triethyl orthoformate: Ethane, 1,1',1''-[methylidynetris(oxy)]tris- ; (122-51-0) **87**, 77,
    350, 362
Triethyl(pent-4-enyloxy)silane: Silane, triethyl(4-penten-1-yloxy)-; (374755-00-7) **87**,
    299
Triethylphosphine; (554-70-1) **87**, 275
Trifluoroacetic acid (76-05-1) **87**, 143; **88**, 317
Trifluoroacetic anhydride; (407-25-0) **85**, 64; **86**, 18
2,2,2-Trifluoroethanol (75-89-8) **85**, 45
2,2,2-Trifluoroethyl *p*-toluenesulfonate: Ethanol, 2,2,2-trifluoro-, 4-
    methylbenzenesulfonate; (433-06-7) **88**, 162
2,2,2-Trifluoroethyl trifluoroacetate; (407-38-5) **88**, 212
Trifluoromethanesulfonic acid; HIGHLY CORROSIVE, Methanesulfonic acid, trifluoro-
    (8, 9); (1493-13-6) **86**, 308; **88**, 197
Trifluoromethanesulfonimide; (82113-65-3) **87**, 253
Trifluoromethylation **87**, 126
1-(Trifluoromethyl)-1,2-benziodoxol-3(1H)-one; (887144-94-7) **88**, 168
1-Trifluoromethyl-1,3-dihydro-3,3-dimethyl-1,2-benziodoxole; (887144-97-0) **88**, 168

470

Trifluoro, 2-(trimethylsilyl)phenyl ester; (88284-48-4) **86**, 161

Triflic anhydride: Trifluoromethanesulfonic acid anhydride; (358-23-6) **85**, 88; **88**, 197, 353

Triisopropyl borate: Boric acid (H₃BO₃), tris(1-methylethyl)ester; (5419-55-6) **88**, 79, 181

Triisopropyl phosphite; (116-17-6) **86**, 36

(Triisopropylsilyl)acetylene: Ethynyltriisopropylsilane; (89343-06-6) **86**, 225

3,4,5-Trimethoxybenzaldehyde; (86-81-7) **88**, 427

3,4,5-Trimethoxybenzoyl chloride: Benzoyl chloride, 3,4,5-trimethoxy-; (4521-61-3) **88**, 427

Trimethylacetaldehyde: Pivaldehyde: Propanal, 2,2-dimethyl; (630-19-3) **85**, 267

Trimethylaluminum: (75-24-1) **87**, 104

Trimethylamine oxide: Methanamine, *N,N*-dimethyl-, *N*-oxide; (1184-78-7) **88**, 162

2,4,6-Trimethylaniline; (88-05-1) **87**, 362

(1*R*,2*R*,3*R*,5*S*)-2,6,6-Trimethylbicyclo[3.1.1]heptan-3-ol: (+)-isopinocampheol; (24041-60-9) **88**, 87

Trimethyl orthoformate; (149-73-5) **86**, 81, 130

Trimethyloxonium tetrafluoroborate; (420-37-1) **87**, 350, 362

1-(Trimethylsilyl)acetylene; (1066-54-2) **85**, 88

4-Trimethylsilyl-2-butyn-1-ol; (90933-84-9) **88**, 296

Trimethylsilyl chloride: Silane, chlorotrimethyl-; (75-77-4) **85**, 209; **88**, 121

1-Trimethylsilyloxybicyclo[3.2.0]heptan-6-one (125302-44-5) **85**, 138

1-Trimethylsilyloxycyclopentene (19980-43-9) **85**, 138

1-Trimethylsilyloxy-7,7-dichlorobicyclo[3.2.0]heptan-6-one (66324-01-4) **85**, 138

2-(Trimethylsilyl)phenyl trifluoromethanesulfonate; (88284-48-4) **87**, 95

Trimethylphosphite: Phosphorous acid, trimethyl ester; (121-45-9) **88**, 152

1-Trimethylsilyl-1-propyne; (6224-91-5) **88***, 377*

O-Trimethylsilylquinidine; **88**, 121

Triphenylacetic acid: Benzeneacetic acid, α,α-diphenyl-; (595-91-5) **86**, 58

Triphenyl borate: Boric acid (H₃BO₃), triphenyl ester; (1095-03-0) **88**, 224

Triphenylphosphine, 99%; (603-35-0) **87**, 161; **88**, 79, 121, 162, 309, 377

Triphenylphospine oxide; (791-28-6) **85**, 248; **86**, 274

Triphosgene: Methanol, 1,1,1-trichloro-, 1,1'-carbonate; (32315-10-9) **86**, 315

Tripotassium phosphate, monohydrate: Phosphoric acid, tripotassium salt, monohydrate (8CI,9CI); (27176-10-9) **87**, 299

Tripropargylamine: 2-Propyn-1-amine, *N,N*-di-2-propynyl: Tri-2-propynylamine; (6921-29-5) **88**, 238

Tri-*iso*-propyl borate; (5419-55-6) **87**, 26

Tri-*iso*-propyl phosphite: Phosphorous acid, tris(1-methylethyl) ester; (116-17-6) **87**, 231

Tri-*iso*-propylsilyl trifluoromethanesulfonate; (80522-42-5) **87**, 253

Tris((1-benzyl-1H-1,2,3-triazolyl)methyl)amine; (510758-28-8) **88**, 238

Tris(dibenzylideneacetone) dipalladium(0): Palladium, tris[μ-[(1,2-η:4,5-η)-(1*E*,4*E*)-1,5-diphenyl-1,4-pentadien-3-one]]di-; (51364-51-3) **86**, 194

Tris(dibenzylideneacetone)-dipalladium(0)-chloroform adduct; (52522-40-4) **86**, 47; **88**, 162

Tris(dibenzylideneacetone)dipalladium(0): Pd₂(dba)₃; (51364-51-3) **87**, 104, 263

1,1,1-Tris(hydroxymethyl)ethane: 1,3-Propanediol, 2-(hydroxymethyl)-2-methyl-; (77-85-0) **88**, 79

1,3,5-[Tris-piperazine]-triazine: 1,3,5-Triazine, 2,4,6-tri-1-piperazinyl-; (19142-26-8) **86**, 141, 151

L-Valine methyl ester hydrochloride; (6306-52-1) **87**, 26; **88**, 14
(*S*)-Valinol: 1-Butanol, 2-amino-3-methyl-, (2*S*)-; (2026-48-4) **86**, 70
Vanadyl acetylacetonate; (3153-26-2) **86**, 121
Vanadium **86**, 121
(*S*)-VANOL: [2,2'-Binaphthalene]-1,1'-diol, 3,3'-diphenyl-, (2*S*)-; (147702-14-5) **88**, 224
4-Vinylbenzophenone: Methanone, (4-ethenylphenyl)phenyl-; (3139-85-3) **86**, 274
4-Vinyl-1-cyclohexene: Cyclohexene, 4-ethenyl-; (100-40-3) **88**, 207
3-Vinylquinoline: Quinoline, 3-ethenyl-; (67752-31-2) **86**, 274

X-Phos: 2-Di-cyclohexylphosphino-2',4',6'-triisopropylbiphenyl; (564483-18-7) **87**, 104

Zinc; (7440-66-6) **87**, 330
Zinc-catalyzed reactions **87**, 310
Zinc (II) chloride; (7646-85-7) **85**, 27, 53
Zinc triflate; (54010-75-2) **86**, 113
Zinc trifluoromethanesulfonate: Methanesulfonic acid, 1,1,1-trifluoro-, zinc salt (2:1);
    (54010-75-2) **87**, 310
Zirconocene chloride hydride: Zirconium chlorobis(η^5-2,4-cyclopentadien-1-yl)hydro-;
    (37342-97-5) **88**, 427

472